Series in
Mathematical Biology
and Medicine

HANDBOOK
OF CANCER MODELS
WITH APPLICATIONS

SERIES IN MATHEMATICAL BIOLOGY AND MEDICINE

Series Editors: **P. M. Auger and R. V. Jean**

Published

Vol. 1: Stochastic Models of Tumor Latency and Their Biostatistical Applications
A. Yu. Yakovlev and A. D. Tsodikov

Vol. 2: Volterra-Hamilton Models in the Ecology and Evolution of Colonial Organisms
P. L. Antonelli and R. H. Bradbury

Vol. 3: The Hierarchical Genome and Differentiation Waves
Novel Unification of Development, Genetics and Evolution
R. Gordon

Vol. 4: Symmetry in Plants
Ed. by R. V. Jean and D. Barabe

Vol. 5: Computational Medicine, Public Health and Biotechnology: Building a Man in the Machine
Proceedings of the First World Congress
Vols. 1 & 3 – Ed. by M. Witten
Vol. 2 – Ed. by D. Joan and M. Witten

Vol. 6: Advances in Mathematical Population Dynamics — Molecules, Cells and Man
Proceedings of the 4th International Conference on Mathematical Population Dynamics
Ed. by O. Arino, D. Axelrod and M. Kimmel

Vol. 7: Neuronal Information Processing
From Biological Data to Modelling and Applications
Ed. by G. Burdet, P. Combe and O. Parodi

Vol. 8: Advances in Bioinformatics and Its Applications
Proceedings of the International Conference
Ed. by M. He, G. Narasimhan and S. Petoukhov

Vol. 9
Series in Mathematical Biology and Medicine

HANDBOOK OF CANCER MODELS WITH APPLICATIONS

edited by

Wai-Yuan Tan
University of Memphis, USA

Leonid Hanin
Idaho State University, USA

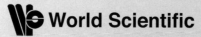

NEW JERSEY · LONDON · SINGAPORE · BEIJING · SHANGHAI · HONG KONG · TAIPEI · CHENNAI

Published by

World Scientific Publishing Co. Pte. Ltd.
5 Toh Tuck Link, Singapore 596224
USA office: 27 Warren Street, Suite 401-402, Hackensack, NJ 07601
UK office: 57 Shelton Street, Covent Garden, London WC2H 9HE

British Library Cataloguing-in-Publication Data
A catalogue record for this book is available from the British Library.

Series in Mathematical Biology and Medicine — Vol. 9
HANDBOOK OF CANCER MODELS WITH APPLICATIONS

Copyright © 2008 by World Scientific Publishing Co. Pte. Ltd.

All rights reserved. This book, or parts thereof, may not be reproduced in any form or by any means, electronic or mechanical, including photocopying, recording or any information storage and retrieval system now known or to be invented, without written permission from the Publisher.

For photocopying of material in this volume, please pay a copying fee through the Copyright Clearance Center, Inc., 222 Rosewood Drive, Danvers, MA 01923, USA. In this case permission to photocopy is not required from the publisher.

ISBN-13 978-981-277-947-2
ISBN-10 981-277-947-7

Typeset by Stallion Press
Email: enquiries@stallionpress.com

Printed in Singapore by Mainland Press Pte Ltd

CONTENTS

Contributors — xvii
Preface — xxiii

1. Oncogenetic Trees — 1
   Aniko Szabo and Kenneth M. Boucher

   1. Introduction — 1
   2. Definitions and Basic Results — 2
      2.1. Description of the Data — 2
      2.2. The Oncogenetic Tree Model — 3
         2.2.1. Error model — 5
   3. Reconstruction — 5
   4. Sample Size Estimation — 11
   5. Parameter Estimation — 13
   6. Example: Renal Carcinoma Development — 14
   7. Properties of the Oncogenetic Tree Estimator: A Simulation Study — 15
      7.1. Simulating Data Based on a Given Tree — 16
      7.2. Probability of Correct Reconstruction — 16
      7.3. Sample Size for High Probability of Reconstruction — 17
   8. Goodness of Fit — 19
      8.1. Bootstrap Estimate of Reconstruction Confidence — 19
      8.2. Analysis of the Stable Portions — 19
   9. Discussion — 22
   References — 23

2. Stochastic Multistage Cancer Models: A Fresh Look
   at an Old Approach   25
   Qi Zheng

   1. Introduction 25
   2. Basic Definitions and Notation 26
   3. Mathematical Details 32
   4. Concluding Remarks 41
   Acknowledgments 42
   References 42

3. Cancer Biology, Cancer Models and Some New
   Approaches to Carcinogenesis   45
   Wai Y. Tan, Chao W. Chen and Li J. Zhang

   1. Introduction 45
   2. Some Recent Cancer Biology for Modeling
      Carcinogenesis 47
      2.1. The Multi-Staging Nature of Carcinogenesis 49
      2.2. The Sequential Nature 50
      2.3. The Genetic Changes and Cancer Genes 53
      2.4. Cell Cycle and Carcinogenesis 55
      2.5. Epigenetic and Cancer 57
      2.6. Telomere, Immortalization and Cancer 60
      2.7. Single Pathway versus Multiple Pathways
           of Carcinogenesis 61
   3. Some General Stochastic Models of Carcinogenesis 64
      3.1. The Extended Multi-Event Model of
           Carcinogenesis 64
      3.2. The Mixed Models of Carcinogenesis 65
   4. Some New Approaches for Analyzing Stochastic
      Models of Carcinogenesis 66
      4.1. Stochastic Differential Equations 67
      4.2. The Probability Distribution of $T(t)$ 69
      4.3. Probability Distribution of the State Variables . . . 69

	5.	A State Space Model for the Extended Multi-Event Model of Carcinogenesis	70
		5.1. The Stochastic System Model, the Augmented State Variables and Probability Distribution	71
		5.2. The Observation Model and the Probability Distribution of Cancer Incidence	72
		5.3. The Posterior Distribution of the Unknown Parameters and State Variables	74
		5.4. The Generalized Bayesian Method for Estimating Unknown Parameters and State Variables	75
	6.	Analysis of British Physician Data of Lung Cancer and Smoking	76
	7.	Conclusions and Summary	83
	Acknowledgments		83
	References		84

4. Modeling the Effects of Radiation on Cell Cycle Regulation and Carcinogenesis — 91
William D. Hazelton

	1.	Introduction	92
		1.1. Multistage Carcinogenesis Models	92
		1.2. Analyses of Environmentally Exposed Cohorts . .	93
	2.	Modeling Biological Mechanisms	95
		2.1. A Combined Cell Cycle and Multistage Clonal Expansion Model	95
	3.	Summary	102
	Acknowledgments		103
	References		103
	Appendix		104

5. Cancer Models, Ionizing Radiation, and Genomic Instability: A Review — 109
Mark P. Little

	1.	Introduction	109
	2.	Armitage-Doll Multi-Stage Model	113

	3. Two-Mutation Model	119
	4. Generalized MVK and Multi-Stage Models	123
	5. Multiple Pathway Models	128
	5.1. Multiple Pathway Models Incorporating Genomic Instability	129
	6. Discussion and Conclusions	133
	Acknowledgments	138
	References	138

6. **Distribution of the Sizes of Metastases: Mathematical and Biomedical Considerations**   149
   Leonid Hanin

   1. Introduction 149
   2. The Model 155
      - 2.1. Tumor Latency 155
      - 2.2. Primary Tumor Growth 155
      - 2.3. Metastasis Formation 155
      - 2.4. Timeline of the Natural History of Metastatic Cancer and Observables 156
      - 2.5. Secondary Metastasis 157
      - 2.6. Metastasis Growth 157
      - 2.7. Metastasis Detection 158
      - 2.8. Effects of Treatment 158
   3. Distribution of the Sizes of Detectable Metastases . . . 158
   4. Distribution of the Sizes of Detectable Metastases for Exponentially Growing Tumors 162
      - 4.1. Model Specification and Results 162
      - 4.2. Model Identification 165
   - Acknowledgments 167
   - References 168

7. **Mathematical Models of Cancer and their Relevant Insights**   173
   Evans Afenya

   1. Introduction 173
   2. Cancer Models 176

| | | 2.1. | Models of Leukemia . . . . | 177 |
| | | 2.2. | Cell Kinetics . . . . | 195 |

	3.	Cancer Treatment Models	200
		3.1. Optimal Control Models	201
	4.	Parameter Estimation	210
	5.	Concluding Remarks	211
		References	213

8. Major Epigenetic Hypotheses of Carcinogenesis Revisited — 225
   King-Thom Chung

   1. Introduction 226
   2. Why are Epigenetic Factors Important? 230
   3. The Warburg's Hypothesis 232
   4. The Linus Pauling Hypothesis: Vitamin C and Cancer . . . 233
   5. Szent-Györgyi (Bioelectronic) Hypothesis 236
   6. Micronutrients and Cancer 238
   7. NAD Deficiency as a Factor in Carcinogenesis 242
   8. GAP Junction Intercellular Communication (GJIC)
      and Cancer 251
   9. Viral Infections and Cancer 254
   10. Other Epigenetic Hypotheses 259
   11. Concluding Remarks and Perspectives 262
   Acknowledgments 265
   References 266

9. Induction and Repair of DNA Damage Formed
   by Energetic Electrons and Light Ions — 291
   Robert D. Stewart and Vladimir A. Semenenko

   1. Dosimetric Quantities and Units 291
      1.1. Absorbed Dose 291
      1.2. Linear Energy Transfer (LET) 292
      1.3. Microdosimetry 294
   2. Induction of DNA Damage 297
      2.1. Classification of DNA Damage 297
      2.2. Mechanisms 297
      2.3. Initial Yield and Characteristics 300

	3.	Repair of Base Damage and Single-Strand Breaks 303
	3.1.	Mechanisms 303
	3.2.	Excision Repair Outcomes and Kinetics 304
	3.3.	Point Mutations Arising from Base Damage and Single-Strand Breaks 307
	4.	Repair of Double-Strand Breaks 308
	4.1.	Mechanisms 308
	4.2.	Repair Kinetics, Chromosome Aberrations and Small-Scale Mutations 310

Acknowledgments 314
References 314
Appendix 319

10. Radiation-Induced Bystander Effects   323
Linda C. DeVeaux

1. Bystander Effects 323
    1.1. Introduction 323
    1.2. Definition of Bystander Effects 325
    1.3. History of Bystander Effects 326
    1.4. Bystander Endpoints 328
    1.5. Transmission of Signal 330
    1.6. Identification of Signal 331
    1.7. Status of Sending and Receiving Cells 333
    1.8. Dependence on Radiation Type 335
    1.9. Bystander Effects and Cancer Risk 336
    1.10. Evolutionary Considerations of Bystander Effects . 336
2. Summary 338

Acknowledgments 338
References 338

11. A Stochastic Model of Human Colon Cancer Involving Multiple Pathways   345
Wai Y. Tan, Li J. Zhang, Chao W. Chen and J. M. Zhu

1. Introduction 345
2. A Brief Summary of Colon Cancer Biology 347

       2.1. The LOH Pathway of Human Colon Cancer
            (The APC-β – Catenin – Tcf – myc Pathway) .. 347
       2.2. The MSI (Micro-Satellite Instability) Pathway
            of Human Colon Cancer 349
  3. The Stochastic Multi-Stage Model of Carcinogenesis ... 352
       3.1. Stochastic Equations of State Variables 353
       3.2. The Expected Number of $I_j(t)$ 354
       3.3. The Probability Distribution of the Number
            of Detectable Tumors 355
  4. A Statistical Model and the Probability Distribution
    of Cancer Incidence Data 356
       4.1. Data Augmentation and the Expanded Model ... 358
       4.2. The Genetic Parameters 359
  5. The State Space Model and the Generalized Bayesian
    Approach for Estimating the Unknown Parameters 361
       5.1. The Prior Distribution of the Parameters 361
       5.2. The Posterior Distribution of the Parameters
            Given $\{Y, Z\}$ 362
       5.3. The Multi-Level Gibbs Sampling Procedure
            for Estimating Parameters 363
       5.4. The Genetic Algorithm 364
  6. Application and Results 364
  7. Conclusions and Discussion 370
Acknowledgments 371
References 371

12. Cancer Risk Assessment of Environmental Agents by
    Stochastic and State Space Models of Carcinogenesis   375
    Wai Y. Tan, Chao W. Chen and Li J. Zhang

  1. Introduction 375
  2. A General Stochastic Model of Carcinogenesis 377
       2.1. The Stochastic Difference Equations for State
            Variables 380
       2.2. The Probability of Developing Cancer Tumors ... 381
       2.3. Probability Distribution of the State Variables ... 381
  3. The Data for Risk Assessment of Environmental Agents . 382

	4.	State Space Models of Carcinogenesis and the Prediction of State Variables	383
	5.	A State Space Model for Cancer Risk Assessment	384
		5.1. The Stochastic System Model and Probability Distributions	384
		5.2. The Observation Model	385
	6.	The Genetic Algorithm and the Predicted Inference Procedures	386
		6.1. The Genetic Algorithm	386
		6.2. The Predictive Inference Procedures	387
	7.	Developing Confidence Intervals for Probabilities of Developing Cancer by Genetic Algorithm	388
	8.	Developing Dose-Response Curves of Environmental Agents by Genetic Algorithm	388
	9.	An Application and Illustration	389
	10.	Conclusions	394
		Acknowledgments	394
		References	394
13.	Stochastic Models for Preneoplastic Lesions and Their Application for Cancer Risk Assessment		397

Annette Kopp-Schneider, Iris Burkholder, Jutta Groos and Lutz Edler

	1.	Introduction	397
	2.	Modeling Preneoplastic Lesions	401
		2.1. The Multistage Model with Clonal Expansion of Intermediate Cells	401
		2.2. A Geometric Model for Colonies of Intermediate Cells	405
		2.3. Comparison of Multistage and Color-Shift Model	406
	3.	Application of Carcinogenesis Models to Preneoplastic Lesion Data	408
		3.1. Mouse Skin Carcinogenesis: Testing Biological Hypotheses about Papilloma and Carcinoma Formation	408

	3.2.	Liver Focal Lesion Data: Testing Hypotheses about FAH Formation and Phenotype Change	411

      3.2. Liver Focal Lesion Data: Testing Hypotheses about
           FAH Formation and Phenotype Change 411
      3.3. Liver Focal Lesion Data: Dose-Response Analyses 412
  4. Discussion 416
Acknowledgments 418
References 418
Appendix A: Basic Ideas of the Color-Shift Model 422
Appendix B: Likelihood Functions 423
      Appendix B(1): Likelihood Function for Skin Papilloma
           and Carcinoma Data 423
      Appendix B(2): Likelihood Function for Liver
           Focal Lesion Data 424

14. Drug Resistance in Cancer Models   425
Jaroslaw Smieja

  1. Introduction 425
  2. Biological Background 426
  3. Preliminaries for Mathematical Models 429
  4. Drug Resistance and a Single Chemotherapeutic Agent .. 431
      4.1. A Simple, Two-Compartmental Model 431
      4.2. Evolution of Drug Resistance Stemming
           from Gene Amplification 434
      4.3. Partial Sensitivity of the Resistant Subpopulation . 439
      4.4. Phase-Specific Chemotherapy 440
      4.5. General Compartmental Model 442
  5. Multidrug Therapy and Drug Resistance 447
      5.1. A Two-Compartmental Model 448
      5.2. A Four-Compartmental Model 448
  6. Concluding Remarks 449
Acknowledgments 451
References 452

15. Bladder Cancer Screening by Magnetic Resonance Imaging   457
Lihong Li, Zigang Wang and Zhengrong Liang

  1. Introduction 458

2. Methods 459
　2.1. MR Image Protocols 459
　2.2. Image Segmentation 461
　2.3. Interactive Visualization System 462
　2.4. Detection of Bladder Lesions 463
3. Results 465
4. Discussion and Conclusions 466
Acknowledgments 467
References 468

16. **Mathematical Framework and Wavelets Applications in Proteomics for Cancer Study**   471
　Don Hong and Yu Shyr

　1. Introduction 471
　2. Mathematical Representation and Preprocessing
　　of Maldi MS Data 475
　　2.1. Mathematical Model for MALDI-TOF MS Data . . 476
　　　2.1.1. Baseline correction and normalization . . 479
　　　2.1.2. Spectra registration and peak alignment . 480
　3. Multiscale Tools 482
　　3.1. Wavelets and WaveSpec Software 483
　　3.2. Diffusion Maps 488
　4. Clustering and Cancer Data Classifications 489
　Acknowledgments 496
　References 496

17. **Advanced Statistical Methods for the Design and Analysis of Tumor Xenograft Experiments**   501
　Ming Tan and Hong-Bin Fang

　1. Introduction 502
　2. Design of Experiments for Combination Studies 504
　　2.1. Fixed-Ratio Design and Ray Design 506
　　2.2. Abdelbasit-Plackett Optimal Experimental Design . 506
　　2.3. Uniform Experimental Design 507
　3. Statistical Analysis for Tumor Growth 509

		3.1.	Statistical Models	510
		3.2.	Parameter Estimation via the ECM Algorithm . . .	512
	4.	Comparison of Treatment Effects		514
		4.1.	Quasi t-Test Based on the EM Algorithm	515
		4.2.	Bayesian Test	516
	5.	Summary and Discussion		516
	Acknowledgments			518
	References			518

18. **Analysis of Occult Tumor Studies**   521
    Shesh N. Rai

    1. Introduction 521
    2. A Review of the Literature 523
    3. Preliminary Considerations 527
        3.1. Constructing the Likelihood Function 530
        3.2. Non-Parametric Settings 531
        3.3. Fitting the Semi-Parametric Model 534
    4. Interval Estimation 539
    5. Testing Tumor Lethality and Carcinogenic Effect 543
    6. Two Examples 546
    7. Discussion 556
    References 560

Index   563

CONTRIBUTORS

Evans Afenya
Department of Mathematics, Elmhurst College
190 Prospect Avenue, Elmhurst, IL 60126, USA

Kenneth M. Boucher
Department of Oncological Sciences
Huntsman Cancer Institute, University of Utah
2000 Circle of Hope Drive
Salt Lake City, UT 64112-5550, USA

Iris Burkholder
Biostatistics-C060, German Cancer Research Center
D-69009 Heidelberg, Germany

Chao W. Chen
National Center of Environmental Assessment
United States Environmental Protection Agency
Washington, DC 20460, USA

King-Thom Chung
Department of Biology, University of Memphis
Memphis, TN 38152, USA

Linda C. DeVeaux
Department of Biological Sciences
Idaho State University
921 South 8th Avenue, Stop 8007
Pocatello, ID 83209-8007, USA

Lutz Edler
Biostatistics-C060, German Cancer Research Center
D-69009 Heidelberg, Germany

Hong-Bin Fang
Division of Biostatistics, Medical School Teaching Facility
University of Maryland, Greenebaum Cancer Center
685 W. Baltimore Street, Suite 261
Baltimore, MD 21201, USA

Jutta Groos
Biostatistics-C060, German Cancer Research Center
D-69009 Heidelberg, Germany

Leonid Hanin
Mathematics Department, Idaho State University
921 S. 8th Avenue, Stop 8085
Pocatello, ID 83209-8085, USA

William D. Hazelton
Senior Staff Scientist
Fred Hutchinson Cancer Research Center
1100 Fairview Ave. N. M2-B500
Seattle, WA 98109, USA

Don Hong
Department of Mathematical Sciences
Tennessee State University
Box 34, Middle
Murfreesboro, TN 37132, USA

Annette Kopp-Schneider
Biostatistics-C060, German Cancer Research Center
D-69009 Heidelberg, Germany

Lihong Li
Department of Engineering Science and Physics
City University of New York/College of Staten Island
2800 Victory Blvd., Room 1N-225
Staten Island, NY 10314, USA

Zhengrong Liang
Department of Radiology
State University of New York at Stony Brook
Health Science Center
L8, Room 067
Stony Brook, NY 11794, USA

Mark P. Little
Department of Epidemiology and Public Health
Division of Epidemiology, Public Health and Primary Care
Faculty of Medicine, Imperial College London
St Mary's Campus, Norfolk Place
London W2 1PG, UK

Shesh N. Rai
Director, Biostatistics Shared Facility
JG Brown Cancer Center
Associate Professor
Department of Bioinformatics and Biostatistics
School of Public Health and Information Sciences
University of Louisville
Louisville, KY 40202, USA

Vladimir A. Semenenko
Department of Radiation Oncology
Medical College of Wisconsin
8701 Watertown Plank Rd.
Milwaukee, WI 53226, USA

Yu Shyr
Vanderbilt University Medical Center
2220 Pierce Avenue, 571 Preston Building
Nashville, TN 37232-6848, USA

Jaroslaw Smieja
Institute of Automatic Control
Silesian University of Technology
Akademicka 16, 44-100 Gliwice, Poland

Robert D. Stewart
Purdue University School of Health Sciences
550 Stadium Mall Drive
West Lafayette, IN 47907-2051, USA

Aniko Szabo
Department of Population Sciences
Division of Biostatistics
Medical College of Wisconsin
8701 Watertown Plank Rd.
Milwaukee, WI 53226, USA

Ming Tan
Division of Biostatistics, Medical School Teaching Facility
University of Maryland Greenebaum Cancer Center
685 W. Baltimore Street, Suite 261
Baltimore, MD 21201, USA

Wai Y. Tan
Department of Mathematical Sciences
University of Memphis
Memphis, TN 38152, USA

Zigang Wang
Molecular Imaging, Siemens Medical Solutions Inc.
810 Innovation Drive
Knoxville, TN 37932-2751, USA

Li J. Zhang
Department of Biostatistics and Bioinformatics
Dana-Farber Cancer Institute
44 Binney St., Boston, MA 02115, USA

Qi Zheng
Department of Epidemiology and Biostatistics
School of Rural Public Health
Texas A&M Health Science Center
1266 TAMU, College Station, TX 77843, USA

J. M. Zhu
Frankfurt Institute for Advanced Studies (FIAS)
Johann Wolfgang Goethe University
Max-von-Laue-Str. 1
60438 Frankfurt am Main, Germany

PREFACE

In the past 15 years, molecular biologists and geneticists have uncovered some of the most basic mechanisms by means of which normal stem cells in a certain organ or tissue develop into cancerous tumors. This biological knowledge serves as a basis for various models of carcinogenesis. Furthermore, in order for biological findings to be tested quantitatively against human epidemiological data and animal experimental data and in order to develop efficient diagnostic, controlling, curative and preventive strategies for cancer, it is essential that these biological theories be transformed into adequate mathematical models supported by relevant methods of statistical data analysis. With this in mind, the goal of this book is to present review papers on most recent and advanced cancer models and their applications from world experts in cancer modelling.

The book contains 18 chapters and is organized into two parts. Part I consists of 11 chapters giving detailed treatment of an assortment of cancer models whereas Part II consists of 7 chapters dealing with applications of various cancer models. In Part I, oncogenetic trees, stochastic multi-stage models of carcinogenesis, models of cancer metastasis, cell cycle regulation by radiation, genomic instability, DNA damage and repair, bystander effects of radiation as well as epigenetic hypotheses and multiple pathways of colon cancer are extensively discussed. Also, Part I reviews the most recent discoveries in cancer biology and the ways they are transformed into stochastic models of carcinogenesis. Among topics dealt with in Part II the reader will find bladder cancer screening by MRI (Magnetic Resonance Imaging), models of drug resistance in cancer chemotherapy, and applications of stochastic and state space models to the assessment of cancer risks associated with environmental agents. Part II of the book also contains chapters on applications of wavelets to proteomics, design and analysis of tumor xenograft experiments, and analysis of occult tumor trial data.

Cancer models presented in the book include deterministic, stochastic, statistical, and state space models. They provide the reader with a state-of-the-art overview of the rapidly evolving field of cancer modelling. Our aspiration when putting together this volume was to bring the best of modern science to bear on biomedical problems related to the study of cancer. We hope that the book will be a useful source of information and references for scientists working in cancer research, biomathematics, biostatistics and bioinformatics as well as for biologists, cancer epidemiologists, clinical investigators and medical doctors at various stages of their career. We also hope that graduate students and their instructors will find it beneficial to their learning and teaching. The chapters of the book should not be viewed as a fixed compendium of methods and results but rather as a source of ideas inspiring further research by those who use (or wish to learn how to use) quantitative methods in cancer studies.

We are well aware that selection of topics for the book reflected our research interests and that our desire to cover a broad range of cancer models and their applications made the book to a certain degree heterogeneous. Through reviewing and editing process we encouraged clarity of exposition; however, it is up to the reader to judge the measure of our success. Finally, the reader should be advised that several chapters require some proficiency in mathematics and statistics.

We would like to express our sincere appreciation to all contributors who have made this book possible. We wish also to thank Ms. J. Quek, senior editor of World Scientific Publication Company, for her assistance in the publication of this book. Finally, it is acknowledged that the work by Wai-Yuan Tan was supported by a research grant from National Cancer Institute of NIH, grant number R15 CA113347-01, and the work by Leonid Hanin was supported by grant #974 awarded on May 1, 2006 by the Faculty Research Committee at Idaho State University.

Wai-Yuan Tan and Leonid Hanin
July 23, 2007

Chapter 1
ONCOGENETIC TREES

Aniko Szabo and Kenneth M. Boucher

Human solid tumors are believed to be caused by a sequence of genetic abnormalities arising in normal and premalignant cells. The understanding of these sequences is important for improving cancer treatment. Models for the occurrence of the abnormalities include linear structure and a recently proposed tree-based structure. We will describe the oncogenetic tree model and an efficient algorithm for its estimation. We also discuss methods for estimating the reliability and goodness-of-fit of this reconstruction. An R package "Oncotree" implementing the described methodology is available from the authors.

Keywords: Oncogenesis, branching, robustness, bootstrap.

1. INTRODUCTION

A seminal effort in describing the steps involved in carcinogenesis was a study of colorectal tumor development by Vogelstein *et al.* (1988). The authors have shown that while the genetic profile of individual tumors varied widely and there was no single mutation present in all tumors, certain changes tended to occur early in the development, and other ones relatively late. In a subsequent paper Fearon and Vogelstein (1990) proposed a linear genetic model for colorectal tumorigenesis as a preferred order of occurrence of the genetic abnormalities while acknowledging the existence of other pathways. Their conclusion was equivocal: "...although a preferred order for the genetic alterations...exist, the data suggest that the progressive accumulation of these alterations is the most consistent feature..." (Fearon and Vogelstein, 1990).

The idea of multiple pathways of cancer progression was introduced in the works of Zelen (Zelen, 1968; Feldstein and Zelen, 1984). Szabo and Yakovlev (2001) considered modeling the process of tumorigenesis as a mixture of all possible pathways; however, not so surprisingly, they showed that given the available data such a model is generally not identifiable in the case of three or more genetic events. Thus one linear pathway does not adequately describe the diversity of cancer development, yet the commonly available data does not contain sufficient information to identify a model with an arbitrarily large number of pathways.

As a compromise between the extremes of too few and too many pathways, Desper *et al.* (1999) introduced the oncogenetic tree model in which the possible pathways form a directed tree structure: they share a common beginning, but can "branch out" at the ends. The linear model of Fearon and Vogelstein (1990) is a special case of the oncogenetic tree model; however the latter is more flexible and appears to be more realistic. This paper is dedicated to describing oncogenetic trees and investigating their properties.

2. DEFINITIONS AND BASIC RESULTS

2.1. Description of the Data

Before defining the oncogenetic tree model, we first describe the data that it is designed to model and that will be used for model fitting. Let $M_1, M_2, \ldots, M_n$ denote the genetic alterations of interest. These could be point mutations, gain or loss of chromosomal regions or other genetic events. N independent specimens ("tumors") are obtained and the presence or absence of the alterations of interest is recorded as a binary vector $\mathbf{x}_j = (x_{j1}, x_{j2}, \ldots, x_{jn})$, where

$$x_{j\ell} = \begin{cases} 0, & \text{if } M_\ell \text{ is absent in the } j\text{th tumor} \\ 1, & \text{if } M_\ell \text{ is pressent in the } j\text{th tumor} \end{cases}, \quad j = 1, \ldots, N; \ell = 1, \ldots, n.$$

The reconstruction algorithm will only use the marginal and pairwise frequencies of occurrence of the alterations, so we introduce the following notations:

- $p_i = P(M_i \text{ occurs}), i = 1, \ldots, n; p_0 = 1$
- $p_{ij} = \begin{cases} P(\text{both } M_i \text{ and } M_j \text{ occur}), i, j = 1, \ldots, n; i \neq j \\ p_i, \qquad\qquad\qquad\qquad\quad i = 1, \ldots, n; j = 0, i \end{cases}$
- $p_{i|j} = P(M_i \text{ occurs given that } M_j \text{ has occurred}), i, j = 1, \ldots, n; i \neq j$
- $p_{i \vee j} = P(M_i \text{ or } M_j \text{ or both occur}), i, j = 1, \ldots, n; i \neq j.$

We will assume that only actually observed alterations are modeled, so $p_i > 0$ always.

2.2. The Oncogenetic Tree Model

In this section we give a short description of an oncogenetic tree and provide some pertinent definitions. For a more complete treatment we refer the reader to Desper *et al.* (1999). An oncogenetic tree models the process of occurrence of genetic alterations in carcinogenesis using a directed tree structure. In this paper we will use both the words *tree* and *branching* to refer to a rooted directed graph T with vertex set $\{M_0\} \cup V = \{M_0, M_1, M_2, \ldots, M_n\}$ such that for every vertex $M_i \in V$ there is a unique directed path from M_0 to M_i along the edges of T. In the literature such a structure is also called an *arborescence*. We will use the common "arrow" notation to denote the edges of the tree: $\overrightarrow{M_i M_j}$ denotes the directed edge from vertex M_i to vertex M_j.

Intuitively, vertex M_0 (the root of the tree) represents the "no alterations" event and each of the vertices of V represent a certain mutation or other genetic alteration. Thus the alteration status of a tumor is described by a set of the vertices that correspond to the alterations that are present in the tumor.

First we give an intuitive description of the oncogenetic tree using a simple example given in Figure 1; here $M_1, M_2, \ldots, M_7$ represent hypothetical alterations of interest. The development of a tumor according to this tree could be the following: the tumor starts as $\{M_0\}$, that is none of the alterations have occurred. Now the events M_1 and M_2 can occur, and their appearance is independent of each other, that is the occurrence of one of them does not change the probability of occurrence for the other one. Suppose M_2 has occurred and so the status of the tumor becomes $\{M_0, M_2\}$. Now in addition to M_1, the alterations M_3, M_4 and

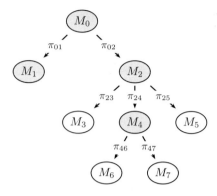

Fig. 1. An example of an untimed oncogenetic tree with seven possible alterations.

M_5 can also occur, so the tumor can move to the status $\{M_0, M_1, M_2\}$, $\{M_0, M_2, M_3\}$, $\{M_0, M_2, M_4\}$ or $\{M_0, M_2, M_5\}$ and so on. The observed status of the tumor depends on the time of the observation. The values π_{ij} on the edges are the probabilities of transition along the given edge by the time of observation. These values allow finding the model-based probability of observing any combination of the alterations in a tumor. For example, $P(\{M_0, M_4\}) = 0$ because, according to the tree, M_2 had to occur before M_4 could; while the probability of the set highlighted with grey is $P(\{M_0, M_1, M_2, M_4\}) = \pi_{01}\, \pi_{02}\, \pi_{24}(1 - \pi_{23})(1 - \pi_{25})(1 - \pi_{46})(1 - \pi_{47})$.

This intuitive description is formalized by the following definitions:

Definition 1. *A pure untimed oncogenetic tree is a tree T with a probability $\pi(e)$ attached to each edge e. This tree generates observations on mutation presence/absence the following way: each edge e is independently retained with probability $\pi(e)$; the set of vertices that are still reachable from M_0 gives the set of the observed genetic alterations.*

A somewhat more realistic model incorporates the progression of time.

Definition 2. *A pure timed oncogenetic tree is a tree T with a rate $\lambda(e)$ attached to each edge and an observation-time distribution φ on $\mathbb{R}^+$. This tree generates observations on mutation presence/absence the following way: first the time of observation t is drawn from φ and the transition time along each edge e is drawn independently from an exponential distribution with rate $\lambda(e)$. The set of vertices that are reachable from M_0 along a path for* which the sum of transition times is less than t gives the set of the observed genetic alterations.

While the above definition of an oncogenetic tree gives a clearly interpretable model for the process of occurrence of genetic events during carcinogenesis, real data never quite follows prescribed models. Thus before a tree model can be fitted, an error structure describing the character of random deviations from the model has to be defined. There are several sources of errors in the context of this model. Some of the observations $x_{j\ell}$ might be incorrect due to the imperfection of the detection technology or the spatial heterogeneity of the tumor. A more fundamental source of "errors" is the truly random occurrence of genetic alteration unrelated to the causal process of carcinogenesis. The error model introduced by Szabo and Boucher (2002) suggests combining the possible errors regardless of their source into two basic types: false positives and false negatives, and base the error model on the probabilities of occurrence of these errors.

2.2.1. *Error model*

- The tumor develops according to the pure oncogenetic tree model.
- The presence/absence of each alteration is independently measured.
- If the alteration is present it is not observed with probability ε_-. If the alteration is absent it is observed with probability ε_+.

3. RECONSTRUCTION

The main goal of the analysis is the reconstruction of the topology of the oncogenetic tree T; the estimation of the edge transition probabilities and error probabilities is of secondary importance. First we will concentrate on the conceptual aspects of reconstruction and assume that there is no sampling error (the sample size $N \to \infty$). One of the main results of the theory of oncogenetic trees is the Reconstruction Algorithm given in Figure 2 that provides an explicit construction method for T (Szabo and Boucher, 2002). This algorithm takes a greedy bottom-up approach: it assigns the parent of each node by finding the maximum-weight in-edge starting from the leaves.

> **Reconstruction algorithm**
>
> (i) Estimate p_i and p_{ij}, $i, j = 0, \ldots, n$ from the marginal frequencies in the data using the definitions (Section 2.1).
>
> (ii) Construct a complete directed graph on vertices $\{M_0, M_1, \ldots, M_n\}$ representing the occurrence of individual events with weight $w(M_i, M_j) = \log \frac{p_{ij}}{p_j(p_i+p_j)}$ for the directed edge $\overrightarrow{M_i M_j}$.
>
> (iii) Build a directed spanning tree (branching) B by defining the ancestor of each vertex the following way:
>
>   (a) Let S denote the set of vertices with assigned parent. Start with $S = \emptyset$.
>
>   (b) Find the vertex $M_i \notin S$ with the smallest probability p_i (in case of a tie, choose randomly).
>
>   (c) Let its parent in B be the vertex $M_j \in S$ such that $w(M_j, M_i)$ is maximal. Set $S = S \cup \{M_i\}$.
>
>   (d) Repeat steps (b)–(c) until all vertices have an assigned parent, that is $S = V$ (vertex M_0 does not need a parent).

Fig. 2. Algorithm for reconstructing the oncogenetic tree from marginal and pairwise joint distribution of alterations.

In the absence of false observations, this algorithm reconstructs the original tree T provided that it is not skewed:

Definition 3. An oncogenetic tree T is skewed if there exist two vertices M_i, M_j with a least common ancestor M_k in T such that

$$p_{i|j} \geq p_{i \vee j | k}. \tag{1}$$

Lemma 0. *An untimed oncogenetic tree is not skewed.*

Proof. In an untimed tree events M_i and M_j are conditionally independent given the status of their least common ancestor M_k, so $p_{i|j} = p_{i|k} < p_{i \vee j | k}$. □

Note that in a timed tree $p_{i|j} > p_{i|k}$, so skewness can occur.

It can be easily seen that in a pure oncogenetic tree the non-skewness condition is equivalent to having

$$p_{i|j} < \frac{p_i + p_j}{p_k + p_j}. \tag{2}$$

This form will be easier to use in the proofs.

Theorem 1 (Reconstruction Theorem). *Let T be a non-skewed oncogenetic tree (timed or untimed) and ε_+, ε_- be the probabilities of, respectively, a false positive and false negative observation, and let $p_{\min} = \min_i p_i$. If $\varepsilon_+ + \varepsilon_- < 1$ and $\varepsilon_+ < (p_{\min})^{1/2}(1 - \varepsilon_+ - \varepsilon_-)$, then the branching B given by the tree reconstruction algorithm is exactly T.*

We will prove this theorem using three lemmas. First note that after incorporating false positives and negatives, the probabilities of observing alterations will become

$$\begin{aligned} p_i^* &= p_i(1 - \varepsilon_-) + (1 - p_i)\varepsilon_+ \\ p_{ij}^* &= p_{ij}(1 - \varepsilon_-)^2 + (p_{i \vee j} - p_{ij})(1 - \varepsilon_-)\varepsilon_+ + (1 - p_{i \vee j})\varepsilon_+^2. \end{aligned} \tag{3}$$

Lemma 1.1. *If M_j is a parent of M_i in T, then $p_j^* > p_i^*$.*

Proof. Since $p_j > p_i$, the statement easily follows from the first equation in Eq. (3): $p_j^* - p_i^* = (p_j - p_i)(1 - \varepsilon_- - \varepsilon_+) > 0$ unless $\varepsilon_- + \varepsilon_+ \geq 1$. □

Lemma 1.2. *If M_j is not an ancestor of M_i in T, then $w(M_k, M_i) > w(M_j, M_i)$, where M_k is the least common ancestor of M_i and M_j.*

Proof. From the definition,

$$w(M_k, M_i) - w(M_j, M_i) = \log \frac{p_{ki}^*(p_i^* + p_j^*)}{p_{ji}^*(p_k^* + p_i^*)}.$$

As M_k is an ancestor of M_i, $p_{ki} = p_i$, so without observation errors the non-skewness assumption Eq. (2) would ensure that the above expression is positive, proving the lemma. We will show that under the assumptions of this theorem the non-skewness inequality is maintained even after the introduction of observational errors.

From Eqs. (3) we have

$$p^*_{ki}(p^*_i + p^*_j) - p^*_{ji}(p^*_k + p^*_i) = [p_i(1 - \varepsilon_-)^2 + (p_k - p_i)(1 - \varepsilon_-)\varepsilon_+$$
$$+ (1 - p_k)\varepsilon^2_+][(p_i + p_j)(1 - \varepsilon_- - \varepsilon_+) + 2\varepsilon_+] - [p_{ij}(1 - \varepsilon_-)^2$$
$$+ (p_{i\vee j} - p_{ij})(1 - \varepsilon_-)\varepsilon_+ + (1 - p_{i\vee j})\varepsilon^2_+][(p_i + p_k)(1 - \varepsilon_- - \varepsilon_+)$$
$$+ 2\varepsilon_+] = (1 - \varepsilon_- - \varepsilon_+)[(p_i^2 - p_i p_{ij} + p_i p_j - p_{ij} p_k)(1 - \varepsilon_- - \varepsilon_+)^2$$
$$+ 2\varepsilon_+(p_i - p_{ij})(1 - \varepsilon_- - \varepsilon_+) + (p_k - p_j)\varepsilon^2_+] > 0.$$

The second equality can be checked by expanding both sides of the equation and the last inequality follows because $1 - \varepsilon_- - \varepsilon_+ > 0$ (assumption of the theorem), $p_i^2 - p_i p_{ij} + p_i p_j - p_{ij} p_k > 0$ (non-skewness assumption Eq. (2)), $p_i > p_{ij}$ (by definition) and $p_k > p_j$ (M_k is an ancestor of M_j). Thus

$$\frac{p^*_i + p^*_j}{p^*_k + p^*_i} > \frac{p^*_{ji}}{p^*_{ki}},$$

so $w(M_k, M_i) > w(M_j, M_i)$, proving the statement. $\square$

Lemma 1.3. *If M_j is the parent of M_i in T and M_k is any other ancestor of M_i in T then $w(M_k, M_i) < w(M_j, M_i)$.*

Proof.

$$w(M_j, M_i) - w(M_k, M_i) = \log \frac{p^*_{ji}(p^*_k + p^*_i)}{p^*_{ki}(p^*_j + p^*_i)}.$$

Without observational errors $p_{ji} = p_{ki} = p_i$ and this expression is positive as $p_k > p_j$. We will show that this statement holds in the presence of the errors as well by invoking the condition $\varepsilon_+ < (p_{\min})^{1/2}(1 - \varepsilon_+ - \varepsilon_-)$.

As M_k and M_j are ancestors of M_i, $p_{ji} = p_{ki} = p_i$ and $p_{k\vee i} = p_k$, so from Eq. (3) we have

$$p^*_{ji}(p^*_k + p^*_i) - p^*_{ki}(p^*_j + p^*_i) = [p_i(1 - \varepsilon_-)^2 + (p_j - p_i)(1 - \varepsilon_-)\varepsilon_+$$
$$+ (1 - p_j)\varepsilon^2_+][(p_i + p_k)(1 - \varepsilon_- - \varepsilon_+) + 2\varepsilon_+] - [p_i(1 - \varepsilon_-)^2$$

$$+ (p_k - p_i)(1 - \varepsilon_-)\varepsilon_+ + (1 - p_k)\varepsilon_+^2][(p_i + p_j)(1 - \varepsilon_- - \varepsilon_+)$$
$$+ 2\varepsilon_+] = (1 - \varepsilon_- - \varepsilon_+)(p_k - p_j)[(1 - \varepsilon_- - \varepsilon_+)^2 p_i - \varepsilon_+^2] > 0.$$

Again, the verification of the second equality is straightforward, while the inequality follows because $1 - \varepsilon_- - \varepsilon_+ > 0$, $p_k > p_j$ (M_k is an ancestor of M_j) and $(1 - \varepsilon_- - \varepsilon_+)^2 p_i - \varepsilon_+^2 > (1 - \varepsilon_- - \varepsilon_+)^2 p_{\min} - \varepsilon_+^2 > 0$ (assumption of the theorem).

Hence $w(M_j, M_i) > w(M_k, M_i)$. □

Proof of the Reconstruction Theorem. Combining together the results of these lemmas, we have proven that the vertex M_i chosen in step (3b) of the Reconstruction Algorithm cannot be the parent of any other vertex in S (Lemma 1.1); and the vertex M_j chosen in step (3c) is its parent in T (Lemmas 1.2, 1.3). Hence B coincides with T. □

The Reconstruction Theorem can be generalized in a variety of ways. For example, Szabo and Boucher (2002) developed sufficient conditions for reconstructing the oncogenetic tree in the case when the rates of false negative errors are allowed to vary as long as they are "almost equal". Here we will not attempt to find the most general statement possible, but rather explore other issues.

The Reconstruction Algorithm provides an intuitively appealing and computationally fast approach for estimating an oncogenetic tree. However its constructivist nature does not provide a good conceptual description of the tree. The following theorem from Desper et al. (1999) gives such a characterization.

Theorem 2. *The oncogenetic tree T is a maximum weight branching spanning all the vertices $\{M_0, M_1, \ldots, M_n\}$.*

Proof. Assume that T is not a maximum weight branching, but instead D is. We will prove that D coincides with T in three steps, each using one of the three lemmas proved in the Reconstruction Theorem. □

Lemma 2.1. *M_0 is the root of the maximum weight branching D.*

Suppose another vertex M_i is the root of D, while the parent of M_0 in D is M_j ($j = i$ is possible). Consider the branching D' obtained by replacing the edge $\overrightarrow{M_jM_0}$ by $\overrightarrow{M_0M_i}$: $D' = D - \overrightarrow{M_jM_0} + \overrightarrow{M_0M_i}$. D' has M_0 as a root. Then D' has a higher weight than D, since

$$w(D') - w(D) = w(M_0, M_i) - w(M_j, M_0) = -\log(1 + p_i^*) - \log(p_j^*)$$
$$+ \log(1 + p_j^*) > -\log(2) + \log(1 + 1/p_j^*) > 0,$$

using the fact that all probabilities are less than one.

Hence if M_0 is not the root, the branching cannot have maximal weight.

Lemma 2.2. *If $\overrightarrow{M_iM_j}$ is an edge in D, then M_i is an ancestor of M_j in T.*

Suppose the statement is false. From all the vertices M_j with parent in D not an ancestor in T, choose the one closest to M_0 in T. Let M_k be the least common ancestor of M_i and M_j in T.

Consider the branching D' obtained by replacing the edge $\overrightarrow{M_iM_j}$ by $\overrightarrow{M_kM_j}$: $D' = D - \overrightarrow{M_iM_j} + \overrightarrow{M_kM_j}$. Since M_k is closer to the root then M_j, the statement of the lemma holds for M_k and its ancestors. Hence M_j cannot be M_k's ancestor and D' is really a branching. Then

$$w(D') - w(D) = w(M_k, M_i) - w(M_i, M_j) = \log \frac{p_{jk}^*(p_i^* + p_j^*)}{p_{ij}^*(p_{ij}^* + p_k^*)} > 0$$

as shown in Lemma 1.2.

Lemma 2.3. *For every edge $\overrightarrow{M_iM_j}$ of D, M_i is the parent of M_j in T as well.*

Again, suppose $M_k \neq M_i$ is the parent of M_j in T. Then from the previous step M_i is an ancestor of M_k in T. Consider the branching $D' = D - \overrightarrow{M_iM_j} + \overrightarrow{M_kM_j}$. The previous step ensures that M_j cannot be an ancestor of M_k, so D' really is a branching. Then

$$w(D') - w(D) = w(M_k, M_i) - w(M_i, M_j) = \log \frac{p_{jk}^*(p_i^* + p_j^*)}{p_{ij}^*(p_{ik}^* + p_k^*)} > 0$$

as proven in Lemma 1.3 of the Reconstruction Theorem.

Note, that unlike finding a maximum weight spanning tree, the problem of finding a maximum weight branching (a directed tree) is not simple. Polynomial-time algorithms have been developed for this problem (Edmonds, 1967; Karp, 1971; Tarjan, 1977), however none are as simple and fast as the Reconstruction Algorithm. The reason for this is that our algorithm uses the special structure of the weights that is not available in the general case.

4. SAMPLE SIZE ESTIMATION

The success of the Reconstruction Algorithm depends on the relative order of the frequencies of occurrence of the mutations and of the edge weights. In the Reconstruction Theorem we have shown that (under certain conditions) the introduction of false positive and negative errors maintains the correct ordering. However these results were proven only for the "true" probabilities p_i^* and p_{ij}^*, ignoring the variability inherent to sampling. In this section we give a lower bound on the sample size that is sufficient for reconstruction with a (large) predefined probability $1 - \xi$.

First, we introduce a few notations. Let $\alpha = \min_{i,j,k} \frac{p_i + p_j}{p_k + p_j} - p_{i|j}$, where the minimum is taken over all triples (i, j, k) such that M_k is the least common ancestor of M_i and M_j, $\beta = \min_i (p_{\text{parent}(i)} - p_i)$, $p_{\min} = \min_i p_i$ and $p_{\max} = \max_i p_i$. Intuitively, α measures the tightness of the non-skewness assumption, and β measures the ability to determine the order of "adjacent" events.

Theorem 3. *Let T be a non-skewed oncogenetic tree (timed or untimed) with n vertices (not including the root M_0) and $\varepsilon_+, \varepsilon_-$ be the probabilities of, respectively, a false positive and false negative observation. If $\varepsilon_+ + \varepsilon_- < 1$, $\chi = \varepsilon_+/(1 - \varepsilon_+ - \varepsilon_-) < p_{\min}^{1/2}$, and the sample size*

$$N \geq \frac{81(p_{\max} + \chi)^3 (\ln[n(n+1)] - \ln(2\xi))}{(1 - \varepsilon_+ - \varepsilon_-)^3 (\min[\alpha p_{\min}^2 + \beta \chi^2, \beta(p_{\min} - \chi^2)])^2}, \quad (4)$$

then with probability at least $1 - \xi$ the branching B given by the Reconstruction Algorithm is exactly T.

Proof. Let $\hat{\delta}_i = \hat{p}_i^* - p_i^*$ and $\hat{\delta}_{ij} = \hat{p}_{ij}^* - p_{ij}^*$, $i, j = 0, \ldots, n$ denote the deviation of the observed frequencies from their theoretical counterparts, and let $\delta = \max(\max_i \hat{\delta}_i, \max_{ij} \hat{\delta}_{ij})$. With some extra work in each of the lemmas in the proof of the Reconstruction Theorem, it can be shown to remain valid if

$$\delta < \frac{1}{9 p_{\max}^*} (1 - \varepsilon_+ - \varepsilon_-)^3 \min[\alpha p_{\min}^2 + \beta \chi^2, \beta(p_{\min} - \chi^2)], \quad (5)$$

where $\chi = \varepsilon_+ / (1 - \varepsilon_+ - \varepsilon_-)$, $p_{\max}^* = \max_i p_i^*$. From the first equation in Eq. (3), $p_{\max}^* = (1 - \varepsilon_+ - \varepsilon_-) p_{\max} + \varepsilon_+$.

The sample size estimation will be based on the Chernoff inequality Ross (2002): if $X \sim Binomial(N, p)$ and $\hat{p}_N = X/N$ denotes the estimated response probability, then for any $u > 0$:

$$P\left(\hat{p}_N - p > \frac{u\sqrt{p}}{\sqrt{N}}\right) \leq e^{-u^2}. \quad (6)$$

Specifically,

$$P\left(\hat{\delta}_i > \frac{u\sqrt{p_{\max}^*}}{\sqrt{N}}\right) < P\left(\hat{\delta}_i > \frac{u\sqrt{p_i^*}}{\sqrt{N}}\right) \leq e^{-u^2}$$

$$P\left(\max_i \hat{\delta}_i > \frac{u\sqrt{p_{\max}^*}}{\sqrt{N}}\right) \leq n e^{-u^2}.$$

Similarly,

$$P\left(\max_{ij} \hat{\delta}_{ij} > \frac{u\sqrt{p_{\max}^*}}{\sqrt{N}}\right) \leq \binom{n}{2} e^{-u^2},$$

hence

$$P\left(\delta > \frac{u\sqrt{p_{\max}^*}}{\sqrt{N}}\right) \leq \frac{n(n+1)}{2} e^{-u^2}.$$

We select u to ensure the desired significance level by setting $e^{-u^2} n(n+1)/2 = \xi$, that is $u^2 = \ln[n(n+1)/(2\xi)]$. On the other side of the inequality, using the limit on the sample size N set in Eq. (4), we can see that the requirement of Eq. (5) is satisfied:

$$\frac{u\sqrt{p^*_{max}}}{\sqrt{N}} \leq \frac{1}{9p^*_{max}}(1 - \varepsilon_+ - \varepsilon_-)^3 \min[\alpha p^2_{min} + \beta\chi^2, \beta(p_{min} - \chi^2)].$$

$\square$

5. PARAMETER ESTIMATION

So far we have addressed only the reconstruction of the topological structure of the oncogenetic tree. Here we will discuss the estimation of the further parameters of the model for untimed oncogenetic trees, that is the estimation of the edge transition probabilities $\pi(e)$ and the error rates $\varepsilon_+, \varepsilon_-$. The estimation of these parameters is linked.

Edge transition probabilities. We propose using a method-of-moments estimator based on the relationship valid in the pure tree: $\pi(\overrightarrow{M_j M_i}) = p_{i|j}$. Equations (3) describe the effect of the error model on this relationship. Thus, in the absence of sampling variability, we can recover the edge weights by solving these equations for $p_{i|j}$:

$$p_{i|j} = \frac{p^*_{ij} - (p^*_i + p^*_j)\varepsilon_+ + \varepsilon_+^2}{(p^*_j - \varepsilon_+)(1 - \varepsilon_+ - \varepsilon_-)}. \tag{7}$$

In practice, we use the observed marginal probabilities $\hat{p}^*_i$ and $\hat{p}^*_{ij}$ to estimate the edge weights.

Error rates. The tree structure, the edge weights and the error probabilities jointly define a distribution of all the possible outcome sets. Thus, we can estimate the error probabilities by fitting this estimated distribution to the observed one. However, obtaining the model-based distribution is computationally expensive (each outcome based on the error-free tree can be "corrupted" in 2^n ways), so we propose fitting the only the marginal probabilities which are much easier to compute.

The tree-based marginal probability of occurrence for event M_i is

$$p^*_i(\varepsilon_+, \varepsilon_-)_T = \varepsilon_+ + (1 - \varepsilon_-)\prod_k p_{j_{k+1}|j_k}, \tag{8}$$

where $0 = j_1, j_2, \ldots, j_d = i$ is the path from the root to M_i.

Thus, combining Eqs. (7) and (8), for every value of the error rates $(\varepsilon_+, \varepsilon_-)$ we can obtain the tree-based marginal probabilities of occurrence and compare them to the observed frequencies. After defining a suitable error-function (in the application we use the squared ℓ_2-distance $\sum_i (\hat{p}_i^* - p_i^*(\varepsilon_+, \varepsilon_-)_T)^2)$, the error rates can be estimated by minimizing the error-function.

6. EXAMPLE: RENAL CARCINOMA DEVELOPMENT

In this section we show an application of the oncogenetic tree model to comparative genomic hybridization data for clear cell renal carcinoma. The comparative genomic hybridization technique (CGH) developed by Kallioniemi et al. (1992) was used on each of the $N = 124$ samples as described in Jiang et al. (1998) to obtain information on chromosome number aberrations (CNAs) on each of the arms of the chromosomes. A more detailed description of the dataset can be found in Desper et al. (1999); Jiang et al. (2000). The human genome consists of 22 autosomal and 2 sex-linked chromosomes. All chromosomes have a long arm q and most (except 13, 14, 15, 21 and 22) have a significant short arm p. The chromosome arms are denoted by attaching the letter p or q to the appropriate chromosome number. The CGH technique uses fluorescent staining to detect abnormal (increased or decreased) number of DNA copies. In contrast to microarray technology, the results cannot be narrowed down to a specific gene, only to a segment of the chromosome, called a band. However when two tumors have abnormalities along a similar region, it is often difficult to tell whether they are based on the same genetic change, so in the renal carcinoma data set the results are reported as a gain or loss on a certain arm, without further distinction for specific bands. Also, as some samples were from females, the Y chromosome was excluded from consideration. This resulted in 82 possible events from 41 locations (both a gain and a loss could occur on different bands of the same chromosomal arm). It is common to denote a change in DNA copy number on a specific chromosome arm by prefixing a "−" sign for decrease and a "+" for increase. Thus, say, $-3q$ denotes abnormally low DNA copy number on the q arm of the 3rd chromosome.

To reduce the total number of events and to keep $p_{\min}$ reasonably large, we selected the seven most frequent events (listed in decreasing order of

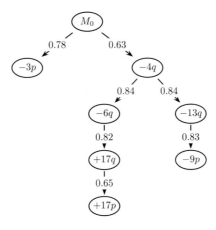

Fig. 3. Oncogenetic tree based on the renal carcinoma CGH data. The edges are labeled with the estimated transition probability $\hat{\pi}_e$.

frequency): $-3p, -4q, -6q, -13q, -9p, +17q, +17p$. Other approaches to the selection of relevant events have also been proposed, including the method due to Brodeur *et al.* (1982) that allows to adjust for prior probabilities of alteration on the chromosome arms. The later approach was used in Jiang *et al.* (2000) and resulted in a list of 12 events; the selected events are included in that list.

Upon applying the Reconstruction Algorithm we obtain the tree shown in Figure 3. This tree will be used as a basis of simulation studies in Section 7, so we introduce the notation $\mathcal{T}_{\text{CGH}}$ for it. The error rate was estimated by fitting the tree-based marginal probabilities of occurrence $p_i(\varepsilon_+, \varepsilon_-)$ to the observed frequencies $\hat{p}_i$ through minimizing the squared ℓ_2-distance $\sum_i (\hat{p}_i - p_i(\varepsilon_+, \varepsilon_-))^2)$ as described in Section 5: $\hat{\varepsilon}_+ = 0.00, \hat{\varepsilon}_- = 0.228$.

Since the combination of the pure oncogenetic tree model with the error process defines a non-zero probability for any possible outcome (given the parameters), maximum likelihood estimation of the error rates is possible.

7. PROPERTIES OF THE ONCOGENETIC TREE ESTIMATOR: A SIMULATION STUDY

In this section we investigate the properties of the Reconstruction Algorithm as an estimator of the true oncogenetic tree structure. The Reconstruction Theorem guarantees that given a sufficiently large sample and with error rates satisfying its restrictions, the correct tree will be reconstructed with a given probability. However, the sufficient sample size provided by Eq. (4) is extremely conservative. For example, for $\mathcal{T}_{\text{CGH}}$ we have $\alpha = 0.032$, $\beta = 0.031$, $\varepsilon_+ = 0.00$, $\varepsilon_- = 0.228$, $p_{\min} = 0.26$, and $p_{\max} = 0.78$, so for a 95% confidence ($\xi = 0.05$) the formula requires $N \approx 210{,}000{,}000$ samples. In practice, it is likely that much smaller sample sizes are sufficient. Also, the dependence of N on the parameters is quite convoluted, their relative significance is hidden. All the investigations are empirical and are based on the oncogenetic tree $\mathcal{T}_{\text{CGH}}$ estimated in Section 6. While many of the specific results depend on the exact structure of the tree, the values of the transition probabilities, etc. the qualitative conclusions are likely to be valid generally.

7.1. Simulating Data Based on a Given Tree

The generation of data from an oncogenetic tree is fairly straightforward based on the definitions of a pure oncogenetic tree and of the error process. First a "clean" observation is generated by retaining each edge with the associated probability and creating the set of occurred alterations from the vertices still reachable from M_0. Then the errors are introduced independently for each alteration: each occurred alteration is not observed with probability ε_-, and each alteration that has not occurred is observed with probability ε_+. This process is repeated until the required sample size is reached.

7.2. Probability of Correct Reconstruction

First, we investigate the dependence of the probability of correct reconstruction of the data generating tree on the error probabilities when the sample size is fixed ($N = 124$). We considered 12 combinations for $(\varepsilon_+, \varepsilon_-)$ with the parameters taking the values 0.00 through 0.10 and 0.20, respectively, and generated 1000 random data sets for each combination. The Reconstruction Algorithm was applied to each random data set and the probability of incorrect reconstruction was estimated as the proportion of sets for which the reconstructed tree had the same structure as $\mathcal{T}_{\text{CGH}}$. To ease interpretation in Figure 4, a logistic surface with quadratic dependence on ε_+ and ε_- was fitted to the resulting estimates.

Oncogenetic Trees 17

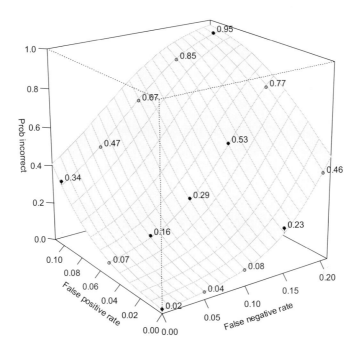

Fig. 4. Effect of error rate on the probability of incorrect reconstruction. Based on the renal carcinoma tree $\mathcal{T}_{\text{CGH}}$ in Fig. 3, 1000 simulated samples of size 124 at each point.

From the results it is evident that the effect of the error probabilities ε_+ and ε_- on the success of reconstruction is quite different. False positive errors appear to have a significantly higher deteriorating effect than false negative errors. Unfortunately, the presence of observations that are false positive with respect to the tree model is an intrinsic feature of the problem and cannot be eliminated by technological improvement. Fortunately, in our data the false negative rate is the one that is estimated to be high. Using simulations, the probability of correct reconstruction (assuming the model is correct) was estimated to be 39%.

7.3. Sample Size for High Probability of Reconstruction

For a researcher planning to collect mutation occurrence data the important question is rather the sample size required to achieve correct reconstruction with a sufficiently high probability. For the same 15 combinations of

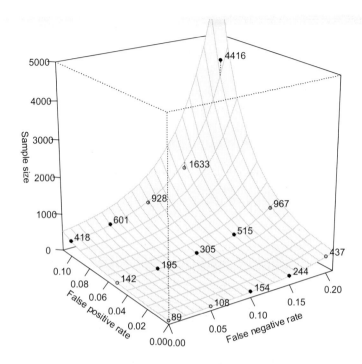

Fig. 5. Effect of error rates on the sample size $N_{0.95}(\varepsilon_+, \varepsilon_-)$ required for 95% confidence reconstruction. Based on the renal carcinoma tree $\mathcal{T}_{\text{CGH}}$ in Fig. 3, 1000 simulated samples of size 124 at each point.

$(\varepsilon_+, \varepsilon_-) \in \{0; 0.05; 0.1\} \times \{0; 0.05; 0.1; 0.15; 0.20\}$ as above, we estimated the sample size $N_{0.95}(\varepsilon_+, \varepsilon_-)$ at which the tree reconstructed from a data set randomly generated from $\mathcal{T}_{\text{CGH}}$ coincided with it 95% of the time. The frequency estimates were based on 1000 simulated data sets. Figure 5 shows the results for the 15 design points with a fitted surface. As previously, false positive errors have a much stronger impact on the required sample size. It is notable that unless the error rates are large, the required samples size is realistic in practice.

In Section 7.2 we found that with error rates $\varepsilon_+ = 0.00$ and $\varepsilon_- = 0.228$, the probability of correct reconstruction with sample size 124 is 39%. We used 1000 simulated data sets to estimate the sample size required for 95% confidence reconstruction of the oncogenetic tree for the renal carcinoma data: $N_{0.95}(\hat{\varepsilon}_+, \hat{\varepsilon}_-) = 658$.

8. GOODNESS OF FIT

In the previous section we used simulation to estimate the confidence of correct reconstruction *if the oncogenetic tree model is correct*. Thus the 39% confidence level is model-based. In this section we use bootstrap techniques to obtain a non-parametric estimate of the reconstruction confidence, and thus examine the goodness of fit of the oncogenetic tree model.

8.1. Bootstrap Estimate of Reconstruction Confidence

The reconstruction confidence is closely related to the variability of the estimator of the structure of the oncogenetic tree. A well-established non-parametric method to evaluate the variability of an estimator is bootstrap resampling. We generated 1000 bootstrap data sets, that is the original data set was sampled with replacement to obtain new data sets of the same size, and an oncogenetic tree $\tilde{T}_i$, $i = 1, \ldots, 1000$ was estimated for each of these data sets. We found a very high variability among the $\tilde{T}_i$'s: 149 different trees were found with the most frequent (T_{CGH}) occurring only 8% of the time. Figure 6 shows the nine most frequently occurring trees.

Our non-parametric estimate of the reconstruction confidence is the proportion of the bootstrap estimates $\tilde{T}_i$ that have the same structure as the tree estimated from the original data T_{CGH}, that is 8%. This is substantially lower than the parametric estimate of 39% obtained in Section 7.2. Such a disagreement raises doubt in the goodness of fit of the oncogenetic tree model.

8.2. Analysis of the Stable Portions

Despite the high variability of the trees based on bootstrap resamples of the data, there are conserved portions that are present in a large proportion of the trees. These pieces and the corresponding probabilities of occurrence are highlighted in Figure 7.

$-3p$ is a direct descendant of the root. According to the oncogenetic tree model, the most frequently occurring event cannot be a consequence of another event, so it has to be a direct descendant of the root. In the renal carcinoma data, the $-3p$ event is by far the most frequent ($\approx 60\%$ while

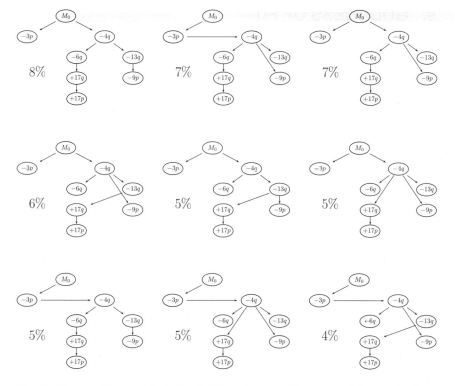

Fig. 6. The most frequent (out of 149 different) trees with associated frequencies from 1000 bootstrap resamples of size 124 of the renal carcinoma data.

for the next most frequent event $P(-4q) \approx 50\%$). Thus the stability of this edge is expected and does not provide any additional insight.

+17p is a child of +17q. This edge is more interesting, especially because the two events occur on the same chromosome and thus an association through the gain/loss of the entire chromosome is not unexpected. The occurrence of $+17p$ jumps from almost none (4%) with normal $17q$ to 50% when there is a gain on $17q$.

−6q and −13q are children of −4q. This cluster is the most complex and unexpected *a priori*. According to the definition of an oncogenetic tree, it implies not only increased probability of occurrence of events $-6q$ and $-13q$ after a loss on $4q$, but also a conditional independence of the "lower" events (given $-4q$). For reference, in the left panel of Figure 8 we present a mosaic plot for the joint distribution of these three events expected under

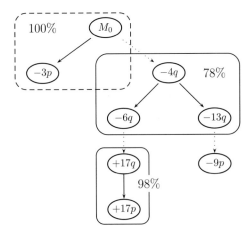

Fig. 7. Highly preserved portions of $\mathcal{T}_{\text{CGH}}$ among the bootstrapped trees with associated frequencies of occurrence. The solid frames mark the "non-trivial" portions; the dashed frame contains an expected edge as it includes the most frequent event (see text for details).

the oncogenetic tree model with the estimated error rates $\hat{\varepsilon}_+ = 0.00$ and $\hat{\varepsilon}_- = 0.228$. In this plot the area of each cell is proportional to the frequency of occurrence of the corresponding combination of the events. When $4q$ is normal (left side), the cell with normal $6q$ and $13q$ dominates; the other cells are due to false positives. However when $-4q$ is present (right side), the probabilities of $-6q$ and $-13q$ become larger, so all the cells increase. Note that due to the presence of false negative errors (but not false positives), the observed data is not expected to follow conditional independence for normal $4q$, only the underlying "true" events would show the grid pattern that is present on the $-4q$ side. The observed mosaic plot is in the right panel of Figure 8. The cells of this plot are coded according to their deviation from the Poisson regression model adjusting for expected frequencies: the fit is very good, none of the cells have residuals with absolute value above 1.3. Comparing this plot to the expected plot on the left side we see a general agreement in the patterns and a clear effect of $-4q$ on the probabilities of occurrence of $-6q$ and $-13q$. While there is some evidence for an excess of positive correlation (cells that are concordant with respect to $-6q$ and $-13q$ are larger), this excess is not statistically significant.

In conclusion of the above discussion, it appears that different analysis of the available data supports the presence of the stable associations found

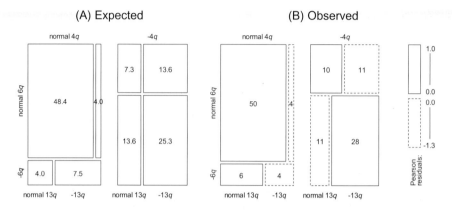

Fig. 8. The expected (A) and observed (B) mosaic plots of the joint distribution of $-4q$, $-6q$ and $-13q$. The area of the cells is proportional to the probability of occurrence; the labels are the expected/observed frequencies, respectively. In (B) the cells are coded according to the Pearson residual for a fit to the expected values in a Poisson regression model (see text).

by the oncogenetic tree model. As a next step, such hypotheses should be validated on an independent data set.

9. DISCUSSION

We have seen that oncogenetic trees provide a flexible, yet not too rich space for modeling genetic alteration data. More development is needed for the estimation of the timing of the alterations; the estimation of the parameters of the timed oncogenetic tree model is an open problem. While time information is not directly available for human tumors, stage and/or tumor size could possibly be used as surrogates.

An alternative tree-based modeling approach has been recently proposed for genetic alteration data (Desper et al., 2000; von Heydebreck et al., 2004). This methodology, that unfortunately is also often referred to as oncogenetic tree building, constructs a phylogenetic tree where the alterations are the leaves. These phylogenetic trees do not have the same mechanistic interpretation as the oncogenetic trees that we described. Additionally, estimation is much more difficult, probabilistic search algorithms have to be used.

References

1. Brodeur GM, Tsiatsis AA, Williams DL, Luthardt FW and Green AA. Statistical analysis of cytogenic abnormalities in human cancer cells. *Cancer Genet. Cytogenet.* 1982; **7**: 137–152.
2. Desper R, Jiang F, Kallioniemi O, Moch H, Papadimitriou C and Schäffer A. Distance-based reconstruction of tree models for oncogenesis. *J. Comput. Biol.* 2000; **7**: 789–803.
3. Desper R, Jiang F, Kallioniemi OP, Moch H, Papadimitriou CH and Schäffer AA. Inferring tree models for oncogenesis from comparative genome hybridization data. *J. Comput. Biol.* 1999; **6**: 37–51.
4. Edmonds J. Optimum branchings. *J. Res. Natl. Bur. Stand.* 1967; **71B**: 233–240.
5. Fearon ER and Vogelstein B. A genetic model for colorectal tumorigenesis. *Cell* 1990; **61**: 759–767.
6. Feldstein M and Zelen M. Inferring the natural time history of breast cancer: implications for tumor growth rate and early detection. *Breast Cancer Res. Treat.* 1984; **4**: 3–10.
7. Jiang F, Desper R, Papadimitriou C, Schäffer A, Kallioniemi O, Richter J, Schraml P, Sauter G, Mihatsch M and Moch H. Construction of evolutionary tree models for renal cell carcinoma from comparative genomic hybridization data. *Cancer Res.* 2000; **60**: 6503–6509.
8. Jiang F, Richter J, Schraml P, Bubendorf L, Gasser T, Mihatsch MJ, Sauter G and Moch H. Chromosomal imbalances in papillary renal cell carcinoma: genetic differences between histological subtypes. *Am. J. Pathol.* 1998; **153**: 1467–1473.
9. Kallioniemi A, Kallioniemi OP, Sudar D, Rutovitz D, Gray JW, Waldman F and Pinkel D. Comparative genomic hybridization for molecular cytogenetic analysis of solid tumors. *Science* 1992; **258**: 818–821.
10. Karp RM. A simple derivation of Edmonds' algorithm on optimum branching. *Networks* 1971; **1**: 265–272.
11. Ross SM. *Probability Models for Computer Science.* Harcourt/ Academic Press, Burlington, MA, 2002.
12. Szabo A and Boucher K. Estimating an oncogenetic tree when false negatives and positives are present. *Math. Biosci.* 2002; **176**: 219–236.
13. Szabo A and Yakovlev A. Preferred sequences of genetic events in carcinogenesis: quantitative aspects of the problem. *J. Biol. Syst.* 2001; **9**: 105–121.
14. Tarjan RE. Finding optimum branchings. *Networks* 1977; **7**: 25–35.

15. Vogelstein B, Fearon ER, Hamilton SR, Kern SE, Preisinger AC, Leppert M, Nakamura Y, White R, Smits AMM and Bos JL. Genetic alterations during colorectal tumor development. *N. Engl. J. Med.* 1988; **319**: 525–532.
16. von Heydebreck A, Gunawan B and Füzesi L. Maximum likelihood estimation of oncogenetic tree models. *Biostatistics* 2004; **5**: 545–556.
17. Zelen M. A hypothesis for the natural time history of breast cancer. *Cancer Res.* 1968; **28**: 207–216.

Chapter 2

STOCHASTIC MULTISTAGE CANCER MODELS: A FRESH LOOK AT AN OLD APPROACH

Qi Zheng

The hazard function is an important tool in epidemiologic research. Mathematical cancer models are often applied in epidemiologic research via the hazard functions induced by such models. This chapter examines some of the existing methods for computing hazard functions induced by a wide class of stochastic cancer models.

Keywords: Hazard rate, birth-and-death process, gamma frailty model, differential equation, initial value problem.

1. INTRODUCTION

Mathematical modeling of carcinogenesis is increasingly viewed as an important means of enhancing understanding of cancer biology. Because cancer is believed to be the result of a series of genetic or epigenetic events occurring in a somatic cell, most of present-day mathematical cancer models embrace this fundamental doctrine of cancer biology (e.g., Tan, 1991). The cancer models that form the subject of this chapter express the multistage doctrine in the language of stochastic birth-and-death processes. This type of model can be traced to Neyman (1961), who focused on the dynamics of the tumor size distribution induced by a stochastic two-stage cancer model. Tumor size distributions remain a daunting mathematical challenge, as progress in this regard has been slow. However, Moolgavkar and Venzon (1979) stimulated widespread and long-lasting interest in the birth-and-death process based cancer model by shifting focus to the survival function (the probability of being tumor free as a function of time) and its closely related hazard function. The impact of this shift in focus can hardly be overestimated, for a considerable portion of mathematical cancer modeling efforts in the ensuing two decades or so were expended on the computation and application of these two types of functions. Although recent years have seen the emergence of different approaches to quantitative modeling of carcinogenesis, e.g., Nowak *et al.* (2002) and Komarova *et al.* (2003), the old models and approaches remain useful in cancer research.

Since its infancy mathematical cancer modeling has been inseparably intertwined with cancer epidemiology. For example, one of the earliest and most influential mathematical cancer models was inspired by epidemiologic observations made by Armitage and Doll (1954) and their contemporary investigators. As tremendous amounts of resources are being continually spent on collecting large-scale epidemiologic data, e.g., the SEER database (www.seer.cancer.gov), epidemiologic methods will play an increasingly important part in unraveling the mystery of carcinogenesis. Because the hazard function is an essential tool in epidemiologic research, the usefulness of a mathematical cancer model in epidemiologic research depends to a large extent on our ability to compute hazard functions induced by the model.

The importance of hazard functions in cancer research has been further highlighted by the recent upsurge of interest in the puzzling observation of a decrease in age-specific cancer incidence at older ages. Explanations of this intriguing phenomenon were offered in slightly different contexts by Carey *et al.* (1992) and by Aalen (1988, 1994), and in the context of cancer epidemiology, by Herrero-Jimenez *et al.* (2000) and by Frank (2004). Since computing hazard rates under various assumptions is essential to testing these and other relevant hypotheses, a better understanding of existing algorithms for computing hazard functions induced by quantitative cancer models will be helpful.

In this chapter I shall explain the basic ideas that yield efficient algorithms for computing hazard functions induced by stochastic multistage cancer models. In doing so, I shall slightly extend and refine some existing methods.

2. BASIC DEFINITIONS AND NOTATION

A stochastic multistage cancer model can be conceptually viewed as a series of connected compartments — each compartment represents the number of mutations a cell residing in it has acquired. A cell in a compartment is allowed to divide, die or mutate. Note that the term death in the present context encompasses cellular necrosis, apoptosis and differentiation. While cellular division and death admit obvious mathematical formulations, the mathematical formulation of mutation is less obvious. Three mutational modes have been proposed in mathematical cancer modeling. Adapting the shorthand notation of Kendall (1960), one can vividly signify the three mutational modes by

$$
\begin{aligned}
&\text{(A)} \quad C_k \to C_{k+1}, \\
&\text{(B)} \quad C_k \to C_k + C_{k+1}, \\
&\text{(C)} \quad C_k \to C_{k+1} + C_{k+1},
\end{aligned}
$$

where C_k denotes a cell carrying k mutations. From a modern biological point of view, mode A corresponds to division-independent mutation, whereas modes B and C represent division-dependent mutation. Because division-independent mutation is still a controversial concept, present-day investigators are increasingly reluctant to adopt mode A in cancer modeling. Historically, many authors used mode A for reasons other than purposefully mimicking division-independent mutation. For example, Armitage and Doll (1954) seemed to use mode A for mathematical convenience. Mode B is becoming more popular mainly because it is consistent with the modern view that most mutations result from DNA replication errors. Mode C does not seem to have been used in cancer modeling. Note that models based on mode A abound in the literature; the coexistence of models based on modes A and B is prone to cause confusion. Another source of confusion is the fact that some investigators simplify the first and last compartments in their cancer models to render the mathematics more tractable. The first compartment, representing "normal" cells carrying no mutations, was often treated as a Poisson process generating cells carrying their first mutations; the last compartment, representing fully malignant cells, was sometimes treated as a "sink" in which cells neither multiply nor die. It is useful to adopt the shorthand notation introduced by Zheng (1997) that allows the precise structure of a multistage model to be succinctly specified. A model based on mode A with k stages is denoted by A_k, and a model based on mode B with k stages is similarly denoted by B_k. The presence of birth-death mechanisms in the starting and final compartments is indicated by

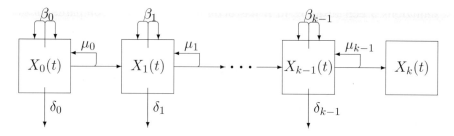

Fig. 1. A $B_k/N/0$ model.

adding two trailing letters (N and M respectively) separated by slashes, and the absence of such a feature is signaled by the numeral zero at the corresponding position. For example, the model depicted in Figure 1 is a $B_k/N/0$ model. The model proposed by Neyman (1961) is an $A_2/0/M$ model, and the two models introduced by Moolgavkar and Venzon (1979) can be denoted by $B_2/N/0$ and $B_2/0/0$, respectively.

I now use the $B_k/N/0$ model depicted in Figure 1 to show how a carcinogenesis model is mathematically defined in the language of stochastic birth-and-death processes. For convenience, a cell carrying i mutations will be called a type i cell. The symbol $X_i(t)$ ($i = 0, 1, \ldots, k$) denotes the number of type i cells living at time t. In general, a type i ($0 \leq i < k$) cell living at time t can undergo one of the following three changes in a small time interval $[t, t + \Delta t]$:

(i) it divides into two type i cells with probability $\beta_i \Delta t + o(\Delta t)$;
(ii) it dies with probability $\delta_i \Delta t + o(\Delta t)$;
(iii) it divides into a type i cell and a type $(i + 1)$ cell with probability $\mu_i \Delta x + o(\Delta t)$.

In the above definition, β_i, δ_i and μ_i are respectively the cell birth rate, cell death rate, and cell mutation rate. They are collectively called cellular kinetic parameters. The foregoing model definition is based on several important assumptions that are shared by most of birth-and-death process based multistage cancer models. These assumptions are usually inconsistent with current understanding of cancer biology, but they help formulate a mathematically tractable model that can still provide insights into cancer biology. For instance, one such assumption is that cells behave independently of one another. In other words, cell communication is ignored.

Despite this and other simplifying assumptions, computing the tumor size distribution, $\Pr[X_k(t) = j]$, remains an open research problem. However, from an epidemiological point of view, this formidable mathematical challenge is not as relevant as it might appear at first sight. Because values of $X_k(t)$ are not readily observable on an epidemiological scale, the usefulness of expressions for $\Pr[X_k(t) = j]$ in epidemiology is not clear. Moolgavkar and Venzon (1979) were among the first to realize that more useful in epidemiology was the zero size tumor probability function, namely, $S(t) = \Pr[X_k(t) = 0]$ regarded as a function of t. The rationale for advocating the use of $S(t)$ is as follows.

In a $B_k/N/0$ model a type k cell is immortal; therefore, the occurrence of the first type k cell will certainly lead to a full-blown tumor. Often, the time needed for the first type k cell to develop into a detectable tumor is considered negligible. If T denotes the random time at which the first type k cell emerges, then the event $\{T > t\}$ is equivalent to the event $\{X_k(t) = 0\}$. Thus, the zero size tumor probability function induces a failure time model, of which advanced statistical theory is available (e.g., Elandt-Johnson and Johnson, 1980). Specifically, the survival function induced by a $B_k/N/0$ model is defined by

$$S(t) = \Pr[T > t] = \Pr[X_k(t) = 0]. \qquad (1)$$

Note that this argument appears awkward when last-stage cells can also die, e.g., type k cells in a $B_k/0/M$ model. For this reason, the algorithms to be discussed here should be confined to models in which last-stage cells are assumed immortal.

The hazard function, more commonly known in epidemiology as the age-specific incidence, is related to the survival function in (1) by

$$h(t) = -\frac{d \log S(t)}{dt} = -\frac{1}{S(t)} \frac{dS(t)}{dt} = \frac{f(t)}{S(t)}, \qquad (2)$$

where $f(t) = -dS(t)/dt$ is the probability density functions of T. Another concept also intimately related to the survival function is the cumulative hazard defined as

$$H(t) = \int_0^t h(u) du.$$

The following two relations involving the cumulative hazard will be useful in the ensuing discussion (e.g., Elandt-Johnson and Johnson, 1980, pp. 51–52):

$$S(t) = \exp(-H(t)),$$
$$H(t) = -\log(S(t)). \tag{3}$$

The intriguing relation between age-specific cancer incidence and chronological age has stimulated an important debate about the long-held assumption that cancer hazard rate increases with age. Analyses of large-scale epidemiologic data began to show that deviations from this received wisdom are not uncommon. There exist several hypotheses in explanation of the observation that cancer hazard rates can decrease at advanced ages. One such hypothesis is the theory of frailty (Aalen, 1988, 1994); a common and simple model based on the frailty theory is the proportional hazard model, under which the hazard function for an individual subject is of the form $zh(t)$. While the unobservable frailty z is assumed to vary across subjects in a population to be studied, the hazard function $h(t)$ is regarded as a baseline hazard common to all subjects in the population. The nonnegative frailty z is considered as a random variable obeying some distributional law having a probability density function $\phi(z)$. A common choice for the distribution of z is a gamma distribution with unity mean and variance v, i.e.,

$$\phi(z) = \left(\frac{1}{v}\right)^{1/v} z^{1/v-1} e^{-z/v} / \Gamma(1/v). \tag{4}$$

Because the frailty z is unobservable, it must be integrated out. Using the density $\phi(z)$ given in (4), one obtains the marginal survival function $S^*(t)$ as follows (e.g., Hougaard 2000, pp. 60–61):

$$S^*(t) = \int_0^\infty e^{-zH(t)} \phi(z) dz = (1 + vH(t))^{-1/v}. \tag{5}$$

Applying (2) to (5) yields the marginal hazard function

$$h^*(t) = \frac{h(t)}{1 + vH(t)} = \frac{h(t)}{1 - v\log S(t)}. \tag{6}$$

This marginal hazard function is characteristic of the so-called gamma frailty model. Izumi and Ohtaki (2004) substituted the Weibull hazard for

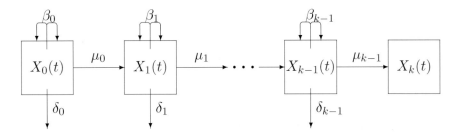

Fig. 2. An $A_k/N/0$ model.

$h(t)$ in (6) and applied the resulting model to radiation-related cancer data. I shall briefly explore (6) by replacing the baseline hazard function $h(t)$ in (6) with hazard functions induced by some typical stochastic multistage cancer models. The gamma frailty model is no doubt a simple one, but nonetheless it may provide useful insight.

The importance of accurately computing hazard functions can be more fully appreciated from a historical perspective. The well-known Armitage-Doll model is an important milestone in the history of mathematical cancer modeling. One can view the Armitage-Doll model as a special case of the $A_k/N/0$ model with all β_i and δ_i equal to zero (see Figure 2). The distribution of T is the sum of k independent exponential variates. Under the simplifying assumption that $\mu_i = \mu$ for all i, T obeys a gamma distribution with shape parameter k and scale parameter μ. Thus the density of T is of the form

$$f(t) = \frac{\mu^k}{(k-1)!} t^{k-1} e^{-\mu t}, \quad (7)$$

from which the hazard function is readily seen to be (e.g., Kalbfleisch and Prentice, 1980, p. 26)

$$h(t) = \frac{\mu^k t^{k-1} e^{-\mu t}}{(k-1)! - \gamma(k, \mu t)}, \quad (8)$$

where $\gamma(x, a) = \int_0^a s^{x-1} e^{-s} ds$ denotes the incomplete gamma function. Because $k \geq 2$ in the present context, the hazard function in (8) is monotone increasing and asymptotes to μ (e.g., Kalbfleisch and Prentice, 1980, p. 27). As (8) clearly indicates, when μt is small, the hazard function $h(t)$ can be approximated by the density function $f(t)$ given in (7). Furthermore, setting

$e^{-\mu t} \approx 1$ (valid when μt is small) yields the famous power law of cancer incidence

$$h(t) \approx \frac{\mu^k}{(k-1)!} t^{k-1}. \tag{9}$$

The aforementioned controversial issue naturally directs one's attention from the approximate hazard function in (9) back to the exact hazard function in (8). When not all μ_i are equal, (8) becomes algebraically cumbersome. When the possibility of cell death has to be taken into consideration, a systematic approach as suggested by Whittemore and Keller (1978, p. 9) is highly desirable. To account for the important effects of cell proliferation on carcinogenesis, one needs well-defined algorithms that also use information about cell proliferation. As (6) indicates, such algorithms are also helpful in applying the frailty theory to cancer research. The following section describes one systematic approach to computing hazard functions induced by a wide class of multistage cancer models.

3. MATHEMATICAL DETAILS

I shall use the $B_k/N/0$ model (Figure 1) to derive an algorithm, providing some mathematical details that are the key to understanding and properly using the algorithm. The basic idea underlying the algorithm has been applied to the $B_2/N/0$ model by Serio (1984), and to the $B_2/0/0$ model by Quinn (1989). The following exposition is adapted from Zheng (1995, 1997) where I refined and extended the methods of Serio and Quinn. The principal idea is essentially an application of the method of the characteristics from the theory of partial differential equations (PDEs) (Carrier and Pearson 1988, pp. 95–97), but the following elementary approach seems to offer a more lucid exposition.

Let G denote the joint probability generating function (p.g.f.) of $X_0(t), \ldots, X_k(t)$ defined by

$$G(z_0, \ldots, z_k; t) = E[z_0^{X_0(t)} z_1^{X_1(t)} \cdots z_k^{X_k(t)}] \tag{10}$$

with $z_0, \ldots, z_k$ being dummy variables. Note, as (1) suggests, the survival function can be obtained from the p.g.f. by the basic relation

$$S(t) = G(1, \ldots, 1, 0; t). \tag{11}$$

with the aid of the so-called "random variable" technique (Bailey, 1964, pp. 70–74), one immediately has the PDE

$$\frac{\partial G}{\partial t} = \sum_{i=0}^{k-1} \left[\beta_i (z_i - 1) z_i + \delta_i (z_i^{-1} - 1) z_i + \mu_i (z_{i+1} - 1) z_i \right] \frac{\partial G}{\partial z_i}.$$

A more convenient form of the above PDE is

$$\frac{\partial G}{\partial t} + \sum_{i=0}^{k-1} \left[-\beta_i z_i^2 + (\beta_i + \delta_i + \mu_i) z_i - \mu_i z_i z_{i+1} - \delta_i \right] \frac{\partial G}{\partial z_i} = 0. \tag{12}$$

In the above PDE all cellular kinetic parameters can be functions of time, and I suppress this parameter time-dependency only for simplicity of notation. It is customarily assumed that there exist only type 0 cells at time $t = 0$, that is, $X_j(0) = 0$ for $j = 1, \ldots, k$. Without loss of generality, I further assume that $X_0(0) = 1$. Therefore, in view of (10), the PDE in (12) is subject to the initial condition

$$G(z_0, \ldots, z_k; 0) = z_0. \tag{13}$$

Let $t_0 > 0$ be an arbitrarily prescribed epoch at which the survival probability $S(t_0)$ and the hazard rate $h(t_0)$ are to be computed. An important step to understanding the algorithm is to view all the dummy variables in the p.g.f. as functions of time. Specifically, consider all z_i as functions of t satisfying a system of ordinary differential equations (ODEs) of the form

$$\begin{aligned} z_i'(t) &= -\beta_i(t) z_i(t)^2 + [\beta_i(t) + \delta_i(t) + \mu_i(t)] z_i(t) \\ &\quad - \mu_i(t) z_i(t) z_{i+1}(t) - \delta_i(t) \quad (i = 0, \ldots, k-2), \\ z_{k-1}'(t) &= -\beta_{k-1}(t) z_{k-1}(t)^2 \\ &\quad + [\beta_{k-1}(t) + \delta_{k-1}(t) + \mu_{k-1}(t)] z_{k-1}(t) - \delta_{k-1}(t) \end{aligned} \tag{14}$$

and the initial conditions

$$z_0(t_0) = \cdots = z_{k-1}(t_0) = 1. \tag{15}$$

Equations (14) and (15) together constitute an initial value problem. The solution of this initial value problem allows one to consider

$G(z_0(t), \ldots, z_{k-1}(t), 0; t)$ as a univariate function of t. In light of the chain rule for differentiation, it is clear that the p.g.f. G and the solution of the initial value problem are related by

$$\frac{d}{dt} G(z_0(t), \ldots, z_{k-1}(t), 0; t) = 0.$$

Integrating with respect to t on $[0, t_0]$ yields

$$G(z_0(t_0), \ldots, z_{k-1}(t_0), 0; t_0) = G(z_0(0), \ldots, z_{k-1}(0), 0; 0). \quad (16)$$

On account of relations (13) and (15), one has

$$G(1, \ldots, 1, 0; t_0) = z_0(0). \quad (17)$$

It then follows from (11) that the survival probability at time t_0 can be extracted from the solution of the initial value problem by setting

$$S(t_0) = z_0(0). \quad (18)$$

Because the initial conditions in (15) involve t_0, in general one needs to solve an initial value problem for each given $t_0 > 0$. To compute the hazard rate $h(t_0)$, one can first compute $S(t)$ for several values of t in a small neighborhood of t_0 and then numerically compute the derivative $S'(t_0)$ (see Eq. (2)). Little (1995) proposed a method that allows one to directly compute $h(t_0)$ by augmenting the initial value problem given in (14) and (15). But I (Zheng, 1998) incorrectly stated that Little's method was applicable only to time-homogeneous cases (where all cellular kinetic parameters are constant).

Note epidemiologists are often interested in time-homogeneous cases; the methods discussed here simplify considerably for time-homogeneous cases. Instead of solving an initial value problem like that given by (14) and (15) for each epoch t_0, one needs to solve only a single initial value problem to obtain the survival function $S(t)$ and the hazard function $h(t)$ for all $t \in [0, C]$, where C is a predetermined positive number. To clarify this claim, denote the solution of the initial value problem defined by (14) and (15) by $z_i(t; t_0)$. Thus, for an arbitrary but fixed t one can rewrite (18) as $S(t) = z_0(s; t)|_{s=0}$. When all cellular kinetic parameters are constant, (14) reduces to an autonomous system. Because an autonomous system is invariant under translations of the independent variable, $z_i(s; t) = z_i(s - t; 0)$ (e.g., Brauer and Nohel, 1969, p. 84). In particular, one has

$$S(t) = z_0(s; t)|_{s=0} = z_0(s - t; 0)|_{s=0} = z_0(-t; 0). \quad (19)$$

By introducing $y_i(t) = z_i(-t; 0)$, one can recast the initial value problem defined by (14) and (15) into

$$y_i'(t) = \beta_i y_i(t)^2 - [\beta_i + \delta_i + \mu_i] y_i(t) + \mu_i y_i(t) y_{i+1}(t) + \delta_i$$
$$(i = 0, \ldots, k - 2), \quad (20)$$

$$y_{k-1}'(t) = \beta_{k-1} y_{k-1}(t)^2 - [\beta_{k-1} + \delta_{k-1} + \mu_{k-1}] y_{k-1}(t) + \delta_{k-1}$$

subject to

$$y_0(0) = \cdots = y_{k-1}(0) = 1. \quad (21)$$

Because (19) holds for arbitrary t, it follows that for all $t \geq 0$

$$S(t) = y_0(t), \quad (22)$$

where $y_0(t)$ is determined by the initial value problem given in (20) and (21). Furthermore, it follows from (22) and (20) that

$$S'(t) = \beta_0 y_0^2(t) - (\beta_0 + \delta_0 + \mu_0) y_0(t) + \mu_0 y_0(t) y_1(t) + \delta_0. \quad (23)$$

On account of (2) one can compute the hazard function by

$$h(t) = \beta_0 + \delta_0 + \mu_0 - \beta_0 y_0(t) - \mu_0 y_1(t) - \delta_0/y_0(t). \quad (24)$$

If the gamma frailty model in (6) is desirable, the hazard function can be computed by

$$h^*(t) = \frac{\beta_0 + \delta_0 + \mu_0 - \beta_0 y_0(t) - \mu_0 y_1(t) - \delta_0/y_0(t)}{1 - v \log(y_0(t))}. \quad (25)$$

Figure 3 depicts three hazard functions. The baseline hazard function is generated by a $B_2/N/0$ model. It is well-known that both $A_k/N/0$ and $B_k/N/0$ models tend to produce bell-shaped hazard curves. As Figure 3 shows, the gamma frailty model preserves this feature, although the shape of a hazard curve can be altered by a gamma frailty model.

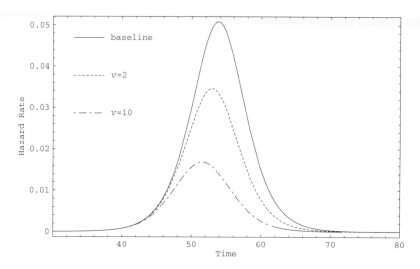

Fig. 3. A $B_2/N/0$ model with $\beta_0 = \beta_1 = 1.0$, $\delta_0 = 0.6$, $\delta_1 = 0.5$, and $\mu_0 = \mu_1 = 10^{-6}$ was used to generate a baseline hazard function; two gamma frailty models with $v = 2$ and $v = 10$ respectively were then applied to the baseline hazard function to obtain marginal hazard functions.

If the initial compartment of a $B_k/N/0$ model degenerates into a Poisson process generating type 1 cells at an intensity rate v, the resultant model is a $B_k/0/0$ model (see Figure 4). This method of simplifying the initial compartment was motivated by mathematical expediency, for early investigators found that a $B_2/0/0$ model was mathematically more tractable than a $B_2/N/0$ model.

Because the initial compartment is reduced to a Poisson stream having intensity function $v(t)$, the number of type 0 cells is no longer of direct interest. Consequently, the p.g.f. G for a $B_k/0/0$ model loses the z_0 dummy variable:

$$G(z_1, \ldots, z_k; t) = E[z_1^{X_1(t)}, \ldots, z_k^{X_k(t)}]. \tag{26}$$

The same random variable technique enables one to have a PDE of the form

$$\frac{\partial G}{\partial t} = \sum_{i=1}^{k-1} \left[\beta_i z_i^2 - (\beta_i + \delta_i + \mu_i) z_i + \mu_i z_i z_{i+1} + \delta_i \right] \frac{\partial G}{\partial z_i} + v(z_1 - 1)G,$$

which can be recast into the following more convenient form

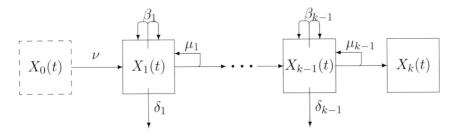

Fig. 4. A $B_k/0/0$ model.

$$\frac{1}{G}\left\{\frac{\partial G}{\partial t} + \sum_{i=1}^{k-1}[-\beta_i z_i^2 + (\beta_i + \delta_i + \mu_i)z_i - \mu_i z_i z_{i+1} - \delta_i]\frac{\partial G}{\partial z_i}\right\}$$
$$= \nu(z_1 - 1). \tag{27}$$

Because $X_1(0) = \cdots = X_k(0)$ is customarily assumed, the above PDE should be subject to the initial condition

$$G(z_1, \ldots, z_k; 0) = 1. \tag{28}$$

For a prescribed $t_0 > 0$ consider the initial value problem

$$\begin{aligned}
z_i'(t) &= -\beta_i(t)z_i(t)^2 + [\beta_i(t) + \delta_i(t) + \mu_i(t)]z_i(t) \\
&\quad - \mu_i(t)z_i(t)z_{i+1}(t) - \delta_i(t) \quad (i = 1, \ldots, k-2), \\
z_{k-1}'(t) &= -\beta_{k-1}(t)z_{k-1}(t)^2 \\
&\quad + [\beta_{k-1}(t) + \delta_{k-1}(t) + \mu_{k-1}(t)]z_{k-1}(t) - \delta_{k-1}(t), \\
z_1(t_0) &= \cdots = z_{k-1}(t_0) = 1.
\end{aligned} \tag{29}$$

In a similar fashion, it can be verified that the solution of the above initial value problem and the p.g.f. G are related by

$$\frac{d}{dt}\log G(z_1(t), \ldots, z_{k-1}(t), 0; t) = \nu(t)(z_1(t) - 1). \tag{30}$$

One can integrate both sides on $[0, t_0]$ to obtain the relation

$$\log G(1, \ldots, 1, 0; t_0) - \log G(z_1(0), \ldots, z_{k-1}(0), 0; 0)$$
$$= \int_0^{t_0} \nu(t)(z_1(t) - 1)dt. \tag{31}$$

Considering the initial condition (28), one obtains

$$S(t_0) = \exp\left(\int_0^{t_0} v(t)\{z_1(t) - 1\}dt\right). \tag{32}$$

As in the case of the $B_k/N/0$ model, computation simplifies when all cellular kinetic parameters are constant. The initial value problem simplifies to

$$y_i'(t) = \beta_i y_i(t)^2 - [\beta_i + \delta_i + \mu_i]y_i(t) + \mu_i y_i(t)y_{i+1}(t) + \delta_i$$
$$(i = 1, \ldots, k-2), \tag{33}$$
$$y_{k-1}'(t) = \beta_{k-1}y_{k-1}(t)^2 - [\beta_{k-1} + \delta_{k-1} + \mu_{k-1}]y_{k-1}(t) + \delta_{k-1},$$
$$y_1(0) = \cdots = y_k(0) = 1.$$

For all $t > 0$ one has

$$S(t) = \exp\left(v \int_0^t (y_1(u) - 1)du\right), \tag{34}$$

and applying (2) yields

$$h(t) = v(1 - y_1(t)). \tag{35}$$

Furthermore, applying the gamma frailty model yields

$$h^*(t) = \frac{v(1 - y_1(t))}{1 + vv \int_0^t (1 - y_1(u))du}. \tag{36}$$

If the survival function $S(t)$ in (34) is also of direct interest, it is sometimes desirable to merge the separate numerical integration step into the initial value problem (33). For notational consistency, let $y_0(t)$ be the same as $S(t)$ in (34). Clearly, $y_0'(t) = v(y_1(t) - 1)y_0(t)$ and $y_0(0) = 1$. Therefore, one can augment (33) to get the following initial value problem:

$$y_0'(t) = v(y_1(t) - 1)y_0(t)$$
$$y_i'(t) = \beta_i y_i(t)^2 - [\beta_i + \delta_i + \mu_i]y_i(t) + \mu_i y_i(t)y_{i+1}(t) + \delta_i$$
$$(i = 1, \ldots, k-2), \tag{33a}$$
$$y_{k-1}'(t) = \beta_{k-1}y_{k-1}(t)^2 - [\beta_{k-1} + \delta_{k-1} + \mu_{k-1}]y_{k-1}(t) + \delta_{k-1},$$
$$y_0(0) = y_1(0) = \cdots = y_k(0) = 1.$$

Thus, (34) becomes

$$S(t) = y_0(t), \tag{34a}$$

(35) remains unaltered, and (36) changes to

$$h^*(t) = \frac{v(1 - y_1(t))}{1 - v\log(y_0(t))}. \tag{36a}$$

The above algorithms can be readily extended to the $A_k/N/0$ and $A_k/0/0$ models. The extension is almost trivial. However, because the well-known Armitage-Doll model can be considered as a special case of the $A_k/N/0$ model, I shall use the $A_k/N/0$ model to outline an algorithm for the Armitage-Doll model that algorithmically generalizes Eq. (8). Conceptually, a k-stage Armitage-Doll model can be obtained by setting $\beta_i = \delta_i = 0$ for all i in an $A_k/N/0$ model as depicted in Figure 2. The PDE satisfied by the p.g.f. for an $A_k/N/0$ model is clearly of the form

$$\frac{\partial G}{\partial t} + \sum_{i=0}^{k-1}[-\beta_i z_i^2 + (\beta_i + \delta_i + \mu_i)z_i - \mu_i z_{i+1} - \delta_i]\frac{\partial G}{\partial z_i} = 0. \tag{37}$$

For time-homogeneous cases, it can be shown in a similar fashion that the survival function satisfies $S(t) = y_0(t)$ with $y_0(t)$ being determined by the initial value problem

$$\begin{aligned}
y_i(t)' &= \beta_i y_i(t)^2 - [\beta_i + \delta_i + \mu_i]y_i(t) + \mu_i y_{i+1}(t) + \delta_i \\
&\qquad (i = 0, \ldots, k-2), \\
y'_{k-1}(t) &= \beta_{k-1}y_{k-1}(t)^2 - [\beta_{k-1} + \delta_{k-1} + \mu_{k-1}]y_{k-1}(t) + \delta_{k-1}, \\
y_0(0) &= \cdots = y_{k-1}(0) = 1.
\end{aligned} \tag{38}$$

Therefore, the hazard function can be computed by

$$h(t) = \beta_0 + \delta_0 + \mu_0 - \beta_0 y_0(t) - \mu_0\frac{y_1(t)}{y_0(t)} - \frac{\delta_0}{y_0(t)}, \tag{39}$$

which generalizes (8).

At first glance one might find the above methods peculiar. It is therefore useful to corroborate (39) by applying it to the two-stage Armitage-Doll model, i.e., an $A_2/N/0$ model with $\beta_i = \delta_i = 0$ for $i = 0, 1$. For simplicity assume that $\mu_1 \neq \mu_2$. Clearly, under this model the survival time T is distributed as the sum of two independent exponential random variables having means μ_0^{-1} and μ_1^{-1} respectively. Therefore, the survival function is

$$S(t) = 1 - \int_0^t \int_0^{t-x} \mu_0 \mu_1 e^{-\mu_0 x - \mu_1 y} dy dx = \frac{\mu_0 e^{-\mu_1 t} - \mu_1 e^{-\mu_0 t}}{\mu_0 - \mu_1}. \quad (40)$$

Applying (2) to the above survival function yields the hazard function

$$h(t) = \mu_0 \mu_1 \frac{e^{\mu_0 t} - e^{\mu_1 t}}{\mu_0 e^{\mu_0 t} - \mu_1 e^{\mu_1 t}}. \quad (41)$$

On the other hand, under the two-stage Armitage-Doll model the initial value problem (38) reduces to

$$y_0'(t) = -\mu_0 y_0(t) + \mu_0 y_1(t),$$
$$y_1'(t) = -\mu_1 y_1(t),$$
$$y_0(0) = y_1(0) = 1.$$

The solution of the above initial value problem is readily seen to be

$$y_0(t) = \frac{\mu_0 e^{-\mu_1 t} - \mu_1 e^{-\mu_0 t}}{\mu_0 - \mu_1},$$
$$y_1(t) = e^{-\mu_1 t}. \quad (42)$$

Thus, $S(t) = y_0(t)$ is validated by (40) and (42). Furthermore, it is straightforward to verify that

$$\mu_0 - \mu_0 \frac{y_1(t)}{y_0(t)} = \mu_0 \mu_1 \frac{e^{\mu_0 t} - e^{\mu_1 t}}{\mu_0 e^{\mu_0 t} - \mu_1 e^{\mu_1 t}}, \quad (43)$$

which confirms (39) in view of (41).

4. CONCLUDING REMARKS

Several thought-provoking hypotheses have been proposed to explain the existence of a supposed or true maximum in age-specific cancer incidence. Although this chapter does not attempt to evaluate these hypotheses, the algorithms discussed here are useful in generating and testing relevant hypotheses. In writing this chapter I made several observations that might be worth further consideration. First, the Armitage-Doll model ($A_k/N/0$) is capable of generating bell-shaped hazard curves; ignoring cell death effects results in monotone increasing hazard rates, and a further approximation leads to the power law of carcinogenesis. Second, a $B_k/N/0$ model in general produces bell-shaped hazard curves; but when the initial compartment is relegated to a Poisson process, the resultant $B_k/0/0$ model yields monotone increasing hazard functions (compare Figure 3 with Figure 5). It thus appears that a mathematical assumption can alter the hazard function in a fundamental way. Third, as Figure 6 shows, the effects of population heterogeneity can be confounded with the effects of cell death. It is hoped

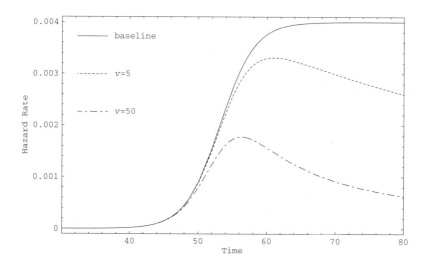

Fig. 5. A $B_2/0/0$ model with $\beta_1 = \beta_2 = 1.0$, $\delta_1 = 0.6$, $\delta_2 = 0.5$, $\mu_1 = \mu_2 = 10^{-6}$ and $v = 0.01$ was used to generate a baseline hazard function; two gamma frailty models with $v = 5$ and $v = 50$ respectively were then applied to the baseline hazard function to obtain marginal hazard functions.

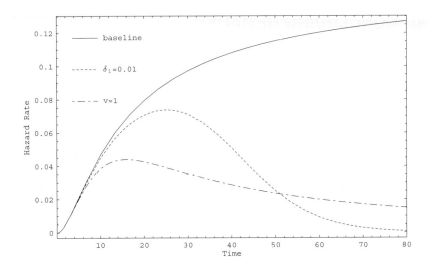

Fig. 6. An $A_2/N/0$ model with $\mu_0 = \mu_1 = \mu_2 = 0.15$ and $\beta_i = \delta_i = 0$ for all i was used to generate a baseline hazard function. One unimodal hazard curve was generated by allowing cellular death ($\delta_1 = 0.01$), and another unimodal hazard curve was generated by applying a gamma frailty model with $v = 1$.

that the computational methods discussed here will facilitate research into important issues in cancer epidemiology.

ACKNOWLEDGMENTS

I am deeply grateful to Dr. L. Hanin for his extensive and valuable suggestions that led to much improved clarity and usefulness of the manuscript.

References

1. Aalen OO. Heterogeneity in survival analysis. *Stat. Med.* 1988; **7**: 1121–1137.
2. Aalen OO. Effects of frailty in survival analysis. *Stat. Methods Med. Res.* 1994; **3**: 227–243.
3. Armitage P and Doll R. The age distribution of cancer and a multi-stage theory of carcinogenesis. *Br. J. Cancer* 1954; **8**: 1–12.
4. Bailey NTJ. *The Elements of Stochastic Processes with Applications to the Natural Sciences*. Wiley, New York, 1964.

5. Brauer F and Nohel JA. *The Qualitative Theory of Ordinary Differential Equations: An Introduction.* W.A. Benjamin, Inc., New York, 1969.
6. Carey TR, Liedo P, Orozco D and Vaupel JW. Slowing of mortality at older ages in large medfly cohorts. *Science* 1992; **258**: 457–461.
7. Carrier GF and Pearson CE. *Partial Differential Equations: Theory and Technique,* 2nd ed. Academic Press, Boston, 1988.
8. Elandt-Johnson RC and Johnson NL. *Survival Models and Data Analysis.* Wiley, New York, 1980.
9. Frank SA. Genetic variation in cancer predisposition: mutational decay of a robust genetic control network. *Proc. Natl. Acad. Sci. USA* 2004; **101**: 8061–8065.
10. Herrero-Jimenez P, Tomita-Mitchell A, Furth EE, Morgenthaler S and Thilly WG. Population risk and physiological rate parameters for colon cancer. The union of an explicit model for carcinogenesis with the public health records of the United States, *Mutat. Res.* 2000; **447**: 73–116.
11. Hougaard P. *Analysis of Multivariate Survival Data.* Springer, New York, 2000.
12. Izumi S and Ohtaki M. Aspects of the Armitage-Doll gamma frailty model for cancer incidence data. *Environmetrics* 2004; **15**: 209–218.
13. Kalbfleisch JD and Prentice RL. *The Statistical Analysis of Failure Time Data.* Wiley, New York, 1980.
14. Kendall DG. Birth-and-death process, and the theory of carcinogenesis. *Biometrika* 1960; **47**: 13–21.
15. Komarova NL, Sengupta A and Nowak MA. Mutation-selection networks of cancer initiation: tumor suppressor genes and chromosomal instability. *J. Theor. Biol.* 2003; **223**: 433–450.
16. Little MP. Are two mutations sufficient to cause cancer? Some generalizations of the two-mutation model of carcinogenesis of Moolgavkar, Venzon, and Knudson, and of the multistage model of Armitage and Doll. *Biometrics* 1995; **51**: 1278–1291.
17. Moolgavkar SH and Venzon DJ. Two-event models for carcinogenesis: incidence curves for childhood and adult tumors. *Math. Biosci.* 1979; **47**: 55–77.
18. Neyman, J. A two-step mutation theory of carcinogenesis. *Bull. Int. Stat. Inst.* 1961; **38**: 123–135.
19. Nowak MA, Komarova NL, Sengupta A, Jallepalli PV, Shih IM, Vogelstein B and Lengauer C. The role of chromosomal instability in tumor initiation. *Proc. Natl. Acad. Sci. USA* 2002; **99**: 16226–16231.
20. Quinn DW. Calculating the hazard function and probability of tumor for cancer risk assessment when the parameters are time-dependent. *Risk Anal.* 1989; **9**: 407–413.

21. Serio G. Two-stage stochastic model for carcinogenesis with time-dependent parameters. *Stat. Probab. Lett.* 1984; **2**: 95–103.
22. Tan WY. *Stochastic Models of Carcinogenesis*. Marcel Dekker, New York, 1991.
23. Whittemore A and Keller JB. Quantitative theories of carcinogenesis. *SIAM Review* 1978; **20**: 1–30.
24. Zheng Q. On the MVK stochastic carcinogenesis model with Erlang distributed cell life lengths. *Risk Anal.* 1995; **15**: 495–502.
25. Zheng Q. A unified approach to a class of stochastic carcinogenesis models. *Risk Anal.* 1997; **17**: 617–624.
26. Zheng Q. To use or not to use? Backward equations in stochastic carcinogenesis models. *Biometrics* 1998; **54**: 384–388.

Chapter 3

CANCER BIOLOGY, CANCER MODELS AND SOME NEW APPROACHES TO CARCINOGENESIS

Wai Y. Tan, Chao W. Chen and Li J. Zhang

In this chapter we survey recent studies by molecular biologists and cancer geneticists and propose some general stochastic models of carcinogenesis. To develop analytic results for these stochastic models, because the traditional Markov theory approach becomes too complicated to be of much use, in this chapter we propose an alternative approach through stochastic equations. Given observed cancer incidence data, we further combine these stochastic models with statistical models to develop state space models for carcinogenesis. By using these state space models, we then develop a generalized Bayesian method and a predicted inference procedure to estimate the unknown parameters and to predict state variables via multi-level Gibbs sampling procedures. In this chapter we use the extended multi-event model and mixture model as examples to illustrate our modeling approach and some basic theories.

Keywords: Generalized Bayesian procedures, observation model, extended multi-event model of carcinogenesis, multi-level Gibbs sampling procedures, mixture model of carcinogenesis, state space model, stochastic differential equations, stochastic system model.

1. INTRODUCTION

It is now universally recognized that each cancer tumor develops through stochastic proliferation and differentiation from a single stem cell which has sustained a series of irreversible genetic changes (Tan, 1991; MacDonald *et al.*, 2004; Weinberg, 2007). Furthermore, the number of stages and the number of pathways of the carcinogenesis process are significantly influenced by environmental factors underlying the individuals (Tan, 1991;

Weinberg, 2007). Recently, it has been demonstrated that carcinogenesis is an evolution process in cell populations referred to as a micro-evolution process; and each cancer tumor is the outcome of growth of a most fitted genetically altered stem cell (Carhill *et al.*, 1999; Hopkin, 1996).

In this chapter we will summarize recent results from cancer biology and propose some general stochastic models of carcinogenesis. For these models, mathematical results by the classical methods are very difficult even under some simplifying assumptions which may not be realistic in the real world; see **Remark 1**. It follows that except possibly for the simplest two-stage model, analytical mathematical results remain to be developed and published. In order to derive analytical mathematical results and to relax some unrealistic assumptions, in this chapter we will provide new approaches through stochastic differential equations to analyze these models. For combining information from different sources and for easing problems of identifiability, we will combine these stochastic models with statistical models to develop some state space models for carcinogenesis. By using these state space models, we will develop a generalized Bayesian method and a predictive inference procedure to estimate the unknown parameters and to predict the state variables.

In Section 2, we will summarize recent results from cancer biology. Based on these cancer biology, in Section 3, we will propose some general stochastic models of carcinogenesis. To derive analytical results and to extend the models, in Section 4, we will propose an alternative approach to analyze these stochastic models through stochastic differential equations. For combining information from stochastic models and statistical models and for fitting the models to cancer data, in Section 5, we will proceed to develop some state space models for the process of carcinogenesis. In Section 6, we will illustrate the application of the models and methods by analyzing the British data from physician's lung cancer and smoking. Finally in Section 7, we will discuss some possible applications of these models and methods.

Remark 1. In almost all models of carcinogenesis in the literature (Tan, 1991; Chu, 1985; Little, 1995; Lubeck and Moolgavkar, 2002), it is assumed that each primary I_k cell grows instantaneously into a malignant cancer tumor in which case one may consider each I_k cell as a cancer tumor.

As pointed out by Yakovlev and Tsodikov (Yakovlev, 1996), however, in many practical situations this assumption may not be realistic.

2. SOME RECENT CANCER BIOLOGY FOR MODELING CARCINOGENESIS

Using tissue culture method, biologists have shown that all organs consist of two types of cells: The differentiated cells which are major components of the organ proper and the stem cells from which cancer tumors develop (Al-Hajj and Clarke, 2004; Zheng and Rosen, 2006). Only stem cells can divide giving rise to new stem cells and new differentiated cells to replace old differentiated cells; the differentiated cells do not divide and are end cells to serve as components of the tissue and to perform specific functions of the tissue. That is, stem cells are subject to stochastic proliferation and differentiation with differentiated cells replacing old cells of the organ.

To understand cancer, notice that in normal individuals, there is a balance between proliferation and differentiation in stem cells and there are devices such as the DNA repair system and apoptosis in the body to protect against possible errors in the metabolism process. Thus, in normal individuals, the proliferation rate and birth rate of stem cells equal to the differentiation rate and death rate of stem cells so that the size of organ is normally not changed. If some genetic changes have occurred in a stem cell to increase the proliferation rate of the cell; then the proliferation rate is greater than the differentiation rate in this genetically altered cell so that this type of genetically altered cells will accumulate; however, with high probability these genetically altered cells will be eliminated by apoptosis or other protection devices unless more genetic changes have occurred in these cells to abrogate apoptosis and to overcome other existing protection devices. Furthermore, it requires at least one round of cell proliferation for a genetic change to be fixed (Farber, 1987; Kalunaga, 1974). Also, since genetic changes are rare events, further genetic changes will occur in at least one of the genetically altered cells only if this type of cells is fixed and established and if the number of these cells is very large. These steps have clearly been demonstrated by cell culture experiments by Barrett and coworkers using rat tracheal epithelial cells and on Syrian hamster embryo fibroblasts

(Nettesherim, 1985); for more detail and some more specific examples, see Chapter 1 in Tan (1991). These results indicate that carcinogenesis in humans and animals is a multi-step random process and that these steps reflect genetic changes and/or epigenetic changes that drive the progressive transformation of normal stem cells into highly malignant ones. These biological results also imply that carcinogenesis is a micro-evolution process in cell population as described in the references (Cahill *et al.*, 1999; Hopkin, 1996).

The above discussion and studies in cancer biology (Baylin, 2006; Hanahan and Weinberg, 2000; Jones and Baylin, 2002; Weinberg, 2007) illustrate that cancer is initiated by some genetic changes or epigenetic changes to increase cell proliferation while decreasing differentiation and cell death. Further genetic changes or epigenetic changes are required to overcome existing protection devises in the body resulting in abrogation of apoptosis, telomere protection (immortalization) and uncontrolled growth as well as angiogenesis and metastasis. Because somatic cell division occurs only through cell division cycle whereas epigenetic changes, gene mutation and genetic changes occur only during cell division, most of the genetic changes and epigenetic changes affect carcinogenesis through the control of cell division cycle. By articulating these findings, Hanahan and Weinberg (2000) have proposed six basic acquired capabilities which each normal stem cell must require to become a malignant cancer tumor cell. These six capabilities are: (1) Self-sufficiency of growth factor signals via genetic changes and/or epigenetic changes. Because cells can only be induced to enter into cell division cycles by growth factor signals (Hanahan and Weinberg, 2000), obviously this capability is required for cancer stem cells to achieve uncontrolled growth. (2) Insensitivity to anti-growth signals mostly via inactivation or silencing of tumor suppressor genes (e.g., the RB gene in chromosome 13). This capability is required of cancer stem cells to eliminate differentiation to complete the cell division cycle yielding daughter cells. (3) Evading apoptosis. When cells grow abnormally and uncontrolled, it will invoke the apoptosis protection devise to kill these cells. When this happens, tumor cells can never increase significantly. For uncontrolled growth, cancer tumor cells must abrogate the apoptosis protection devise via genetic changes and/or epigenetic changes (e.g. silencing by epigenetic mechanisms or inactivation or mutation of *p53* gene in chromosome

17 (Baylin, 2006; Chen et al., 2005). (4) Unlimited replicative potential (immortalization) via the activation of telomerase; see Section 2.6. (5) Sustained angiogenesis to draw nutrition and oxygens from the blood vessels for cancer cell growth. (6) Tissue invasion and metastasis to overcome limitations of space and nutrition. The first four capabilities are required to establish uncontrollable growth of stem cells (avascular carcinogenesis) whereas the last two are for the development of cancer spread and metastasis of cancer cells (vascular carcinogenesis). Each of the above capabilities involves at least one or many genetic and/or epigenetic changes although in some cases some genetic changes may invoke more than one capabilities. To understand carcinogenesis, in what follows we further discuss some important issues in carcinogenesis.

2.1. The Multi-Staging Nature of Carcinogenesis

The discussion above indicates that for a normal stem cell to develop into a malignant cancer tumor cell, it must accumulate many gene mutations or genetic changes. Because gene mutations and genetic changes are rare events and can occur only during cell division, it is a statistical near-impossibility that all mutation and genetic changes can occur simultaneously during a single cell division. It follows that different gene mutations or genetic changes must occur in different cell division at different times. This also leads to the observation that all steps in the carcinogenesis process must occur in sequence.

Notice that micro-array analysis have indicated that in most human cancers, a large number of cancer genes are involved. However, only a few of the genes are stage and rate limiting, leading to a finite number of stages in the multi-stage model of carcinogenesis; see Renan (1993). The age-dependent cancer incidence data for many human cancers imply four to seven rate-limiting stages from normal stem cells to malignant cancer tumors in most of human cancers (Renan, 1993). These stages are reflected by observable pathological lesions and the transition from one stage to the next higher stage may involve several genetic changes and/or epigenetic changes. For example, in FAP or most sporadic human colon cancer, the first lesion that can be observed are the dysplastic aberrant cript foci (ACF) due to the inactivation or loss or mutation of *APC* genes in 5q (see Fodde et al., 2001); the ACF's further grow into dysplastic adenomas, promoted

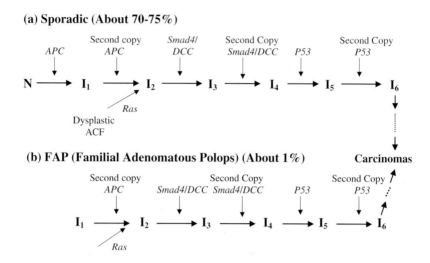

Fig. 1. LOH (Loss of heterozygosity) pathway.

probably to mutation or activation of the *ras* gene in 11p or the *src* gene in 20q. These adenomas grow to a maximum size of about 10 mm^3; further growth and malignancy require the abrogation of growth inhibiting factors and apoptosis which are facilitated by the inaction or loss or mutation of the *Smad2/Smad4* in chromosome 18q and the inactivation or loss or mutation of the *p53* gene in 17p. This multi-stage pathway is represented schematically by Figure 1. Notice that in this multi-stage model, the *APC*, the *Smad2/Smad4* and the *p53* genes are rate limiting genes whereas the oncogenes *H-ras* and *src* are not rate limiting but only speed up these transitions by promoting the proliferation rates of the respective intermediate initiated cells (Jessup *et al.*, 2002).

2.2. The Sequential Nature

While any genetic change can take place at any time, only certain order of genetic changes can lead to a successful completion of the cascade of carcinogenesis to generate cancer tumors. This follows from the observation that carcinogenesis is a micro-evolution process in the cell population so that only those genetic changes or epigenetic changes which increase the fitness of the cells can be fixed and established in the cell population (Cahill *et al.*, 1999; Hopkin, 1996).

To serve as an example and illustration, consider the human colon cancer. In this cancer, observe that a cell with a mutated *ras* oncogene but no other genetic changes would eventually be eliminated from the cell population. It follows that in human colon cancer, mutation of *ras* had never been observed as an initiating early event. From molecular biology a clear explanation is that although the mutated *ras* enables the cell to enter cell division cycle continuously without growth signals (autocrine theory) and can also evoke the MAPK pathway and the PI3k-Akt pathway to increase transcription of many genes in the nucleus (Osada and Takahashi, 2002; Weinberg, 2007), it can also induce the suppressor gene $p14^{ARF}$ to activate the *p53* gene via the ARF-MDM2-p53 pathway (Osada and Takahashi, 2002; Weinberg, 2007), leading to apoptosis of the cell; see Figure 2. On the other hand, if a copy of the *APC* gene in chromosome 5q has been mutated or inactivated, it would generate chromosome instability and LOH (loss of heterozygosity) because the *APC* gene affects the G_2 checking point dominantly during the mitosis stage by interfering with the microtubule and hence centrosome causing

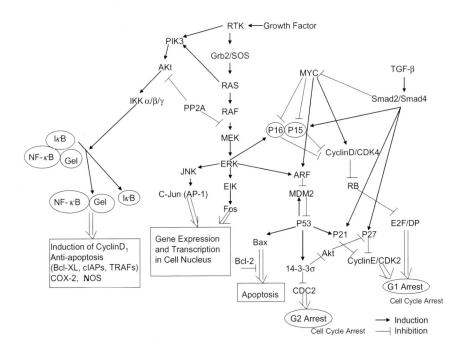

Fig. 2. Signal transduction and pathways of RAS.

aberrant chromosomal segregation and hence aneuploidy and polyploidy daughter cells (Fodde *et al.*, 2001; Green and Kaplan, 2003); this would increase the fitness of the cell and also speed up the mutation or inactivation of the second copy of the *APC* gene. These biological observations may help explain why in FAP and in most sporadic human colon cancer, the first event leading to the cancer phenotype is the inactivation or loss or mutation of the *APC* gene at 5q whereas the mutation or activation of the oncogene genes *ras* and *src*, and the mutation or inactivation of the suppressor *p53* appear to be relatively late (Fodde *et al.*, 2001; Jessup *et al.*, 2002).

To further explain the order of genetic changes in Figure 1, notice that the TGFβ signaling inhibits the *myc* oncogene through binding of *Smad2/3* with p107 and E2F and inhibit the heterodimers cyclin D/CDk4/6 and cyclin E/CDk2 by activating $p15^{INK4b}$ and $p21^{cip1}$ and/or $p27^{kip1}$ respectively (Derynck and Zhang, 2003; Wakefield and Roberts, 2002; Weinberg, 2007). The LOH of *Smad2/Smad4* would abrogate all these negative effects of the TGFβ signaling. Recent studies (Derynck and Zhang, 2003; Wakefield and Roberts, 2002) have also revealed that the *ras* oncogene can also inhibit the effects of *Smad* genes, thus changing the negative effects of the TGFβ signaling into positive effects of cell proliferation and cell invasiveness. Finally the apoptosis effects of *ras* and *myc* and other oncogenes are abrogated by the inactivation or loss or mutation of the *p53* gene in 17p. These biological studies clearly indicate that the order of genetic changes as given in Figure 1 can lead to a successful completion of the cascade of carcinogenesis to generate cancer tumors in human colons.

To serve as another example, consider the human lung cancer. As reported by Fong and Sekido (2002), Osaka and Takahashi (2002), and Wistuba *et al.* (2002), the loss of the suppressor genes (i.e. *FHIT* and *VHL*) in 3p through LOH are the early event, followed by the loss of the gene $p16^{INK4a}$ in 9p through LOH, the loss of *p53* in 17p through LOH and the mutation of the oncogene *ras*. The reason why LOH in 3p is the early event is that this region is a fragile site and is most likely to subject to breakage than other sites (Osada and Takahashi, 2002; Wistuba *et al.*, 2002). Notice also that because epigenetic changes are much more frequent than genetic changes, it is very likely that most cancers are initiated by epigenetic changes. In fact, Zochbauer-Miller *et al.* (2001) had observed multiple methylation of many genes in non-small cell lung cancer. Recently, Jones and Baylin (2002) and Baylin and Ohm (2006) have observed that epigenetic changes often lead to LOH and gene mutations, which may underline the importance of epigenetic changes in the initiation and progression of carcinogenesis.

2.3. The Genetic Changes and Cancer Genes

Carcinogenesis is initiated either by genetic changes (Hesketh, 1997; Loeb and Loeb, 2000; MacDonald *et al.*, 2004; Osada and Takahashi, 2002; Pharoah *et al.*, 2004; Smolinski and Meltzer, 2002; Tan, 1991; Welch and Chrysogelos, 2002) or by epigenetic change through activation of oncogene product or silencing effects of suppressor genes (Baylin, 2005; Belinsky, 2004; Breivik and Gaudernack, 1999; Esteller, 2000; Feinberg and Tycko, 2004; Flintoft, 2005; Holm *et al.*, 2005; Jaffe, 2003; Lund and Lohuizen, 2004; Ohlsson *et al.*, 2003; Robertson, 2005; Ushijima, 2005). The genetic changes may either be as small as point gene mutations, or as large as some chromosomal aberrations such as deletion of chromosomal segments, chromosome inversion and chromosomes translocation leading to mutation or deletion of some cancer genes, or activation of some dominant cancer genes, or inactivation of some recessive cancer genes. The cancer genes which contribute to the creation of cancer phenotype are the oncogenes (dominant cancer genes), the suppressor genes (recessive cancer genes) and the mis-match repair genes (MMR) which are involved in DNA synthesis and repair and/or chromosomal segregation. (As the suppressor genes, MMR genes are recessive genes.) Some of the cancer genes may be considered as both a suppressor gene and an oncogene (e.g. the APC gene in colon cancer (Fodde *et al.*, 2001; Green and Kaplan, 2003)).

The oncogenes, the suppressor genes and the MMR genes are the major genes for the creation of the cancer phenotype although some other modifying genes may also contribute to cancer through its interaction with proteins of oncogenes and/or suppressor genes or its interference with some cancer pathways. To date, about 250 oncogenes and about 75 suppressor genes have been identified.

Oncogenes are highly preserved dominant genes which regulate development and cell division. When these genes are activated or mutated, normal control of cell growth is unleashed, leading to the cascade of carcinogenesis. Specifically, some of the oncogenes such as the *Ras* oncogene induces

$G_0 \to G_1$ by functioning as a signal propagator from signal receptor at the cell membrane to the transcription factors in the cell nucleus in the signal transduction process. Some of the oncogenes serve as transcription factors (e.g., *myc, jun* and *fos, etn*) to affect DNA synthesis during the *S* stage while some other oncogenes serve as anti-apoptosis (e.g. bcl-2) agents. As illustrated in Figure 2, the growth factor signals, or the RTK (Receptor Tyrosine Kinase) or cytokine or integrin may activate and induce expression of the oncogene *ras* which evoke the MAPK pathway and the PI3K-Akt signals to induce expression of the oncogenes *myc, jun* and *fos* in the nucleus to promote cell proliferation (Osada and Takahashi, 2002; Weinsberg, 2007).

Suppressor genes are recessive genes whose inactivation or mutation lead to uncontrolled growth. Mutation or deletion of MMR genes (suppressor genes) lead to microsatellite repeats and create a mutator phenotype, predisposing the affected cells to genetic instability and to increase mutation rates of many relevant cancer genes. Many of the suppressor genes either function to control the gap stages (G_1 and G_2) or by abrogating the apoptosis process or function to control the activation of an oncogene such as myc. For example, the protein of the suppressor gene *RB* forms a complex with E2F and some differentiating-related proteins (DP) to block transition from $G_1 \to S$; when the RB gene protein is phosphorylated or the *RB* gene inactivated or mutated, E2F is unleashed to push the cell cycle from the G_1 phase to the *S* phase; see **Remark 2**. The protein products of the suppressor gene $p16^{INK4a}$ at 9p21 inhibit the function of cyclin D1 and CDk4 proteins which phosphorylate the RB gene product to release E2F. The inactivation of many suppressor genes such as the *p53* gene abrogates or suppresses the apoptosis process. In colon cancer, the mutation or deletion of both copies of the suppressor gene *APC* at 5q lead to increased expression level of the *myc* gene and *D1* gene in the nucleus. Recent studies have shown that the *APC* gene may also affect the G_2 checking point dominantly by interfering with the microtube and hence centrosome causing aberrant chromosomal segregation and hence aneuploidy and polyploidy daughter cells (Fodde *et al.*, 2001; Green and Kaplan, 2003). (In this sense, the *APC* gene in chromosome 5q act both as a recessive gene and a dominant gene.) Recently, Blasco (Blasco, 2005) has shown that mutation or epigenetic silencing of the *RB* gene can also induce immortalization by elongating telomere length.

Carcinogenesis can be initiated either by mutation or activation of an oncogene, or inactivation or loss or mutation of a suppressor gene, or by mutation or epigenetic silencing of some MMR gene leading to microsatellite genomic instability. For example, the mutation of the *ras* oncogene at codon 61 has initiated skin cancer in mouse treated with dimethylbenzanthraccine (DMBA). The deletion or epigenetic inactivation or mutation of the *RB* gene at 13q14 has initiated the human retinoblastoma cascade. The familial adenomatous polyposis (FAP) colon cancer is initiated by the mutation of the *APC* gene at chromosome 5q21; the hereditary non-polyposis colorectal cancer (HNPCC) is initiated by the epigenetic silencing or mutation of the MMR gene *hMSH2* at 2p16 or the MMR gene *hMLH1* at 3p21.

Remark 2. As demonstrated by Cheng (2004), when the *RB* gene is phosphorylated, E2F-1 is partially released from an inhibited state and it turns on a series of genes including cyclin A and cyclin E. Cyclin E forms a complex with CDK2 and cdc25A. Cdc25A can remove the inhibitory phosphates from CDK2 and the resulting cyclin E/CDK2 complex, thus further phosphorylating RB, leading to a complete release of E2F and the transcription of a series genes essential for *S*-phase progression and DNA synthesis.

2.4. Cell Cycle and Carcinogenesis

Cancer evolves through promotion of cell proliferation, abrogation of apoptosis and differentiation, immortalization and acquiring angiogenesis and metastasis, all these being achieved by accumulation of irreversible genetics and/or epigenetic changes. Cell proliferation is generated by the completion of somatic cell divisions and genetic changes can occur only during cell division whereas somatic cell division starts by entering and then completing the cell division cycle:

$$G_0 \to G_1 \to S \to G_2 \to M \to G_0,$$

where G_0 is the resting stage, S the synthesis stage during which proteins and DNA are synthesized, M the mitosis stage during which the cell divides to produce two daughter cells, and G_1 and G_2 are the gap steps which serve as control steps of the cycle; see Figure 3.

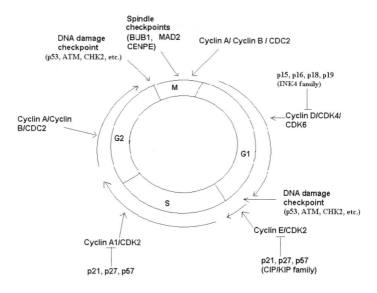

Fig. 3. Mitotic cell cycle and checkpoints.

Recent biological findings indicate that the epigenetic changes and cancer genes affect carcinogenesis mostly through the control of cell cycle. To illustrate this, observe first that the progression of G_1, S, G_2, M stages and from $G_1 \to S$ are controlled by heterodimers of cyclins and cyclin-dependent kinases (CDK's) cyclin D/CDK4, cyclin A_1/CDK2, cyclin A/cyclin B/CDC2, cyclin A/cyclin B/CDC2 and cyclin E/CDK2 respectively. Furthermore in the cell cycle, there are DNA damage checkpoints between G_1 stage and S stage and between G_2 stage and M stage; and there are spindle checkpoints in prometaphase and between metaphase and analysis during the M stage; see Figure 3. These checkpoints are controlled by suppressor genes (*p53*, *ATM*, *BUB1*, *MAD2*, etc.). Thus, if the suppressor genes $p15^{INK4b}$ or $p16^{INK4a}$ or $p18^{INK4c}$ or $p19^{INK4d}$ is activated, then the heterodimer cyclin D/CDk4 is inhibited to cause arrest at the G_1 stage; similarly, if the suppressor genes $p21^{cip1}$ or $p27^{kip1}$ or $p57^{kip2}$, then the heterodimer cyclin E/CDk2 is inhibited to cause arrest at the S stage or at the transition between G_1 and S stage. Similarly, if DNA is damaged, then damage signals activate the tumor suppressor genes *ATM* and *p53* causing apoptosis or cell division arrest at G_1 and/or G_2 stages; see Figure 4. If *p53* is inactivated or mutated, then the processes of apoptosis and cell arrest are impaired; thus cell proliferation is promoted. This may help explain why *p53* is inactivated or mutated in most of human cancers.

To illustrate how some oncogenes control cell cycle, notice that under normal conditions, cells can enter cell division cycle only through induction by a growth signal (growth factor, hormones, cytokines, integrins, etc; see Weinberg (2007) and Hanahan and Weinberg (2000)). Hence, for cancer tumor cells to achieve uncontrolled growth, it must achieve the capability of self-sufficiency (or independence) of growth signals. This is achieved by many of the oncogenes such as *hst, int-2, erb, ras* and *src*. To illustrate, consider the oncogene *H-ras* which encodes a G-protein p21. In normal and resting cells, this G-protein binds with GDP (guanosine di-phosphate) to form the complex p21-GDP and is inactive. In the event of a growth factor signaling, p21-GDP changes into p21-GTP (guanosine tri-phosphate) and is active through growth factor receptors, GRB2 and SOS (Weinsberg, 2007). When the rat breast was exposed to NMU (N-nitroso-N-methylurea), it induced a point mutation in the *ras* gene through a $G \to A$ base change at the second nucleotide of codon 12 (Tan, 1991; Zarl *et al.*, 1985). It appeared that this mutated $p21^{Ras}$ was unable to hydrolyze GTP to yield GDP, thus G-GTP remains to be active and excited, leading to continuous entering into cell division cycle independent of growth signals and processing signal transduction to transcript many genes in the nucleus (Tan, 1991; Zarl *et al.*, 1985).

2.5. Epigenetic and Cancer

Cancer initiation and progression are achieved and controlled by gene mutations and genetic changes. However, these genetic effects can also be achieved by changes of functions of these gene products through non-genetic avenues without affecting the nucleotide sequences in DNA molecules (Baylin, 2005; Belinsky, 2004; Breivik and Gaudernack, 1999; Esteller, 2000; Feinberg and Tycko, 2004; Flintoft, 2005; Holm *et al.*, 2005; Jaffe, 2003; Lund and Lohuizen, 2004; Ohlsson *et al.*, 2003; Robertson, 2005; Ushijima, 2005). These are called epigenetic changes which mainly involve activation of oncogenes products or silencing of suppressor genes proteins through DNA methylation of cytosine at C_pG base pair islands (Baylin, 2005; Belinsky, 2004; Breivik and Gaudernack, 1999; Esteller, 2000; Feinberg and Tycko, 2004; Flintoft, 2005; Holm *et al.*, 2005; Jaffe,

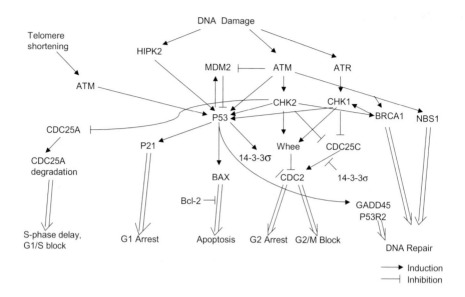

Fig. 4. DNA damage-invoked pathways and consequences.

2003; Lund and Lohuizen, 2004; Ohlsson et al., 2003; Robertson, 2005; Ushijima, 2005) or histone acetylation (Ohlsson et al., 2003), or loss of imprinting (LOI) (Holm et al., 2005; Robertson, 2005), or tissue disorganization and gap junction disruption (Jones and Baylin, 2002; Baylin, 2005). For example, Ferreira et al. (2001) have showed that besides genetic inactivation or mutation of the *RB* gene, the process that the *RB* gene represses E2F-regulated genes in differentiated cells can also be achieved by an epigenetic mechanism linked to heterochromatin and involving histone H3 and promoter DNA methylation. In human colon cancer, Breivik and Gaudernack (1999) showed that either methylating carcinogens or hypermethylation at C_pG islands would lead to G/T mismatch which in turn leads to MMR gene deficiency or epigenetic silencing of the MMR genes and hence MSI (microsatellite instability); alternatively, either hypo-methylation, or bulky-adduct forming (BAF) carcinogens such as alkylating agents, UV radiation and oxygen species promote chromosomal rearrangement via activation of mitotic check points (MCP), thus promoting CIS (chromosomal instability). These data clearly suggest that the epigenetic changes and/or interaction between genetic and epigenetic changes are equally important as genetic changes in generating the cancer phenotype. The following biological results further underline the importance of epigenetic mechanisms in cancer initiation and progression.

(1) Hypomethylation in non-promoting regions of the gene results in increased transcription of the oncogene and genomic instability (Feinberg and Tykco (2004), Yamada et al. (2005)).

(2) Hypermethylation of the CpG islands of the promoter region of the gene would silence tumor suppressor genes. This often leads to allele losses and LOH; see (Jones and Baylin (2002), Baylin and Ohm (2006)). Notice that LOH of suppressor genes is the major avenue by means of which most cancers of human beings are initiated (e.g. colon cancer, lung cancer, liver cancer).

(3) Because 5-methylcytosine is mutagenic, aberrant methylation of the cytosine in CpG region may often lead to gene mutation (Jones and Baylin, 2002; Baylin, 2006; Breivik and Gaudernack, 1999). For example, promoter methylation of the MGMT (O^6-methylguanine-DNA methyltransferase) leads to G $\rightarrow$ A mutation (Jones and Baylin, 2002). Cytosine methylation in the coding region of the gene increase mutation rate (C $\rightarrow$ T mutation) because of the spontaneous hydrolytic deamination of the methylated cytosine, which causes C $\rightarrow$ T mutation (Jones and Baylin, 2002). Methylation of CpG changes the absorption wavelength of the cytosine into the range of incidence sunlight to enhance UV absorption, resulting in CC $\rightarrow$ TT mutation (Jones and Baylin, 2002). Methylation enhances carcinogen binding with DNA to form DNA adducts and induces G $\rightarrow$ T transversion mutation (Jones and Baylin, 2002).

(4) Many human cancers appeared to be initiated by epigenetic mechanisms. For example, in 75% cases of the lung cancer, carcinogenesis are initiated by gene silencing and hence LOH of the RASSF1 and other genes in chromosome 3p (Baylin, 2006; Osada and Takahashi, 2002). In human liver with hepatitis B virus infection, in 70% cases, HCC is initiated by reversible epigenetic event and aberrant methylation of genes as methyltranferase causing elevated level of the *TGFα* gene or elevated level of the *IGF − 2* gene (Thorgeirsson, 2002). In colon cancer, silencing of the gene *SFRP* (secreted frizzled-related protein) activates the Wnt signal pathway to initiate the colon carcinogenesis process (Baylin, 2006).

(5) Hypermethylation of the CpG islands of the promoter region of the gene never occur in mutants or chromosomes involving genetic changes (Jones and Baylin, 2002; Baylin, 2006).
(6) If one allele is mutated, hypermethylation is often seen as the second inactivating change (Jones and Baylin, 2002; Baylin, 2006). For example, in liver cancer, promoter hypermethylation often precedes mutations and genetic changes (Thorgeirsson, 2002).

2.6. Telomere, Immortalization and Cancer

It is well-documented that normal stem cells have finite life span and can divide only a finite number of times whereas cancer tumor cells can divide indefinitely (i.e. immortalized) (Blasco, 2005; Hahn and Meyerson, 2001; Shay et al., 2001). Biological studies have shown that this is related to telomeres which make up the ends of chromosomes to protect it from recombination and degradation activities. Telomeres are special chromatin structures and are composed of tandem repeats of TTAGGG sequences and a single stranded overhang of the G-rich strand. When each normal stem cell divides, telomeres shorten by 50–200 bp, due to the fact that the lagging strand of DNA synthesis is unable to replicate the extreme $3'$ end of the chromosome. When telomeres are sufficiently shortened, cells enter an irreversible growth arrest called cellular senescence; when the length of the telomeres have shortened below some critical points resulting in loss of telomere protection of the chromosomes, then the cells will die or lead to chromosomal instability. In cancer cells, the telomerase helps to stabilize telomere length so that cancer cells become immortalized and can divide indefinitely.

Telomerase is a reverse transcriptase and is encoded by the TERT (telomerase reverse transcriptase) gene. This gene recognizes the $3'$-OH group of the end of the G-strand overhang of telomere. It elongates telomeres by extending from this group using the RNA, which is encoded by the TERT, as a template. Blasco (2005) has shown that besides being substrate for telomerase and the telomere repeat-binding factors, the telomeres are also bound and regulated by many chromatin regulators and related proteins, including TRF1, TRF2, TERT, TERC, DKC1, SUV39H1, SUV39H2, SUV20H1, HP1α, HP1β, HP1γ and the retinoblastoma family of proteins (RB1, RBL1,

RBL2). This implies that the telomere length and function are also regulated by many chromatin and regulator proteins as given above. For example, if the retinoblastoma gene has been inactivated so that the RB1 function is lost, then trimethylation of H4-K20 is down, leading to abnormally long telomeres; as shown by Blasco (2005), this telomere length elongation can also be achieved by epigenetic regulation of telomeric chromatin. Henson *et al.* (2002) and Lundbald (2001) have shown that lengthening of telomere and hence immortalization can also be achieved by telomerase-independent mechanisms.

2.7. Single Pathway versus Multiple Pathways of Carcinogenesis

In some type of cancers such as retinoblastoma, cancer tumor is derived by a single pathway (Tan, 1991; Weinberg, 2007).

In many other cancers, however, the same cancer may arise from different carcinogenic pathways. This include skin cancers, liver cancers and mammary gland in animals, the melanoma development in skin cancer in human beings, breast cancer, colon cancer, liver cancer and lung cancer in human beings.

To serve as an example, consider the colon cancer of human beings. For this cancer, genetic studies have indicated that there are two major avenues by means of which colon cancer is developed (de la Chapelle, 2004; Fodde, Smit and Clevers, 2001; Fodde *et al.*, 2001; Green and Kaplan, 2003; Hawkins and Ward, 2001; Jessup *et al.*, 2002; Lauren-Puig *et al.*, 1999; Luebeck and Moolgavkar, 2002; Peltomaki, 2001; Potter, 1999; Sparks *et al.*, 1998; Ward *et al.*, 2001): the chromosomal instability (CIN) and the micro-satellite instability (MSI). The CIN pathway involves loss or mutation of the suppressor genes — the *APC* gene in chromosome 5q, the *Smad4/DCC* genes in chromosome 18q and the *p53* gene in chromosome 17p. This pathway accounts for about 75–80% of all colon cancers and has been referred to as the LOH (loss of heterozygosity) pathway because it is often characterized by aneuploidy /or loss of chromosome segments (chromosomal instability); it has also been referred to as the APC-β-catenin-Tcf-myc pathway because it involves β-catenin, Tcf (T-cell factor) and the *myc* oncogene; see **Remark 3**. The MSI pathway involves microsatellite MMR

(a) Sporadic (About 10-15%)

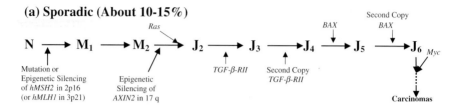

(b) HNPCC (Hereditary Non-Polyposis Colon Cancer) (≤ 5%)

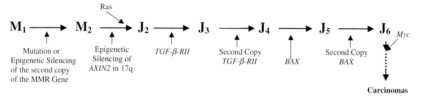

Fig. 5. MSI-H (high level of MSI) pathway.

genes, *hMLH1*, *hMSH2*, *hPMS1*, *hPMS2*, *hMSH6* and *hMSH3*. (Mostly *hMLH1* and *hMSH2*.) This pathway accounts for about 10–15% of all colon cancers and appears mostly in the right colon. It has been referred to as the MSI (micro-satellite instability) pathway or the mutator phenotype pathway because it is often characterized by the loss or mutations in the mismatch repair genes creating a mutator phenotype to significantly increase the mutation rates of many critical genes.

The CIS and MSI pathways of human colon cancer are represented by Figures 1 and 5, respectively. The model in Figure 1 is a six-stage model. However, because of the haplo-insufficiency of the *Smad4* gene (Alberici et al., 2006) and the haplo-insufficiency of the *p53* gene (Lynch and Milner, 2006), one may reduce this six-stage model into a four-stage model by combining the third stage and the fourth stage into one stage and by combining the fifth stage and the sixth stage into one stage. This may help explain why the four-stage model fits the human colon cancer better than other stage models (Luebeck and Moolgavkar, 2002).

Recent biological studies by Green and Kaplan (2003) and others have also shown that the mutation or inactivation or deletion of the *APC* gene can cause defects in microtubule plus-end attachment during mitosis dominantly, leading to aneuploidy and chromosome instability. This would speed up the mutation or inactivation of the second copy of the *APC* gene and increase fitness of the *APC*-carrying cells in the micro-evolution process of cancer progression. This could also help explain why the APC LOH pathway is more frequent than other pathways. Recent biological studies (Baylin, 2002; Koinuma *et al.*, 2006) have indicated that both the CIS and the MSI pathways involve the Wnt signalling pathway, the TGF-β signalling and the p53-Bax signalling but different genes in different pathways are affected in these signalling processes. (In the CIS pathway, the affected gene is the *APC* gene in the Wnt signalling, the *Smad4* in the TGF-β signalling and *p53* gene in the p53-Bax signalling; on the other hand, in the MSI pathway, the affected gene is the *Axin 2* gene in the Wnt signalling, the *TGF-β-Receptor II* in the TGF-β signalling and the *Bax* gene in the p53-Bax signalling.) Because the probability of genetic change and point mutation of genes are in general very small compared to epigenetic changes, one may speculate that colon cancer may actually be initiated by some epigenetic mechanisms. For example, Breivik and Gaudernack (1999) demonstrated that either methylating carcinogens or hypermethylation at C_pG islands would lead to G/T mismatch which in turn leads to MMR gene deficiency or epigenetic silencing of the MMR genes and hence MSI; alternatively, either hypo-methylation, or bulky-adduct forming (BAF) carcinogens such as alkylating agents, UV radiation and oxygen species promote chromosomal rearrangement via activation of mitotic check points (MCP), thus promoting CIS (chromosomal instability).

Remark 3. In the APC-β-catenin−Tcf−myc pathway, the *APC* gene forms a complex with β-catenin, Axin1/2 and GSK−3β (glycogen synthase kinases 3-β) to degrade the β-catenin protein. When both copies of *APC* gene is lost or mutated, the β-catenin protein then accumulates to form a complex with Tcf to promote cell proliferation, usually via the elevated level of the *myc* gene and/or cyclin D1. Additionally, the β-catenin protein also binds with E-cadherin and α−catenin to disrupt cell junction function, promoting mobility of the cell (Bienz and Hamada, 2004).

3. SOME GENERAL STOCHASTIC MODELS OF CARCINOGENESIS

Based on the above biological mechanisms of carcinogenesis, it is clear that carcinogenesis develops by a series of steps each of which involves epigenetic changes and/or genetic changes followed by stochastic proliferation and differentiation of newly generated cells. This leads to the following stochastic multi-stage models to be referred to as extended multi-event models.

3.1. The Extended Multi-Event Model of Carcinogenesis

The most general model for a single pathway is the extended k-stage ($k \geq 2$) multi-event model proposed by Tan and Chen (1998, 2005) and Tan et al. (2004). This is an extension of the multi-event model first proposed by Chu (1985) and studied by Tan (1991) and Little (1995). It views carcinogenesis as the end point of k ($k \geq 2$) discrete, heritable and irreversible events (mutations, genetic changes or epigenetic changes) with intermediate cells subjected to stochastic proliferation and differentiation. It takes into account cancer progression by following Yang and Chen (1991) to postulate that cancer tumors develop from primary I_k cells by clonal expansion (i.e. stochastic birth-death process), where a primary I_k cell is an I_k cell which arise directly from an I_{k-1} cell.

Let N denote normal stem cells, T the cancer tumors and I_j the j-th stage initiated cells arising from the $(j-1)$-th stage initiated cells ($j = 1, \ldots, k$) by mutation or some genetic changes. Then the model assumes $N \to I_1 \to I_2 \to \cdots \to I_k$ with the N cells and the I_j cells subject to stochastic proliferation (birth) and differentiation (death). The cancer tumors develop from primary I_k cells by clonal expansion.

As examples, given in Figure 1 is the multi-stage model for the APC$-\beta-$Catenin$-$Tcf pathway for human colon cancer. This is a major pathway which accounts for 80% of all colon cancers. In Figure 5, we present another multistage model for human colon cancer involving mis-match repair genes (mostly *hMLH1* and *hMSH2*) and the *Bax* gene in chromosome 19q and the *TGFβR₁I* gene in chromosome 3p. This pathway accounts for about 15% of all human colon cancer. In Figure 6, we present a multi-stage pathway for

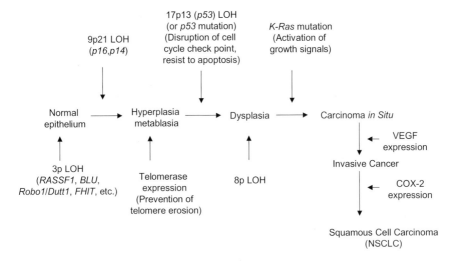

Fig. 6. Histopathology lesions and genetic pathway of squamous cell carcinoma of non-small cell lung cancer (NSCLC).

the squamous NSCLC (non-small cell lung cancer) as proposed by Osaka and Takahashi (2002) and Wituba *et al.* (2002).

3.2. The Mixed Models of Carcinogenesis

In the population, for the same type of cancer, different individual may involve different pathways or different number of stages (Tan, 1988, 1991, 2002; Tan and Chen, 1998, 2000, 2005; Tan *et al.*, 2001). These models have been referred by Tan (1990, 1991) and Tan and Singh (1990) as mixed models of carcinogenesis. These models are basic consequences of the observations: (1) Different individuals are subject to different environmental conditions, (2) the mutation of critical cancer genes can occur in either germline cells or in somatic cells, and (3) As shown in the previous section, the same cancer can be derived by several different pathways, referred to as multiple pathways.

To serve as an example, consider again the human colon cancer. The multiple pathways for human colon cancer as described in the previous section then leads to the following five different pathways: The sporadic LOH (about 70%, see Figure 1), the familial LOH (FLOH, about 10–15%), the FAP (familial adenomatous polys, about 1%), the sporadic MSI (about

10–15%, see Figure 5) and the HNPCC (hereditary non-polyposis colon cancer, about 4–5%). For sporadic pathways, the individuals at birth are normal individuals and do not carry any mutated or inactivated suppressor genes. For FAP, the individual has inherited a mutated *APC* gene in chromosome 5 at birth. For HNPCC, the individuals has inherited a mutated mismatch gene *hMLH1* or *hMSH2*. For the familial colon cancer, the individuals have inherited a low penetrance mutated gene such as *APCI1307K* at birth. Hence, FAP and FLOH are special cases of the APC–β-catenin–Tcf–myc pathway and HNPCC a special case of the MSI pathway.

The above indicates that from the population perspective, the human colon cancer can best be described by a mixture of five pathways. Assume that each individual in the population develops colon cancer by following the same biological mechanism independently of other individuals. Let ω_i be the proportion for the i-th pathway and $P_i(j)$ the probability that a normal person would develop colon cancer during the j-th age group $[t_{j-1}, t_j)$ by the i-th pathway. Then the probability that this person would develop colon cancer during the j-th age group is given by $Q(j) = \sum_{i=1}^{5} \omega_i(j) P_i(j)$. Let Y_j be the total number of people who develop colon cancer during the j-th age group and let n_j be the number of people at risk for colon cancer. Then the probability density of Y_j given n_j is:

$$P(Y_j \mid n_j) = \binom{n_j}{Y_j} [Q(j)]^{Y_j} [1 - Q(j)]^{n_j - Y_j},$$

where $\binom{n_j}{Y_j} = \frac{n_j!}{(Y_j!)[(n_j - Y_j)!]}$. When n_j is very large and $Q(j)$ very small so that $\lambda_j = n_j Q(j) < \infty$, $P(Y_j \mid n_j)$ is closely approximated by a Poisson distribution with mean λ_j.

4. SOME NEW APPROACHES FOR ANALYZING STOCHASTIC MODELS OF CARCINOGENESIS

To develop mathematical theories for the stochastic models of carcinogenesis, the traditional approach is by way of Markov theories (Tan, 1991). The basic approach along this line consists of the following four basic steps: (a) Deriving the probability generating function (PGF) of the number of cancer tumors, (b) deriving the incidence function of cancer tumors, (c) deriving the probability distribution of time to tumor onset, and (d) deriving the probabilities of the number of cancer tumors. This approach has been described and illustrated in detail in Tan (1991).

Using the above approach, theoretically one may derive some useful information for stochastic models of carcinogenesis. However, a careful scrutiny would reveal that the above approach suffers from several drawbacks: (1) The process may not be Markov so that the above approach is not applicable. For example, if one can not ignore cancer progression, then the number of cancer tumors is not Markov since it depends on the time when the last stage initiated cell is generated (Tan, 2002; Tan and Chen, 1998, 2005; Tan *et al.*, 2001, 2004); see also **Remark 1**. (2) As illustrated in Tan (1991), it is mathematically manageable only for a two-stage model under very restrictive assumptions; further many of these assumptions have significant impacts on cancer incidence (Tan, 1991, 2002; Tan and Chen, 1998, 2000, 2005; Tan *et al.*, 2001, 2004; Yang and Chen, 1991). (3) It is extremely difficult, if not impossible, to fit and to adapt to cancer data, especially beyond the simplest MVK two-stage model. (4) The cancer stages and many of the parameters are not identifiable when three or more stages are involved. In fact, as shown by Hanin and Yakovlev (1996), even for the simple homogeneous two-stage MVK model, it is not possible to estimate $\{b_1, d_1, \lambda_0\}$ and α_1 separately; hence, not all parameters are estimable by using the above Markov approach unless some other data and some further external information about the parameters is available.

Because of the above difficulties, we have developed an alternative approach to developed stochastic models of carcinogenesis (Tan, 2002; Tan and Chen, 1998, 2005; Tan *et al.*, 2001, 2004). As shown by Tan and Chen (1998) through pgf (probability generation function) method, this alternative approach is equivalent to the above approach but is more powerful and can get more information. In this section we illustrate how to derive some basic results by using an extended k-stage multi-event model of carcinogenesis with ($k \geq 2$) carcinogenesis as an example.

4.1. Stochastic Differential Equations

For the extended k-stage multi-event model, it is assumed that the number $N_0(t) = I_0(t)$ of normal stem cells are deterministic functions of time since these numbers are usually very large. Hence, for this model, the state variables are $(I_i(t), i = 1, \ldots, k-1, T(t))$, where $I_i(t) =$ number of I_i cells at time t, and $T(t) =$ number of cancer tumors at time t.

To derive stochastic differential equations for $I_j(t)$, $j = 1, \ldots, k-1$, note: (i) The numbers of $\{I_j, j = 1, \ldots, k-1\}$ cells at time $t+\Delta t$ derive from the numbers of $\{I_j, j = 0, 1, \ldots, k-1\}$ cells at time t through stochastic birth and death processes and mutation processes. (ii) The birth-death-mutation processes during the small interval with length Δt is equivalent to multinomial distributions.

Define for $j = 1, \ldots, k-1$:

- $B_j(t) =$ Number of new I_j cells generated by stochastic cell proliferation (birth) of I_j cells during $(t, t + \Delta t]$,
- $D_j(t) =$ Number of death of I_j cells during $(t, t + dt]$,
- $M_j(t) =$ Number of new I_j cells arising from I_{j-1} cells by mutation or some genetic changes during $(t, t + \Delta t]$, $j = 1, 2, \ldots, k$.

Then, for $j = 0, 1, \ldots, k-1$, the above principle leads to:

$$M_1(t) \sim Poisson\{\lambda_0(t)\Delta t\},$$

where $\lambda_0(t) = I_0(t)\alpha_0(t)$, and for $j = 1, \ldots, k-1$,

$$[B_j(t), D_j(t), M_{j+1}(t)] | I_j(t) \sim ML[I_j(t); b_j(t)\Delta t, d_j(t)\Delta t, \alpha_j(t)\Delta t].$$

It follows that $EM_1(t) = \lambda_0(t)\Delta t$ and for $j = 1, \ldots, k-1$,

$$E\{B_j(t) \mid I_j(t)\} = I_j(t)b_j(t)\Delta t,$$
$$E\{D_j(t) \mid I_j(t)\} = I_j(t)d_j(t)\Delta t,$$
$$E\{M_{j+1}(t) \mid I_j(t)\} = I_j(t)\alpha_j(t)\Delta t.$$

By the conservation law, we have then for $i = 1, \ldots, k-1$,

$$I_i(t + \Delta t) = I_i(t) + M_i(t) + B_i(t) - D_i(t).$$

The stochastic differential equations for the state variables are:

$$\Delta I_j(t) = I_j(t + \Delta t) - I_j(t) = M_j(t) + B_j(t) - D_j(t)$$
$$= \{I_{j-1}(t)\alpha_{j-1}(t) + I_j(t)\gamma_j(t)\}\Delta t + \epsilon_j(t)\Delta t,$$
$$j = 1, \ldots, k-1, \qquad (1)$$

where $\gamma_j(t) = b_j(t) - d_j(t)$, $j = 0, 1, \ldots, k-1$ and for $j = 1, \ldots, k-1$,
$e_j(t)\Delta t = [M_j(t) - I_{j-1}(t)\alpha_{j-1}(t)\Delta t] + [B_j(t) - I_j(t)b_j(t)\Delta t] - [D_j(t) - I_j(t)d_j(t)\Delta t]$.

4.2. The Probability Distribution of $T(t)$

To develop probability distribution for $T(t)$, observe that cancer tumors develop from primary I_k cells by following a stochastic birth-death process with birth rate $b_T(s, t)$ and death rate $d_T(s, t)$, where s is the time the primary tumor cell was generated. Hence we can derive the probability distribution for number of detectable cancer tumors at time t ($T(t)$).

Then as shown in Tan (2002, Chapter 8), the conditional probability distribution of $T(t)$ given $\{I_{k-1}(s), s \leq t\}$ is:

$$T(t) \mid \{I_{k-1}(s), s \leq t\} \sim \text{Poisson}(\Lambda_T(t)) \tag{2}$$

where $\Lambda_T(t) = \int_{t_0}^{t} I_{k-1}(x)\alpha_{k-1}(x) P_T(x, t) dx$, and the $P_T(s, t)$ is given by:

$$P_T(s, t) = \frac{1}{h_T(s, t) + g_T(s, t)} \left(\frac{g_T(s, t)}{h_T(s, t) + g_T(s, t)} \right)^{N_T - 1} \tag{3}$$

where N_T is the number of tumor cells for the tumor to be detectable,

$$h_T(s, t) = \exp\left\{ -\int_s^t [b_T(s, y) - d_T(s, y)] dy \right\}$$

and

$$g_T(s, t) = \int_s^t b_T(s, y) h_T(y, t) dy.$$

4.3. Probability Distribution of the State Variables

Chose some fixed small interval for Δt as 1 time unit (i.e. $\Delta t \sim 1$) and denote by $\widetilde{X}(t) = \{I_i(t), i = 1, \ldots, k-1\}$. Let $g_0\{j; \lambda_0(t)\}$ denote the probability $M_1(t) = j$ from the Poisson distribution $M_1(t) \sim \text{Poisson}\{\lambda_0(t)\}$; for $r = 1, \ldots, k-1$, let $f_r\{i, j \mid I_r(t)\}$ denote the probability $\{B_r(t) = i, D_r(t) = j\}$ from the multinomial distribution $\{B_r(t), D_r(t)\} \sim ML[I_r(t); b_r(t), d_r(t)]$ and let $h_r\{j \mid i_r, j_r, I_r(t)\}$ be the probability of $(M_r(t) = j)$ from the binomial distribution $M_r(t) \sim Binomial\{I_r(t) - i_r - j_r; \frac{\alpha_r(t)}{1 - b_r(t) - d_r(t)}\}$. Then, from results in Section 4.1, the probability density function of $X = \{\underset{\sim}{X}(1), \ldots, \underset{\sim}{X}(t_M)\}$ is

$$P(X) = \prod_{t=1}^{t_M} \{P[\underset{\sim}{X}(t) \mid \underset{\sim}{X}(t-1)]\}$$

and for $t = 0, 1, \ldots, t_M - 1$,

$$P\{\underset{\sim}{X}(t+1) \mid \underset{\sim}{X}(t)\} = \sum_{i_1=0}^{I_1(t)} \sum_{j_1=0}^{I_1(t)-i_1} f_1\{i_1, j_1 \mid I_1(t)\} g_0\{a_1(t), \lambda_0(t)\}$$

$$\times \sum_{i_2=0}^{I_2(t)} \sum_{j_2=0}^{I_2(t)-i_2} f_2\{i_2, j_2 \mid I_2(t)\} h_1\{a_2(t) \mid I_2, j_2, I_2(t)\}$$

$$\times \cdots \times \sum_{i_{k-1}=0}^{I_{k-1}(t)} \sum_{j_{k-1}=0}^{I_{k-1}(t)-i_{k-1}} f_{k-1}\{i_{k-1}, j_{k-1} \mid I_{k-1}(t)\}$$

$$\times h_{k-2}\{a_{k-1}(t) \mid i_{k-1}, j_{k-1}, I_{k-1}(t)\}, \quad (4)$$

where $a_r(t) = \text{Max}(0, I_r(t+1) - I_r(t) - i_r + j_r)$.

5. A STATE SPACE MODEL FOR THE EXTENDED MULTI-EVENT MODEL OF CARCINOGENESIS

State space model is a stochastic models which consists of two sub-models: The stochastic system model which is the stochastic model of the system and the observation model which is a statistical model based on available observed data from the system. Hence it takes into account the basic mechanisms of the system and the random variation of the system through its stochastic system model and incorporate all these into the observed data from the system; furthermore, it validates and upgrades the stochastic model through its observation model and the observed data of the system. Thus the state space model adds one more dimension to the stochastic model and to the statistical model by combining both of these models into one model.

As illustrated in Tan (2002, Chapters 8–9), the state space model has many advantages over both the stochastic model and the statistical model when used alone since it combines information and advantages from both of these models.

As an example, in this section we will illustrate how to develop a state space model for the extended multi-event model given in Section 3.1 with the observation model being based on the observed number of cancer incidence over time. For this state space model, the stochastic system model is specified by the stochastic equations given by (1) with the probability distribution of state variables being given in Section 4.3. The observation model is a statistical model based on the number of cancer cases $(y_i(j), i = 1, \ldots, m, j = 1, \ldots, n)$ over n different age groups and m different exposure levels.

5.1. The Stochastic System Model, the Augmented State Variables and Probability Distribution

The probability distribution for the state variables in Eq. (4) is extremely complicated involving many summations. For implementing the Gibbs sampling procedures to estimate the unknown parameters and the state variables, we thus expand the model by augmenting the dummy un-observable variables $\underset{\sim}{U}(t) = \{B_r(t), D_r(t), r = 1, \ldots, k-1\}$ and put $\boldsymbol{U} = \{\underset{\sim}{U}(t), t = 0, \ldots, t_M - 1\}$. Then, from the distribution results in Section (4.3), we have:

$$P\{\underset{\sim}{U}(t) \mid \underset{\sim}{X}(t)\} = \prod_{i=1}^{k-1} f_i\{B_i(t), D_i(t) \mid I_i(t)\};$$

$$P\{\underset{\sim}{X}(t+1) \mid \underset{\sim}{U}(t), \underset{\sim}{X}(t)\} = g_0\{c_1(t), \lambda_0(t)\} \prod_{i=2}^{k-1} h_{i-1}\{c_i(t) \mid B_i(t), D_i(t)\},$$

(5)

where for $i = 1, \ldots, k-1$, $c_i(t) = I_i(t+1) - I_i(t) - B_i(t) + D_i(t)$.

The joint density of $\{\boldsymbol{X}, \boldsymbol{U}\}$ is

$$P\{\boldsymbol{X}, \boldsymbol{U}\} = \prod_{t=1}^{t_M} P\{\underset{\sim}{X}(t) \mid \underset{\sim}{X}(t-1), \underset{\sim}{U}(t-1)\} P\{\underset{\sim}{U}(t-1) \mid \underset{\sim}{X}(t-1)\}.$$

5.2. The Observation Model and the Probability Distribution of Cancer Incidence

The observation model is based on y_{ij}, where y_{ij} is the observed number of new cancer cases in the j-th age group $[t_{j-1}, t_j)$ under exposure to the carcinogen with dose level s_i. Let $n_i(j)$ be the number of normal people from whom the y_{ij} are generated. To derive the probability distribution of y_{ij} given $n_i(j)$ and given the state variables, let $\{I_{k-1}(t; i, r)$ be the number of $I_{k-1}(t)$ cells in the r-th individual who was exposed to the carcinogen with dose level s_i and let $\alpha_{k-1}(i)$ be the rate of the transition $I_{k-1} \to I_k$ under exposure to the carcinogen with dose level s_i. Among people who have been exposed to the carcinogen with dose level s_i, let $P_r(i, j)$ denote the conditional probability given the state variables that the r-th individual would develop cancer during the j-th age group. Then, as shown in Tan (2002, Chapter 8), $P_r(i, j)$ is given by:

$$P_r(i, j) = \exp\left\{ -\sum_{t=0}^{t_{j-1}-1} I_{k-1}(t; i, r)\alpha_{k-1}(i) \right\} (1 - e^{-R(i,j,r)\alpha_{k-1}(i)}),$$

where $R(i, j, r) = \sum_{t=t_{j-1}}^{t_j-1} I_{k-1}(t; i, r)$.

From the above it follows that the conditional probability density of y_{ij} given $n_i(j)$ and given the state variables $\underset{\sim}{I}_{k-1}(i, j) = \{I_{k-1}(t; i, r), t \le t_j, r = 1, \ldots, n_i(j)\}$ is

$$P\{y_{ij} \mid n_i(j), \underset{\sim}{I}_{k-1}(i, j)\} = \binom{n_i(j)}{y_{ij}} \prod_{r=1}^{y_{ij}} P_r(i, j) \prod_{u=y_{ij}+1}^{n_i(j)} [1 - P_u(i, j)]. \quad (6)$$

Let $I_{k-1}(t; i)$ denote the number of I_{k-1} cells at time t under dose level s_i. When $n_i(j)$ and $n_i(j) - y_i(j)$ are very large and when $n_i(j)P_r(i, j)$ are finite for all r, the above probability is closely approximated by:

$$P\{y_{ij} \mid n_i(j), \underset{\sim}{I}_{k-1}(i, j)\} = \frac{1}{y_{ij}!} \exp\{-\lambda_i(j)\} \prod_{r=1}^{y_{ij}} [n_i(j)P_r(i, j)], \quad (7)$$

where $\lambda_i(j) = n_i(j)EP_T(i, j) = n_i(j)EP_r(i, j), r = 1, \ldots, n_i(j)$ and where

$$P_T(i,j) = \exp\left\{-\sum_{t=0}^{t_{j-1}-1} I_{k-1}(t;i)\alpha_{k-1}(i) + \log(1 - e^{\{-R(i,j)\alpha_{k-1}(i)\}})\right\},$$

with $R(i,j) = \sum_{t=t_{j-1}}^{t_j-1} I_{k-1}(t;i)$. (For Proof, see Tan (2002), Tan et al. (2004).)

From Eq. (7), the conditional likelihood of the parameters given data $Y = \{y_{ij}, i=1,\ldots,m, j=1,\ldots,n\}$ and given the state variables is

$$L\{\Theta \mid Y, \underset{\sim}{I}_{k-1}\} = \prod_{i=1}^{m}\prod_{j=1}^{n} P\{y_{ij} \mid n_i(j), \underset{\sim}{I}_{k-1}(i,j)\}, \qquad (8)$$

Also, since $\alpha_{k-1}(i)$ is very small, it can be shown (Tan et al., 2004) that with $\bar{\alpha}_{k-1}(i) = 10^6 \alpha_{k-1}(i)$,

$$E[P_T(i,j)] = \exp\left\{-\frac{1}{10^6}\sum_{t=0}^{t_{j-1}-1} E I_{k-1}(t;i)\bar{\alpha}_{k-1}(i)\right.$$

$$\left. + E\log(1 - e^{-\frac{1}{10^6}R(i,j)\bar{\alpha}_{k-1}(i)})\right\}$$

$$\approx B_i(j)\bar{\alpha}_{k-1}(i)\exp\{-A_i(j)\bar{\alpha}_{k-1}(i)\},$$

where $A_i(j) = \frac{1}{10^6}\{\sum_{t=0}^{t_j-1} E I_{k-1}(t;i) - \frac{1}{2}ER(i,j)\}$ and $B_i(j) = \frac{1}{10^6}ER(i,j)$.

From the above distribution results, it is obvious that the joint density of $\{X, U, Y\}$ is

$$P\{X, U, Y\} = \prod_{i=1}^{m}\prod_{j=1}^{n} P\{y_{ij} \mid n_i(j), \underset{\sim}{I}_{k-1}(i,j)\}$$

$$\times \prod_{t=t_{j-1}+1}^{t_j} P\{\underset{\sim}{X}(t) \mid \underset{\sim}{X}(t-1), \underset{\sim}{U}(t-1)\}$$

$$\times P\{\underset{\sim}{U}(t-1) \mid \underset{\sim}{X}(t-1)\}. \qquad (9)$$

Notice that in the above equation, the birth rates, death rate and mutation rates $\{\lambda_0(t), b_r(t), d_r(t), \alpha_r(t), r = 1, \ldots, k-1\}$ are functions of the dose level s_i.

The above distribution will be used to derive the conditional posterior distribution of the unknown parameters Θ given $\{X, U, Y\}$. Notice that because the number of parameters is very large, the classical sampling theory approach by using the likelihood function $P\{Y \mid X, U\}$ is not possible without making assumptions about the parameters; however, this problem can easily be avoided by new information from the stochastic system model and the prior distribution of the parameters.

5.3. The Posterior Distribution of the Unknown Parameters and State Variables

In many practical situations, one may assume that the birth rates, death rates and mutation rates are time !"homogeneous. For the i-th dose level, denote these rates by $\{\lambda_i, \alpha_{ji}, b_{ji}, d_{ji}, j = 1, \ldots, k-1\}$. Then the set of unknown parameters are $\Theta = \{\lambda_i, \alpha_{ji}, b_{ji}, d_{ji}, j = 1, \ldots, k-1, i = 1, \ldots, m\}$. To derive the posterior distribution of Θ given $\{X, U, Y\}$, let $P\{\Theta\}$ be the prior distribution of Θ and for the i-th dose level, denote the $\{\underset{\sim}{X}(t), \underset{\sim}{U}(t)\}$ by $\underset{\sim}{X}^{(i)}(t) = \{I_j^{(i)}(t), j = 1, \ldots, k-1\}$ and $\underset{\sim}{U}^{(i)}(t) = \{B_j^{(i)}(t), D_j^{(i)}(t), j = 1, \ldots, k-1\}$. From equations (8)–(9), the conditional posterior distribution $P\{\Theta \mid X, U, Y\}$ of Θ given $\{X, U, Y\}$ is:

$$P\{\Theta \mid X, U, Y\} \propto P\{\Theta\} \prod_{i=1}^{m} \lambda_i^{\{\sum_{t=0}^{t_M-1} M_1(t)\}} e^{\{-t_M \lambda_i\}} \prod_{j=1}^{k-1} [b_{ji}]^{\{\sum_{t=0}^{t_M-1} B_j^{(i)}(t)\}}$$

$$\times [d_{ji}]^{\{\sum_{t=0}^{t_M-1} D_j^{(i)}(t)\}} \alpha_{ji}^{\{\sum_{t=0}^{t_M-1} R_j^{(i)}(t)\}}$$

$$\times (1 - b_{ji} - d_{ji} - \alpha_{ji})^{\{\sum_{t=1}^{t_M} [I_j(t) - B_j^{(i)}(t) - D_j^{(i)}(t) - R_j^{(i)}(t)]\}},$$

(10)

where $R_j^{(i)}(t) = I_j^{(i)}(t+1) - I_j^{(i)}(t) - B_j^{(i)}(t) + D_j^{(i)}(t)$.

For the prior distribution of the unknown parameters, we will assume that *a priori* the parameters in Θ are independently distributed of one another. Furthermore, we will assume natural conjugate priors for all the parameters. That is, we assume:

$$P\{\Theta\} \propto \prod_{i=1}^{m} \lambda_i^{p_i-1} \exp\{-\lambda_i q_i\} \prod_{j=1}^{k-1} [b_{ji}]^{u_{ji}-1} [d_{ji}]^{v_{ji}-1} \alpha_{ji}^{r_{ji}-1}$$
$$\times (1 - b_{ji} - d_{ji} - \alpha_{ji})^{w_{ji}-1}, \qquad (11)$$

where the hyperparameters $\{p_i, q_i, u_{ji}, v_{ji}, w_{ji}, r_{ji}\}$ are positive real numbers. These hyperparameters can be estimated from previous studies. In the event that prior studies and information are not available, we will follow Box and Tiao (1973) to assume that $P\{\lambda_i, i = 1, \ldots, m\} \propto \prod_{i=1}^{m} (\lambda_i)^{-1}$ and that all other parameters are uniformly distributed to reflect the fact that our prior information are vague and imprecise.

5.4. The Generalized Bayesian Method for Estimating Unknown Parameters and State Variables

Using the above distribution results, the multi-level Gibbs sampling procedures for estimating the unknown parameters Θ and the state variables X are given by the following loop:

(1) Given the parameter values, we will use the stochastic Eq. (1) and the associated probability distributions to generate a large sample of $\{X, U\}$. Then, by combining this large sample with $P\{Y \mid X, U\}$, we select $\{X, U\}$ from this sample through the weighted Bootstrap method due to Smith and Gelfant (1992). This selected $\{X, U\}$ is then a sample generated from $P\{X, U \mid \Theta, Y\}$ although the latter density is unknown (for proof, see Tan 2002, Chapter 3). Call the generated sample $\{X^{(*)}, U^{(*)}\}$.
(2) On substituting $\{U^{(*)}, X^{(*)}\}$ which are generated numbers from the above step, generate Θ from the conditional density $P\{\Theta \mid X^{(*)}.U^{(*)}, Y\}$ given by Eq. (10).
(3) With Θ being generated from Step 2 above, go back to Step 1 and repeat the above [1]–[2] loop until convergence.

The convergence of the above algorithm has been proved in Tan (2002, Chapter 3). At convergence, one then generates a random sample of $\{X, U\}$

from the conditional distribution $P\{X, U \mid Y\}$ of $\{X, U\}$ given Y, independent of Θ and a random sample of Θ from the posterior distribution $P\{\Theta \mid Y\}$ of Θ given Y, independent of $\{X, U\}$. Repeat these procedures one then generates a random sample of size N of $\{X, U\}$ and a random sample of size M of Θ. One may then use the sample means to derive the estimates of $\{X, U\}$ and Θ and use the sample variances as the variances of these estimates. Alternatively, one may also use Efron's bootstrap method (Efron, 1982) to derive estimates of the standard errors of the estimates.

6. ANALYSIS OF BRITISH PHYSICIAN DATA OF LUNG CANCER AND SMOKING

It has long been recognized that smoking can cause lung cancer (Fong and Sekido, 2002; Osada and Takahashi, 2002; Wistuba et al., 2002) in most cases. To reveal the basic mechanisms of how tobacco nicotine cause lung cancer, in this section we will apply the above state space model to analyze the British physician smoking data given in Doll and Peto (1978). Given in Table 1 is the British physician data extracted from the paper by Doll and Peto (1978).

In the data set in Table 1, we have included only the age groups between 40 years old and 80 years old because in this data set, lung cancer incidence are non-existent before 40 years old and are also rare among people who are older than 80 years old.

For data in Table 1, observe that there are eight dose levels represented by the number of cigarettes smoked per day and there are eight age groups each with a period of five years. Because lung cancer incidence were reported for a five-year period, as in Luebeck and Moolgavkar (2002) and Tan et al. (2004) we will assume that the initiated cells in the last stage grow instantaneously into malignant tumors, unless otherwise stated. To implement the procedures in Section 5, we let $\Delta t \sim 1$ correspond to a period of three months.

To analyze data given in Table 1, we will use the state space model with the observation model being given by the number (y_{ij}) of total lung cancer incidence. For the stochastic system model we will entertain four extended k-stage multi-event models: (a) a time non-homogeneous two-stage model,

Table 1. British physician lung cancer data with smoking information.

Age (years)	Cigarettes/day (range and mean)								
	0 / 0	1–4 / 2.7	5–9 / 6.6	10–14 / 11.3	15–19 / 16	20–24 / 20.4	25–29 / 25.4	30–34 / 30.2	35–40 / 38
40–44	17846.5[1] / 0[2] / 0[3]	1216 / 0 / 0	2041.5 / 0 / 0	3795.5 / 1 / 0	4824 / 0 / 0	7046 / 1 / 1	2523 / 0 / 0	1715.5 / 1 / 1	892.5 / 0 / 0
45–49	15832.5 / 0 / 0	1000.5 / 0 / 0	1745 / 0 / 0	3205 / 1 / 0	3995 / 0 / 0	6460.5 / 1 / 2	2565.5 / 2 / 1	2123 / 2 / 2	1150 / 0 / 1
50–54	12226 / 1 / 1	853.5 / 0 / 0	1562.5 / 0 / 0	2727 / 2 / 1	3278.5 / 1 / 2	5583 / 6 / 4	2620 / 3 / 3	2226.5 / 3 / 3	1281 / 3 / 2
55–59	8905.5 / 2 / 1	625 / 1 / 0	1355 / 0 / 0	2288 / 1 / 2	2466.5 / 2 / 2	4357.5 / 8 / 6	2108.5 / 5 / 4	1923 / 6 / 5	1063 / 4 / 4
60–64	6248 / 0 / 0	509.5 / 1 / 0	1068 / 1 / 0	1714 / 1 / 2	1829.5 / 2 / 2	2863.5 / 13 / 9	1508.5 / 4 / 5	1362 / 11 / 7	826 / 7 / 6
65–69	4351 / 0 / 0	392.5 / 0 / 0	843.5 / 1 / 0	1214 / 2 / 3	1237 / 2 / 2	1930 / 12 / 10	974.5 / 5 / 6	763.5 / 9 / 7	515 / 9 / 7
70–74	2723.5 / 1 / 1	242 / 1 / 0	696.5 / 2 / 0	862 / 4 / 4	683.5 / 4 / 2	1055 / 10 / 8	527 / 7 / 7	317.5 / 2 / 4	233 / 5 / 5
75–79	1772 / 2 / 0	208.5 / 0 / 0	517.5 / 0 / 0	547 / 4 / 4	370.5 / 5 / 2	512 / 7 / 7	209.5 / 4 / 4	130 / 2 / 3	88.5 / 2 / 3

Notes: [1]–population [2]–observed lung cancer incidence [3]–predicted lung cancer incidence based on four-stage homogeneous model.

(b) a time homogeneous two-stage model, (c) a time homogeneous three-stage model, (d) a time homogeneous four-stage model, and (e) a time homogeneous five-stage model. In models in (b)–(e), the mutation rates, the birth rates and the death rates are assumed to be independent of time. In model (a), while the mutation rates are independent of time, we assume the birth rate and the death rate on initiated cells are two-step piece-wise non-homogeneous with $t_1 = 60$ years old as the cut-off time point.

To assess effects of dose level, we let $x_i = \log(1 + u_i)$, where u_i is the mean dose of the i-th dose level. Then, based on some preliminary analysis (Tan et al., 2004) we assume a Cox regression model for the mutation rates and assume linear regression models for the birth rates and the death rates. Thus, we let $\lambda_0(i, t) = a_{00}(t)e^{a_{01}(t)x_i}$, $\alpha_j(i) = a_{j0}e^{a_{j1}x_i}$ for all models; but let $\{b_j(i) = b_{j0} + b_{j1}x_i, d_j(i) = d_{j0} + d_{j1}(i)x_i\}$ for the time homogeneous models and let $\{b_1(i, s) = b_{10}(s) + b_{11}(s)x_i, d_1(i) = d_{10}(s) + d_{11}(s)x_i\}$ with $s = 1$ for $t \leq t_1$ and $s = 2$ for $t > t_1$ for the two-stage time non-homogeneous model.

Applying the procedures in Section 5, we have estimated the parameters and fitted the data in Table 1. The AIC and BIC values of all models as well as the p-values for testing goodness of fit of the models are given in Table 2. The p-values are computed using the approximate probability distribution results $\sum_{i=1}^{m}\sum_{j=1}^{n}(y_{ij} - \hat{\lambda}_{ij})^2/\hat{\lambda}_{ij} \sim \chi^2(mn - k)$, for large mn, where k is the number of parameters estimated under the model.

The estimates for the unknown parameters in the four-stage model are given in Table 3. From the p-values of the models, apparently that all models except the five-stage model fit the data well, but the values of AIC and BIC suggested that the four-stage model is more appropriate for the data. This

Table 2. BIC, AIC and loglikelihood values for two-, three- and four-stage models.

Model	BIC	AIC	$\log L(\hat{\Theta} \mid Y)$	p-value
Homogeneous two-stage	188.93	182.07	−87.04	0.10
Non-homogeneous two-stage	177.48	167.20	−77.60	0.42
Homogeneous three-stage	151.42	139.42	−62.71	0.98
Homogeneous four-stage	150.76	133.61	−56.81	0.99
Homogeneous five-stage	211.67	189.38	−81.69	0.06

Notes: (1) The p-value is based on $\sum_{i=1}^{m}\sum_{j=1}^{n}\{y_{ij} - \hat{\lambda}_{ij}\}^2/\hat{\lambda}_{ij} \sim \chi^2(72 - k)$, ($k =$ the number of parameters under the model).
(2) $\log L(\hat{\Theta} \mid Y) = \sum_{i=1}^{m}\sum_{j=1}^{n}\{-\hat{\lambda}_{ij} + y_{ij}\log\hat{\lambda}_{ij} - \log(y_{ij}!)\}$, $\hat{\lambda}_{ij} = n_{ij}p_i(j)$.

Table 3. Estimates of parameters for the four-stage homogeneous model with predictive.

	Cigarettes/day (range and mean)				
Parameters	0 0	1–4 2.7	5–9 6.6	10–14 11.3	15–19 16
$\lambda_0(i)$	218.03 ±12.81	247.95 ±15.79	286.41 ±16.79	285.07 ±16.04	295.57 ±16.73
$b_1(i)$	$5.93E-02$ $\pm 8.77E-05$	$6.72E-02$ $\pm 5.40E-05$	$7.17E-02$ $\pm 3.43E-05$	$7.47E-02$ $\pm 1.46E-05$	$7.66E-02$ $\pm 1.16E-05$
$d_1(i)$	$6.18E-02$ $\pm 9.68E-05$	$6.02E-02$ $\pm 4.78E-05$	$5.99E-02$ $\pm 2.90E-05$	$5.95E-02$ $\pm 1.06E-05$	$5.93E-02$ $\pm 1.50E-05$
$b_1(i) - d_1(i)$	$-1.31E-03$ $\pm 5.60E-05$	$1.78E-03$ $\pm 6.07E-05$	$3.95E-03$ $\pm 6.64E-05$	$5.21E-03$ $\pm 4.12E-05$	$6.21E-03$ $\pm 3.17E-05$
$\alpha_1(i)$	$5.29E-05$ $\pm 2.67E-06$	$6.74E-05$ $\pm 1.67E-06$	$5.48E-05$ $\pm 1.12E-06$	$5.68E-05$ $\pm 3.96E-07$	$5.52E-05$ $\pm 3.65E-07$
$b_2(i)$	$8.08E-02$ $\pm 1.59E-03$	$8.92E-02$ $\pm 7.34E-04$	$8.97E-02$ $\pm 5.04E-04$	$9.14E-02$ $\pm 2.10E-04$	$9.34E-02$ $\pm 2.62E-04$
$d_2(i)$	$8.84E-02$ $\pm 1.99E-03$	$8.63E-02$ $\pm 6.82E-04$	$9.00E-02$ $\pm 5.52E-04$	$9.19E-02$ $\pm 2.15E-04$	$9.37E-02$ $\pm 2.15E-04$
$b_2(i) - d_2(i)$	$-7.60E-03$ $\pm 2.55E-03$	$2.90E-03$ $\pm 1.00E-03$	$-3.00E-04$ $\pm 7.47E-03$	$-5.00E-04$ $\pm 3.01E-04$	$-3.00E-04$ $\pm 3.39E-04$
$\alpha_2(i)$	$4.01E-05$ $\pm 3.86E-05$	$5.55E-06$ $\pm 4.99E-06$	$3.28E-06$ $\pm 3.70E-06$	$3.84E-05$ $\pm 4.75E-06$	$1.61E-05$ $\pm 3.37E-06$
$b_3(i)$	$1.08E-02$ $\pm 1.07E-02$	$9.32E-02$ $\pm 2.53E-04$	$4.79E-02$ $\pm 1.65E-02$	$9.84E-02$ $\pm 7.63E-04$	$1.01E-01$ $\pm 4.89E-04$

(Continued)

Table 3. (Continued)

Parameters	\multicolumn{6}{c}{Cigarettes/day (range and mean)}						
	0 / 0	1–4 / 2.7	5–9 / 6.6	10–14 / 11.3	15–19 / 16
$d_3(i)$	$7.18E-02$	$6.27E-02$	$1.10E-01$	$8.59E-02$	$8.16E-02$
	$\pm 2.45E-02$	$\pm 1.68E-04$	$\pm 2.45E-02$	$\pm 5.60E-04$	$\pm 4.49E-04$
$b_3(i)-d_3(i)$	$-2.21E-03$	$7.88E-03$	$1.46E-02$	$1.97E-02$	$2.11E-02$
	$\pm 2.42E-03$	$\pm 2.71E-03$	$\pm 1.84E-03$	$\pm 4.78E-04$	$\pm 5.79E-04$
$\alpha_3(i) = \alpha_3$	\multicolumn{5}{c}{$9.54E-06 \pm 2.51E-08$}				

Parameters	\multicolumn{5}{c}{Cigarettes/day (range and mean)}					
	20–24 / 20.4	25–29 / 25.4	30–34 / 30.2	35–40 / 38
$\lambda_0(i)$	313.02	318.24	330.79	341.41
	± 16.86	± 16.40	± 18.61	± 19.33
$b_1(i)$	$7.81E-02$	$7.94E-02$	$8.04E-02$	$8.18E-02$
	$\pm 7.85E-06$	$\pm 8.83E-06$	$\pm 9.26E-06$	$\pm 7.29E-06$
$d_1(i)$	$5.92E-02$	$5.91E-02$	$5.90E-02$	$5.89E-02$
	$\pm 7.83E-06$	$\pm 7.30E-06$	$\pm 7.99E-06$	$\pm 6.56E-06$
$b_1(i)-d_1(i)$	$6.79E-03$	$7.30E-03$	$7.86E-03$	$8.30E-03$
	$\pm 2.09E-05$	$\pm 2.66E-05$	$\pm 2.15E-05$	$\pm 3.21E-05$
$\alpha_1(i)$	$5.52E-05$	$5.44E-05$	$5.56E-05$	$5.51E-05$
	$\pm 2.03E-07$	$\pm 2.24E-07$	$\pm 2.28E-07$	$\pm 1.85E-07$
$b_2(i)$	$9.49E-02$	$9.61E-02$	$9.70E-02$	$9.80E-02$
	$\pm 1.49E-04$	$\pm 1.34E-04$	$\pm 1.30E-04$	$\pm 1.26E-04$
$d_2(i)$	$9.42E-02$	$9.40E-02$	$9.46E-02$	$9.53E-02$
	$\pm 1.54E-04$	$\pm 1.42E-04$	$\pm 1.67E-04$	$\pm 1.24E-04$

(Continued)

Table 3. (Continued)

Parameters	Cigarettes/day (range and mean)			
	20–24 20.4	25–29 25.4	30–34 30.2	35–40 38
$b_2(i) - d_2(i)$	$7.00E-04$ $\pm 2.13E-04$	$2.10E-03$ $\pm 1.95E-04$	$2.40E-03$ $\pm 2.12E-04$	$2.70E-03$ $\pm 1.77E-04$
$\alpha_2(i)$	$1.42E-05$ $\pm 2.12E-06$	$7.39E-06$ $\pm 1.16E-06$	$1.48E-05$ $\pm 2.11E-06$	$1.20E-05$ $\pm 1.50E-06$
$b_3(i)$	$1.01E-01$ $\pm 9.55E-04$	$1.05E-01$ $\pm 4.39E-04$	$1.04E-01$ $\pm 6.81E-04$	$1.07E-01$ $\pm 2.70E-04$
$d_3(i)$	$8.16E-02$ $\pm 8.36E-04$	$7.90E-02$ $\pm 4.00E-04$	$7.99E-02$ $\pm 6.56E-04$	$8.02E-02$ $\pm 2.06E-04$
$b_3(i) - d_3(i)$	$2.48E-02$ $\pm 2.58E-04$	$2.69E-02$ $\pm 2.98E-04$	$2.74E-02$ $\pm 1.85E-04$	$2.98E-02$ $\pm 3.85E-04$
$\alpha_3(i) = \alpha_3$		$9.54E-06 \pm 2.51E-08$		

Notes:

$\lambda_{00} = 215.17 \pm 4.90$ $\lambda_{01} = 0.12 \pm 7.90E-03$
$b_{10} = 0.0592 \pm 0.0001$ $b_{11} = 0.0062 \pm 0.0001$
$d_{10} = 0.0615 \pm 0.0002$ $d_{11} = -0.0008 \pm 0.0001$
$\alpha_{10} = 5.82E\text{-}05 \pm 3.63E\text{-}06$ $\alpha_{11} = -0.01 \pm 0.02$
$b_{20} = 0.0814 \pm 0.0008$ $b_{21} = 0.0044 \pm 0.0003$
$d_{20} = 0.0862 \pm 0.0012$ $d_{21} = 0.0024 \pm 0.0005$
$\alpha_{20} = 3.30E\text{-}05 \pm 1.16E\text{-}05$ $\alpha_{21} = -0.30 \pm 0.18$
$b_{30} = 0.0265 \pm 0.0153$ $b_{31} = 0.0239 \pm 0.0057$
$d_{30} = 0.0752 \pm 0.0107$ $d_{31} = 0.0025 \pm 0.0040$

four-stage model seems to fit the following molecular biological model for squamous cell lung carcinoma proposed recently by Wistuba et al. (2002): normal epithelium → hyperplasia (3p/9p LOH, genomic instability) → dysplasia (telomerase dysregulation) → *in situ* carcinoma (8p LOH, *FHIT* gene inactivation, gene methylation) → invasive carcinoma (*p53* gene inactivation, *k-ras* mutation).

From Table 3, we observe the following interesting results:

(1) The estimates of $\lambda_0(i)$ increases as the dose level increases. This indicates the tobacco nicotine is an initiator. From molecular biological studies, this initiation process may either be associated with the LOH (loss of heterozygosity) of some suppressor genes from chromosomes 3p or silencing of these genes by epigenetic actions (Jones and Baylin, 2002; Baylin, 2006; Zochbauer-Muller *et al.*, 2001).

(2) The estimates of the mutation rates $\alpha_j(i) = \alpha_j$ for $j = 1, 2, 3$ are in general independent of the dose level x_i. These estimates are of order 10^{-5} and do not differ significantly from one another.

(3) The results in Table 3 indicate that for non-smokers, the death rates $d_j(0)$ ($j = 1, 2, 3$) are slightly greater than the birth rates $b_j(0)$ so that the proliferation rates $\gamma_j(0) = b_j(0) - d_j(0)$ are negative. For smokers, however, the proliferation rates $\gamma_j(i) = b_j(i) - d_j(i)(i > 0)$ are positive and increases as dose level increases (the only exception is $\gamma_2(3)$). This is not surprising since most of the genes are tumor suppressor genes which are involved in cell differentiation and cell proliferation and apoptosis (e.g., *p53*) (Osada and Takahashi, 2002; Fong and Sekido, 2002; Wistuba *et al.*, 2002).

(4) From Table 3, we observed that the estimates of $\gamma_3(i)$ are of order 10^{-2} which are considerably greater that the estimates of $\gamma_1(i)$ respectively. The estimates of $\gamma_1(i)$ are of order 10^{-3} and are considerably greater than the estimates of $\gamma_2(i)$ respectively. The estimates of $\gamma_2(i)$ are of order 10^{-4} and assume negative values for $i = 3, 4, 5$. One may explain these observations by noting the results: (1) significant cell proliferation may trickle apoptosis leading to increased cell death unless the apoptosis gene (*p53*) has been inactivated and (2) the inactivation of the apoptosis gene (*p53*) occurred in the very last stage; see Osaka and Takahashi (2002) and Wistuba *et al.* (2002).

7. CONCLUSIONS AND SUMMARY

Based on most recent biological studies, in this chapter we have presented some stochastic models for carcinogenesis. To develop mathematical analysis for these models, the traditional approach based on theories of Markov process is extremely difficult and has some serious drawbacks. To get around these difficulties, in this chapter we have proposed an alternative approach through stochastic differential equations and state space models for carcinogenesis. This provides an unique approach to combine information from both stochastic models and statistical models of carcinogenesis. By using state space models, we have developed a general procedure via multiple Gibbs sampling method to estimate the unknown parameters. In this paper we have used the multi-event model as an example to illustrate the basic approach and our new modelling ideas.

To illustrate some applications of results of this chapter, we have applied the model and method to the British physician data on lung cancer and smoking. Our analysis has shown that a four-stage homogeneous stochastic model fits the data well. This model appears to be consistent with the molecular biological model of squamous cell lung carcinoma proposed by Osaka and Takahashi (2002) and Wistuba *et al.* (2002).

By assuming a four-stage model for the data we have obtained the following results:

(1) The tobacco nicotine is both an initiator and promoter.
(2) The mutation rates can best be described by the Cox regression model so that $\alpha_j(i, t) = \alpha_{j0}(t) \exp\{\alpha_{j1}(t)x_i\}$, $j = 0, 1$; similarly, the birth rates $b_j(i, t)$ and the death rates $d_j(i, t)$ can best be described by linear regression models.
(3) The estimates of the mutation rates and proliferation rates appear to be consistent with biological observations.

ACKNOWLEDGMENTS

The research of this paper is supported by a research grant from NCI/NIH, grant number R15 CA113347-01.

References

1. Al-Hajj M and Clarke MF. Self-renewal and solid tumor stem cells. *Oncogene* 2004; **23**: 7274–7288.
2. Alberici P, Jagmohan-Changur S, De Pater E, *et al*. Smad4 haplo-insufficiency in mouse models for intestinal cancer. *Oncogene* 2006; **25**: 1841–1851.
3. Baylin SB and Ohm JE. Epigenetic silencing in cancer — a mechanism for early oncogenic pathway addiction. *Nat. Rev. Cancer* 2006; **6**: 107–116.
4. Baylin SB. DNA methylation and gene silencing in cancer. *Nat. Clin. Pract. Oncol.* 2005; **2**: S4–S11.
5. Belinsky SA. Gene-promoter hypermethylation as a biomarker in lung cancer. *Nat. Rev. Cancer* 2004; **4**: 707–717.
6. Bienz M and Hamada F. Adenomatous polyposis coli proteins and cell adhesion. *Curr. Opin. Cell Biol.* 2004; **16**: 528–535.
7. Blasco MA. Telemeres and human diseases: aging, cancer and beyond. *Nat. Rev. Genet.* 2005; **6**: 611–622.
8. Breivik J and Gaudernack G. Genomic instability, DNA methylation, and natural selection in colorectal carcinogenesis. *Semin. Cancer Biol.* 1999; **9**: 245–254.
9. Box GEP and Tiao GC. *Bayesian Inferences in Statistical Analysis*. Addison-Wesley, Reading, MA, 1973.
10. Cahill DP, Kinzler KW, Vogelstein B and Lengauer C. Genetic instability and Darwinian selection in tumors. *Trends Cell Biol.* 1999; **9**: M57–60.
11. Chen W, *et al*. Tumor suppressor HIC1 directly regulates SIRT1 and modulates p53-dependent apoptotic DNA damage responses. *Cell* 2005; **123**: 437–448.
12. Cheng T. Cell cycle inhibitors in normal and tumor stem cells. *Oncogene* 2004; **23**: 7256–7266.
13. Chu KC. Multi-event model for carcinogenesis: a model for cancer causation and prevention. In: *Carcinogenesis: A Comprehensive Survey Volume 8: Cancer of the Respiratory Tract — Predisposing Factors*. Mass MJ, Ksufman DG, Siegfied JM, Steel VE and Nesnow S (eds.) Raven Press, New York, 1985, pp. 411–421.
14. de la Chapelle A. Genetic predisposition to colorectal cancer. *Nat. Rev. Cancer* 2004; **4**: 769–780.
15. Derynck R and Zhang YE. Smad-dependent and smad-independent pathways in TGF–β family signalling. *Nature* 2003; **425**: 577–584.
16. Doll R and Peto R. Cigarette smoking and bronchial carcinoma: dose and time relationships among regular smokers lifelong non-smokers. *J. Epidemiol. Community Health* 1978; **32**: 303–313.

17. Efron B. *The Jackknife, the Bootstrap and Other Resampling Plans.* SIAM, Philadelphia, PA, 1982.
18. Esteller M. Epigenetic lesions causing genetic lesions in human cancer: promoter hypermethylation of DNA repair genes. *Eur. J. Cancer* 2000; **36**: 2294–2300.
19. Farber E. Experimental induction of hepatocellular carcinoma as a paradigm for carcinogenesis. *Clin. Physiol. Biochem.* 1987; **5**: 152–159.
20. Feinberg AP and Tycko B. The history of cancer epigenetics. *Nat. Rev. Cancer* 2004; **4**: 143–153.
21. Ferreira R, Naguibneva I, Pritchard LL, Ait-Si-Ali S and Harel-Bellan A. The Rb/chromatin connection and epigenetic control: opinion. *Genes Oncogene* 2001; **20**: 3128–3133.
22. Flintoft L. Silent transmission. *Nat. Rev. Genet.* 2005; **5**: 720–721.
23. Fodde R, Kuipers J, Rosenberg C, *et al.* Mutations in the APC tumor suppressor gene cause chromosomal instability. *Nat. Cell Biol.* 2001; **3**: 433–438.
24. Fodde R, Smit R and Clevers H. APC, signal transduction and genetic instability in colorectal cancer. *Nat. Rev. Cancer* 2001; **1**: 55–67.
25. Fong KM and Sekido Y. The molecular biology of lung carcinogenesis. In: *The Molecular Basis of Human Cancer.* Coleman WB and Tsongalis GJ (eds.) Humana Press, Totowa, NJ, 2002, Chapter 17, pp. 379–405.
26. Green RA and Kaplan KB. Chromosomal instability in colorectal tumor cells is associated with defects in microtubule plus-end attachments caused by a dominant mutation in APC. *J. Cell Biol.* 2003; **163**: 949–961.
27. Hahn WC and Meyerson M. Telomere activation, cellular immortalization and cancer. *Ann. Med.* 2001; **33**: 123–129.
28. Hanahan D and Weinberg RA. The hallmarks of cancer. *Cell* 2000; **100**: 57–70.
29. Hanin LG and Yakovlev AY. A nonidentifiability aspect of the two-stage model of carcinogenesis. *Risk Anal.* 1996; **16**: 711–715.
30. Hawkins NJ and Ward RL. Sporadic colorectal cancers with microsatellite instability and their possible origin in hyperplastic polyps and serrated adenomas. *J. Natl. Cancer Inst.* 2001; **93**: 1307–1313.
31. Henson JD, Neumann AA, Yeager TR and Reddel RR. *Oncogene* 2002; **21**: 589–610.
32. Hesketh R. *The Oncogene and Tumor Suppressor Gene Facts Book*, 2nd ed. Academic Press, San Diego, CA, 1997.
33. Holm TM, *et al.* Global loss of imprinting leads to widespread tumorigenesis in adult mice. *Cancer Cell* 2005; **8**: 275–285.

34. Hopkin K. Tumor evolution: survival of the fittest cells *J. NIH Res.* 1996; **8**: 37–41.
35. Jaffe LF. Epigenetic theory of cancer initiation. *Adv. Cancer Res.* Elsevier Inc., USA, 2003.
36. Jessup JM, Gallic GG and Liu B. The molecular biology of colorectal carcinoma: importance of the Wg/Wnt signal transduction pathway. In: *The Molecular Basis of Human Cancer.* Coleman WB and Tsongalis GJ (eds.) Humana Press, Totowa, NJ, 2002, Chapter 13, pp. 251–268.
37. Jones PA and Baylin SB. The fundamental role of epigenetic events in cancer. *Nat. Rev. Genet.* 2002; **3**: 415–428.
38. Kalunaga T. Requirement for cell replication in the fixation and expression of the transformed state in mouse cells treated with 4-nitroquinoline-1-oxide. *Int. J. Cancer* 1974; **14**: 736–742.
39. Koinuma K, Yamashita Y, Liu W, Hatanaka H, Kurashina K, Wada T, et al. Epigenetic silencing of AXIN2 in colorectal carcinoma with microsatellite instability. *Oncogene* 2006; **25**: 139–146.
40. Laurent-Puig P, Blons H and Cugnenc PH. Sequence of molecular genetic events in colorectal tumorigenesis. *Eur. J. Cancer Prevent.* 1999; **8**: S39–S47.
41. Little MP. Are two mutations sufficient to cause cancer? Some generalizations of the two-mutation model of carcinogenesis of Moolgavkar, Venson and Knudson, and of the multistage model of Armitage and Doll. *Biometrics* 1995; **51**: 1278–1291.
42. Loeb KR and Loeb LA. Significance of multiple mutations in cancer. *Carcinogenesis* 2000; **21**: 379–385.
43. Luebeck EG and Moolgavkar SH. Multistage carcinogenesis and colorectal cancer incidence in SEER. *Proc. Natl. Acad. Sci. USA* 2002; **99**: 15095–15100.
44. Lund AH and Lohuizen Van M. Epigenetics and cancer. *Genes Dev.* 2004; **18**: 2315–2335.
45. Lundbald V. Telemere maintenance without telomerase. *Oncogene* 2001; **21**: 522–531.
46. Lynch CJ and Milner J. Loss of one p53v allele results in four-fold reduction in p53 mRNA and protein: a basis for p53 haplo-insufficiency. *Oncogene* 2006; **25**: 3463–3470.
47. MacDonald F, Ford CHJ and Casson AG. *Molecular Biology of Cancer.* Taylor and Frances Group, New York, 2004.
48. Nettesheim P and Barrett JC. *In vitro* transformation of rat tracheal epithelial cells as a model for the study of multistage carcinogenesis. In: *Carcinogenesis,*

Vol. 9. Barrett JC and Tennant RW (eds.) Raven Press, New York, 1985, pp. 283–292.
49. Ohlsson R, Kanduri C, Whitehead J, *et al.* Epigenetic variability and evolution of human cancer. In: *Advances in Cancer Research.* Elsevier Inc., USA, 2003.
50. Osada H and Takahashi T. Genetic alterations of multiple tumor suppressor genes and oncogenes in the carcinogenesis and progression of lung cancer. *Oncogene* 2002; **21**: 7421–7434.
51. Peltomaki P. Deficient DNA mismatch repair: a common etiologic factor for colon cancer. *Human Mol. Genet.* 2001; **10**: 735–740.
52. Pharoah PDP, Dunning AM, Ponder BAJ and Easton DF. Association studies for finding cancer-susceptaibility genetic variants. *Nat. Rev. Cancer* 2004; **4**: 850–860.
53. Potter JD. Colorectal cancer: molecules and population. *J. Natl. Cancer Inst.* 1999; **91**: 916–932.
54. Renan MJ. How many mutations are required for tumorigenesis? Implications from human cancer data. *Mol. Carcinog.* 1993; **7**: 139–146.
55. Robertson T. DNA methylation and human diseases. *Nat. Rev. Genet.* 2005; **6**: 597–610.
56. Shay JW, Zou Y, Hiyama E and Wright WE. Telomerase and cancer. *Mol. Genet.* 2001; **10**: 677–685.
57. Smith AFM and Gelfand AE. Bayesian statistics without tears: a sampling and resampling perspective. *Am. Stat.* 1992; **46**: 84–88.
58. Smolinski KN and Meltzer SJ. Inactivation of negative regulators during neoplastic transformation. In: *The Molecular Basis of Human Cancer.* Coleman WB and Tsongalis GJ (eds.) Humana Press, Totowa, NJ, 2002, Chapter 5, pp. 81–111.
59. Sparks AB, Morin PJ, Vogelstein B and Kinzler KW. Mutational analysis of the APC/beta-catenin/Tcf pathway in colorectal cancer. *Cancer Res.* 1998; **58**: 1130–1134.
60. Tan WY. *Stochastic Models of Carcinogenesis.* Marcel Dekker, New York, 1991.
61. Tan WY. Some mixed models of carcinogenesis. *Math. Comp. Model.* 1988; **10**: 765–773.
62. Tan WY. *Stochastic Models with Applications to Genetics, Cancers, AIDS and Other Biomedical Systems.* World Scientific, Singapore and River Edge, NJ, 2002.
63. Tan WY and Chen CW. A multiple pathway model of carcinogenesis involving one stage models and two-stage models. In: *Mathematical Population*

Dynamics. Arino O, Axelrod DE and Kimmel M (eds.) Marcel Dekker, New York, 1991, Chapter 31, pp. 469–482.
64. Tan WY and Chen CW. Assessing effects of changing environment by a multiple pathway model of carcinogenesis. *Math. Comp. Model.* 2000; **32**: 229–250.
65. Tan WY and Chen CW. Stochastic modeling of carcinogenesis: some new insight. *Math. Comp. Model.* 1998; **28**: 49–71.
66. Tan WY and Chen CW. Cancer stochastic models. In: *Encyclopedia of Statistical Sciences*, revised edition. John Wiley and Sons, New York, 2005.
67. Tan WY, Chen CW and Wang W. Stochastic modeling of carcinogenesis by state space models: a new approach. *Math. Comp. Model.* 2001; **33**: 1323–1345.
68. Tan WY and Singh KP. A mixed model of carcinogenesis — with applications to retinoblastoma. *Math. Biosci.* 1990; **98**: 201–211.
69. Tan WY, Zhang LJ and Chen CW. Stochastic modeling of carcinogenesis: state space models and estimation of parameters. *Disc. Cont. Dynam. Syst. B* 2004; **4**: 297–322.
70. Thorgeirsson SS and Grisham JW. Molecular pathogenesis of human hepatocellular carcinoma. *Nat. Rev. Genet.* 2002; **31**: 339–346.
71. Ushijima T. Detection and intepretation of altered methylation patterns in cancer cells. *Nat. Rev. Genet.* 2005; **5**: 223–231.
72. Wakefield LM and Roberts AB. TGF-β signaling: positive and negative effects on tumorigenesis. *Curr. Opin. Genet. Dev.* 2002; **12**: 22–29.
73. Ward R, Meagher A, Tomlinson I, O'Connor T, *et al.* Microsatellite instability and the clinicopathological features of sporadic colorectal cancer. *Gut* 2001; **48**: 821–829.
74. Weinberg RA. *The Biology of Human Cancer*. Garland Sciences, Taylor and Frances Group, New York, 2007.
75. Welch JN and Chrysogelos SA. Positive mediators of cell proliferation in neoplastic transformation. In: *The Molecular Basis of Human Cancer*. Coleman WB and Tsongalis GJ (eds.) Humana Press, Totowa, NJ, 2002, Chapter 4, pp. 65–79.
76. Wistuba II, Mao L and Gazdar AF. Smoking molecular damage in brochial epithelium. *Oncogene* 2002; **21**: 7298–7306.
77. Yang GL and Chen CW. A stochastic two-stage carcinogenesis model: a new approach to computing the probability of observing tumor in animal bioassays. *Math. Biosci.* 1991; **104**: 247–258.
78. Yamada Y, *et al.* Opposing effects of DNA hypomethylation on intestinal and liver carcinogenesis. *Proc. Natl. Acad. Sci. USA* 2005; **102**: 13580–13585.

79. Yakovlev AY and Tsodikov AD. *Stochastic Models of Tumor Latency and Their Biostatistical Applications*. World Scientific, Singapore and River Edge, NJ, 1996.
80. Zarl H, *et al.* Direct mutagenesis of Ha-ras-1 oncogenes by N-nitro-N-methylurea during initiation of mammary carcinogenesis in rats. *Nature* 1985; **315**: 382–385.
81. Zheng M and Rosen JM. Stem cells in the etiology and treatment of cancer. *Curr. Opin. Genet. Dev.* 2006; **16**: 60–64.
82. Zochbauer-Muller S, Fong KM, Virmani AK, *et al.* Promoter methylation of multiple genes in non-small cell lung cancers. *Cancer Res.* 2001; **61**: 249–255.

Chapter 4

MODELING THE EFFECTS OF RADIATION ON CELL CYCLE REGULATION AND CARCINOGENESIS

William D. Hazelton

Epidemiological analyses using the two-stage clonal expansion (TSCE) model indicate that low-dose radiation and exposure to some chemicals, such as arsenic or tobacco smoke, operate through several mechanisms during carcinogenesis. Typically, the most significant effect of these exposures is to increase the promotion of initiated cells. The estimated increases in promotion are small in absolute terms, yet highly significant, suggesting that initiated cells are almost, but not completely, under homeostatic regulation that serves to maintain the size of adult tissues and organs. To understand these effects better, we introduce a combined cell cycle and multistage carcinogenesis model with both deterministic and stochastic components. The model incorporates homeostatic regulation of cell cycle progression based on the difference between the total mean (deterministic) number and target number of stem cells. There are several reasons for extending carcinogenesis models to represent cell cycle states. First, the early mutations in many cancers appear to involve proteins associated with checkpoint control or other aspects of cell cycle regulation. Second, the sensitivity of cells exposed to radiation or chemicals changes markedly throughout the cell cycle. Modeling of damage and repair processes along with tissue homeostasis may be key to understanding the mechanisms that underlie promotion and other exposure-response phenomena. For example, loss of a checkpoint protein in initiated cells may shorten the cell cycle period and provide a slight growth advantage compared to normal cells in replacing cells damaged by endogenous processes, radiation, or chemical exposure.

Keywords: Multistage carcinogenesis, cell cycle, checkpoint regulation, ionizing radiation (IR), chemical exposure.

1. INTRODUCTION

1.1. Multistage Carcinogenesis Models

Solid tumors appear to develop through several phenomenological stages described as initiation, promotion, and malignant transformation. The two-stage clonal expansion (TSCE) model is perhaps the simplest mathematical model that represents this process (Moolgavkar *et al.*, 1981, 1988, 1990). It assumes a deterministic normal stem cell compartment and stochastic intermediate and malignant compartments. Initiation of normal cells is described by a filtered Poisson process, with initiated stem cells assumed to undergo birth (cell division), death (loss through apoptosis or differentiation), or transformation to the malignant state, modeled as appearance of the first malignant cell. A fixed lag or lag time distribution is used to represent the time from first malignant cell to incidence or mortality. A generating function approach is used to represent the evolving distribution of initiated cells in calculating the hazard and survival functions.

It is possible to extend the multistage clonal expansion approach to represent additional stages in the carcinogenic process (Little, 1995), although this requires estimation of additional parameters and increased numerical computation. Extensions of the TSCE model with three, four, and five stages for colorectal cancer were compared using data from the Surveillance, Epidemiology, and End Results (SEER) registry for 1973–1996 (Luebeck *et al.*, 2002), with the four stage model having the highest likelihood. A slightly different approach was used to estimate dose-response on multiple sequential mutation stages prior to a clonal expansion stage in analysis of the effects of low-dose sparsely ionizing radiation on lung cancer incidence in the Canadian National Dose Registry cohort (Hazelton *et al.*, 2005a). This analysis compared models using from one to ten sequential initiating stages, with three stages prior to clonal expansion giving the highest likelihood.

In this chapter, extensions of multistage methods are introduced that provide a more detailed biological representation of processes that may be important in the response of organisms to radiation and chemical exposure. We illustrate this approach by describing a model that includes early and late G1, S, G2, and M cell cycle states, simple and complex damage, slow and fast repair, checkpoint regulation, apoptosis, differentiation triggered by a light-dark cycle, and mutations. Complicated models such as this generally have too many parameters to allow estimation of all parameters from epidemiological data, but may be useful for exploration of biological hypotheses.

1.2. Analyses of Environmentally Exposed Cohorts

Epidemiological analysis of the effects of chemical or radiation exposure require extensive and reliable data, typically from a large cohort with exposure and outcome histories at the individual or group level, a simple cancer model such as the TSCE model, and a flexible dose-response model that allows each exposure to potentially modify each of the parameters of the cancer model. Maximum likelihood techniques are used to optimize the dose response parameters to best represent the pattern of outcomes in the cohort.

These methods have been applied to analysis of lung (Luebeck *et al.*, 1999; Stevens *et al.*, 1984; Hazelton *et al.*, 2001, 2005a, 2005b) colon (Luebeck *et al.*, 2002), breast (Moolgavkar *et al.*, 1980) and other cancers (Stevens *et al.*, 1984; Moolgavkar *et al.*, 1981). A series of TSCE and extended model analyses focusing on the effects of radiation, tobacco and other exposures on lung cancer incidence and mortality is summarized below.

The TSCE model was used to study the effects of radon and smoking on lung cancer mortality in a cohort of 3238 white Colorado uranium miners (Luebeck *et al.*, 1999). Radon and smoking were both found to significantly increase promotion. Initiation was significantly increased by smoking, but radon had a non-significant effect. Neither exposure significantly increased the rate of malignant conversion. Although the dose-response of smoking and radon affecting initiation and promotion was additive, the risk associated with these exposures showed an interaction that was between additive and multiplicative. Radon exposure was found to have an inverse dose rate effect, conferring a higher risk when a given dose was protracted over a longer time interval.

The effects of radon, arsenic, cigarettes, and combined long pipe and water pipe consumption on lung cancer mortality were analyzed using the TSCE model with data from a cohort of 12,011 tin miners from Yunnan province in China (Hazelton *et al.*, 2001). Tobacco and arsenic exposures each accounted for approximately 20% of the attributable risk, with about

5% attributed to radon, 9% to background, and the balance to interactions between exposures. The interaction between tobacco and arsenic was large, accounting for approximately 20% of cases in the cohort. Dose-response functions for each exposure were found to have distinct effects on the malignant conversion and promotion rates. The estimated promotional effect of radon and arsenic was seen as an effect on the net cell proliferation (cell division minus death) rate, whereas tobacco influenced the cell division rate.

The effect of low-dose sparsely ionizing (primarily X-ray and gamma) radiation on lung cancer incidence was studied in a cohort of over 190,000 Canadian medical, dental, industrial, and nuclear power workers (Hazelton et al., 2005a). Among males, these occupational exposure were found to significantly affect promotion and malignant conversion. The estimated dose-response for females was smaller but qualitatively similar to that for males. The estimated effect on promotion using the TSCE model did not diminish when using models with one to ten initiation stages prior to the clonal expansion stage. Models without promotion had poor likelihoods. The estimated non-linear risk as a function of dose appeared generally consistent with estimates of risk from the generally higher, acute doses experienced by Japanese atomic bomb survivors. This study suggests that linear extrapolation from high acute doses may underestimate low-dose protracted radiation risk.

The effects of cigarette smoking were studied among never and current smokers using the TSCE model in an analysis of lung cancer mortality in three large cohorts (Hazelton et al., 2005b). The cohort data includes 20 years follow-up for the British Doctor's cohort, 12 years for the American Cancer Society's Cancer Prevention Studies I (CPS-I) cohort, and 6 years for the Cancer Prevention Studies II (CPS-II) cohort. The CPS-I and -II studies each included over a million individuals. Smoking related promotion was found to be the dominant etiological mechanism in all cohorts. Smoking-related initiation was less important than promotion in the earlier contemporaneous British doctors and CPS-I cohorts. The CPS-II cohort had non-significant smoking related initiation, but higher smoking related promotion than the earlier cohorts. Predicted risk among former smokers was found to be in good agreement with risks observed among ex-smokers derived from other studies. The increased overall risk and shift to higher rates for adenocarcinoma in the CPS-II cohort may be associated with changes in cigarette composition and/or deeper inhalation required to achieve a satisfactory nicotine delivery among smokers of the more recent low-nicotine cigarettes.

2. MODELING BIOLOGICAL MECHANISMS

2.1. A Combined Cell Cycle and Multistage Clonal Expansion Model

Many genes associated with carcinogenesis are regulators of cell cycle progression, checkpoint control, repair, or apoptosis. Understanding the role of these genes in carcinogenesis requires a model that bridges from these intracellular processes to the increasingly larger temporal and spatial scales of cells, tissues, and organisms where small differences in promotion or other aspects of cell cycle regulation may contribute significantly to carcinogenesis. We are attempting to model these mechanisms using an extension of the multistage model that includes cell cycle and DNA damage states, cell cycle checkpoints, repair, apoptosis, and mutations leading from normal, to initiated, to malignant phenotypes.

The combined cell cycle and carcinogenesis model includes states that are identified by the cell cycle status, damage status, and cancer progression status. Processes include rates for cell cycle progression, endogenous and IR damage (both simple and complex), checkpoint delays, fast and slow repair, apoptosis, mutation, and differentiation. Differentiation rates are assumed to depend on the light-dark cycle (circadian rhythm). This model is currently being used to study the radiation response of mouse intestinal crypts (see Figure 1).

The cell-cycle status is assumed to progress sequentially taking on eight values representing early and late G1, S, G2, and M phases. The damage status takes on three values, normal, simple damage, or complex damage, as illustrated in Figure 1. See Hazelton *et al.* (2005) for a description of an earlier version of this model. Three values for cancer progression status are assumed, e.g., normal, initiated, or malignant stem cell. Altogether, this requires $8 \times 3 \times 3 = 72$ states. Three additional states are used to represent the number of cells undergoing apoptosis, the number of differentiated cells before they are sloughed off, and the total number of stem cells, including apoptotic cells before they are cleared.

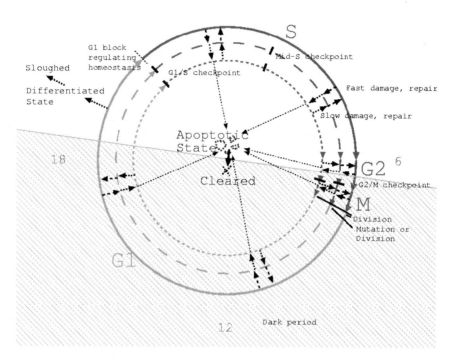

Fig. 1. A schematic representation of the combined cell cycle and carcinogenesis model, with states identified by early and late G1, S, G2, and M cell cycle status, with no damage (solid), simple damage (dashed) or complex damage (dotted). A light-dark cycle is assumed, with light turning on at 19 hours and off at 7 hours that controls differentiation out of G1. Endogenous and IR cause normal to simple damage transitions, and IR is assumed to cause simple to complex damage transitions. Cell cycle checkpoints are between late G1 and early S, at mid-S, and between late G2 and early M. Mutations occur during cell division, with one simple damaged cell progressing from normal to initiated, or from initiated to malignant. Fast repair moves cells from simple damage to normal. Slow repair moves cells from complex to simple damage. Apoptosis is assumed to occur from complex damaged cells, with apoptotic cells eventually cleared.

A number of processes are assumed to link these states, including cell cycle progression, endogenous and ionizing radiation (IR) damage, fast repair between simple damage and normal status, slow repair between complex and simple damage status, checkpoint delays for both simple and complex damaged cells, apoptosis for cells with complex damage, and mutation for cells with simple damage status. These processes are described in more detail below.

Cell cycle progression: The time spent in G1 is assumed variable depending on the total stem cell count in a crypt. We assume a target number of 16 stem cells. If the total stem cell count is below target, at most about 4–6 cells are triggered to divide, spending approximately 9 hours in S, 2 hours in G2, and 1 hour in M phase (Ijiri *et al.*, 1990).

Endogenous damage: The model parameters are set to represent approximately 50 endogenous double strand breaks (EDSB) during S phase due to conversion of endogenous single strand lesions (Vilenchik *et al.*, 2003), with an average of 0.05 EDSB/hr in other cell cycle states, but increasing proportional to the DNA content during cell cycle progression. In the absence of IR, transitions from normal to simple damage status are assumed to occur on average at these EDSB rates.

IR damage: The model assumes IR causes transitions from normal to simple damage status, and from simple to complex damage status at rates corresponding to approximately 30 DSB/Gy on average, with rates increasing proportional to DNA content during cell cycle progression.

Fast repair from simple damage to normal status is modeled with a half-life of 0.4 hours. Slow repair from complex to simple damage status is modeled with a half-life of 4 hours (van Rongen *et al.*, 1995).

Mutation is assumed to occur when a M phase cell divides, with one cell fixing a mutation that causes a transition from normal stem cell status with simple damage to an initiated cell, or from an initiated cell with simple damage to a malignant cell.

Apoptosis is assumed to be governed by a rate for transition of a cell with complex damage status to enter the apoptotic state. Experimental data for mouse intestinal stem cells suggests that the apoptotic transition rate is large for these ultra-sensitive stem cells, where most such cells that are actively cycling undergo apoptosis at sub-Gy radiation doses (Ijiri *et al.*, 1990). We assume a rate for clearance of the apoptotic cells, leading them to eventually disappear. Experiments indicate that the number of apoptotic cells is maximal for radiation at the end of G2/M, reaching a maximum of about 4 cells (Ijiri *et al.*, 1990).

Differentiation is assumed to be governed by a transition rate for a cell in G1 phase to enter the differentiated state. Experimental data for mouse intestinal stem cells suggest that differentiation is linked to the circadian light-dark cycle. We model this by assuming a 12 hour on, 12 hour off light cycle influencing the differentiation rate out of G1. We do not model further division or progressive differentiation that may occur. We assume a rate for sloughing of the differentiated cells so that the number does not continually increase.

Checkpoint delays: We assume that transition rates at G1/S, mid-S, and G2/M boundaries are slowed significantly. We begin by assuming rates are slowed by a factor of 2 for simple damage, and a factor of 4 for complex damage, compared with normal cell cycle progression rates.

Figures 2(a)–2(g) below are calculated using model equations that provide a deterministic description of the mean behavior of the cells. They show the calculations for growth beginning with one stem cell under homeostatic regulation in a mouse intestinal crypt with a target of 16 cells. There is a photoperiod effect on differentiation out of G1 phase, which causes the oscillating pattern of growth. At 600 hours, there is an 18 minute exposure

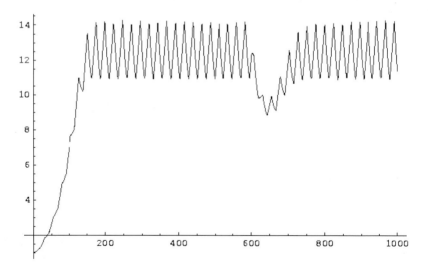

Fig. 2(a). Stem cell number. The y-axis shows the number of stem cells beginning with one cell at $t = 0$, and the x-axis is time in hours. At $t = 600$ hours, a dose of 1 Gy of radiation is applied. Homeostatic regulation is targeted at 16 cells, with the oscillation in cell number caused by differentiation of cells out of the G1 cell cycle state during the light photoperiod.

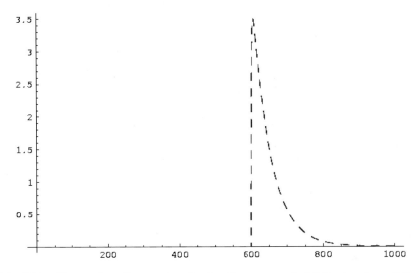

Fig. 2(b). The number of apoptotic cells rises from zero to 3–4 following the radiation dose at $t = 600$ hours.

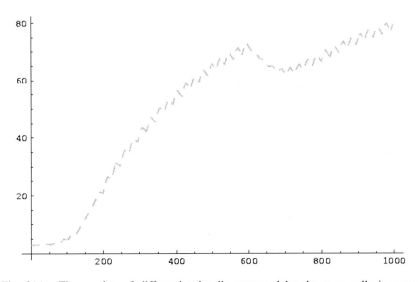

Fig. 2(c). The number of differentiated cells generated by the stem cells increases following crypt growth, with a dip following the radiation dose at 600 hours.

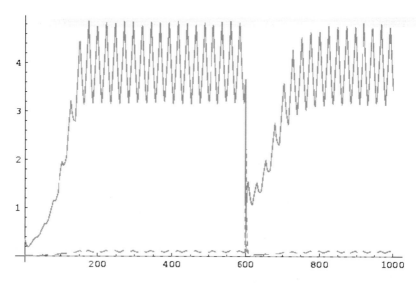

Fig. 2(d). The number of cells in G1 in the normal undamaged state (solid line, top), with simple damage (dashed), and complex damage (dotted). The radiation at $t = 600$ hours causes a spike in the number of complex damaged cells.

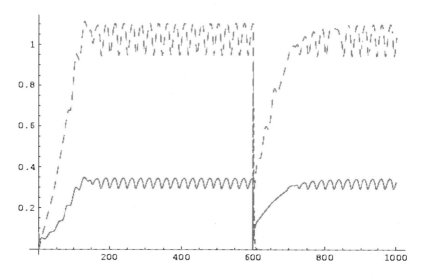

Fig. 2(e). The number of S-phase cells. There is a significant probability of simple damage due to conversion of endogenous simple strand lesions to double strand breaks during S phase (dashed line, top). The IR causes additional simple damage and conversion of simple to complex damage (dotted). The number of undamaged S-phase cells is shown by the middle (solid) line.

Modeling the Effects of Radiation on Cell Cycle Regulation and Carcinogenesis

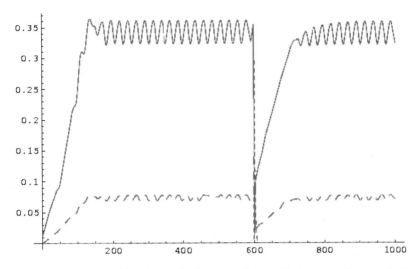

Fig. 2(f). The number of G2 phase cells that are undamaged is shown by the top solid line, with simple damaged cells (dashed line) arising from unrepaired S-phase cells, and a spike of complex damaged cells (dotted) following radiation at $t = 600$ hours.

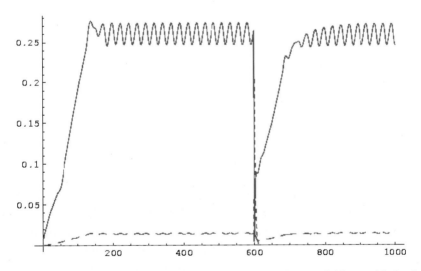

Fig. 2(g). The number of M-phase cells that are undamaged (top solid line), with simple damage (dashed), and with complex damage (dotted).

to 1 Gy of IR. This leads to apoptosis of 3–4 cells, followed by clearance and recovery of homeostasis.

We have also implemented a stochastic version of the model that is solved following solution of the deterministic equations. A probability generating function approach is used to represent the time dependent probability distribution for cells in all states. The continuous time coupled Markov system representing this joint distribution satisfies a partial differential equation (pde). Time dependent survival and hazard functions are found through numerical solution of the characteristic equations for the pde.

Homeostatic regulation (controlling the rate of cell cycle progression out of late G1 phase) is based on the deterministic model estimate of the time varying G1/S transition rate that regulates the total number of stem cells. The stochastic equations allow modeling of carcinogenic progression where the stochastic effects of clonal growth and possible extinction of small clones are important.

3. SUMMARY

The multistage clonal expansion model of carcinogenesis is generalized to include cell cycle and DNA damage states for normal and initiated stem cells. Initiated cells may undergo transformation to a malignant state, eventually leading to cancer incidence or death. The model allows oxidative or radiation induced DNA damage, checkpoint delay, DNA repair, apoptosis, and transformation rates to depend on the cell cycle state or DNA damage state of normal and initiated cells.

A deterministic version of the model tracks the mean number of cells in each state. These equations can be supplemented with equations representing a stochastic version of the model. The stochastic version uses a probability generating function to represent the time dependent probability distribution for cells in all states. Time dependent survival and hazard functions are found through numerical solution of the equations describing this process.

ACKNOWLEDGMENTS

We would like to acknowledge support from the Department of Energy (DOE) under grant number DE-FG02-03ER63675, and the National Institutes of Health (NIH) under grant number RO1 CA119224-01.

References

1. Hazelton WD, Luebeck EG, Heidenreich WF and Moolgavkar SH. Analysis of a historical cohort of Chinese tin miners with arsenic, radon, cigarette smoke, and pipe smoke exposures using the biologically based two-stage clonal expansion model. *Radiat. Res.* 2001; **156**: 78–94.
2. Hazelton WD, Moolgavkar SH, Curtis SB, Zielinski JM, Ashmore JP and Krewski D. Biologically based analysis of lung cancer incidence in a large Canadian occupational cohort with low-dose ionizing radiation exposure, and comparison with Japanese atomic bomb survivors. *J. Toxicol. Environ. Health Part A* 2006; **69**(11): 1013–1038.
3. Hazelton WD, Curtis SB and Moolgavkar SH. Analysis of radiation effects using a combined cell cycle and multistage carcinogenesis model. *Adv. Space Res.* 2006, **37**: 1809–1812.
4. Hazelton WD, Clements MS and Moolgavkar SH. Multistage carcinogenesis and lung cancer mortality in three cohorts. *Cancer Epidemiol. Biomark. Prev.* 2005; **14**(5): 1171–1181.
5. Ijiri K and Potten CS. The circadian rhythm for the number and sensitivity of radiation-induced apoptosis in the crypts of mouse small intestine. *Int. J. Radiat. Biol.* 1990; **56**: 165–175.
6. Little MP. Are two mutations sufficient to cause cancer? Some generalizations of the two-mutation model of carcinogenesis of Moolgavkar, Venzon, and Knudson, and of the multistage model of Armitage and Doll. *Biometrics* 1995; **51**: 1278–1291.
7. Luebeck EG and Moolgavkar SH. Multistage carcinogenesis and the incidence of colorectal cancer. *Proc. Natl. Acad. Sci. USA* 2002; **99**: 15095–15100.
8. Luebeck EG, Heidenreich WF, Hazelton WD, Paretzke HG and Moolgavkar SH. Biologically-based analysis of the Colorado uranium miners cohort data: age, dose and dose-rate effects. *Radiat. Res.* 1999; **152**: 339–351.
9. Moolgavkar S and Knudson A. Mutation and cancer: A model for human carcinogenesis. *J. Natl. Cancer Inst.* 1981; **66**: 1037–1052.

10. Moolgavkar SH, Dewanji A and Venzon DJ. A stochastic two-stage model for cancer risk assessment. I. The hazard function and the probability of tumor. *Risk Anal.* 1988; **8**: 383–392.
11. Moolgavkar SH and Luebeck EG. Two-event model for carcinogenesis: Biological, mathematical, and statistical considerations. *Risk Anal.* 1990; **10**: 323–341.
12. Moolgavkar SH, Day NE and Stevens RG. Two-stage model for carcinogenesis: Epidemiology of breast cancer in females. *J. Natl. Cancer Inst.* 1980; **65**: 559–569.
13. Moolgavkar SH and Stevens RG. Smoking and cancers of bladder and pancreas: risks and temporal trends. *J. Natl. Cancer Inst.* 1981; **67**: 15–23.
14. Stevens RG and Moolgavkar SH. A cohort analysis of lung cancer and smoking in British males. *Am. J. Epidemiol.* 1984; **119**: 624–641.
15. Stevens RG and Moolgavkar SH. Malignant melanoma: dependence of site-specific risk on age. *Am. J. Epidemiol.* 1984; **119**: 890–895.
16. van Rongen E, Travis EL and Thames HD Jr. Repair rate in mouse lung after clinically relevant radiation doses per fraction. *Radiat. Res.* 1995; **141**: 74–78.
17. Vilenchik NM and Knudson AG. Endogenous DNA double-strand breaks: production, fidelity of repair, and induction of cancer. *Proc. Natl. Acad. Sci. USA* 2003; **100**: 12871–12876.

APPENDIX

Let $\bar{n}_{i,j}(t), \bar{s}_{i,j}(t), \bar{c}_{i,j}(t)$ represent the mean number of normal, simple damaged, and complex damaged stem cells in cell cycle state i and mutation state j at time t. Indices $i = 0, 1$ represent early and late G1-phase, $i = 2, 3$ represent S-phase, $i = 4, 5$ represent G2-phase, and $i = 6, 7$ represent M-phase. The index i is cyclic. Let $j = 0$ represent normal stem type, $j = 1, \ldots, n_m - 1$ successive pre-malignant cell types, and $j = n_m$ represent the malignant type. Let $\bar{d}(t)$ represent the mean number of differentiated cells, $\bar{a}(t)$ the number of apoptotic cells, and $\bar{h}(t)$ the total number of living undifferentiated cells under homeostatic regulation at time t. Let $cc_{i,j}(t)$ represent rates regulating progression through the cell cycle, $ks_{i,j}(t)$ rates regulating progression through the cell cycle for simple damaged cells including checkpointing (slowing of progression) at end of G1, mid-S, and end of G2, and $kc_{i,j}(t)$ rates regulating progression for complex damaged cells with checkpoints as for simple damaged cells, and additional checkpointing representing M-phase arrest. Let $ds_{i,j}(t)$ represent rates for damage from normal to simple damage status, and $dc_{i,j}(t)$ rates for damage from simple to complex damage status, $rs_{i,j}(t)$ repair rates from simple damage to normal status, and $rc_{i,j}(t)$ rates for repair from complex to simple damage status. Let $df_{i,j}(t)$ represent rates for differentiation of normal cells (assumed to only occur during G1-phase), and $sl(t)$ the rate for sloughing of differentiated cells. Let $ap_{i,j}(t)$ represent rates of apoptosis for complex damaged cells, and $ph(t)$ the rate for phagocytosis of apoptotic cells. Mutation is assumed to occur at rate $\mu_j(t)$ during division of a simple damaged cell of type j in late M-phase, creating a new simple damaged cell and new mutant normal cell of type $j+1$, both in early G1-phase. The Kronecker delta function $\delta(i, k)$ (equaling 1 for $i = k$, and 0 otherwise) is used in the following equations to represent addition of cell during division or mutation, with a cell in state $i = 7$ transforming to two cells in state $i = 0$.

Deterministic equations for combined cell cycle and multistage cancer model

Normal cells:

$$\frac{d}{dt}\bar{n}_{i,j}(t) = -cc_{i,j}(t)\bar{n}_{i,j}(t) - ds_{i,j}(t)\bar{n}_{i,j}(t) - df_{i,j}(t)\bar{n}_{i,j}(t)$$
$$+ (1 + \delta(i, 0))cc_{i-1,j}(t)\bar{n}_{i-1,j}(t) + rs_{i,j}(t)\bar{s}_{i,j}(t)$$
$$+ \delta(i, 0)mu_{j-1}(t)\bar{s}_{i-1,j-1}(t).$$

Simple damaged cells:

$$\frac{d}{dt}\bar{s}_{i,j}(t) = -ks_{i,j}(t)\bar{s}_{i,j}(t) - dc_{i,j}(t)\bar{s}_{i,j}(t) - \delta(i, 7)mu_j(t)\bar{s}_{i,j}(t)$$
$$- rs_{i,j}(t)\bar{s}_{i,j}(t) + (1 + \delta(i, 0))ks_{i-1,j}(t)\bar{s}_{i-1,j}(t) + ds_{i,j}(t)\bar{n}_{i,j}(t)$$
$$+ rc_{i,j}(t)\bar{c}_{i,j}(t) + \delta(i, 0)mu_j(t)\bar{s}_{i-1,j}(t).$$

Complex damaged cells:

$$\frac{d}{dt}\bar{c}_{i,j}(t) = -kc_{i,j}(t)\bar{c}_{i,j}(t) - ap_{i,j}(t)\bar{c}_{i,j}(t) - rc_{i,j}(t)\bar{c}_{i,j}(t)$$
$$+ (1 + \delta(i, 0))kc_{i-1,j}(t)\bar{c}_{i-1,j}(t) + dc_{i,j}(t)\bar{s}_{i,j}(t).$$

Apoptotic cells:

$$\frac{d}{dt}\bar{a}(t) = -ph(t)\bar{a}(t) + \sum_{i,j} ap_{i,j}(t)\bar{c}_{i,j}(t).$$

Differentiated cells (loss through sloughing, gain through differentiation of normal cells):

$$\frac{d}{dt}\bar{d}(t) = -sl(t)\bar{d}(t) + \sum_{i,j} df_{i,j}(t)\bar{n}_{i,j}(t).$$

Total number of cells under homeostatic regulation

$$\frac{d}{dt}\bar{h}(t) = -ph(t)\bar{a}(t) - \sum_{i,j} df_{i,j}(t)\bar{n}_{i,j}(t)$$
$$+ \sum_{i,j}[\delta(i,0)(cc_{i-1,j}(t)\bar{n}_{i-1,j}(t) + (mu_j(t) + ks_{i-1,j}(t))\bar{s}_{i-1,j}(t)$$
$$+ kc_{i-1,j}(t)\bar{c}_{i-1,j}(t))].$$

Homeostatic control with target number $h0$ cells per crypt (regulates exit from G1 cell cycle phase)

$$cc_{i,j}(t) = cc0_{i,j}(1 - \delta(i,1)) + \delta(i,1)cc0_{i,j}\max[0, (h0 - \bar{h}(t))/h0].$$

Checkpoint for simple damage (slows exit from G1, mid-S, end of G2)

$$ks_{i,j}(t) = cc_{i,j}(t)(1 - ks0\,\delta(i,1))(1 - ks0\,\delta(i,2))(1 - ks0\,\delta(i,5)).$$

Checkpoint for complex damage (slows exit from G1, mid-S, end of G2, and causes arrest in M)

$$kc_{i,j}(t) = cc_{i,j}(t)(1 - kc0\,\delta(i,1))(1 - kc0\,\delta(i,2))$$
$$\times (1 - kc0\,\delta(i,5))(1 - kc0\,\delta(i,6)).$$

Differentiation only out of G1 phase, influenced by circadian rhythm:

$$df_{i,j}(t) = \delta(i,1)df0_j cir(t)$$
$$cir(t) = IF[Mod[t,24] < 6 || Mod[t,24] > 18, 0, 1].$$

Relative DNA content in early and late G1, S, G2, and M phases:

$$dna_i = (1,\ 1,\ 1.25, 1.75,\ 2,\ 2,\ 2,\ 2).$$

Simple damage rates combine endogenous damage (occurring mostly in S phase) and ionizing radiation damage, assumed proportional to DNA content and dose rate:

$$ds_{i,j}(t) = en_{i,j} + irs_{i,j}\, dna_i\, dose(t).$$

Simple repair rates are assumed proportional to DNA content:

$$rs_{i,j}(t) = rs0_{i,j}\, dna_i.$$

Complex damage rates assume conversion of simple to complex damage is proportional to dose rate:

$$dc_{i,j}(t) = irc_{i,j}\, dna_i\, dose(t).$$

Repair rates from complex to simple damage are assumed proportional to DNA content:

$$rc_{i,j}(t) = rc0_{i,j}\, dna_i.$$

Characteristic equations for stochastic cell cycle and multistage cancer survival model:

Let $\widehat{\eta}_{i,j}(u)$, $\widehat{\sigma}_{i,j}(u)$, $\widehat{\chi}_{i,j}(u)$ represent conditional probability generating functions (pgf)s representing normal, simple damaged, and complex damaged cell types, that are conditioned on no malignant cell occurring at any time s prior to t. We use a change of variables $u = t - s$ to solve the pgfs backwards in time, with survival boundary conditions. However, we must first solve the deterministic equations above to find the time dependence of cell cycle and checkpoint rates that maintain homeostasis. Using these time dependent rates, we solve the following characteristic equations.

Normal cells:

$$\frac{d}{du}\widehat{\eta}_{i,j}(u) = cc_{i,j}(t-u)\left[\widehat{\eta}_{i+1,j}^{(1+\delta(i,7))}(u) - \widehat{\eta}_{i,j}(u)\right]$$
$$+ ds_{i,j}(t-u)[\widehat{\sigma}_{i,j}(u) - \widehat{\eta}_{i,j}(u)]$$
$$+ df_{i,j}(t-u)[1 - \widehat{\eta}_{i,j}(u)].$$

Simple damaged cells:

$$\frac{d}{du}\widehat{\sigma}_{i,j}(u) = ks_{i,j}(t-u)\left[\widehat{\sigma}_{i+1,j}^{(1+\delta(i,7))}(u) - \widehat{\sigma}_{i,j}(u)\right]$$
$$+ dc_{i,j}(t-u)[\widehat{\chi}_{i,j}(u) - \widehat{\sigma}_{i,j}(u)]$$
$$+ \delta(i,7)mu_j(t-u)[\widehat{\sigma}_{i+1,j}(u)\eta_{i+1,j+1}(u)$$
$$- \widehat{\sigma}_{i,j}(u)] + rs_{i,j}(t-u)[\eta_{i,j}(u) - \widehat{\sigma}_{i,j}(u)].$$

Complex damaged cells:

$$\frac{d}{du}\widehat{\chi}_{i,j}(u) = kc_{i,j}(t-u)\left[\widehat{\chi}_{i+1,j}^{(1+\delta(i,7))}(u) - \widehat{\chi}_{i,j}(u)\right]$$
$$+ ap_{i,j}(t-u)[1 - \widehat{\chi}_{i,j}(u)]$$
$$+ rc_{i,j}(t-u)[\widehat{\sigma}_{i,j}(u) - \widehat{\chi}_{i,j}(u)].$$

The cancer survival probability, $S(t)$, is calculated by assuming nc crypts per individual with $h0$ cells per crypt that begin as normal stem cells in beginning G1 phase at time 0,

$$S(t) = \widehat{\eta}_{0,0}^{h0*nc}(t),$$

using survival boundary conditions for an n_m-stage model,

$$S(0) = 1, \quad \widehat{\eta}_{i,j}(0) = 1, \quad \widehat{\sigma}_{i,j}(0) = 1, \quad \widehat{\chi}_{i,j}(0) = 1 \text{ (for } j < n_m\text{)}.$$
$$\widehat{\eta}_{i,n_m}(0) = 0, \quad \widehat{\sigma}_{i,n_m}(0) = 0, \quad \widehat{\chi}_{i,n_m}(0) = 0.$$

Chapter 5

CANCER MODELS, IONIZING RADIATION, AND GENOMIC INSTABILITY: A REVIEW

Mark P. Little

A variety of quasi-mechanistic models of carcinogenesis are reviewed, and in particular, the multi-stage model of Armitage and Doll and the two-mutation model of Moolgavkar, Venzon, and Knudson, in the light of the known effects of radiation on cancer risk. Both the latter models, and various generalizations of them also, are capable of describing at least qualitatively many of the observed patterns of excess cancer risk following ionizing radiation exposure. However, there are certain inconsistencies with the biological and epidemiological data both for the multi-stage model and the two-mutation model. In particular, there are indications that the two-mutation model is not totally suitable for describing the pattern of excess risk for solid cancers that is often seen after exposure to radiation, although leukemia may be better fitted by this type of model. Generalizations of the model of Moolgavkar, Venzon, and Knudson which require three or more mutations, and models allowing for genomic instability, are easier to reconcile with the epidemiological and biological data relating to solid cancers.

Keywords: Cancer, Armitage-Doll model, two-mutation model, multiple pathway models, transmissible genomic instability, ionizing radiation.

1. INTRODUCTION

One of the principal uncertainties that surround the calculation of population cancer risks from epidemiological data results from the fact that few radiation-exposed cohorts have been followed up to extinction. For example, 50 years after the atomic bombings of Hiroshima and Nagasaki, about half of the survivors were still alive (Preston *et al.*, 2003). In attempting to calculate lifetime population cancer risks it is therefore important to predict how risks might vary as a function of time after radiation exposure, in particular for that group for whom the uncertainties in projection of risk to the end of life are most uncertain, namely those who were exposed in childhood.

In what follows we shall use *cancer risk* to mean the instantaneous cancer hazard function $h(t)$ at time t, defined by $h(t) = -\frac{d}{dt} \ln[P[\text{cancer occurs at age} > t]$. This can be defined either with respect to mortality from cancer or cancer incidence. One way to model the variation in cancer risk is to use empirical models incorporating adjustments for a number of variables (e.g. age at exposure, time since exposure, sex) and indeed this approach has been used in the Fifth Report of the Biological Effects of Ionizing Radiations (BEIR V) Committee (National Research Council, 1990) in its analyses of data from the Japanese atomic bomb survivor Life Span Study (LSS) cohort and various other irradiated groups. Recent analyses of solid cancers for these groups have found that the radiation-induced excess risk can be described fairly well by a relative risk model (ICRP, 1991). The *time-constant relative risk model* assumes that if a dose of radiation is administered to a population, then, after some latent period, t_{lat}, there is an increase in the cancer rate, the excess rate being proportional to the underlying cancer rate in an unirradiated population. That is to say that if $h(t, D, a)$ is the hazard function at age t corresponding to a radiation dose D administered at age a, then $h(t, D, a) = h(t, 0, a) \cdot [1 + ERR(D) \cdot 1_{t-a>t_{lat}}]$, where $ERR(D)$ is the *excess relative risk*. For leukemia, this model provides an unsatisfactory fit, consequently a number of other models have been used for this group of malignancies, including one in which the excess cancer rate resulting from exposure is assumed to be constant, i.e. the *time-constant additive risk model* (UNSCEAR, 1988). That is to say, using the notation above, the hazard function at age t corresponding to a radiation dose D administered at age a, is given by $h(t, D, a) = h(t, 0, a) + EAR(D) \cdot 1_{t-a>t_{lat}}$, where $EAR(D)$ is the *excess absolute risk*.

It is well known that for all cancer subtypes (including leukemia) the excess relative risk (ERR) diminishes with increasing age at exposure (UNSCEAR, 2000). For those irradiated in childhood there is evidence of a reduction in the ERR of solid cancer 25 or more years after exposure (Little *et al.*, 1991; Little, 1993; Thompson *et al.*, 1994; Pierce *et al.*, 1996; UNSCEAR, 2000). For solid cancers in adulthood the ERR is more nearly constant, or perhaps even increasing over time (Little and Charles, 1991; Little, 1993; UNSCEAR, 2000), although there are some indications to the contrary (Weiss *et al.*, 1994). Clearly then, even in the case of solid cancers various factors have to be employed to modify the ERR.

Associated with the issue of projection of cancer risk over time is that of projection of cancer risk between two populations with differing underlying susceptibilities to cancer. Analogous to the relative risk time projection model one can employ a *multiplicative transfer* of risks, in which the ratio of the radiation-induced excess cancer rates to the underlying cancer rates in the two populations might be assumed to be identical. Similarly, akin to the additive risk time projection model one can use an *additive transfer* of risks, in which the radiation-induced excess cancer rates in the two populations might be assumed to be identical. The data that are available suggests that there is no simple solution to the problem (UNSCEAR, 1994). For example, there are weak indications that the relative risks of stomach cancer following radiation exposure may be more comparable than the absolute excess risks in populations with different background stomach cancer rates (UNSCEAR, 1994). Comparison of breast cancer risks observed in the Japanese atomic bomb survivor incidence data and those in various medically exposed populations, many from North America and Europe, where underlying breast cancer rates are higher than in Japan, suggests that ERRs are rather higher in the LSS than those in the medically irradiated groups, but (time- and age-adjusted) EARs are more similar (Little and Boice, 1999; Preston *et al.*, 2002). The observation that gender differences in solid tumor ERR are generally offset by differences in gender-specific background cancer rates (UNSCEAR, 1994) might suggest that EARs are more alike than ERRs. Taken together, these considerations suggest that in various circumstances relative or absolute transfers of risk between populations may be advocated or, indeed, the use of some sort of hybrid approach such as that employed by Muirhead and Darby (1987) and Little *et al.* (1999).

The exposed populations that are often used for deriving cancer risks, e.g. the Japanese atomic bomb survivors, were exposed to ionizing radiation at high doses and high dose rates. However, it is the possible risks arising from low dose and low dose-rate exposure to ionizing radiation which are central to the setting of standards for radiological protection. The

International Commission on Radiological Protection (ICRP) (1991) recommended application of a dose and dose-rate effectiveness factor of 2 to scale cancer risks from high dose and high dose-rate exposure to low dose and low dose-rate exposure on the basis of animal data, the shape of the cancer dose-response in the bomb survivor data and other epidemiological data. Although the linear-quadratic dose-response model (with upward curvature) found for leukemia is perhaps the most often employed departure from linearity in analyses of cancer in radiation-exposed groups (Pierce and Vaeth, 1991; Pierce et al., 1996), other shapes are possible for the dose-response curve (UNSCEAR, 1993). While for most tumor types in the Japanese data linear-quadratic curvature adequately describes the shape of the dose-response curve, for non-melanoma skin cancer (NMSC) there is evidence for departures from linear-quadratic curvature. The NMSC dose-response in the Japanese cohort is consistent with a dose threshold of $\approx$ 1 Sv (Little and Muirhead, 1996; Little and Charles, 1997) or with an induction term proportional to the fourth power of dose, with in each case an exponential cell sterilization term to reduce NMSC risk at high doses (>3 Sv).

Arguably, models which take account of the biological processes leading to the development of cancer can provide insight into these related issues of projection of cancer risk over time, transfer of risk across population and extrapolation of risks from high doses and dose-rates to low doses and dose-rates. For example, Little and Charles (1991) have demonstrated that a variety of mechanistic models of carcinogenesis predict an ERR which reduces with increasing time after exposure for those exposed in childhood, while for those exposed in adulthood the ERR might be approximately constant over time. Mechanistic considerations also imply that the interactions between radiation and the various other factors that modulate the process of carcinogenesis may be complex (Leenhouts and Chadwick, 1994), so that in general one would not expect either relative or absolute risks to be invariant across populations.

There is much biological data suggesting that the initiating lesion in the multistage process leading to cancer might be one involving a destabilization of the genome resulting in elevation of mutation rates, reviewed by Morgan (2003a, 2003b). This phenomenon has implications both for the shape of the dose response curve, and for projection of cancer risk over time. There have been a few previous attempts to incorporate transmissible genomic instability (GI) in mechanistic carcinogenesis models (Mao et al., 1998; Ohtaki and Niwa, 2001).

In this article we review the epidemiology of cancer in relation to the action of radiation exposure, and assess the consequences for quasi-mechanistic cancer models, in particular for the overall form of the model and the way in which radiation might be considered to modify the parameters of these models. We will pay particular attention to models that allow for transmissible GI.

2. ARMITAGE-DOLL MULTI-STAGE MODEL

Mechanistic models of carcinogenesis were originally developed to explain phenomena other than the effects of ionizing radiation. One of the more commonly observed patterns in the age-incidence curves for epithelial cancers is that the cancer incidence rate varies approximately as $C \cdot [\text{age}]^\beta$ for some constants C and β. The so-called multi-stage model of carcinogenesis of Armitage and Doll (1954) was developed in part as a way of accounting for this approximately log-log variation of cancer incidence with age. The model supposes that at age t an individual has a population of $X(t)$ completely normal (stem) cells and that these cells acquire one mutation at a rate $M(0)(t)$. The cells with one mutation acquire a second mutation at a rate $M(1)(t)$, and so on until at the $(k-1)$th stage the cells with $(k-1)$ mutations proceed at a rate $M(k-1)(t)$ to become fully malignant. The model is illustrated schematically in Figure 1.

It can be shown that when $X(t)$ and the $M(i)(t)$ are constant, a model with k stages predicts a cancer incidence rate that is approximately given by the expression $C \cdot [\text{age}]^{k-1}$ with $C = M(0) \cdot M(1) \cdot \ldots \cdot M(k-1)/(1 \cdot 2 \cdot \ldots \cdot (k-1))$ (Armitage and Doll, 1954; Moolgavkar, 1978).

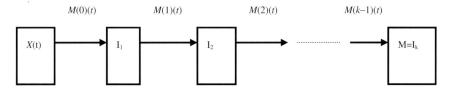

Fig. 1. Schematic diagram of the Armitage-Doll multi-stage model.

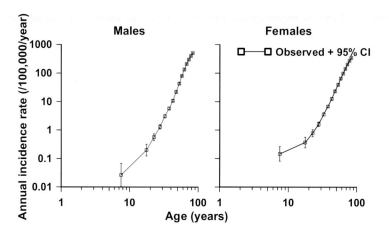

Fig. 2. SEER (2005) colon cancer data, and observed data (with 95% confidence intervals (CI), adjusted for overdispersion (McCullagh and Nelder 1989)). The use of double logarithmic (log-log) axes shows that except for the youngest age group (<10 years) the age-incidence relationship is well described by $C \cdot [\text{age}]^{k-1}$.

As can be seen from Figure 2, for colon cancer the age-incidence relationship is remarkably well described by a power of age, as predicted by this model.

Departures from this form of relationship are only apparent at very young ages (<10 years) (Figure 2). For many common epithelial cancers in adulthood this function, $C \cdot [\text{age}]^{k-1}$, fits the age-incidence and age-mortality relationships well, with the implied number of rate-limiting stages, k, between 5 and 7 (Doll, 1971). In the intervening 30 years, there has accumulated substantial biological evidence that cancer is a multi-step process involving the accumulation of a number of genetic and epigenetic changes in a clonal population of cells. This evidence is reviewed by the United Nations Scientific Committee on the Effects of Atomic Radiation (UNSCEAR) (1993, 2000). However, there are certain problems with the model proposed by Armitage and Doll (1954) associated with the fact that, as noted above, to account for the observed age incidence curve $C \cdot [\text{age}]^{\beta}$, between five and seven rate-limiting stages are needed. For colon cancer there is evidence that six stages might be required (Fearon and Vogelstein, 1990). However, for other cancers there is little evidence that there are as many rate-limiting stages as this. BEIR V (National Research Council, 1990) surveyed evidence for all cancers and found that two or three stages might be justifiable, but not a much larger number. To this extent the large number of stages predicted by the Armitage-Doll model appears, for certain cancer sites, to be verging on the biologically unlikely. Related to the large number of stages required by the Armitage-Doll multi-stage model is the high mutation rates predicted by the model. Moolgavkar and Luebeck (1992) fitted the Armitage-Doll multi-stage model to datasets describing the incidence of colon cancer in a general population and in patients with familial adenomatous polyposis. Moolgavkar and Luebeck (1992) found that Armitage-Doll models with five or six stages gave good fits to these datasets, but that both of these models implied mutation rates that were too high by at least two orders of magnitude. The discrepancy between the predicted and experimentally measured mutation rates might be eliminated, or at least significantly reduced, if account were to be taken of the fact that the experimental mutation rates are locus-specific. A "mutation" in the sense in which it is defined in this model might result from the "failure" of any one of a number of independent loci, so that the "mutation" rate would be the sum of the failure rates at each individual locus.

Notwithstanding these problems, much use has been made of the Armitage-Doll multi-stage model as a framework for understanding the time course of carcinogenesis, particularly for the interaction of different carcinogens (Peto, 1977). Day and Brown (1980) discuss in a qualitative way the pattern of variation of risk in a number of animal and human groups exposed to a variety of chemical carcinogens (as well as radiation). Freedman and Navidi (1989) also assess the fits of such models to various animal and human datasets, and as a result of fitting the model to three cohorts of smokers, in which they allow the mutation rates $M(i)(t)$ for early and late stages to be changed by the administration of tobacco smoke, conclude that none of the models adequately describes all the features of the data; this finding is to some extent contradicted by the fitting of similar models (allowing the first and penultimate mutation rates to be affected) to smoking data by Brown and Chu (1987). Brown and Chu (1983) also fitted the Armitage-Doll model to a dataset of copper smelter workers occupationally exposed to arsenic, as did Mazumdar *et al.* (1989); in both cases evidence was found that arsenic might act at both early and late mutation rates of such a multi-step process. Crump and Howe (1984) fitted the model to rats exposed to ethylene dibromide and found a satisfactory fit with a single affected stage (the first out of six stages). Thomas (1983) considered a cohort exposed to asbestos for which information on smoking was also available and found that the effects of both of these factors could each be modelled adequately by assuming that a single mutation rate was affected (stage 4 for asbestos and stage 5 for smoking in a model with a total of six stages). Thomas (1990) has fitted the Armitage-Doll model with one and two radiation-affected stages to the solid cancer data in the Japanese LSS Report 11 cohort of bomb survivors. Thomas (1990) found that a model with a total of five stages, of which either stages 1 and 3 or stages 2 and 4 were radiation-affected, fitted significantly better than models with a single radiation-affected stage. Little et al. (1992, 1994) also fitted the Armitage-Doll model with up to two radiation-affected stages to the Japanese LSS Report 11 dataset and also to data on various medically exposed groups, using a slightly different technique to that of Thomas (1990). Little et al. (1992, 1994) found that the optimal solid cancer model for the Japanese data had three stages, the first of which was radiation affected, while for the Japanese leukemia data the best fitting model had three stages, the first and second of which were radiation affected. A version of the Armitage-Doll has also been fitted to the LSS solid tumor incidence data by Pierce and Mendelsohn (1999). Pierce and Mendelsohn (1999) found that a model with five or six stages gave the best fit to this data.

Both the paper of Thomas (1990) and those of Little et al. (1992, 1994) assumed the ith and the jth stages or mutation rates ($M(i-1)$, $M(j-1)$) ($j > i$) in a model with k stages to be (linearly) affected by radiation and the transfer coefficients (other than $M(i-1)$ and $M(j-1)$) to be constant (as is the stem cell population $X(t)$). In these circumstances it can be shown (Little et al., 1992) that if an instantaneously administered dose of radiation d is given at age a, then at age $t(>a)$ the cancer rate is approximately:

$$\mu \cdot t^{k-1} + \alpha \cdot d \cdot a^{i-1} \cdot [t-a]^{k-i-1} + \beta \cdot d \cdot a^{j-1} \cdot [t-a]^{k-j-1}$$
$$+ \gamma \cdot d^2 \cdot a^{i-1} \cdot [t-a]^{k-j-1} \quad (1)$$

for some positive constants μ, α and β, and where γ is given by:

$$\gamma = \frac{\alpha \cdot \beta \cdot \Gamma(k-i) \cdot \Gamma(j)}{2 \cdot \mu \cdot \Gamma(k)} \quad \text{if } j = i+1$$
$$= 0 \quad \text{if } j > i+1 \quad (2)$$

and $\Gamma(.)$ is the gamma function (Abramowitz and Stegun, 1964). The first term ($\mu \cdot t^{k-1}$) in expression (1) corresponds to the cancer rate that would be observed in the absence of radiation, while the second term ($\alpha \cdot d \cdot a^{i-1} \cdot [t-a]^{k-i-1}$) and the third term ($\beta \cdot d \cdot a^{j-1} \cdot [t-a]^{k-j-1}$) represent the separate effects of radiation on the ith and jth stages, respectively. The fourth term ($\gamma \cdot d^2 \cdot a^{i-1} \cdot [t-a]^{k-j-1}$), which is quadratic in dose d, represents the consequences of interaction between the effects of radiation on the ith and the jth stages and is only non-zero when the two radiation-affected stages are adjacent ($j = i+1$). Thus if the two affected stages are adjacent, a quadratic (dose plus dose-squared) relationship will occur, whereas the relationship will be approximately linear if the two affected stages have at least one intervening stage. Another way of considering the joint effects of radiation on two stages is that for a brief exposure, unless the two radiation-affected stages are adjacent, there will be insignificant interaction between the cells affected by radiation in the earlier and later of the two radiation-affected cell compartments. This is simply because very few cells will move between the two compartments in the course of the radiation exposure. If the ith and the jth stages are radiation-affected the result of a brief dose of radiation will be to cause some of the cells which have already accumulated ($i - 1$) mutations to acquire an extra mutation and move from the ($i - 1$)th to the ith compartment. Similarly, it will cause some of the cells which have already acquired ($j - 1$) mutations to acquire an extra mutation and so move from the ($j - 1$)th to the jth compartment. It should be noted that the model does not require that the same cells be hit by the radiation at the ith and jth stages, and in practice for low total doses, or whenever the two radiation-affected stages are separated by an additional unaffected stage or stages, an insignificant proportion of the same cells will be hit (and mutated) by the radiation at both the ith and the jth stages. The result is that, unless the radiation-affected stages are adjacent, for a brief exposure the total effect on cancer rate is approximately the sum of the effects, assuming radiation were to act on each of the radiation-affected stages alone. One interesting implication of models with two or more radiation-affected stages is that as a result of interaction between the effects of radiation at the various stages, protraction of dose in general results in an increase in cancer rate, i.e. an inverse dose-rate effect (Little et al., 1992). However, it can be shown that in practice the resulting increase in cancer risk is likely to be small (Little et al., 1992).

The variant of the Armitage-Doll model fitted by Pierce and Mendelsohn (1999) is unusual in that it assumes that radiation equally affects all k mutation rates in the model except the last. (In the last stage radiation is not assumed to have any effect.) This assumption distinguishes their use of this model from the approaches of Little *et al.* (1992) or Thomas (1990), both of whom assumed that radiation affected at most two of the mutation rates (and did not constrain the effects of radiation to be equal in these stages). There are some technical problems with the paper of Pierce and Mendelsohn (1999) arising from the authors failure to take account of interactions between the effects of radiation on the $(k-2)$ pairs of adjacent stages, and which contribute significantly, by adding a quadratic term in the dose-response. These cannot be ignored, even to a first order approximation. The fact that in general there is little evidence for upward curvature in the solid cancer dose-response in the LSS (Pierce and Vaeth, 1991; Little and Muirhead, 1996, 1998, 2000) argues that if proper account were taken of these interaction terms the model of Pierce and Mendelsohn (1999) would not fit the data well. Moreover, one implication of the model of Pierce and Mendelsohn (1999) is that the ERR will be proportional to $1/a$, i.e. the inverse of attained age. However, this is known to provide a poor description of the ERR of solid cancer, even within the LSS cohort (Little *et al.*, 1997a, 1999). For these reasons, there are some grounds for regarding the model of Pierce and Mendelsohn (1999) as providing a poor description of the pattern of excess risk of solid tumors within the LSS cohort. Other problems with the model of Pierce and Mendelsohn (1999) are discussed by Heidenreich *et al.* (2002).

The optimal leukemia model found by Little *et al.* (1992, 1994), having adjacent radiation-affected stages, predicts a linear-quadratic dose-response, in accordance with the significant upward curvature which has been observed in the Japanese dataset (Pierce and Vaeth, 1991; Preston *et al.*, 1994; Pierce *et al.*, 1996). This leukemia model, and also that for solid cancer, predicts the pronounced reduction of ERR with increasing age at exposure (see Figure 3) which has been seen in the Japanese atomic bomb survivors and other datasets (UNSCEAR, 2000).

The optimal Armitage-Doll leukemia model predicts a reduction of ERR with increasing time after exposure for leukemia. At least for those exposed in childhood, the optimal Armitage-Doll solid cancer model also predicts

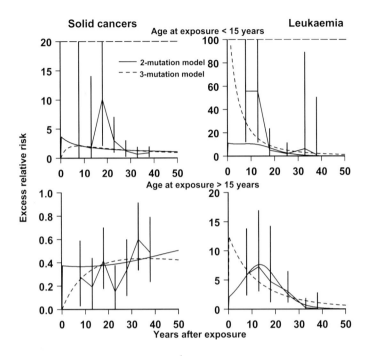

Fig. 3. Observed excess relative risk Sv^{-1} (+90% CI) for solid cancers and leukemia and fitted excess relative risk (evaluated at 1 Sv) using optimal two-mutation and three-mutation generalized MVK models (Little, 1996).

a reduction in ERR with time for solid cancers. These observations are consistent with the observed pattern of risk in the Japanese and other datasets (Little, 1993; UNSCEAR, 2000). Nevertheless, there are indications that the Armitage-Doll model may not provide an adequate fit to the Japanese data (Little *et al.*, 1995). For this reason, and because of the other problems with the Armitage-Doll model discussed above, one needs to consider a slightly different class of models.

3. TWO-MUTATION MODEL

In order to reduce the arguably biologically implausible number of stages required by their first model, Armitage and Doll (1957) developed a further model of carcinogenesis, which postulated a two-stage probabilistic process whereby a cell following an initial transformation into a pre-neoplastic state (initiation) was subject to a period of accelerated (exponential) growth. At some point in this exponential growth a cell from this expanding population might undergo a second transformation (promotion) leading quickly and directly to the development of a neoplasm. Like their previous model, it satisfactorily explained the incidence of cancer in adults, but was less successful in describing the pattern of certain childhood cancers.

The two-mutation model developed by Knudson (1971) to explain the incidence of retinoblastoma in children took account of the process of growth and differentiation in normal tissues. Subsequently, the stochastic two-mutation model of Moolgavkar and Venzon (1979) generalized Knudson's model, by taking account of cell mortality at all stages as well as allowing for differential growth of intermediate cells. The two-stage model developed by Tucker (1967) is very similar to the model of Moolgavkar and Venzon but does not take account of the differential growth of intermediate cells. The two-mutation model of Moolgavkar, Venzon and Knudson (MVK) supposes that at age t there are $X(t)$ susceptible stem cells, each subject to mutation to an intermediate type of cell at a rate $M(0)(t)$. The intermediate cells divide at a rate $G(1)(t)$; at a rate $D(1)(t)$ they die or differentiate; at a rate $M(1)(t)$ they are transformed into malignant cells. The model is illustrated schematically in Figure 4.

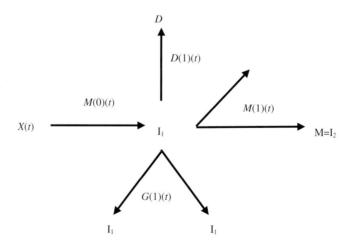

Fig. 4. Schematic diagram of the two-mutation (MVK) model.

In contrast with the case of the (first) Armitage-Doll model, there is a considerable body of experimental biological data supporting this initiation-promotion type of model (see e.g. Moolgavkar and Knudson, 1981; Tan, 1991). The model has recently been developed to allow for time-varying parameters at the first stage of mutation (Moolgavkar *et al.*, 1988). A further slight generalization of this model (to account for time varying parameters at the second stage of mutation) was presented by Little and Charles (1991), who also demonstrated that the ERR predicted by the model, when the first mutation rate was subject to instantaneous perturbation, decayed at least exponentially for a sufficiently long time after the perturbation. Moolgavkar *et al.* (1990), Luebeck *et al.* (1996) and Heidenreich *et al.* (1999, 2000) have used the two-mutation model to describe the incidence of lung cancer in rats exposed to radon, and in particular to model the inverse dose-rate effect that has been observed in this data. Moolgavkar *et al.* (1993), Luebeck *et al.* (1999), Leenhouts (1999), Hazelton *et al.* (2001), Little *et al.* (2002) and Heidenreich *et al.* (2004) have applied the model to describe the interaction of radon, smoking and other agents causing lung cancer in various miner cohorts. The two-mutation model has also been utilised to describe lung, stomach, and colon cancer in the Japanese atomic bomb survivor incidence data (Kai *et al.*, 1997), and to fit to liver cancer data from a cohort of Swedish Thorotrast-exposed patients (Heidenreich *et al.*, 2003).

It is vital in fitting these and other models to take account of problems of parameter identifiability. It has been known for some time that there is redundancy in the parameterization of the two-mutation model, so that only three combinations of the five available combinations of model parameters $(X, M(0), M(1), G(1), D(1))$ can be estimated from knowledge of the hazard function (Hanin and Yakovlev, 1996; Heidenreich, 1996; Hanin, 2002), i.e. two combinations of parameters cannot be estimated. There is a large literature on this, the most important parts of which can be found in the articles of Heidenreich *et al.* (1997) and Hanin (2002). More general material on parameter identifiability can be found in the papers by Jacquez and Perry (1990) and Feng and DiStefano (1995). Little *et al.* (2008) have extended the results of Heidenreich (1996) and Heidenreich *et al.* (1997), showing that for the class of models considered by Little and Wright (2003), that includes the two-mutation model as a special case, two parameter combinations cannot be estimated.

A curious finding in many analyses of lung cancer in relation to radon-daughter exposure using the two-mutation model is that there is significant radon action on intermediate cell proliferation. This has been observed both in radon-exposed rats (Heidenreich *et al.*, 1999, 2000), in the Colorado Plateau uranium miners (Luebeck *et al.*, 1999; Little *et al.*, 2002) and in the Chinese tin miners (Hazelton *et al.*, 2001). This is very much associated with fits of the two-mutation model, and may reflect the limited number of parameters that can be modified in this model. Analyses of rat data using a three-mutation generalized MVK model (see below) did not find any indications of an effect of radon daughter exposure on intermediate cell proliferation (Heidenreich *et al.*, 2000). Likewise, analysis of the Colorado Plateau miners (the same dataset analyzed by Luebeck *et al.* (1999)) using a three-mutation MVK generalized MVK model (see below) did not find any effect of radon daughter exposure on intermediate cell proliferation rates (Little *et al.*, 2002), and the fit of the three-mutation model was somewhat better than that of the two-mutation model (see Figure 5).

Moolgavkar and Luebeck (1992) have used models with two or three mutations to describe the incidence of colon cancer in a general population and in patients with familial adenomatous polyposis. They found that both models gave good fits to both datasets, but that the model with two mutations implied mutation rates that were biologically implausibly low, by at least two orders of magnitude. The three-mutation model, which predicted mutation rates more in line with biological data, was therefore somewhat preferable. The problem of implausibly low mutation rates implied by the two-mutation model is not specific to the case of colon cancer, and is discussed at greater length by Den Otter *et al.* (1990) and Derkinderen *et al.* (1990), who argue that for most cancer sites a model with more than two stages is required. A possible way round the problem of implausibly low mutation rates, at least for colon cancer, is suggested by the model of Nowak *et al.* (2003), who showed that by "washing out" pre-malignant cells in the intestinal lumen a relatively high mutation rate at the cellular level may translate into a much lower apparent mutation rate at the tissue (intestinal crypt) level.

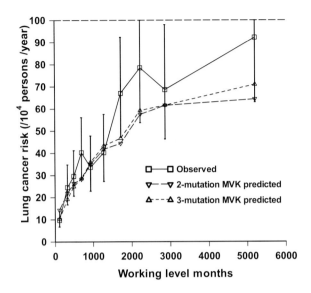

Fig. 5. Observed absolute risk of lung cancer mortality (+95% CI) and predicted risk associated with the optimal two-mutation and three-mutation models fitted to the Colorado Plateau uranium miner data as a function of cumulative radon-daughter exposure, taken from Little *et al.* (2002).

4. GENERALIZED MVK AND MULTI-STAGE MODELS

A number of generalizations of the Armitage-Doll and two- and three-mutation models have been developed (Tan, 1991; Little, 1995; Little and Wright, 2003). In particular, two closely related models have been developed, whose properties have been described in the paper of Little (1995). The first model is a generalization of the two-mutation model of Moolgavkar, Venzon, and Knudson and so will be termed the *generalized MVK model*. The second model generalizes the multi-stage model of Armitage and Doll and will be referred to as the *generalized multi-stage model*. For the generalized MVK model it may be supposed that at age t there are $X(t)$ susceptible stem cells, each subject to mutation to a type of cell carrying an irreversible mutation at a rate of $M(0)(t)$. The cells with one mutation divide at a rate $G(1)(t)$; at a rate $D(1)(t)$ they die or differentiate.

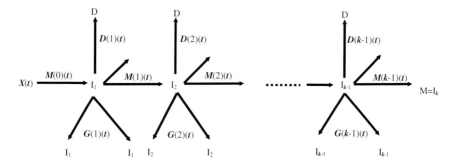

Fig. 6. Schematic diagram of the generalized MVK model.

Each cell with one mutation can also divide into an equivalent daughter cell and another cell with a second irreversible mutation at a rate $M(1)(t)$. For the cells with two mutations there are also assumed to be competing processes of cell growth, differentiation, and mutation taking place at rates $G(2)(t)$, $D(2)(t)$, and $M(2)(t)$, respectively, and so on until at the $(k-1)$th stage the cells which have accumulated $(k-1)$ mutations proceed at a rate $M(k-1)(t)$ to acquire another mutation and become malignant. The model is illustrated schematically in Figure 6.

The two-mutation model of Moolgavkar, Venzon, and Knudson corresponds to the case $k = 2$. The generalized multi-stage model differs from the generalized MVK model only in that the process whereby a cell is assumed to split into an identical daughter cell and a cell carrying an additional mutation is replaced by the process in which only the cell with an additional mutation results, i.e. an identical daughter cell is not produced. The classical Armitage-Doll multi-stage model corresponds to the case in which the intermediate cell proliferation rates $G(i)(t)$ and the cell differentiation rates $D(i)(t)$ are all zero.

It can be shown (Little, 1995) that the ERR for either model following a perturbation of the parameters will tend to zero as the attained age tends to infinity. One can also demonstrate that perturbation of the parameters $M(k-2)$, $M(k-1)$, $G(k-1)$, and $D(k-1)$ will result in an almost instantaneous change in the cancer rate (Little, 1995).

Generalized MVK models have been fitted to the Japanese atomic bomb survivor LSS Report 11 mortality data (Little, 1996, 1997). Both for leukemia and solid cancers the only models with a single radiation-affected parameter which give at all satisfactory fit are those in which radiation is assumed to affect $M(0)$ (Little, 1996). Both for leukemia and for solid cancer generalized two- and three-mutation MVK models fit equally well. For leukemia, the three-mutation model provides at all satisfactory a fit only when $M(0)$ and $M(1)$ are assumed affected by radiation. For solid cancer and leukemia there are indications of lack of fit to the youngest age at exposure group for the three-mutation model; there is also some lack of fit of the optimal solid cancer three-mutation model to this age at exposure group (Figure 3). Little *et al.* (1996) also showed that the age-incidence relationship for lymphocytic leukemia incidence in the UK population could be adequately described by models with either two or three stages. Little *et al.* (2002) modelled lung cancer mortality in the Colorado Plateau uranium miner cohort using generalizations of the MVK model, and demonstrated that models with three mutations provided a superior fit to models with two mutations.

For solid cancer only $M(0)$ is (linearly) affected by radiation for two- or three-mutation generalized MVK models. In contrast to the solid cancer models, both leukemia models assume a linear-quadratic dose-dependence of the $M(i)$. The non-linearity found in the leukemia $M(i)$ dose-response reflects known curvature in the leukemia dose-response in the Japanese (National Research Council, 1990; Pierce and Vaeth, 1991). There is some evidence, e.g. for chromosome aberrations that the mutation induction curve is linear-quadratic at least for low linear energy transfer (LET) radiation such as X-rays, γ-rays or β particles, although linearity is generally observed for high LET radiation such as α particles and neutrons (Lloyd and Edwards, 1983). Little (2000) has shown that the degree of (upward) curvature in the chromosome-aberration dose response in various *in vitro* datasets is statistically compatible with that observed in the Japanese atomic bomb survivor leukemia incidence dose response.

Despite the indications of lack of fit discussed above, the variation of ERR with time since exposure and age at exposure predicted by the optimal two- and three-mutation models for solid cancer (Figure 3) is in qualitative agreement with the variation seen in the Japanese bomb survivors and in other irradiated groups (UNSCEAR, 2000). In particular, the optimal models demonstrate the progressive reduction in ERR with increasing age at exposure seen in many datasets (UNSCEAR, 2000), together with the marked reduction in ERR with increasing time since exposure observed in various groups exposed in childhood (Little *et al.*, 1991; Pierce *et al.*, 1996; UNSCEAR, 2000).

Figure 3 reinforces the theoretical predictions of a previous paper (Little, 1995), and shows that immediately after perturbing $M(0)$ in the two-mutation model the ERR for solid cancers and leukemia quickly increases. However, there are no data in the first five years of follow-up in the Japanese cohort (Pierce *et al.*, 1996), so that it is difficult to test the predictions made in a previous paper (Little, 1995) concerning the variation in risk shortly after exposure using that dataset.

There is a suggestive elevation in the ERR of cancers other than leukemia and colon cancer in the UK ankylosing spondylitis patients <5 years after first treatment (the first two datapoints in the top-left panel of Figure 7), but the authors caution against interpreting this as the effect of the X-irradiation (Darby *et al.*, 1987). There are no strong indications of an increase in risk in the first five years after radiotherapy for cancers other than leukemia and of the reproductive organs in a study of women followed up for second cancer after radiotherapy for cervical cancer (Boice *et al.*, 1985). This corresponds to the first two datapoints in the bottom panel of Figure 7. (Lung cancers are also excluded from the International Radiation Study of Cervical Cancer (IRSCC) data shown in the lower left panel of Figure 7 because of indications of above-average smoking rates in this cohort (Boice *et al.*, 1985).) In general there are no strong indications of an elevation in solid cancer risk soon after irradiation in other exposed groups (UNSCEAR, 2000). To this extent there are indications of inconsistency for solid cancers between the predictions of the two-mutation model and the observed variation in risks shortly after exposure.

Moolgavkar *et al.* (1993) partially overcome the problem posed by this instantaneous rise in the hazard after perturbation of the two-mutation model parameters in their analysis of the Colorado uranium miners data by assuming a fixed period (3.5 years) between the appearance of the first malignant cell and the clinical detection of malignancy. However, the use of such a fixed latent period only translates a few years into the future the sudden step-change in the hazard. To achieve the observed gradual increase in ERR shortly after exposure, a stochastic process must be used to model the transition from the first malignant cell to detectable cancer, such as is provided by the final stage(s) in the three- or four-mutation generalized

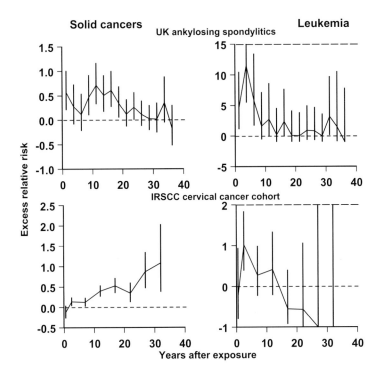

Fig. 7. Excess relative risk (+90% CI) for solid cancers (cancers other than leukemia, colon in spondylitics; cancers other than leukemia, lung, breast, ovary in the International Radiation Study of Cervical Cancer (IRSCC)) and leukemia (=acute non-lymphocytic leukemia in the IRSCC) in the UK ankylosing spondylitics (Darby *et al.*, 1987) and in the IRSCC cervical cancer cohort (Boice *et al.*, 1985).

MVK models used in the analysis of Little (1996). In particular, an exponentially growing population of malignant cells could be modelled by a penultimate stage with $G(k-1) > 0$ and $D(k-1) = 0$, the probability of detection of the clone being determined by $M(k-1)$. In their analysis of lung, stomach and colon cancer in the Japanese atomic bomb survivor incidence data Kai *et al.* (1997) did not assume any such period of latency, perhaps because of the long period after the bombings (12.4 years) before solid cancer incidence follow-up began in the LSS. There are other ways in which an observed gradual increase in tumor risk after parameter perturbation could be achieved, in particular by assuming a random tumor growth rate, or by using a quantal response rate, relating probability of tumor detection to size, as outlined by Bartoszynski *et al.* (2001).

The evidence with respect to the variation in ERR shortly after exposure for leukemias is rather different from that for solid cancers. In the UK ankylosing spondylitis patients (Darby *et al.*, 1987) there is significant excess risk even in the period <2.5 years after first treatment (first datapoint in top-right panel of Figure 7). The IRSCC data (Boice *et al.*, 1985) shows a significant excess risk for acute non-lymphocytic leukemia in the period 1–4 years after first treatment (the second datapoint in the lower-right panel of Figure 7), and this pattern is observed in many other groups (UNSCEAR, 2000). More detailed analysis of UK leukemia incidence data indicate that the age-incidence curves for all subtypes of lymphocytic leukemia can be adequately modelled by two- and three-mutation generalized MVK models (Little *et al.*, 1996, 1997b), although the two-mutation models for acute lymphocytic leukemia (ALL) imply a very small number of stem cells ($<10^4$ cells) if the model is not to yield implausibly low mutation rates (Little *et al.*, 1997b).

5. MULTIPLE PATHWAY MODELS

Little *et al.* (1995) fitted a generalization of the Armitage-Doll model to the Japanese atomic bomb survivor and IRSCC leukemia data which allowed for two cell populations at birth, one consisting of normal stem cells carrying no mutations, the second a population of cells each of which has been subject to a single mutation. The leukemia risk predicted by such a model is equivalent to that resulting from a model with two pathways between the normal stem cell compartment and the final compartment of malignant cells, the second pathway having one fewer stage than the first. This model fitted the Japanese and IRSCC leukemia datasets significantly better, albeit with biologically implausible parameters, than a model which assumed just a single pathway (Little *et al.*, 1995). A number of other such models are described by Tan (1991), who also discusses at some length the biological and epidemiological evidence for such models of carcinogenesis.

We now discuss what may appear to be a special case of these multiple pathway models, but which are of sufficient flexibility to embrace most categories of multiple pathway models.

5.1. Multiple Pathway Models Incorporating Genomic Instability

There is much biological data suggesting that the initiating lesion in the multistage process leading to cancer might be one involving a destabilization of the genome resulting in elevation of mutation rates (Morgan, 2003a, 2003b). In particular, the findings of Kadhim *et al.* (1992, 1994), that exposure of mammalian haemopoietic stem cells to alpha particles could result in a general elevation of mutation rates to very much higher than normal levels, implies, if these findings are at all relevant to carcinogenesis, that there might be multiple pathways in the progression from normal stem cells to malignant cells. A carcinogenesis model based on transmissible GI and clonal selection was proposed by Nowell (1976). More recently Loeb (1991, 2001) has presented evidence that an early step in carcinogenesis is mutation in a gene controlling genome stability. Stoler *et al.* (1999) showed that there are 11,000 mutations per carcinoma cell for a number of different cancer types, again implying that genomic destabilization is an early event in carcinogenesis. In particular, there is strong evidence of such an early genomic destabilization event for colon cancer (Loeb, 1991, 2001; Stoler *et al.*, 1999).

There have been a few attempts to incorporate GI in mechanistic carcinogenesis models (Mao *et al.*, 1998; Ohtaki and Niwa, 2001), although in general these models have not been fitted to data in a statistically rigorous manner. Little and Wright (2003) developed a stochastic carcinogenesis model which allowed for genome destabilization, very close in spirit to the model of Mao *et al.* (1998), and generalizing the class of generalized MVK models developed by Little (1995, 1996, 1997), which in turn therefore generalize the two-mutation model of Moolgavkar, Venzon and Knudson (Knudson, 1971; Moolgavkar and Venzon, 1979). Little and Wright (2003) fitted the model to Surveillance, Epidemiology and End Results (SEER) population-based Caucasian colon cancer incidence data (SEER, 2005).

The model assumes that cells can acquire two sorts of mutation, those associated with progression to a malignant phenotype ("cancer-stage" mutations), and those associated with successive destabilization of the genome ("destabilizing" mutations). With acquisition of successively more destabilizing mutations the cancer-stage mutation rates are generally higher, corresponding to the genome destabilization that is characteristic of GI.

Specifically, the model supposes that at age t there are $X(t)$ susceptible stem cells, each subject to mutation to a type of cell carrying an irreversible cancer-stage mutation at a rate of $M(0,0)(t)$. The cells in the stem cell compartment can also acquire a destabilizing mutation at a rate $A(0,0)(t)$. Thereafter the cells in compartment $I_{(i,j)}$ with i cancer-stage mutation and j destabilizing mutations divide into two such cells at a rate $G(i, j)(t)$; at a rate $D(i, j)(t)$ they die or differentiate. Each such cell can also divide into an equivalent daughter cell and another cell with an additional cancer-stage mutation at a rate $M(i, j)(t)$. In addition, each such cell can also divide into an equivalent daughter cell and another cell with an additional destabilizing mutation, at a rate $A(i, j)(t)$. There are assumed to be a total of k cancer-stage mutations required for a cell to become malignant. Likewise, there are assumed to be m destabilizing mutations. Once a cell has acquired all m such destabilizing mutations it is assumed to remain at the mth destabilizing mutation level. This model is illustrated schematically in Figure 8. The acquisition of carcinogenic (cancer-stage) mutations amounts to moving horizontally (left to right) in Figure 8, whereas acquisition of destabilizing mutations amounts to moving vertically (top to bottom) in this figure. The asymmetric cell divisions associated with most of the cancer-stage and destabilizing mutations (all except $(i, j) = (0, 0)$), in which each cell produces a daughter cell identical to the parent and another carrying an additional mutation, should be contrasted with the symmetric cell divisions associated with the cell proliferation processes (with rates $G(i, j)$), in which each cell produces two identical daughter cells. The two-mutation MVK model corresponds to the case $k = 2, m = 0$, while the generalized MVK model with K stages developed by Little (1995, 1996, 1997) amounts to the case $k = K, m = 0$. In fits to the SEER colon cancer data models with two cancer-stage mutations and one destabilizing mutation, with three cancer-stage mutations and one destabilizing mutation, and with five cancer-stage mutations and two destabilizing mutations all gave good fit (Little and Wright, 2003).

A recent paper by Nowak *et al.* (2002) proposed a formulation of a stochastic carcinogenesis model that incorporates genomic instability, again applied to (although not actually fitted to) colon cancer. In contrast, Luebeck

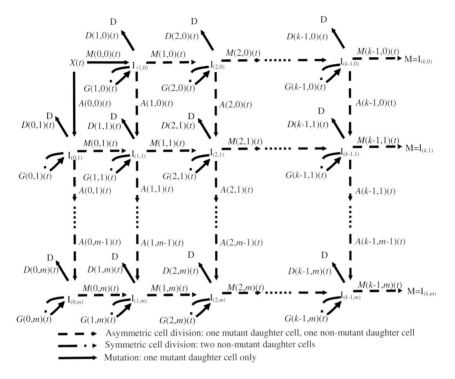

Fig. 8. Schematic diagram of the generalized MVK model with k cancer-stage mutations and m destabilizing mutations.

and Moolgavkar (2002) have recently proposed a four-stage stochastic model positing inactivation of the adenomatous polyposis coli (APC) gene followed by a high frequency event, possibly positional in nature, an extension of the two-stage clonal expansion model of Moolgavkar and Venzon (1979) and Knudson (1971); this model does not assume genomic instability. These models are illustrated schematically in Figures 9 and 10. The model of Little and Wright (2003) includes as special cases the models proposed by Luebeck and Moolgavkar (2002) and Nowak et al. (2002).

Little and Li (2007) have compared the fits of all five of these models (Little and Wright two cancer-stage mutations + one destabilizing mutation, Little and Wright three cancer-stage mutations + one destabilizing mutation, Little and Wright five cancer-stage mutations + two destabilizing mutations three-mutation, Luebeck and Moolgavkar; Nowak et al.) to a slight extension (by three more years of follow-up) of the colon cancer

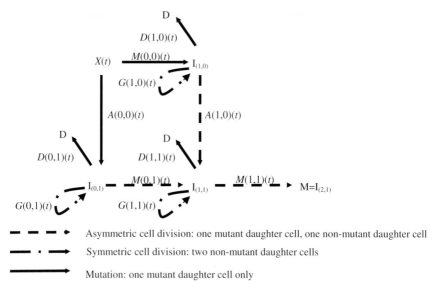

Fig. 9. Schematic diagram of the model of Nowak *et al.* (2002), similar to a generalized MVK model with two cancer-stage mutations and one destabilizing mutation.

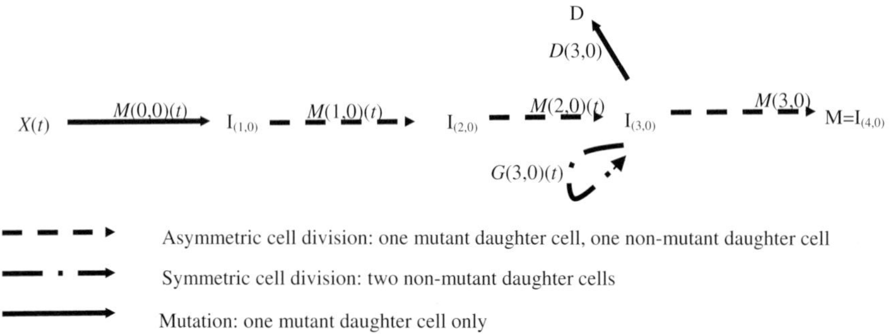

Fig. 10. Schematic diagram of the model of Luebeck and Moolgavkar (2002), similar to a generalized MVK model with four cancer-stage mutations and no destabilizing mutations.

data used by Little and Wright (2003). If the number of stem cells is fixed at a biologically plausible value, 10^8 cells (Potten *et al.*, 2003), the best fitting model is that of Nowak *et al.* (2002), with the two-stage model of Little and Wright (2003) and the four stage model of Luebeck and Moolgavkar (2002) not markedly inferior, as shown in Figures 11 and 12. The fit of the three-stage model of Little and Wright (2003) is somewhat worse than these

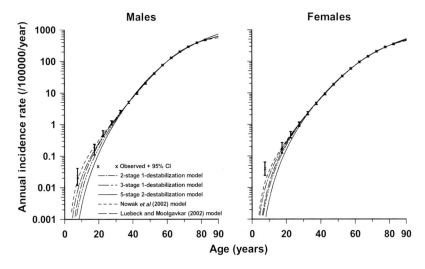

Fig. 11. Cancer hazards predicted by models of Nowak *et al.* (2002) (with two cancer-stage mutations and one destabilizing mutation), of Luebeck and Moolgavkar (2002) (with four cancer-stage mutations and no destabilizing mutations), and of Little and Wright (2003) (with two cancer-stage mutations and one destabilizing mutation, three cancer-stage mutations and one destabilizing mutation, five cancer-stage mutations and two destabilizing mutations), with stem cell population fixed to 10^8 cells, fitted to SEER (2005) colon cancer data, and observed data (with 95% confidence intervals (CI), adjusted for overdispersion (McCullagh and Nelder 1989)) (taken from Little and Li (2007)).

two, even more so that of the five stage model of Little and Wright (2003), as shown in Figures 11 and 12. Comparison of the predictions of the two-stage models of Little and Wright (2003) and Nowak *et al.* (2002), given in Figures 13 and 14, with patterns of excess risk in the Japanese atomic bomb survivor colon cancer incidence data, shown in Figure 15, indicate that radiation might act on cell proliferation rates in the model, and at least for the model of Little and Wright also on one of the parameters governing progression to genomic destabilization.

6. DISCUSSION AND CONCLUSIONS

We have seen that the classical multi-stage model of Armitage and doll and the two-mutation model of Moolgavkar, Venzon, and Knudson, and various generalizations of them also, are capable of describing, at least qualitatively,

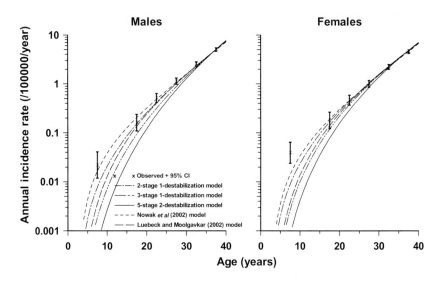

Fig. 12. As for Figure 11, but showing cancer rates and model fits up to age 40 (taken from Little and Li, 2007).

many of the observed patterns of excess cancer risk following ionizing radiation exposure. However, there are certain inconsistencies with the biological and epidemiological data for both the multi-stage and two-mutation models. In particular, there are indications that the two-mutation model is not totally suitable for describing the pattern of excess risk for solid cancers that is often seen after exposure to ionizing radiation, although leukemia may be better fitted by this type of model. Generalized MVK models which require three or more mutations, in particular ones with multiple pathways associated with genomic destabilization, are easier to reconcile with biological and epidemiological data relating to solid cancers.

A common assumption of most carcinogenesis models is that cells are statistically conditionally independent (conditional on the parental lineage and exogenous exposures), so that the cell populations may be described by a branching process. This is assumed for analytic tractability, but it is difficult to test. To the extent that it is known that cells communicate with each other via cell surface markers and otherwise, it is unlikely to be precisely true. One tissue in which, because of its spatial structure, this assumption may break down is the colon. The colon and small intestine are structured into crypts, each crypt containing some thousands of cells, and organized so that the

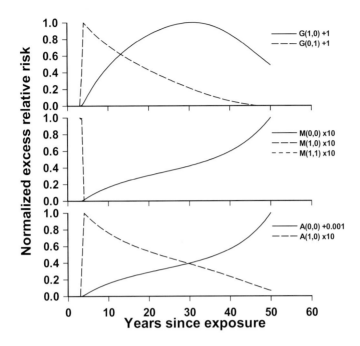

Fig. 13. Normalized excess relative hazard following perturbations of $G(i, j)$, $M(i, j)$ and $A(i, j)$ in generalized MVK model with two cancer-stage mutations and one destabilizing mutation of Little and Wright (2003), with stem cell population fixed to 10^8 cells, fitted to male SEER colon cancer incidence data (models as fitted in Little and Li (2007)). In the absence of perturbation the model is as described in Figures 8 and 11. A 3-year period of latency is assumed between the appearance of the first carcinogenic cell and clinically overt cancer. The parameters $G(i, j)$ are increased by 1 y^{-1} at the age of 25, for 1 year, the parameters $M(i, j)$ and $A(i, j)$ are generally multiplied by 10 at the age of 25 ($A(0, 0)$ is augmented by $10^{-3} y^{-1}$), for 1 year.

stem cells are at the bottom of the crypt (Potten and Loeffler, 1987; Nowak et al., 2003). There is evidence that there may be more than one stem cell at the bottom of each crypt (Bach et al., 2000). The progeny of stem cells migrate up the crypt and continue to divide, becoming progressively more differentiated. The differentiated cells eventually reach the top of the crypt where they are shed into the intestinal lumen. Potten and Loeffler (1987) and Nowak and colleagues (Nowak et al., 2003; Michor et al., 2004) have postulated similar models for cancers of the small intestine and colon taking account of the linear structure of the crypts, and in which necessarily the assumption of conditional independence breaks down.

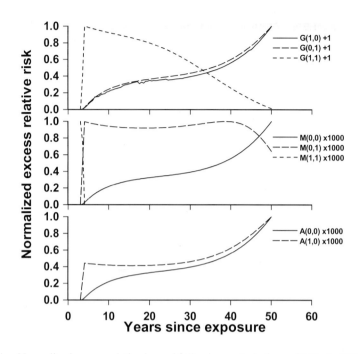

Fig. 14. Normalized excess relative hazard following perturbations of $G(i, j)$, $M(i, j)$ and $A(i, j)$ in generalized MVK model with two cancer-stage mutations and one destabilizing mutation of Nowak *et al.* (2002), with stem cell population fixed to 10^8 cells, fitted to male SEER colon cancer incidence data (models as fitted in Little and Li (2007)). In the absence of perturbation the model is as described in Figures 9 and 11. A 3-year period of latency is assumed between the appearance of the first carcinogenic cell and clinically overt cancer. The parameters $G(i, j)$ are increased by 1 y^{-1} at the age of 25, for 1 year, the parameters $M(i, j)$ and $A(i, j)$ are multiplied by 1000 at the age of 25, for 1 year.

There are known to be at least two distinct phenotypes underlying GI. Chromosomal instability (CIN) is the dominating phenotype of cancer cells, and is characterized by numerical and structural aberrations of the genome. Cell cycle checkpoint genes such as *TP53* (Livingstone *et al.*, 1992), oncogenes such as *ras* and *myc* (Denko *et al.*, 1994; Felsher and Bishop, 1999) and mitotic-spindle checkpoint genes such as *hBUB1* (Cahill *et al.*, 1998) have all been implicated in generating this phenotype. The phenotype of microsatellite instability (MIN) on the other hand, is characterized by a profusion of short DNA sequences, called microsatellites, scattered throughout the genome. It is thought to be caused by defects in DNA mismatch repair genes such as *MSH2* or *MLH1* (Peltomaki and de la Chapelle, 1997). There

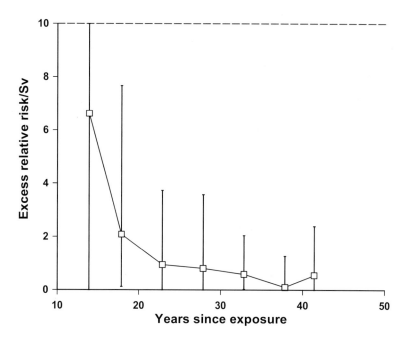

Fig. 15. Excess relative risk (per Sv) and 95% confidence intervals for male colon cancer incidence as a function of years since exposure in the Japanese atomic bomb survivor data of Thompson et al. (1994) (taken from Little and Wright (2003)).

is evidence that the GI pathway, CIN or MIN, a cell goes down may be determined by the selective pressures exerted by specific carcinogens (Bardelli et al., 2001). The GI model devised by Little and Wright (2003) can incorporate such a multiplicity of GI pathways, and indeed this model formally embraces many of the other multiple pathway models that have been proposed (Tan, 1991; Little et al., 1995), for example by setting certain mutation rates to 0 and others to very large values.

It has been generally accepted that most biological damage produced by ionizing radiation occurs when radiation interacts directly with DNA in the cell nucleus or indirectly through the action of free radicals (UNSCEAR, 2000). However, in the last ten or so years there have been a number of reports of cells exposed experimentally to α-particle radiation in which more cells showed damage than were traversed by α particles (Nagasawa and Little, 1992, 1999; Azzam et al., 1998, 2000; Belyakov et al., 2001; Huo et al., 2001; Sawant et al., 2001; Zhou et al., 2001; Little et al., 2003), i.e. a *bystander effect*. Similar effects have been observed in cells exposed to culture medium from cells irradiated *in vitro* with a variety of types of ionizing radiation (Mothersill and Seymour, 2001, 2002). This is observed for a number of end points, including cell killing, micronucleus induction, and mutation induction, as recently reviewed by Iyer and Lehnert (2000) and Morgan (2003a, 2003b). It is also clear, from the use of microbeam approaches, that direct DNA damage from energy deposition is not required to trigger the effect (Shao *et al.*, 2004). There is evidence that the phenomena of the bystander effect and transmissible GI may be linked. In particular, the work of Lorimore *et al.* (1998) suggests that chromosomal instability may develop in unirradiated haemopoietic stem cells in proximity to similar cells that are irradiated with α particles. If this is confirmed in other systems, and in particular *in vivo*, it has important implications for future models of the bystander effect and carcinogenesis, implying the need to incorporate information on the spatial location of individual cells in any model. A recently developed model of the bystander effect that incorporates information on spatial location may indicate the way forward in this respect (Little *et al.*, 2005).

ACKNOWLEDGMENTS

This work has been partially funded by the European Commission under contract FI6R-CT-2003-508842 (RISC-RAD).

References

1. Abramowitz M and Stegun IA. *Handbook of Mathematical Functions*. National Bureau of Standards, Washington, DC, 1964.
2. Armitage P and Doll R. The age distribution of cancer and a multi-stage theory of carcinogenesis. *Br. J. Cancer* 1954; **8**: 1–12.
3. Armitage P and Doll R. A two-stage theory of carcinogenesis in relation to the age distribution of human cancer. *Br. J. Cancer* 1957; **9**: 161–169.
4. Azzam EI, de Toledo SM, Gooding T and Little JB. Intercellular communication is involved in the bystander regulation of gene expression in human cells exposed to very low fluences of alpha particles. *Radiat. Res.* 1998; **150**: 497–504.

5. Azzam EI, de Toledo SM, Waker AJ and Little JB. High and low fluences of α-particles induce a G_1 checkpoint in human diploid fibroblasts. *Cancer Res.* 2000; **60**: 2623–2631.
6. Bach SP, Renehan AG and Potten CS. Stem cells: the intestinal stem cell as a paradigm. *Carcinogenesis* 2000; **21**: 469–476.
7. Bardelli A, Cahill DP, Lederer G, Speicher MR, Kinzler KW, Vogelstein B and Lengauer C. Carcinogen-specific induction of genetic instability. *Proc. Natl. Acad. Sci. USA* 2001; **98**: 5770–5775.
8. Bartoszynski R, Edler L, Hanin L, Kopp-Schneider A, Pavlova L, Tsodikov A, Zorin A and Yakovlev AYu. Modeling cancer detection: tumor size as a source of information on unobservable stages of carcinogenesis. *Math. Biosci.* 2001; **171**: 113–142.
9. Belyakov OV, Malcolmson AM, Folkard M, Prise KM and Michael BD. Direct evidence for a bystander effect of ionizing radiation in primary human fibroblasts. *Br. J. Cancer* 2001; **84**: 674–679.
10. Boice JD, Jr., Day NE, Andersen A, Brinton LA, Brown R, Choi NW, Clarke EA, Coleman MP, Curtis RE, Flannery JT, Hakama M, Hakulinen T, Howe GR, Jensen OM, Kleinerman RA, Magnin D, Magnus K, Makela K, Malker B, Miller AB, Nelson N, Patterson CC, Pettersson F, Pompe-Kirn V, Primic-Žakelj M, Prior P, Ravnihar B, Skeet RG, Skjerven JE, Smith PG, Sok M, Spengler RF, Storm HH, Stovall M, Tomkins GWO and Wall C. Second cancers following radiation treatment for cervical cancer. An international collaboration among cancer registries. *J. Natl. Cancer Inst.* 1985; **74**: 955–975.
11. Brown CC and Chu CC. A new method for the analysis of cohort studies: implications of the multistage theory of carcinogenesis applied to occupational arsenic exposure. *Environ. Health Perspect.* 1983; **50**: 293–308.
12. Brown CC and Chu CC. Use of multistage models to infer stage affected by carcinogenic exposure: example of lung cancer and cigarette smoking. *J. Chron. Dis.* 1987; **40**: 171S–179S.
13. Cahill DP, Lengauer C, Yu J, Riggins GJ, Willson JKV, Markowitz SD, Kinzler KW and Vogelstein B. Mutations of mitotic checkpoint genes in human cancers. *Nature* 1998; **392**: 300–303.
14. Crump KS and Howe RB. The multistage model with a time-dependent dose pattern: applications to carcinogenic risk assessment. *Risk Anal.* 1984; **4**: 163–176.
15. Darby SC, Doll R, Gill SK and Smith PG. Long term mortality after a single treatment course with X-rays in patients treated for ankylosing spondylitis. *Br. J. Cancer* 1987; **55**: 179–190.

16. Day NE and Brown CC. Multistage models and primary prevention of cancer. *J. Natl. Cancer Inst.* 1980; **64**: 977–989.
17. Den Otter W, Koten JW, van der Vegt BJH, Beemer FA, Boxma OJ, Derkinderen DJ, de Graaf PW, Huber J, Lips CJM, Roholl PJM, Sluijter FJH, Tan KEWP, van der Heyden K, van der Ven L and van Unnik JAM. Oncogenesis by mutations in anti-oncogenes: a view. *Anticancer Res.* 1990; **10**: 475–488.
18. Denko NC, Giaccia AJ, Stringer JR and Stambrook PJ. The human Ha-*ras* oncogene induces genomic instability in murine fibroblasts within one cell cycle. *Proc. Natl. Acad. Sci. USA* 1994; **91**: 5124–5128.
19. Derkinderen DJ, Boxma OJ, Koten JW and Den Otter W. Stochastic theory of oncogenesis. *Anticancer Res.* 1990; **10**: 497–504.
20. Doll R. The age distribution of cancer: implications for models of carcinogenesis. *J. R. Stat. Soc. Ser. A* 1971; **132**: 133–166.
21. Fearon ER and Vogelstein B. A genetic model for colorectal tumorigenesis. *Cell* 1990; **61**: 759–767.
22. Felsher DW and Bishop JM. Transient excess of *MYC* activity can elicit genomic instability and tumorigenesis. *Proc. Natl. Acad. Sci. USA* 1999; **96**: 3940–3944.
23. Feng D and DiStefano JJ, III. An algorithm for identifiable parameters and parameter bounds for a class of cascaded mammillary models. *Math. Biosci.* 1995; **129**: 67–93.
24. Freedman DA and Navidi WC. Multistage models for carcinogenesis. *Environ. Health Perspect.* 1989; **81**: 169–188.
25. Hanin LG and Yakovlev AYu. A nonidentifiability aspect of the two-stage model of carcinogenesis. *Risk Anal.* 1996; **16**: 711–715.
26. Hanin LG. Identification problem for stochastic models with application to carcinogenesis, cancer detection and radiation biology. *Disc. Dynam. Nat. Soc.* 2002; **7**: 177–189.
27. Hazelton WD, Luebeck EG, Heidenreich WF and Moolgavkar SH. Analysis of a historical cohort of Chinese tin miners with arsenic, radon, cigarette smoke, and pipe smoke exposures using the biologically based two-stage clonal expansion model. *Radiat. Res.* 2001; **156**: 78–94.
28. Heidenreich WF. On the parameters of the clonal expansion model. *Radiat. Environ. Biophys.* 1996; **35**: 127–129.
29. Heidenreich WF, Luebeck EG and Moolgavkar SH. Some properties of the hazard function of the two-mutation clonal expansion model. *Risk Anal.* 1997; **17**: 391–399.

30. Heidenreich WF, Jacob P, Paretzke HG, Cross FT and Dagle GE. Two-step model for the risk of fatal and incidental lung tumors in rats exposed to radon. *Radiat. Res.* 1999; **151**: 209–217.
31. Heidenreich WF, Brugmans MJP, Little MP, Leenhouts HP, Paretzke HG, Morin M and Lafuma J. Analysis of lung tumour risk in radon-exposed rats: an intercomparison of multi-step modeling. *Radiat. Environ. Biophys.* 2000; **39**: 253–264.
32. Heidenreich WF, Luebeck EG, Hazelton WD, Paretzke HG and Moolgavkar SH. Multistage models and the incidence of cancer in the cohort of atomic bomb survivors. *Radiat. Res.* 2002; **158**: 607–614.
33. Heidenreich WF, Nyberg U and Hall P. A biologically based model for liver cancer risk in the Swedish Thorotrast patients. *Radiat. Res.* 2003; **159**: 656–662.
34. Heidenreich WF, Luebeck EG and Moolgavkar SH. Effects of exposure uncertainties in the TSCE model and application to the Colorado miners data. *Radiat. Res.* 2004; **161**: 72–81.
35. Huo L, Nagasawa H and Little JB. HPRT mutants induced in bystander cell by very low fluences of alpha particles result primarily from point mutations. *Radiat. Res.* 2001; **156**: 521–525.
36. International Commission on Radiological Protection (ICRP). *1990 Recommendations of the International Commission on Radiological Protection*, Annals of the ICRP 21 (1–3). Pergamon Press, Oxford, 1991.
37. Iyer R and Lehnert BE. Effects of ionizing radiation in targeted and nontargeted cells. *Arch. Biochem. Biophys.* 2000; **376**: 14–25.
38. Jacquez JA and Perry T. Parameter estimation: local identifiability of parameters. *Am. J. Physiol.* 1990; **258**: E727–E736.
39. Kadhim MA, Lorimore SA, Hepburn MD, Goodhead DT, Buckle VJ and Wright EG. α-particle-induced chromosomal instability in human bone marrow cells. *Lancet* 1994; **344**: 987–988.
40. Kadhim MA, Macdonald DA, Goodhead DT, Lorimore SA, Marsden SJ and Wright EG. Transmission of chromosomal instability after plutonium α-particle irradiation. *Nature* 1992; **355**: 738–740.
41. Kai M, Luebeck EG and Moolgavkar SH. Analysis of the incidence of solid cancer among atomic bomb survivors using a two-stage model of carcinogenesis. *Radiat. Res.* 1997; **148**: 348–358.
42. Knudson AG, Jr. Mutation and cancer: statistical study of retinoblastoma. *Proc. Natl. Acad. Sci. USA* 1971; **68**: 820–823.

43. Leenhouts HP. Radon-induced lung cancer in smokers and non-smokers: risk implications using a two-mutation carcinogenesis model. *Radiat. Environ. Biophys.* 1999; **38**: 57–71.
44. Leenhouts HP and Chadwick KH. A two-mutation model of radiation carcinogenesis: application to lung tumours in rodents and implications for risk evaluation. *J. Radiol. Prot.* 1994; **14**: 115–130.
45. Little JB, Nagasawa H, Li GC and Chen DJ. Involvement of the nonhomologous end joining DNA repair pathway in the bystander effect for chromosomal aberrations. *Radiat. Res.* 2003; **159**: 262–267.
46. Little MP. Risks of radiation-induced cancer at high doses and dose rates. *J. Radiol. Prot.* 1993; **13**: 3–25.
47. Little MP. Are two mutations sufficient to cause cancer? Some generalizations of the two-mutation model of carcinogenesis of Moolgavkar, Venzon, and Knudson, and of the multistage model of Armitage and Doll. *Biometrics* 1995; **51**: 1278–1291.
48. Little MP. Generalisations of the two-mutation and classical multi-stage models of carcinogenesis fitted to the Japanese atomic bomb survivor data. *J. Radiol. Prot.* 1996; **16**: 7–24.
49. Little MP. Are two mutations sufficient to cause cancer? Modelling radiation-induced cancer in the Japanese atomic bomb survivors using generalizations of the two-mutation model of Moolgavkar, Venzon and Knudson and of the multi-stage model of Armitage and Doll. In: *Health Effects of Low Dose Radiation: Challenges of the 21st Century*, Proceedings of the conference organized by the British Nuclear Energy Society and held in Stratford-upon-Avon, UK, on 11–14 May 1997. British Nuclear Energy Society, London, 1997, pp. 169–174.
50. Little MP. A comparison of the degree of curvature in the cancer incidence dose-response in Japanese atomic bomb survivors with that in chromosome aberrations measured *in vitro*. *Int. J. Radiat. Biol.* 2000; **76**: 1365–1375.
51. Little MP and Li G. Stochastic modelling of colon cancer: is there a role for genomic instability? *Carcinogenesis* 2007; **28**: 479–487.
52. Little MP and Boice JD, Jr. Comparison of breast cancer incidence in the Massachusetts tuberculosis fluoroscopy cohort and in the Japanese atomic bomb survivors. *Radiat. Res.* 1999; **151**: 218–224.
53. Little MP and Charles MW. Time variations in radiation-induced relative risk and implications for population cancer risks. *J. Radiol. Prot.* 1991; **11**: 91–110.

54. Little MP and Charles MW. The risk of non-melanoma skin cancer incidence in the Japanese atomic bomb survivors. *Int. J. Radiat. Biol.* 1997; **71**: 589–602.
55. Little MP, Li G and Heidenreich WF. Parameter identifiability in a general class of stochastic carcinogenesis models. *Biostatistics* 2008; submitted.
56. Little MP and Muirhead CR., Evidence for curvilinearity in the cancer incidence dose-response in the Japanese atomic bomb survivors. *Int. J. Radiat. Biol.* 1996; **70**: 83–94.
57. Little MP and Muirhead CR. Curvature in the cancer mortality dose response in Japanese atomic bomb survivors: absence of evidence of threshold. *Int. J. Radiat. Biol.* 1998; **74**: 471–480.
58. Little MP and Muirhead CR. Derivation of low-dose extrapolation factors from analysis of curvature in the cancer incidence dose response in Japanese atomic bomb survivors. *Int. J. Radiat. Biol.* 2000; **76**: 939–953.
59. Little MP and Wright EG. A stochastic carcinogenesis model incorporating genomic instability fitted to colon cancer data. *Math. Biosci.* 2003; **183**: 111–134.
60. Little MP, Hawkins MM, Shore RE, Charles MW and Hildreth NG. Time variations in the risk of cancer following irradiation in childhood. *Radiat. Res.* 1991; **126**: 304–316.
61. Little MP, Hawkins MM, Charles MW and Hildreth NG. Fitting the Armitage-Doll model to radiation-exposed cohorts and implications for population cancer risks. *Radiat. Res.* 1992; **132**: 207–221.
62. Little MP, Hawkins MM, Charles MW and Hildreth NG. Corrections to the paper "Fitting the Armitage-Doll model to radiation-exposed cohorts and implications for population cancer risks" (letter). *Radiat. Res.* 1994; **137**: 124–128.
63. Little MP, Muirhead CR, Boice JD, Jr and Kleinerman RA. Using multistage models to describe radiation-induced leukemia. *J. Radiol. Prot.* 1995; **15**: 315–334.
64. Little MP, Muirhead CR and Stiller CA. Modelling lymphocytic leukemia incidence in England and Wales using generalisations of the two-mutation model of carcinogenesis of Moolgavkar, Venzon and Knudson. *Stat. Med.* 1996; **15**: 1003–1022.
65. Little MP, de Vathaire F, Charles MW, Hawkins MM and Muirhead CR. Variations with time and age in the relative risks of solid cancer incidence after radiation exposure. *J. Radiol. Prot.* 1997a; **17**: 159–177.
66. Little MP, Muirhead CR and Stiller CA. Modelling acute lymphocytic leukemia using generalizations of the MVK two-mutation model of

carcinogenesis: implied mutation rates and the likely role of ionising radiation. In: *Microdosimetry: An Interdisciplinary Approach.* Goodhead DT, O'Neill P and Menzel HG (eds.) Royal Society of Chemistry, Cambridge, 1997b, pp. 244–247.
67. Little MP, Muirhead CR and Charles MW. Describing time and age variations in the risk of radiation-induced solid tumour incidence in the Japanese atomic bomb survivors using generalized relative and absolute risk models. *Stat. Med.* 1999; **18**: 17–33.
68. Little MP, Haylock RGE and Muirhead CR. Modelling lung tumour risk in radon-exposed uranium miners using generalizations of the two-mutation model of Moolgavkar, Venzon and Knudson. *Int. J. Radiat. Biol.* 2002; **78**: 49–68.
69. Little MP, Filipe JAN, Prise KM, Folkard M and Belyakov OV. A model for radiation-induced bystander effects, with allowance for spatial position and the effects of cell turnover. *J. Theor. Biol.* 2005; **232**: 329–338.
70. Livingstone LR, White A, Sprouse J, Livanos E, Jacks T and Tlsty TD. Altered cell cycle arrest and gene amplification potential accompany loss of wild-type p53. *Cell* 1992; **70**: 923–935.
71. Lloyd DC and Edwards AA. Chromosome aberrations in human lymphocytes: effect of radiation quality, dose, and dose rate. In: *Radiation-Induced Chromosome Damage in Man.* Ishihara T and Sasaki MS (eds.) Alan Liss, New York, 1983, pp. 23–49.
72. Loeb LA. Mutator phenotype may be required for multistage carcinogenesis. *Cancer Res.* 1991; **51**: 3075–3079.
73. Loeb LA. A mutator phenotype in cancer. *Cancer Res.* 2001; **61**: 3230–3239.
74. Lorimore SA, Kadhim MA, Pocock DA, Papworth D, Stevens DL, Goodhead DT and Wright EG. Chromosomal instability in the descendants of unirradiated surviving cells after α-particle irradiation. *Proc. Natl. Acad. Sci. USA* 1998; **95**: 5730–5733.
75. Luebeck EG, Curtis SB, Cross FT and Moolgavkar SH. Two-stage model of radon-induced malignant lung tumors in rats: effects of cell killing. *Radiat. Res.* 1996; **145**: 163–173.
76. Luebeck EG, Heidenreich WF, Hazelton WD, Paretzke HG and Moolgavkar SH. Biologically based analysis of the data for the Colorado uranium miners cohort: age, dose and dose-rate effects. *Radiat. Res.* 1999; **152**: 339–351.
77. Luebeck EG and Moolgavkar SH. Multistage carcinogenesis and the incidence of colon cancer. *Proc. Natl. Acad. Sci. USA* 2002; **99**: 15095–15100.
78. Mao JH, Lindsay KA, Balmain A and Wheldon TE. Stochastic modelling of tumorigenesis in p53 deficient mice. *Br. J. Cancer* 1998; **77**: 243–252.

79. Mazumdar S, Redmond CK, Enterline PE, Marsh GM, Constantino JP, Zhou SJY and Patwardhan RN. Multistage modeling of lung cancer mortality among arsenic-exposed copper-smelter workers. *Risk Anal.* 1989; **9**: 551–563.
80. McCullagh P and Nelder JA. *Generalized Linear Models*, 2nd ed. Chapman and Hall, London, 1989.
81. Michor F, Isawa Y, Rajagopalan H, Lengauer C and Nowak MA. Linear model of colon cancer initiation. *Cell Cycle* 2004; **3**: 358–362.
82. Moolgavkar SH. The multistage theory of carcinogenesis and the age distribution of cancer in man. *J. Natl. Cancer Inst.* 1978; **61**: 49–52.
83. Moolgavkar SH and Knudson AG, Jr. Mutation and cancer: a model for human carcinogenesis. *J. Natl. Cancer Inst.* 1981; **66**: 1037–1052.
84. Moolgavkar SH and Luebeck EG. Multistage carcinogenesis: population-based model for colon cancer. *J. Natl. Cancer Inst.* 1992; **84**: 610–618.
85. Moolgavkar SH and Venzon DJ. Two-event models for carcinogenesis: incidence curves for childhood and adult tumors. *Math. Biosci.* 1979; **47**: 55–77.
86. Moolgavkar SH, Cross FT, Luebeck G and Dagle GE. A two-mutation model for radon-induced lung tumors in rats. *Radiat. Res.* 1990; **121**: 28–37.
87. Moolgavkar SH, Dewanji A and Venzon DJ. A stochastic two-stage model for cancer risk assessment. I. The hazard function and the probability of tumor. *Risk Anal.* 1988; **8**: 383–392.
88. Moolgavkar SH, Luebeck EG, Krewski D and Zielinski JM. Radon, cigarette smoke, and lung cancer: a re-analysis of the Colorado plateau uranium miners' data. *Epidemiology* 1993; **4**: 204–217.
89. Morgan WF. Non-targeted and delayed effects of exposure to ionizing radiation: I. Radiation-induced genomic instability and bystander effects *in vitro*. *Radiat. Res.* 2003a; **159**: 567–580.
90. Morgan WF. Non-targeted and delayed effects of exposure to ionizing radiation: II. Radiation-induced genomic instability and bystander effects *in vivo*, clastogenic factors and transgenerational effects. *Radiat. Res.* 2003b; **159**: 581–596.
91. Mothersill C and Seymour C. Radiation-induced bystander effects: past history and future directions. *Radiat. Res.* 2001; **155**: 759–767.
92. Mothersill C and Seymour CB. Bystander and delayed effects after fractionated radiation exposure. *Radiat. Res.* 2002; **158**: 626–633.
93. Muirhead CR and Darby SC. Modelling the relative and absolute risks of radiation-induced cancers. *J. R. Stat. Soc. Ser. A* 1987; **150**: 83–118.

94. Nagasawa H and Little JB. Induction of sister chromatid exchanges by extremely low doses of α-particles. *Cancer Res.* 1992; **52**: 6394–6396.
95. Nagasawa H and Little JB. Unexpected sensitivity to the induction of mutations by very low doses of alpha-particle radiation: evidence for a bystander effect. *Radiat. Res.* 1999; **152**: 552–557.
96. National Research Council. *Health Effects of Exposure to Low Levels of Ionizing Radiation* (BEIR V). National Academy Press, Washington, DC, 1990.
97. Nowak MA, Komarova NL, Sengupta A, Jallepalli PV, Shih I-M, Vogelstein B and Lengauer C. The role of chromosomal instability in tumor initiation. *Proc. Natl. Acad. Sci. USA* 2002; **99**: 16226–16231.
98. Nowak MA, Michor F and Isawa Y. The linear process of somatic evolution. *Proc. Natl. Acad. Sci. USA* 2003; **100**: 14966–14969.
99. Nowell PC. The clonal evolution of tumor cell populations. *Science* 1976; **194**: 23–28.
100. Ohtaki M and Niwa O. A mathematical model of radiation carcinogenesis with induction of genomic instability and cell death. *Radiat. Res.* 2001; **156**: 672–677.
101. Peltomaki P and de la Chapelle A. Mutations predisposing to hereditary nonpolyposis colorectal cancer. *Adv. Cancer Res.* 1997; **71**: 93–119.
102. Peto R. Epidemiology, multistage models, and short-term mutagenicity tests. In: *Origins of Human Cancer*. Hiatt HH and Winsten JA (eds.) Cold Spring Harbor Laboratory, Cold Spring Harbor, 1977, pp. 1403–1428.
103. Pierce DA and Mendelsohn ML. A model for radiation-related cancer suggested by atomic bomb survivor data. *Radiat. Res.* 1999; **152**: 642–654.
104. Pierce DA and Vaeth M. The shape of the cancer mortality dose-response curve for the A-bomb survivors. *Radiat. Res.* 1991; **126**: 36–42.
105. Pierce DA, Shimizu Y, Preston DL, Vaeth M and Mabuchi K. Studies of the mortality of atomic bomb survivors. Report 12, part I. Cancer: 1950–1990. *Radiat. Res.* 1996; **146**: 1–27.
106. Potten CS and Loeffler M. A comprehensive model of the crypts of the small intestine of the mouse provides insight into the mechanisms of cell migration and the proliferation hierarchy. *J. Theor. Biol.* 1987; **127**: 381–391.
107. Potten CS, Booth C and Hargreaves D. The small intestine as a model for evaluating adult tissue stem cell drug targets. *Cell Prolif.* 2003; **36**: 115–129.
108. Preston DL, Kusumi S, Tomonaga M, Izumi S, Ron E, Kuramoto A, Kamada N, Dohy H, Matsui T, Nonaka H, Thompson DE, Soda M and Mabuchi K. Cancer incidence in atomic bomb survivors. Part III: leukemia, lymphoma and multiple myeloma, 1950–1987. *Radiat. Res.* 1994; **137**: S68–S97.

109. Preston DL, Mattsson A, Holmberg E, Shore R, Hildreth NG and Boice JD, Jr. Radiation effects on breast cancer risk: a pooled analysis of eight cohorts. *Radiat. Res.* 2002; **158**: 220–235.
110. Preston DL, Shimizu Y, Pierce DA, Suyama A and Mabuchi K. Studies of mortality of atomic bomb survivors. Report 13: solid cancer and noncancer disease mortality: 1950–1997. *Radiat. Res.* 2003; **160**: 381–407.
111. Sawant SG, Randers-Pehrson G, Geard CR, Brenner DJ and Hall EJ. The bystander effect in radiation oncogenesis: I. Transformation in C3H 10T$^{1/2}$ cells *in vitro* can be initiated in the unirradiated neighbors of irradiated cells. *Radiat. Res.* 2001; **155**: 397–401.
112. Shao C, Folkard M, Michael BD and Prise KM. Targeted cytoplasmic irradiation induces bystander responses. *Proc. Natl. Acad. Sci. USA* 2004; **101**: 13495–13500.
113. Stoler DL, Chen N, Basik M, Kahlenberg MS, Rodriguez-Bigas MA, Petrelli NJ and Anderson GR. The onset and extent of genomic instability in sporadic colorectal tumor progression. *Proc. Natl. Acad. Sci. USA* 1999; **96**: 15121–15126.
114. Surveillance, Epidemiology and End Results (SEER) Registry Public-use data, 1973–2002, Cancer Statistics Branch, Surveillance Research Program, Division of Cancer Control and Population Sciences, National Cancer Institute, Bethesda, MD, USA (http://seer.cancer.gov/), 2005.
115. Tan W-Y. *Stochastic Models of Carcinogenesis*. Marcel Dekker, New York, 1991.
116. Thomas DC. Statistical methods for analyzing effects of temporal patterns of exposure on cancer risks. *Scand. J. Work Environ. Health* 1983; **9**: 353–366.
117. Thomas DC. Statistical methods for analyzing effects of temporal patterns of exposure on cancer risks. *Scand. J. Work Environ. Health* 1983; **9**: 353–366.
118. Thomas DC. A model for dose rate and duration of exposure effects in radiation carcinogenesis. *Environ. Health Perspect.* 1990; **87**:163–171.
119. Thompson DE, Mabuchi K, Ron E, Soda M, Tokunaga M, Ochikubo S, Sugimoto S, Ikeda T, Terasaki M, Izumi S and Preston DL. Cancer incidence in atomic bomb survivors. Part II: solid tumors, 1958–1987. *Radiat. Res.* 1994; **137**: S17–S67.
120. Tucker HG. A stochastic model for a two-stage theory of carcinogenesis. In: *Fifth Berkeley Symposium on Mathematical Statistics and Probability*. University of California Press, Berkeley, 1967, pp. 387–403.
121. United Nations Scientific Committee on the Effects of Atomic Radiation (UNSCEAR). *Sources, Effects and Risks of Ionizing Radiation*. United Nations, New York, 1988.

122. United Nations Scientific Committee on the Effects of Atomic Radiation (UNSCEAR). *Sources and Effects of Ionizing Radiation.* United Nations, New York, 1993.
123. United Nations Scientific Committee on the Effects of Atomic Radiation (UNSCEAR). *Sources and Effects of Ionizing Radiation.* United Nations, New York, 1994.
124. United Nations Scientific Committee on the Effects of Atomic Radiation (UNSCEAR). *Sources and Effects of Ionizing Radiation. Volume II: Effects.* United Nations, New York, 2000.
125. Weiss HA, Darby SC and Doll R. Cancer mortality following X-ray treatment for ankylosing spondylitis. *Int. J. Cancer* 1994; **59**: 327–338.
126. Zhou H, Suzuki M, Randers-Pehrson G, Vannais D, Chen G, Trosko JE, Waldren CA and Hei TK. Radiation risk to low fluences of α particles may be greater than we thought. *Proc. Natl. Acad. Sci. USA* 2001; **98**: 14410–14415.

Chapter 6

DISTRIBUTION OF THE SIZES OF METASTASES: MATHEMATICAL AND BIOMEDICAL CONSIDERATIONS

Leonid Hanin

The chapter deals with mechanistic biologically motivated modeling of metastasis. A general methodology for such a modeling is outlined, and a comprehensive mathematical model of individual natural history of metastatic cancer allowing for interaction between the primary tumor and metastases is formulated. This model is applied to computing the distribution of the sizes of detectable (or all) metastases in a given host site at any time post-diagnosis. A parametric version of the model for exponentially growing primary and secondary tumors and exponentially distributed metastasis promotion times is fully developed, and identifiability properties of this model are established.

Keywords: Cancer natural history, metastasis, model identifiability, parameter estimation, Poisson process, primary tumor.

1. INTRODUCTION

It has been long surmised that the main reason for the limited success with which treatment of many types of invasive cancer had met so far is the presence of occult (undetectable) metastases at the time of initial diagnosis or start of treatment for the primary disease (see e.g. Paget, 1889 and Douglas, 1971). For example, it has been estimated (Barbour and Gotley, 2003) that more than 70% of cancer patients have metastases at the time of presentation. This suggests that metastatic invasion and dissemination is a very early event in the natural history of cancer. If confirmed, this hypothesis would mean that improvement in the survival of patients afflicted with advanced cancer requires oncologists to start anti-metastatic therapy concurrently with, or shortly after, the treatment of the primary disease. Availability of modern methods of metastasis ablation such as conformal stereotactic hypofractionated radiosurgery (Schell *et al.*, 1995), radioimmunotherapy (Bernhardt *et al.*, 2001; Goddu *et al.*, 1994; O'Donohue, 2000) and treatment with oncolytic viruses (Harrison *et al.*, 2001; Kaplan, 2005; Kasuya *et al.*, 2005; Kaufman *et al.*, 2005; Kirn *et al.*, 1998; Kirn and McCormick, 1996; Khuri *et al.*, 2000; McCormick, 2003; Nemunaitis *et al.*, 2001; Parato *et al.*, 2005; Reid *et al.*, 2002; Shah *et al.*, 2003; Shen and Nemunaitis, 2006; Thorne *et al.*, 2006) makes this approach feasible. Furthermore, the latter two curative means allow one to treat, at least in principle, both detectable and occult metastases of any size and localization. A treatment plan, however, can only be as good as the information about the number and sizes of all secondary tumors. Obtaining such information with any reasonable accuracy is a daunting task, given the presence of undetectable metastases and various uncertainties associated both with the process of metastasis formation and with reading of medical images.

Overcoming this challenge calls upon comprehensive mathematical models of cancer natural history. These models are the ultimate tool of choice for relating unobservable critical microevents, such as formation of the first clonogenic cancer cell and inception of metastases in host organs or tissues, to observable outcomes including detection of the primary tumor or metastases.

Mathematical models of cancer progression involve a multitude of parameters indispensable for describing the natural history of the disease in a given patient. The most important among these parameters are:

(1) the age of onset of the primary disease;
(2) parameters descriptive of the law of primary tumor growth;
(3) intensity of metastasis formation;
(4) parameters characterizing metastasis promotion time (that is, the time between separation of a metastasis from the primary tumor and its inception, i.e. the start of active proliferation, in a host site);
(5) parameters built into the law of growth of metastases at various host organs and tissues; and
(6) the rate of formation of secondary metastases.

Unfortunately, all these parameters are unobservable. To estimate them for a given patient, a mathematical model of the individual natural history of cancer should be fit to the data available for the patient in a given clinical setting. Such observations typically fall into two categories: (1) clinical characteristics of the disease at primary diagnosis, and (2) observations that result from follow-up studies aimed at detecting primary tumor recurrence and/or metastatic spread of the disease. Observations in the first group usually include age at primary diagnosis, stage of the disease, localization and size of the primary tumor, its various histological indicators and biochemical markers, and information on the familial history of the disease. Those observations in the second group that are relevant to metastasis usually consist of the number and sizes of detectable metastases and information on the treatment for the primary tumor and secondary lesions.

Once parameters of the model are estimated, the model can be utilized to assess the distributional characteristics of the number and sizes of occult and detectable metastases at primary diagnosis, start of treatment or any other time point of interest. With the information on the individual natural history of cancer at hand, an oncologist can design a therapeutic intervention plan (including modes, dosage, and time courses of various treatments) that fits the patient best and maximizes the probability of cure or the residual lifetime.

Furthermore, knowledge of the natural history of cancer makes it possible to obtain definitive answers to a number of questions that are vital for understanding the interaction between the primary tumor and metastases and for the design of effective anti-metastatic therapies. Some of these questions are:

(1) Does extirpation of the primary tumor give a boost to metastasis formation?
(2) Does it increase the rate of metastasis growth and shorten their promotion times?

A number of authors hypothesized that the answer to both questions is positive (Retsky *et al.*, 2003, 2004; Smolle *et al.*, 1997). One of the practical outcomes of establishing such a relationship between the primary tumor and metastases would be identification of categories of patients for whom resection of the primary tumor is not conducive to the improved survival.

The goal of this chapter is to lay a theoretical groundwork for development and analysis of mathematical models of the individual natural history of invasive cancer including interaction between primary tumor and metastases. The focus of our modeling effort is on metastasis formation, promotion and growth in a given host site. In particular, we will derive a formula for the distribution of the sizes of all metastases (or all detectable metastases) in a host site at any time before or after the treatment of the primary tumor. The model discussed below is an extension of the model formulated and analyzed by Hanin *et al.*, 2006a where no interaction between the primary tumor and its secondary lesions was assumed. Estimation of parameters for such a model requires availability of the site specific number and sizes (or volumes) of metastases at a certain time. Computational and statistical feasibility of parameter estimation from these data for an individual patient was ascertained by Hanin *et al.* (2006a). Obtaining such data is a laborious task that requires careful reading of CT, PET or other images. An appealing alternative way of model validation and parameter estimation is to make use of the wealth of autopsy data (Douglas, 1971; Schulz and Borchard, 1992; Schulz and Hort, 1981; Yamanati *et al.*, 1999).

The processes of metastasis formation are extremely complex, heterogeneous and selective (Barbour and Gotley, 2003; Chambers *et al.*, 1995; Evans, 1991; Fidler, 1990, 1991, 1997, 2003). To form a micrometastasis in a host site, a tumor cell has to separate itself from the primary tumor, penetrate a blood or lymph channel, traverse the circulation network, evade attacks of the immune system, extravasate, invade a host site, survive through the dormancy period, start to proliferate, and induce angiogenesis. As a result of this multi-stage selection process, only a tiny fraction of cells shed by the primary tumor gives rise to viable metastases (Barbour and Gotley, 2003).

Studies of breast cancer (Pantel and Otte, 2001), lung cancer (Jiao and Krasna, 2002; Sugio *et al.*, 2002), melanoma (Fodstad *et al.*, 2001) and esophageal cancer (Jiao and Krasna, 2002) patients showed that large quantities of circulating tumor cells can be present without overt metastases. This suggests that inception in the host site is the most critical rate limiting step in the multi-stage process of metastasis formation. High sensitivity of tumor cells to the conditions of the host microenvironment and the resulting selective affinity of tumor cells to certain specific organs and tissues has been known since the end of the 19th century under the name of "seed and soil hypothesis" (Paget, 1889). But even if seeded into a fertile "soil," a tumor cell may remain dormant for a very long period of time. As one example, recurrence of breast cancer was reported to occur after 20 to 25 years of disease-free period (Karrison *et al.*, 1999). Yet another critical step on the pathway to formation of clinically manifest metastases is induction of angiogenesis. It was estimated by Demicheli (2001) that only about 4–10% of actively growing metastases eventually develop a capillary network which makes their further growth possible.

Because formation of metastases involves substantial variability and random fluctuations in the characteristics of metastatic cells and host microenvironment (Barbour and Gotley, 2003; Kendal, 2005) and is impelled through a number of sporadically occurring critical microevents, it can be best described using stochastic models. This approach to modeling metastatic progression of cancer was applied, mostly in the form of Monte Carlo simulation studies, in Bernhardt *et al.* (2001) and Kendal (2005). A statistical estimation of the empirical distribution of the sizes of metastases resulting from autopsy studies was performed by Kendal (2001). A semi-stochastic description of the kinetics of the number and sizes of metastases based on the von Förster equation was obtained by Iwata *et al.* (2000). A comparison of deterministic and stochastic approaches to modeling distant metastases resulting from primary and locally recurring prostate cancer can be found in Yorke *et al.* (1993).

An attempt at developing a mechanistic, biologically motivated, stochastic model of metastasis was made in Bartoczyński *et al.* (2001), where the probability that at the time of cancer detection the primary tumor has not yet metastasized was computed. One of the basic ideas entertained in Bartoczyński *et al.* (2001) was to relate the rate of metastases formation to the size of the primary tumor within the framework of quantal response models (Puri, 1967, 1971; Puri and Centuria, 1972). This approach was implemented in Hanin *et al.* (2006a) and will be followed in the present work as well.

In order to assess the extent of cancer progression, a patient diagnosed with primary tumor is given an imaging procedure which, after reading of the images, allows the number of detected metastases and their volumes to be determined. These volume data can be converted into the size estimates based on the density of 10^9 cells/cm^3 typical for most solid tumors or on the published more accurate density data for a specific type of solid cancer. Therefore, the first step in developing a comprehensive model of the individual natural history of metastasis consists in deriving a formula for the conditional joint distribution of the sizes of detectable metastases given their number. Such a formula is a key to statistical estimation of some of the unobservable parameters of the natural history of the disease. This leads to an important question as to what characteristics of occult metastases (such as the distributions of their number, sizes and total volume) can be estimated on the basis of data available for detectable metastases.

When dealing with the sizes of metastases observed in a certain host site at a given time one faces a considerable mathematical and methodological challenge in that the sizes of metastases cannot in general be thought of as resulting from a sequence of independently repeated trials, and thus they do not form a random sample from a probability distribution. The reasons for this are two-fold. First, metastases that were shed by the primary tumor later tend to have smaller sizes at the surveying time. Second, the rate of metastasis shedding depends on the primary tumor size which increase typically causes acceleration of the process of metastasis formation. Therefore, although one can always construct a frequency distribution (histogram) of the sizes of detectable metastases at any time post-diagnosis, it is generally not true that it represents an empirical counterpart of the distribution of the "size of detectable metastasis at a given time." In fact, the latter random variable is not well-defined! Labeling (or numbering) metastases represents yet another concomitant problem. The most natural way to label detectable metastases is through ordering their sizes taken at a certain time from the smallest to the largest (or vice versa). However, with such a labeling the sizes of metastases are represented by random variables that are neither independent nor identically distributed. Thus, the data on the number and sizes of metastases is not generally amenable to standard methods of statistical analysis and parameter estimation.

A way to circumvent this difficulty was proposed by Hanin *et al.* (2006a) in the form of a mechanistic model of metastasis based on certain biologically plausible assumptions. Based on this model it was shown in Hanin *et al.* (2006a) that the joint distribution of the sizes of detectable metastases conditional on their number coincides with that of the vector of order statistics derived from some probability distribution and, furthermore, a formula for the latter distribution was obtained. Although the mechanism of sampling from this distribution is elusive, for many purposes, including parameter estimation in the maximum likelihood setting, this distribution may serve as a surrogate of the distribution of the "size of detectable metastasis." In particular, fitting this distribution to the observed sizes of metastases makes it possible to estimate many parameters descriptive of the natural history of the disease and gain an insight into the processes of metastasis formation and growth.

2. THE MODEL

The natural history of invasive cancer is commonly divided into the periods of tumor latency, primary tumor growth, and metastatic progression. These periods and relevant model assumptions are described below.

2.1. Tumor Latency

The latent period starts with the birth of an individual and ends with the appearance of the first malignant clonogenic cell. This event is termed the *onset of the disease*.

2.2. Primary Tumor Growth

The size of the primary tumor (that is, the number of cells comprising the tumor) at any time w counted from the onset of the disease will be denoted by $\Phi(w)$. It is assumed that Φ is a strictly increasing continuous function such that $\Phi(0) = 1$. The function Φ may depend on one or several parameters that can be deterministic or random. We denote by φ the function inverse to Φ.

2.3. Metastasis Formation

Following Bartoczyński *et al.* (2001) we will assume that metastasis shedding is governed by a non-homogeneous Poisson process with intensity μ proportional to some power of the current size of the primary tumor. Thus

$$\mu(w) = \alpha \Phi^\theta(w), \tag{1}$$

where $\alpha > 0$ and $\theta \geq 0$ are constants, and time w is counted from the age t of the disease onset. Note that model (1) with $\theta = 0$ describes stationary metastasis shedding governed by a homogeneous Poisson process with constant rate α. It is further assumed that metastases shed by the primary tumor give rise to clinically detectable secondary tumors in a given host site independently of each other with the same probability q. Therefore (see e.g. Ross, 1997, pp. 257–259), production of viable metastases in the host site is governed by a Poisson process with the intensity $\nu = q\mu$. Each viable metastasis is assumed to spend some random time between detachment from the primary tumor and inception in the host site (which may include a period of dormancy) after which it starts irreversible proliferation. We assume that these promotion times for different metastases are independent and identically distributed with some probability density function (p.d.f.) f and the corresponding cumulative distribution function (c.d.f.) F. It is well-known (see e.g. Hanin and Yakovlev, 1996) that the resulting delayed Poisson process is again a Poisson process with the rate

$$\lambda(w) = \int_0^w \nu(s) f(w-s) ds. \qquad (2)$$

2.4. Timeline of the Natural History of Metastatic Cancer and Observables

Suppose that at age u a patient was diagnosed with primary cancer and that the primary tumor size at diagnosis was S. It follows from our assumptions in Sections 2.1 and 2.2 that the age t of the disease onset is given by

$$t = u - \varphi(S). \qquad (3)$$

Suppose also that at age v, $v \geq u$, the primary tumor was resected, and that at age τ, $\tau \geq u$, the patient developed n detectable metastases localized in the same host site with the observed sizes $x_1, x_2, \ldots, x_n$, where $x_1 < x_2 < \cdots < x_n$. Note that tumor resection after time τ has no bearing on the sizes of metastases measured at time τ so that in this case (as well as in the case of an untreated primary tumor) we can set, for the purpose of computing the distribution of the sizes of detectable metastases, $v = \tau$. Thus, we will assume without loss of generality that

$$0 \leq t \leq u \leq v \leq \tau,$$

see Figure 1.

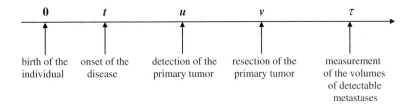

Fig. 1. Timeline of the natural history of metastatic cancer.

2.5. Secondary Metastasis

To retain mathematical tractability of the model, we assume that secondary metastasizing (that is, formation of "metastasis of metastasis") to a given site both from other sites and from within is negligible.

2.6. Metastasis Growth

After inception in the host site the growth of a viable metastasis is irreversible and its size at time w counted from the inception is denoted by $\Psi_T(w)$, where T is the metastasis inception time relative to the onset of the disease. Note that in the simpler setting studied in Hanin et al. (2006a) the growth processes for the primary tumor and metastases were assumed independent, in which case $\Psi_T(w)$ is a function of w alone independent of T. However, if the rate of metastasis growth may be affected by the removal of the primary tumor this assumption is not true anymore. Specifically, we assume that prior to primary tumor resection the growth of metastases is governed by a function Ψ_0 while after the resection the size of metastases is growing according to another function Ψ_1 which acts multiplicatively on the size of metastasis at the time of primary tumor resection. Thus,

$$\Psi_T(w) = \begin{cases} \Psi_0(w), & \text{if } T + w \leq v - t \\ \Psi_0(v - t - T)\Psi_1(w - (v - t - T)), & \text{if } T + w > v - t \text{ and } T < v - t, \\ \Psi_1(w), & \text{if } T \geq v - t. \end{cases}$$
(4)

The functions Ψ_i are assumed to be strictly increasing and differentiable, and to satisfy the condition $\Psi_i(0) = 1, i = 0, 1$. Additionally, they may depend on some deterministic or random parameters.

2.7. Metastasis Detection

The volume of a metastasis becomes measurable when the size of the metastasis reaches some threshold value m. The value of m and the accuracy of volume measurement are determined by the sensitivity of imaging technology. In the case of PET/CT imaging involved in the study by Hanin et al. (2006a) the threshold volume was $0.5 \, \text{cm}^3$, and the accuracy of volume determination was one pixel, which is approximately $0.065 \, \text{cm}^3$.

2.8. Effects of Treatment

After surgical removal of the primary tumor its size is set to zero. Then, in accordance with formula (1), if the primary tumor was resected at age v then $\mu(w) = 0$ for $w > v - t$. Because the rate of secondary metastasizing is assumed to be negligible, the process of new metastasis formation is stopped at the time of primary tumor extirpation. Finally, chemotherapeutic or hormonal treatment of metastases is assumed to affect them only through the rate of their growth and the distribution of their promotion times.

3. DISTRIBUTION OF THE SIZES OF DETECTABLE METASTASES

We intend to compute the joint conditional p.d.f. of the observed sizes of detectable metastases given that their number equals n. Let X be the size of a detectable metastasis with inception time T (relative to the onset of the disease) that was surveyed at age τ. Then according to Eq. (4)

$$X = \Psi_T(\tau - t - T) = \begin{cases} \Psi_0(v - t - T)\Psi_1(\tau - v), & \text{if } T < v - t \\ \Psi_1(\tau - t - T), & \text{if } T \geq v - t \end{cases},$$

where $t - t - T$ is the metastasis growth time from inception to detection. Observe that function Ψ_T actually does not depend on T:

$$X = \Psi(\tau - t - T), \tag{5}$$

where

$$\Psi(y) = \begin{cases} \Psi_1(y), & \text{if } 0 \leq y \leq \tau - v \\ \Psi_0(y - (\tau - v))\Psi_1(\tau - v), & \text{if } \tau - v < y \leq \tau - t \end{cases}. \tag{6}$$

Note also that function Ψ is strictly increasing, continuous, piecewise differentiable, and satisfies $\Psi(0) = 1$. Its maximum value is

$$M = \Psi_0(v - t)\Psi_1(\tau - v), \qquad (7)$$

and the inverse function, $\psi := \Psi^{-1}$, is given by

$$\psi(x) = \begin{cases} \Psi_1^{-1}(x), & \text{if } 1 \leq x \leq \Psi_1(\tau - v) \\ \Psi_0^{-1}\left(\dfrac{x}{\Psi_1(\tau - v)}\right) + \tau - v, & \text{if } \Psi_1(\tau - v) < x \leq M \end{cases}. \qquad (8)$$

The fact that function Ψ in relation (5) between random variables X and T is independent of T makes the argument in Hanin et al. (2006a), that was used for derivation of the joint distribution of the sizes of n detectable metastases in a given site at age t, applicable to the more general case studied in this work. This leads to the following result.

Theorem 1. *The sizes $X_1 < X_2 < \cdots < X_n$ of metastases in a certain host site that are detectable at age τ are equidistributed, given their number n, with the vector of order statistics for a random sample of size n drawn from the distribution with the p.d.f. defined by*

$$p(x) = \omega(\tau - t - \psi(x))\psi'(x), \quad m \leq x \leq M, \qquad (9)$$

and $p(x) = 0$ for $x \notin [m, M]$, where tumor onset time t is given by (3), function ψ is defined in (8), M is specified in (7), and

$$\omega(w) = \frac{\int_0^{\min\{w, v-t\}} \Phi^\theta(s) f(w - s) ds}{\int_0^{\min\{\tau - t - \psi(m), v-t\}} \Phi^\theta(s) F(\tau - t - \psi(m) - s) ds},$$

$$0 \leq w \leq \tau - t - \psi(m). \qquad (10)$$

Observe that if duration of the metastasis promotion time is negligible then formula (10) is reduced to

$$\omega(w) = \frac{\Phi^\theta(w)}{\int_0^{\min\{\tau - t - \psi(m), v-t\}} \Phi^\theta(s) ds}, \quad 0 \leq w \leq \min\{\tau - t - \psi(m), \ v - t\}. \qquad (11)$$

Remark 1. Formula (9) with $m = 1$ describes the site-specific distribution underlying the sizes of all (detectable and occult) metastases. In this case function ω in (10) and (11) takes on, respectively, the following simpler forms

$$\omega(w) = \frac{\int_0^{\min\{w, v-t\}} \Phi^\theta(s) f(w - s) ds}{\int_0^{v-t} \Phi^\theta(s) F(\tau - t - s) ds}, \quad 0 \leq w \leq \tau - t,$$

and

$$\omega(w) = \frac{\Phi^\theta(w)}{\int_0^{\tau-t} \Phi^\theta(s) ds}, \quad 0 \leq w \leq \tau - t,$$

Remark 2. A remarkable feature of the p.d.f. p given by Eq. (9) is that it is independent of n and is also free of the parameter $q\alpha$ that characterizes the intensity of metastasis seeding. The latter parameter, however, is indispensable for determining the distribution of the *number* of metastases in the site in question at any time pre- or post-diagnosis. In fact, it follows from Eqs. (1) and (2) that the distribution of the site specific number of detectable metastases at time τ is Poisson with parameter

$$q\alpha \int_0^{\min\{\tau - \psi(m), v\} - t} \Phi^\theta(s) F(\tau - t - \psi(m) - s) ds,$$

where t is given by Eq. (3). The same formula with $m = 1$ gives the distribution of the total number of metastases at the site at time τ. Thus, this distribution is Poisson with parameter

$$q\alpha \int_0^{v-t} \Phi^\theta(s) F(\tau - t - s) ds.$$

Remark 3. It follows from Theorem 1 that the joint likelihood of the observed sizes $x_1, x_2, \ldots, x_n$ of detectable metastases is given by the formula

$$L(x_1, x_2, \ldots, x_n) = n! \prod_{i=1}^n p(x_i).$$

Thus, apart from the factor $n!$, the likelihood has the same form it would take should the observations $x_1, x_2, \ldots, x_n$ be independent. This shows that identifiable parameters of a suitably parameterized model of disease natural history described in Section 2 can in principle be estimated using maximum likelihood methodology, see Section 4 for further discussion.

Remark 4. If the primary tumor size S at presentation is unknown then the onset time t cannot be computed through Eq. (3) and should be treated as a random variable. In this case an additional integration in Eq. (9) with respect to the distribution of the onset time is required. This distribution can be obtained by utilizing one of the established mechanistic models of tumor latency such as the two-stage clonal expansion model (also termed Moolgavkar-Venzon-Knudson model) (Moolgavkar et al., 1988; Moolgavkar and Knudson, 1981; Moolgavkar and Luebeck, 1990; Moolgavkar and Venzon, 1979) or Yakovlev-Polig model (Yakovlev and Polig, 1996). Alternatively, one can assume that the tumor latency time follows a distribution from a flexible parametric family (e.g. of gamma or Weibull distributions). For methodological approaches to parameter estimation for such models from population data on cancer incidence, see Hanin et al. (2006b); Luebeck and Moolgavkar (2002); and Zorin et al. (2005).

Remark 5. If the laws of primary tumor and/or metastasis growth contain random parameters then Eq. (9) should be additionally integrated with respect to their distribution. More generally, if growth of the primary tumor is governed by a stochastic process then formula (9) can be applied to any of its sample paths $\Phi(w)$ and then integrated with respect to the corresponding probability measure on the space of sample paths of the process (which is typically extremely hard to obtain). In the case where growth of metastases is governed by a stochastic process this approach is feasible only if sample paths of the process are all increasing and never cross each other. These assumptions were used in a very essential way in the derivation of Eq. (4), see Hanin et al. (2006a). The main difficulty that arises for stochastic processes whose sample paths are not necessarily increasing or may intersect is that metastases that were shed earlier may have smaller sizes at the time of detection τ than those shed later or even remain occult at or become extinct by time τ.

In the next section, we will explicate formulas (9)–(11) in the special case of non-random exponential growth of both primary tumor and its metastases combined with an exponentially distributed metastasis promotion time.

4. DISTRIBUTION OF THE SIZES OF DETECTABLE METASTASES FOR EXPONENTIALLY GROWING TUMORS

4.1. Model Specification and Results

Suppose primary tumor grows exponentially with constant rate $\beta > 0$:

$$\Phi(w) = e^{\beta w}, \quad 0 \leq w \leq v - t,$$

where w is time counted from the age t of tumor onset, and v is the age of primary tumor resection. Then for φ, the inverse function for Φ, we have $\varphi(y) = \ln(y)/\beta$. Hence

$$t = u - \frac{\ln(S)}{\beta}. \tag{12}$$

One of the limitations of the work Hanin et al. (2006a) is that the assumed exponential growth of metastases did not allow for distinct growth rates prior to and after the resection of the primary tumor. To account for such a possibility, we will assume that metastases grow exponentially with rates $\gamma_0, \gamma_1 > 0$ before and after extirpation of the primary tumor, respectively: $\Psi_i(w) = e^{\gamma_i w}$, $i = 0, 1$. Then

$$\Psi(y) = \begin{cases} e^{\gamma_1 y}, & \text{if } 0 \leq y \leq \tau - v \\ e^{\gamma_0 y + (\gamma_1 - \gamma_0)(\tau - v)}, & \text{if } \tau - v < y \leq \tau - t \end{cases},$$

and

$$\psi(x) = \begin{cases} \dfrac{\ln x}{\gamma_1}, & \text{if } 1 \leq x \leq e^{\gamma_1(\tau - v)} \\ \dfrac{\ln x}{\gamma_0} + \left(1 - \dfrac{\gamma_1}{\gamma_0}\right)(\tau - v), & \text{if } e^{\gamma_1(\tau - v)} < x \leq e^{\gamma_0(v - t) + \gamma_1(\tau - v)} \end{cases},$$

compare with Eqs. (6) and (8).

Suppose additionally that metastasis promotion time is exponentially distributed with the expected value ρ. Then applying formulas (9) and (10) to the resulting parametric model we obtain, after some computational effort, explicit formulas for the p.d.f. p given in Theorem 1. To formulate these results, denote

$$M = e^{\gamma_0(v - t) + \gamma_1(\tau - v)} \tag{13}$$

(compare with Eq. (7)), and

$$A = e^{\gamma_1(\tau-v)}, \quad a_0 = \frac{\beta\theta}{\gamma_0}, \quad b_0 = \frac{1}{\rho\gamma_0}, \quad b_1 = \frac{1}{\rho\gamma_1}. \quad (14)$$

Theorem 2. (1) If $A \leq m$ then

$$p(x) = (C_1 x)^{-1} \left[\left(\frac{M}{x}\right)^{a_0} - \left(\frac{x}{M}\right)^{b_0} \right], \quad m \leq x \leq M, \quad (15)$$

where

$$C_1 = a_0^{-1} \left[\left(\frac{M}{m}\right)^{a_0} - 1 \right] - b_0^{-1} \left[1 - \left(\frac{m}{M}\right)^{b_0} \right].$$

(2) If $A > m$ then

$$p(x) = \begin{cases} \dfrac{\gamma_0}{\gamma_1}(C_2 x)^{-1} \left[\left(\dfrac{M}{A}\right)^{a_0} - \left(\dfrac{A}{M}\right)^{b_0} \right] \left(\dfrac{x}{A}\right)^{b_1}, & m \leq x < A \\[2ex] (C_2 x)^{-1} \left[\left(\dfrac{M}{x}\right)^{a_0} - \left(\dfrac{x}{M}\right)^{b_0} \right], & A \leq x \leq M \end{cases}, \quad (16)$$

where

$$C_2 = b_0^{-1} \left[\left(\frac{M}{A}\right)^{a_0} - \left(\frac{A}{M}\right)^{b_0} \right] \left[1 - \left(\frac{m}{A}\right)^{b_1} \right] + a_0^{-1} \left[\left(\frac{M}{A}\right)^{a_0} - 1 \right]$$

$$- b_0^{-1} \left[1 - \left(\frac{A}{M}\right)^{b_0} \right].$$

The above expressions C_1 and C_2 are normalization constants that make the respective p.d.f.s (15) and (16) proper probability distributions. Note that in the case $\gamma_0 = \gamma_1$ formulas (15) and (16) reduce to the model introduced and studied in Hanin *et al.* (2006a). Observe also that model (15) is a limiting case of model (16) arising when $A = m$. From biological standpoint the case $A \leq m$ is characterized by the condition that metastases which inception in the host site occurred after the time of primary tumor resection will not reach the minimum detectable size m by time τ. In particular, this happens when the primary tumor was not resected by the time of metastases surveying ($v = \tau$). Finally, model (16) with $m = 1$ describes the site-specific distribution of *all* (detectable and occult) metastases at time τ.

Models (15) and (16) with $a_0 = 0$ will be called *homogeneous*. They represent the distribution of the sizes of detectable metastases at time τ under stationary shedding of metastases by the primary tumor ($\theta = 0$). Specifically, the following results for the homogeneous model follow from Theorem 2:

(1) If $A \leq m$ then

$$p(x) = (C_3 x)^{-1} \left[1 - \left(\frac{x}{M} \right)^{b_0} \right], \quad m \leq x \leq M, \tag{17}$$

where

$$C_3 = \ln \frac{M}{m} - b_0^{-1} \left[1 - \left(\frac{m}{M} \right)^{b_0} \right].$$

(2) If $A > m$ then

$$p(x) = \begin{cases} \dfrac{\gamma_0}{\gamma_1} (C_4 x)^{-1} \left[1 - \left(\dfrac{A}{M} \right)^{b_0} \right] \left(\dfrac{x}{A} \right)^{b_1}, & m \leq x < A \\ (C_4 x)^{-1} \left[1 - \left(\dfrac{x}{M} \right)^{b_0} \right], & A \leq x \leq M \end{cases} \tag{18}$$

where

$$C_4 = \ln \frac{M}{A} + b_0^{-1} \left[1 - \left(\frac{A}{M} \right)^{b_0} \right] \left(\frac{m}{A} \right)^{b_1}.$$

Yet another limiting case of models (15) and (16) arises when metastasis promotion time is negligible ($\rho = 0$). These models will be termed *instantaneous seeding* models. A computation based on Eq. (11) leads to

Distribution of the Sizes of Metastases

the following formulas for the p.d.f. p from Theorem 1:

(1) If $A \leq m$ then

$$p(x) = C_5 x^{-a_0-1}, \quad m \leq x \leq M, \tag{19}$$

where

$$C_5 = \frac{a_0}{m^{-a_0} - M^{-a_0}}.$$

(2) If $A > m$ then

$$p(x) = C_6 x^{-a_0-1}, \quad A \leq x \leq M, \tag{20}$$

where

$$C_6 = \frac{a_0}{A^{-a_0} - M^{-a_0}}.$$

Notice that the instantaneous seeding model (19)–(20) arises as a limiting case of the more general model (15)–(16). In fact, setting $b_0 = \infty$ in Eq. (15) converts it into (19) while setting $b_0 = b_1 = \infty$ in (16) produces Eq. (20).

4.2. Model Identification

The utility of models (15) and (16) and their limiting forms (17)–(20) depends on whether parameters β, θ, γ_0, γ_1, ρ of the natural history of cancer are identifiable from the site-specific distribution of the sizes of detectable metastases (for an extensive discussion of identifiability of stochastic models, see Hanin, 2002). We start with identification of the parameters M, A, a_0, b_0, b_1 of the most general model (16).

Theorem 3. *Suppose that $A > m$. Then parameters M, A, a_0, b_0, b_1 of model (16) are jointly identifiable from the p.d.f. p.*

Proof. Observe that $x = M$ is the only point on the interval $[m, M]$ with the property $p(x) = 0$. Therefore, parameter M is uniquely determined by a given function p. Furthermore, denoting $q(x) := xp(x)$, $m \leq x \leq M$, we conclude from (16) that function q increases on $[m, A]$ and decreases on $[A, M]$. Thus, A is the unique point of maximum of function q on $[m, M]$

and hence is completely determined by function p. Next, it follows from the first formula in (16) that

$$b_1 = m\frac{q'(m)}{q(m)},$$

which makes parameter b_1 identifiable. From the second equation in (16) we find that

$$q'(x) = -(C_2 x)^{-1}\left[a_0\left(\frac{M}{x}\right)^{a_0} + b_0\left(\frac{x}{M}\right)^{b_0}\right], \quad A \leq x \leq M. \quad (21)$$

Therefore,

$$q'(M) = -\frac{a_0 + b_0}{C_2 M}.$$

Also, for the function $r(x) := x^2 q'(x)$ we obtain using (21)

$$r'(M) = -\frac{b_0(b_0 + 1) - a_0(a_0 - 1)}{C_2}$$

and

$$r''(M) = -\frac{a_0^2(a_0 - 1) + b_0^2(b_0 + 1)}{C_2 M}.$$

This implies

$$\frac{r'(M)}{Mq'(M)} = \frac{b_0(b_0 + 1) - a_0(a_0 - 1)}{b_0 + a_0} = b_0 - a_0 + 1$$

and

$$\frac{r''(M)}{q'(M)} = \frac{b_0^2(b_0 + 1) + a_0^2(a_0 - 1)}{b_0 + a_0} = (b_0 - a_0)^2 + (b_0 - a_0) + a_0 b_0.$$

It follows from the latter two equations that parameter combinations $b_0 - a_0$ and $a_0 b_0$ are identifiable from function p. Then the same is true for parameters a_0 and b_0. This completes the proof of Theorem 3.

Remark 6. A similar but simpler argument would show that parameters M, a_0, b_0 of model (15), parameters M, A, b_0, b_1 of the homogeneous version (18) of model (16), and parameters M, b_0 of the homogeneous version (17) of model (15) are identifiable. The same is true for parameters M, A, a_0 of the instantaneous seeding model (20) and parameters M, a_0 of its analogue (19).

Finally, we address the question of identifiability of the natural history of cancer from the data on primary tumor size at presentation and sizes of detectable metastases in a given host site. Equations (12) and (13) yield

$$M = S^{\gamma_0/\beta} e^{\gamma_0(v-u) + \gamma_1(\tau-v)}.$$

Combining this formula with Eqs. (14) we arrive at the following conclusions:

(1) All parameters β, θ, γ_0, γ_1, ρ of model (16) are identifiable from the known function p. Furthermore, if the primary tumor size at detection is unknown then parameters γ_0, γ_1, ρ are still identifiable. However, in this case it is only the product $\beta\theta$ of parameters β, θ that can be identified from the p.d.f. p.
(2) Similarly, parameters β, γ_0, γ_1, ρ (or γ_0, γ_1, ρ if S is unknown) are identifiable within the homogeneous model (18).
(3) Parameters β, θ, γ_0 are generally not identifiable within the instantaneous seeding model (20). However, parameter γ_1 is identifiable and, surprisingly enough, so is parameter θ in the case where primary tumor is resected shortly after diagnosis ($v = u$).
(4) Parameters of the natural history of metastatic cancer cannot be identified from the reduced models (15), (17) and (19).
(5) Finally, the distribution of the sizes of detectable metastases in a given host site at time τ given by models (16), (18) or (20) uniquely determines the distribution of the sizes of *all* metastases in the same site at time τ represented by the same respective models with $m = 1$.

ACKNOWLEDGMENTS

This research was supported in part by grant # 974 awarded by the Faculty Research Committee at Idaho State University.

References

1. Barbour A and Gotley DC. Current concepts of tumour metastasis. *Ann. Acad. Med. Singapore* 2003; **32**: 176–184.
2. Bartoczyński R, Edler L, Hanin L, Kopp-Schneider A, Pavlova L, Tsodikov A, Zorin A and Yakovlev A. Modeling cancer detection: tumor size as a source of information on unobservable stages of carcinogenesis. *Math. Biosci.* 2001; **171**: 113–142.
3. Bernhardt P, Forssell-Aronsson E, Jacobsson L and Skarnemark G. Low-energy electron emitters for targeted radiotherapy of small tumors. *Acta Oncol.* 2001; **40**: 602–608.
4. Chambers AF, Macdonald IF, Schmidt E, Koop S, Morris VL, Khokha R and Groom AC. Steps in tumor metastasis: new concepts from intravital videomicroscopy. *Cancer Metastasis Rev.* 1995; **14**: 279–301.
5. Demicheli R. Tumour dormancy: findings and hypotheses from clinical research on breast cancer. *Semin. Cancer Biol.* 2001; **11**: 297–306.
6. Douglas JRS. Significance of the size distribution of bloodborne metastases. *Cancer* 1971; **27**: 379–390.
7. Evans CW. *The Metastatic Cell: Behavior and Biochemistry*. Chapman & Hall, London, 1991.
8. Fidler IJ. Critical factors in the biology of human cancer metastasis. 28th G.H.A. Clowes memorial award lecture. *Cancer Res.* 1990; **50**: 6130–6138.
9. Fidler IJ. The biology of human cancer metastasis. 7th Jan Waldenstrom lecture. *Acta Oncol.* 1991; **30**: 668–675.
10. Fidler IJ. Molecular biology of cancer: invasion and metastasis. In: *Cancer Principles and Practice of Oncology*, 5th ed. DeVita VT, Hellman S and Rosenberg SA (eds.) Lippincott Raven Publishers, Philadelphia, 1997, pp. 135–152.
11. Fidler IJ. The pathogenesis of cancer metastasis: the "seed and soil" hypothesis revisited. *Nat. Rev. Cancer* 2003; **3**: 453–458.
12. Fodstad O, Faye R, Hoifodt HK, Skovlund E and Aamdal S. Immunobead-based detection and characterization of circulating tumor cells in melanoma patients. *Recent Results Cancer Res.* 2001; **158**: 40–50.
13. Goddu SM, Rao DV and Howell RW. Multicellular dosimetry for micrometastases: dependence of self-dose versus cross-dose to cell nuclei on type and energy of radiation and subcellular distribution of radionuclides. *J. Nucl. Med.* 1994; **35**: 521–530.
14. Hanin LG. Identification problem for stochastic models with application to carcinogenesis, cancer detection and radiation biology. *Disc. Dynam. Nat. Soc.* 2002; **7**: 177–189.

15. Hanin LG, Rose J and Zaider M. A stochastic model for the sizes of detectable metastases. *J. Theor. Biol.* 2006; **243**: 407–417.
16. Hanin LG, Miller AB, Yakovlev AY and Zorin AV. The University of Rochester model of breast cancer detection and survival. In: *Journal of the National Cancer Institute*, Monograph 36. *The Impact of Mammography and Adjuvant Therapy on U.S. Breast Cancer Mortality (1975–2000): Collective Results from the Cancer Intervention and Surveillance Modeling Network*, 2006, pp. 86–95.
17. Hanin LG and Yakovlev AY. A nonidentifiability aspect of the two-stage model of carcinogenesis. *Risk Anal.* 1996; **16**: 711–715.
18. Harrison D, Sauthoff H, Heitner S, Jagirdar J, Rom WN and Hay JG. Wild-type adenovirus decreases tumor xenograft growth, but despite viral persistence complete tumor responses are rarely achieved — deletion of the viral E1b-19-kD gene increases the viral oncolytic effect. *Hum. Gene Ther.* 2001; **12**: 1323–1332.
19. Iwata K, Kawasaki K and Shigesada N. A dynamical model for the growth and size distribution of multiple metastatic tumors. *J. Theor. Biol.* 2000; **203**: 177–186.
20. Jiao X and Krasna MJ. Clinical significance of micrometastasis in lung and esophageal cancer: a new paradigm in thoracic oncology. *Ann. Thorac. Surg.* 2002; **74**: 278–284.
21. Kaplan JM. Adenovirus-based cancer gene therapy. *Curr. Gene Ther.* 2005; **5**: 595–605.
22. Karrison TG, Ferguson DJ and Meier P. Dormancy of mammary carcinoma after mastectomy. *J. Natl Cancer Inst.* 1999: **19**: 80–85.
23. Kasuya H, Takeda S, Nomoto S and Nakao A. The potential of oncolytic virus therapy for pancreatic cancer. *Cancer Gene Ther.* 2005; **12**: 725–736.
24. Kaufman HL, Deraffele G, Mitcham J, Moroziewicz D, Cohen SM, Hurst-Wicker KS, Cheung K, Lee DS, Divito J, Voulo M, Donovan J, Dolan K, Manson K, Panicali D, Wang E, Horig H and Marincola FM. Targeting the local tumor microenvironment with vaccinia virus expressing B7.1 for the treatment of melanoma. *J. Clin. Invest.* 2005; **115**: 1903–1912.
25. Kendal WS. The size distribution of human hematogenous metastases. *J. Theor. Biol.* 2001; **211**: 29–38.
26. Kendal WS. Chance mechanisms affecting the burden of metastases. *BMC Cancer* 2005; **5**: 138–146.
27. Khuri FR, Nemunaitis J, Ganly I, Arseneau J, Tannock IF, Romel L, Gore M, Ironside J, MacDougall RH, Heise C, Randlev B, Gillenwater AM, Bruso P,

Kaye SB, Hong WK and Kirn DH. A controlled trial of intratumoral ONYX-015, a selectively-replicating adenovirus, in combination with cisplatin and 5-fluorouracil in patients with recurrent head and neck cancer. *Nat. Med.* 2000; **6**: 879–885.
28. Kirn D, Hermiston T and McCormick F. ONYX-015: clinical data are encouraging. *Nat. Med.* 1998; **4**: 1341–1342.
29. Kirn DH and McCormick F. Replicating viruses as selective cancer therapeutics. *Mol. Med. Today* 1996; **2**: 519–527.
30. Luebeck EG and Moolgavkar SH. Multistage carcinogenesis and the incidence of colorectal cancer. *Proc. Natl. Acad. Sci. USA* 2002; **99**: 15095–15100.
31. McCormick F. Cancer-specific viruses and the development of ONYX-015. *Cancer Biol. Ther.* 2003; **2**: S157–160.
32. Moolgavkar SH, Dewanji A and Venzon DJ. A stochastic two stage model for cancer risk assessment. I. The hazard function and the probability of tumor. *Risk Anal.* 1988; **8**: 383–392.
33. Moolgavkar SH and Knudson AG. Mutation and cancer: a model for human carcinogenesis. *J. Natl. Cancer Inst.* 1981; **66**: 1037–1052.
34. Moolgavkar SH and Luebeck EG. Two-event model for carcinogenesis: biological, mathematical and statistical considerations. *Risk Anal.* 1990; **10**: 323–341.
35. Moolgavkar SH and Venzon DJ. Two event model for carcinogenesis: incidence curves for childhood and adult tumors. *Math. Biosci.* 1979; **47**: 55–77.
36. Nemunaitis J, Khuri F, Ganly I, Arseneau J, Posner M, Vokes E, Kuhn J, McCarty T, Landers S, Blackburn A, Romel L, Randlev B, Kaye S and Kirn D. Phase II trial of intratumoral administration of ONYX-015, a replication-selective adenovirus, in patients with refractory head and neck cancer. *J. Clin. Oncol.* 2001; **19**: 289–298.
37. O'Donoghue JA. Dosimetric principles of targeted radiotherapy. In: *Radioimmunotherapy of Cancer*. Abrams PG and Fritzberg A (eds.) Marcel Dekker, New York, 2000, pp. 1–20.
38. Paget S. The distribution of secondary growths in cancer of the breast. *Lancet* 1889; **1**: 571–573.
39. Pantel K and Otte M. Occult micrometastases: enrichment, identification and characterization of single disseminated tumour cells. *Semin. Cancer Biol.* 2001; **11**: 327–337.
40. Parato KA, Senger D, Forsyth PA and Bell JC. Recent progress in the battle between oncolytic viruses and tumours. *Nat. Rev. Cancer* 2005; **5**: 965–976.
41. Puri PS. A class of stochastic models of response after infection in the absence of defense mechanism. In: *Proceedings of the Fifth Berkeley Symposium on*

Mathematical Statistics and Probability, Vol. 4. University of California Press, Berkeley and Los Angeles, 1967, pp. 511–535.
42. Puri PS. A quantal response process associated with integrals of certain growth processes. In: *Mathematical Aspects of Life Sciences*, Queen's Papers in Pure and Applied Mathematics, No. 26, Wasan MT (ed.) Queen's University, Kingston, Ontario, Canada, 1971.
43. Puri PS and Senturia J. On a mathematical theory of quantal response assays. In: *Proceedings of the Sixth Berkeley Symposium on Mathematical Statistics and Probability*, Vol. 4. University of California Press, Berkeley and Los Angeles, 1972, pp. 231–247.
44. Reid T, Warren R and Kirn D. Intravascular adenoviral agents in cancer patients: lessons from clinical trials. *Cancer Gene Ther.* 2002; **9**: 979–986.
45. Retsky M, Bonadonna G, Demicheli R, Folkman J, Hrushesky W and Valagussa P. Hypothesis: induced angiogenesis after surgery in premenopausal node-positive breast cancer patients is a major underlying reason why adjuvant chemotherapy works particularly well for those patients. *Breast Cancer Res.* 2004; **6**: R372–374.
46. Retsky M, Demicheli R and Hrushesky W. Breast cancer screening: controversies and future directions. *Curr. Opin. Obstet. Gynecol.* 2003; **15**: 1–8.
47. Ross SM. *Introduction to Probability Models*, 6th ed. Academic Press, San Diego, 1997.
48. Schell MC, Bova FJ, Larson DA, Leavitt DD, Lutz WR, Podgorsak EB and Wu A. TG-42 Report on stereotactic external beam irradiation. *AAPM Report* No. 54, Stereotactic Radiosurgery, 1995.
49. Schulz W and Borchard F. Größe der Lebermetastasen bei geringer Metastasenzahl. Eine quantitative Studie an postmortalen Lebern. *Fortschr. Röntgenstr.* 1992; **156**: 320–324.
50. Schulz W and Hort W. The distribution of metastases in the liver. *Virchows Arch. Pathol. Anat.* 1981; **394**: 89–96.
51. Shah AC, Benos D, Gillespie GY and Markert JM. Oncolytic viruses: clinical applications as vectors for the treatment of malignant gliomas. *J. Neurooncol.* 2003; **65**: 203–226.
52. Shen Y and Nemunaitis J. Herpes simplex virus 1 (HSV-1) for cancer treatment. *Cancer Gene Ther.* 2006; **13**: 975–992.
53. Smolle J, Soyer HP, Smolle-Juttner FM, Rieger E and Kerl H. Does surgical removal of primary melanoma trigger growth of occult metastases? An analytical epidemiological approach. *Dermatol. Surg.* 1997; **23**: 1043–1046.
54. Sugio K, Kase S, Sakada T, Yamazaki K, Yamaguchi M, Ondo K, *et al.* Micrometastasis in the bone marrow of patients with lung cancer associated

with reduced expression of E-cadherin and beta-catenin: risk assessment by immunohistochemistry. *Surgery* 2002; **131**: S226–231.
55. Thorne SH, Negrin RS and Contag CH. Synergistic antitumor effects of immune cell-viral biotherapy. *Science* 2006; **311**: 1780–1784.
56. Yakovlev AY and Polig E. A diversity of responses displayed by a stochastic model of radiation carcinogenesis allowing for cell death. *Math. Biosci.* 1996; **132**: 1–33.
57. Yamanati H, Chiba R, Kobari M, Matsuno S and Takahashi T. Total number and size distribution of hepatic metastases of carcinoma. *Anal. Quant. Cytol. Histol.* 1999; **21**: 216–226.
58. Yorke ED, Fuks Z, Norton L, Whitmore W and Ling CC. Modeling the development of metastases from primary and locally recurrent tumors: comparison with a clinical data base for prostatic cancer. *Cancer Res.* 1993; **53**: 2987–2993.
59. Zorin AV, Hanin LG, Edler L and Yakovlev AY. Estimating the natural history of breast cancer from bivariate data on age and tumor size at diagnosis. In: *Quantitative Methods for Cancer and Human Health Risk Assessment*. Edler L and Kitsos CP (eds.) John Wiley & Sons, Chichester, 2005, pp. 317–327.

Chapter 7

MATHEMATICAL MODELS OF CANCER AND THEIR RELEVANT INSIGHTS

Evans Afenya

Recent developments in the prevention, detection, treatment, and management of cancer are highlighted. The significance and importance of mathematical modeling of cancer is discussed. Some mathematical models of cancer are discussed, reviewed, appraised, and related to clinical studies, with a special focus on the cancers of the disseminated type. Special emphasis is placed on mathematical models of acute leukemia, being a dangerous type of cancer. Insights and predictions brought on by the models are discussed and placed within clinical context. These insights and predictions are also placed within the framework of the growing relevance of mathematical models to clinical studies and investigations. The models are used to demonstrate and spotlight the importance of looking at cancer treatment as a formal optimization problem.

Keywords: Mathematical models, cancer, leukemia, multiple myeloma.

1. INTRODUCTION

The American Cancer Society (ACS) recently reported that about 3000 fewer people died from cancer in the United States from 2003 to 2004. It said the big decrease shows that not only has the death rate from cancer been reversed, but it has been reversed so much that fewer people are dying, even though the population of elderly people, who are most susceptible to cancer, is growing (ACS homepage, 2007). The society projected there will be 559,650 deaths from cancer in 2007 and also predicted there will be 1,444,920 new cases of cancer in this same year; 766,860 among men and 678,060 among women. Obviously, the last two decades have witnessed significant advances in the detection, prevention, treatment, and maintenance of cancer. Improvements have been made in the detection of minimal residual disease through the use of polymerase chain reaction techniques that can detect one malignant cell in 100,000 cells (Schrier, 1995; Campana et al., 2001). The use of recombinant hematopoietic growth factors, for example, to stimulate the growth of normal cells while ensuring an appreciably high malignant cell kill holds great promise for the future (Bernstein, 1993; Kalaycio et al., 2001). Despite the advances, however, cancer presently remains one of the challenges posed to society as a whole. It is the second leading cause of death in the United States (ACS homepage). It is predicted that nearly half of all men and a little over one third of all women in the United States will develop cancer during their lifetime (ACS homepage).

Cancer involves the growth of abnormal cells that do not obey or respond to normal regulatory mechanisms. Such cells normally sustain DNA damage that the damage repair system is not able to deal with and tend to proliferate uncontrollably leading to their spread across various parts of the body where they replace normal tissue and cause hematopoietic malfunction, as is characteristic in the acute leukemias. Cancer can be divided into two broad groups. These are solid tumor cancers and dispersed or disseminated cancers. Solid tumor cancers include those of the breast, lungs, liver, pancreas, prostate, and colon, to mention a few. The leukemias, myelomas, and lymphomas make up the disseminated ones. In solid tumors, an abnormal clone of cells forms as a tumor at a specific location in the human body. If the tumor happens to be malignant or cancerous then it proliferates uncontrollably and metastasizes to other parts of the body. On the other hand benign tumors do not possess metastatic properties and are rarely life threatening. In disseminated cancers such as the leukemias a malignant cell may emerge in the bone marrow, proliferate, and replace normal cells in the marrow, peripheral blood, and other body and blood-forming organs and, this may lead to fatalities. Advances that have been made so far in cancer treatment are basically aimed at mitigating fatalities and creating the necessary conditions for cure.

The most effective kind of treatment currently being used against cancer is chemotherapy, which is aimed at inducing remissions and creating the necessary conditions that will make marrow transplantation possible. Chemotherapeutic induction of bone marrow recovery depends to a large extent on the differential regrowth pattern of normal and abnormal cells. In this setting, it is expected that normal cells may repopulate the hematopoietic system more rapidly and assume dominance over any small and latent malignant subpopulation while efforts are made to wipe out this hidden population. The recovery of normal hematopoiesis, therefore, constitutes a major problem in chemotherapy. In association with this problem are others such as low remission rates, longer time periods for achieving remissions, drug-resistant strains of malignant cells, severe and prolonged marrow failure, common early deaths, and unusual long-term survivors. Clinicians have long been grappling with these problems and as a consequence, a number of treatment strategies have arisen out of *in vitro* and *in vivo* clinical studies, with varying degrees of success or failure in different situations. These strategies include combination therapies such as the use of multiple drug treatments, radiation plus chemotherapeutic treatments, and immunotherapeutic plus chemotherapeutic treatments.

In seeking the best treatment strategies and options, clinicians (Citron *et al.*, 2003) are now employing insights brought on by mathematical models. The work of Citron and his co-workers in 2003 was entirely based on the insights and predictions offered by mathematical models proposed by Norton and Simon in 1986. Norton and Simon (1986) postulated that dose densification could possibly be a viable treatment option against cancer but it took quite a reasonable amount of time to test this option in the clinic. Dose densification involves the delivery of heavy concentrations of anti-cancer drugs over time intervals that are shorter than the standard treatment time intervals. Piccart-Gebhart (2003) points out that it took more than 15 years to test the concept of dose densification in the clinic partly due to concerns about the safety of such an approach. He also explains that much of the energy of the oncology community in the last two decades has been driven by specific drug questions, to the neglect of most of the other key variables of chemotherapy that might turn out to be of utmost importance. These include the timing of chemotherapy in relation to tumor resection and initiation of endocrine therapy, the duration of chemotherapy, and the schedule of drug administration. Nonetheless, the testing of the Norton-Simon hypothesis (1986) in the clinic is an indication that the biomedical community is welcoming and opening another significant front in the fight against cancer and other diseases.

Arising from the aforementioned, the objective of this chapter is to review mathematical models of cancer and the insights they yield that are of relevance to diagnosis and treatment. Since cancer is a very broad area of research that encompasses solid and disseminated tumors, the focus here will be on mathematical modeling of disseminated cancers, an area that has witnessed a rather low level of modeling activity compared to solid tumor cancers. Mathematical modeling of cancer began a number of years ago but has intensified and become an important enterprise over the past few decades as the struggle to find a cure for this disease continues. As alluded to earlier, some of the models have provided useful insights that are now being tested in the clinic. Even though a reasonable gulf still exists between work at the level of mathematical modeling of cancer on the one hand and clinical investigations of this disease on the other hand, a coalescence of these two areas is becoming possible as is evidenced by the testing of the work of Norton and Simon (1986) in the clinic. This coalescence is also being made possible by the introduction of more biologically and biomedically-driven and insightful models that clinicians are finding to be useful and easier to understand and employ. It is becoming increasingly clear that as clinicians face the future challenges of ethical questions related to experiments involving animals and humans, mathematical models could be looked at as providing important insights into treatment issues that even though may hold the key to breakthroughs in the clinic, could be considered to be dangerous when experimented in human clinical trials. It is within this context that this chapter is written.

We conduct an appraisal of existing mathematical models of cancer in Section 2 with a focus on cancers of the disseminated type. In Section 3 we review some treatment models. Issues related to parameter estimation are discussed in Section 4 and concluding remarks can be found in Section 5.

2. CANCER MODELS

In this section we consider some models of disseminated cancer that have appeared in the literature over the past few decades. A survey of the literature reveals that it is reasonably replete with modeling activity related to solid tumors (Bajzer, 1999; Bajzer *et al.*, 1997; Bassukas, 1994; Calderón and

Kwembe, 1991; Casey, 1934; Frenzen and Murray, 1986; Gyllenberg and Webb, 1989, Laird, 1965; Norton, 1988; Kendal, 1985; Simpson-Herren and Lloyd, 1970; Steel, 1977) but the same cannot be said of the dispersed tumors. One of our aims in this chapter is to highlight the importance of modeling the disseminated cancers as a way of filling a significant void that currently exists between the extensive work going on in biomedical investigations of diseases such as leukemia, lymphoma, and myeloma on one hand and biomathematical investigations of these diseases on the other hand.

2.1. Models of Leukemia

A survey of the literature shows that the few mathematical models that have focused on the disseminated cancers have been on chronic or acute leukemias, to date. Thus, our review here would spotlight models of leukemia and insights generated by these models, as a representative of the disseminated cancers. Leukemia is by itself divided into many types depending on the type of abnormal cell contributing mainly to the malignancy and can arise in lymphoid or myeloid cell lines. The chronic types progress and worsen slowly while the acute types progress and worsen very rapidly. Our special emphasis in this discourse will be on acute leukemia, being the most dangerous among the leukemia types. The most common types of leukemia are chronic lymphocytic leukemia (CLL), chronic myeloid leukemia (CML), acute lymphocytic leukemia (ALL), and acute myeloid leukemia (AML). Leukemia is characterized by a disorganization of the hematopoietic system in which abnormal cells replace normal cells in the bone marrow and peripheral blood and cause a malfunction of body organs leading to fatalities.

The challenges presented by leukemia led to the introduction of a model of acute myeloblastic leukemia by Rubinow and Lebowitz in 1976. Their model was based on Clarkson's (1972) viewpoint that there exist side by side in leukemia two cell populations: the normal neutrophil cell system and a population of leukemic cells. They assumed that the normal and leukemic populations had active, resting, and blood compartments and possessed similar growth characteristics but the leukemic cells had an aberrant set of kinetic parameters. Another assumption was that the normal cells had a maturation compartment but the leukemic cells had none. The structure of these cell populations is captured by considering the age-time and maturity-time representations of such entities. To understand the age-time and maturity-time formalisms, we first note that in a cell population, each cell can be considered to have a characteristic chronological age a, the time elapsed since its birth. The population of cells of any age could be assumed to be large and as such can be represented as varying continuously. Consequently, the cell density function given by $n(a, t)$, which is the proportion of cells of a given age in the population, describes the continuous functional nature of the aging population. In this case, $n(a, t)da$ describes the number of cells at time t that are aging in the age interval a to $a + da$ where da represents a change in age. The age of a cell is in general not an observable quantity but many cell properties such as mitosis and cell division that are observable are age-dependent and this makes the age-time formalism plausible. For the maturity-time formalism we note that some of the possibilities that exist for defining cell maturity are based on the concepts of cell volume and the amount of DNA in a cell. In simple cells such as *E. coli*, DNA synthesis proceeds from the moment of birth at a linear rate, which makes the amount of DNA contained in a cell a good measure of its state of maturity. Thus, cells may go through a relatively finite interval of maturation but the cell aging process would proceed from the stage of birth over an extended period that surpasses the maturation interval. Accordingly, in a similar fashion, the cell density function given by $n(\mu, t)$, which is the proportion of cells at a given stage of maturity, describes the continuous functional nature of the maturing population and $n(\mu, t)d\mu$ represents the number of cells at time t that are maturing in the maturity interval μ to $\mu + d\mu$.

As is characteristic with biological phenomena, the rate at which the proportion of cells of a given age varies takes place in relation to the same proportion available. This yields the equation given by:

$$\frac{d}{dt}n(a, t) = \gamma n(a, t) \qquad (1)$$

where γ is a proportionality constant. Application of the chain rule to Eq. (1) produces the von Foerster (1959) equation given by:

$$\frac{\partial n}{\partial t} + \frac{\partial n}{\partial a} = -\lambda n \qquad (2)$$

with the initial condition

$$n(a, 0) = f(a) \qquad (3)$$

where $f(a)$ is a given age distribution function and the boundary condition at $a = 0$ is

$$n(0, t) = 2 \int_0^\infty \lambda n(a', t) da'. \tag{4}$$

The parameter γ in Eq. (1) assumes the value $-\lambda$ in Eq. (2) with the minus sign describing cell death or disappearance or division and λ represents the fractional probability per unit time for such cell death, disappearance, or division. The factor of two that appears in front of the integral sign quantifies the assumption that cell division results in two new cells being formed. When $n = n(\mu, t)$, an analogous form of Eq. (2) results with an equation of the form

$$\frac{\partial n}{\partial t} + \frac{\partial}{\partial \mu}(v\mu) = -\lambda n \tag{5}$$

defined over a finite maturation interval $\mu_0 < \mu \leq \mu_1$ where v is the velocity of maturation and is a prescribed function and μ_0 and μ_1 are constants. The parameter λ represents cell disappearance due to death or other means but excluding cell division. The cell birth process is represented by the boundary condition

$$n(\mu_0, t)v(\mu_0) = 2n(\mu_1, t)v(\mu_1) \tag{6}$$

and as in Eq. (3) a similar initial condition is given by

$$n(\mu, 0) = g(\mu) \tag{7}$$

with $g(\mu)$ being a given maturation distribution.

To obtain further understanding of how Eqs. (1)–(7) play their roles in quantifying the assumptions of Rubinow and Lebowitz (1976), we will first consider the natural history of a particular type of leukocyte called the neutrophil or granulocyte that is produced in the bone marrow and which could become malignant and give rise to different types of disseminated cancer. The natural history of the neutrophil supposedly starts with the pluripotent stem cell that proliferates and divides into the myeloblast. Myeloblasts also proliferate and divide into promyelocytes that in turn divide into myelocytes that are also proliferative. After becoming myelocytes, a stage beyond which all granulocytes are non-proliferative, cells undergo a period of maturation and then get expelled from the marrow into the blood through some mechanisms where they may experience random death or disappearance. Since enough is not known about the kinetic details of the individual cell types, we lump the cells with proliferative properties into active (A) and resting (G_0) compartments. The maturation (M_n) compartment comprises cells that are only maturing and are non-proliferative with the maturing cells entering the blood compartment (B). We note that in the active compartment of the normal marrow, cells that proliferate in active cycle are defined by the G_1 (pre-DNA synthesis), S (DNA synthesis), G_2(post-DNA synthesis or pre-mitotic), and M (mitosis) stages in that order. In this active compartment, cells also mature over a finite time interval and divide after going through mitosis (M). Cells that are not in active phase stay in a resting or dormant state (G_0) and may just age. Cells in active cycle could become dormant and enter the resting phase and *vice versa*. Cells in the resting phase move into a phase of maturation and the cells that go through the stage of maturation arrive in the blood to perform phagocytic and other functions. Cells performing such functions in the blood just go through a process of aging. A schematic description of the processes in the active, resting, maturation, and blood pools is given in Figure 1.

We now translate the descriptions given above into mathematical terms by letting $n(\mu, t)$, $g(a, t)$, $m(\mu, t)$, and $b(a, t)$ represent the cell density

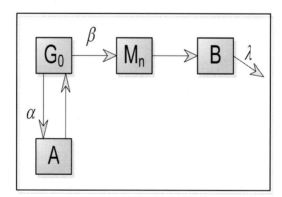

Fig. 1. Schematic description of the normal neutrophil production system. The compartments A, G_0, M_n, and B represent the active, resting, maturation, and blood states, respectively.

functions in the active, resting, maturation, and blood compartments, respectively. Consequently, along the lines described by Eqs. (1)–(7), we obtain the following model system that describes cell dynamics in the various compartments:

$$\frac{\partial n(\mu, t)}{\partial t} + \frac{\partial n(\mu, t)}{\partial \mu} = 0, \quad 0 < \mu \leq T_A \tag{8}$$

$$\frac{\partial g(a, t)}{\partial t} + \frac{\partial g(a, t)}{\partial a} = -(\alpha + \beta)g(a, t), \quad 0 < a \tag{9}$$

$$\frac{\partial m(\mu, t)}{\partial t} + \frac{\partial m(\mu, t)}{\partial \mu} = 0, \quad 0 < \mu \leq T_M \tag{10}$$

$$\frac{\partial b(a, t)}{\partial t} + \frac{\partial b(a, t)}{\partial a} = -\lambda b, \quad 0 < a. \tag{11}$$

The right hand sides in these equations represent cell loss and in this case cell loss from the active and maturation compartments are assumed to be negligible prompting the right hand sides of Eqs. (8) and (10) to be represented by zeros. In this model, feedback control is exercised by making the parameters α and β depend on the total population of cells in all the compartments, $N(t)$. The specific functional dependence of α and β on $N(t)$ is taken to be as follows:

$$\alpha(t) = \alpha_0 + \alpha_1 \left[\left(\frac{\overline{N}}{N(t)} \right)^\nu - 1 \right], \tag{12}$$

$$\beta(t) = \alpha_0 + \beta_1 \left[\left(\frac{\overline{N}}{N(t)} \right)^\nu - 1 \right], \tag{13}$$

where $\alpha_0, \alpha_1, \beta_1, \overline{N}$, and ν are positive constants, and $\alpha_1 > \beta_1$. $\overline{N}$ represents the steady state population of the system. To demonstrate the nature of the feedback control that guarantees the process of homeostasis, a consideration of the functional dependence chosen in Eqs. (12) and (13) shows that when underproduction of cells occurs, $N(t) < \overline{N}$, and the resting compartment responds to depletions in the total population by increasing production and sending more cells to mature since $\frac{\overline{N}}{N(t)} > 1$ in both equations yielding $\alpha > \alpha_0$ and $\beta > \alpha_0$. On the other hand, a state of overproduction of cells brings about $N(t) > \overline{N}$, making $\frac{\overline{N}}{N(t)} < 1$ and resulting in less cells being produced or sent to maturity. The fractional loss rate in the blood, λ, is assumed to be a constant. Following Eqs. (3), (4), (6), and (7), the boundary conditions are given by;

$$n(0, t) = \alpha N_0(t), \quad g(0, t) = 2n(T_A, t),$$
$$m(0, t) = \beta N_0(t), \quad b(0, t) = \gamma N_M(t), \tag{14}$$

and the initial conditions are:

$$n(\mu, 0) = f_A(\mu), \quad g(a, 0) = f_G(a),$$
$$m(\mu, 0) = f_M(\mu), \quad b(a, 0) = f_B(a). \tag{15}$$

Here, $N_0(t)$ represents the total population in the resting compartment, $N_M(t)$ represents the total population of cells in the maturation compartment, the factor of two present in the second equation of Eq. (14) represents the fact that two cells are produced by division of cells in the active compartment when they reach the age T_A, and the quantities appearing on the right sides of equations making up Eq. (15) are all given distribution functions in the active, resting, maturation, and blood compartments, respectively. Analogous to $N_0(t)$ and $N_M(t)$, quantities $N_A(t)$ and $N_B(t)$ would represent the total population of cells in the active and blood compartments, respectively. The total compartment populations at any time can be defined in terms of the cell density functions as follows:

$$N_A(t) = \int_0^{T_A} n(\mu, t) d\mu, \quad N_0(t) = \int_0^{\infty} g(a, t) da,$$
$$N_M(t) = \int_0^{T_M} m(\mu, t) d\mu, \quad N_B(t) = \int_0^{\infty} b(a, t) da, \tag{16}$$

with $N(t) = N_A(t) + N_0(t) + N_M(t) + N_B(t)$.

Integrating Eq. (8) over μ and utilizing Eq. (16), we obtain

$$\frac{dN_A(t)}{dt} + n(T_A, t) - n(0, t) = 0. \tag{17}$$

From the method of characteristics (Haberman, 1997), the general solution of an equation such as Eq. (8) without regard to initial or boundary conditions is an arbitrary function of $t - \mu$. Thus,

$$n(\mu, t) = \begin{cases} n(\mu - t, 0), & t \leq \mu, \\ n(0, t - \mu), & t > \mu. \end{cases} \quad (18)$$

For $t < \mu$, in view of the initial condition in Eq. (15), we obtain

$$n(\mu, t) = f_A(\mu - t), \quad t \leq \mu. \quad (19)$$

Accordingly, the function $n(T_A, t)$ appearing in Eq. (17) is given by the expression

$$n(T_A, t) = \begin{cases} f_A(T_A - t), & t \leq T_A, \\ n(0, t - T_A), & t > T_A. \end{cases} \quad (20)$$

By substituting the expression in Eq. (20) and the boundary conditions in Eqs. (14) into Eq. (17) and performing similar manipulations of Eqs. (9) to (11), we obtain the following system of ordinary differential-difference equations for the total number of cells in each compartment.

$$\frac{dN_A}{dt} = \alpha(t)N_0(t) - \begin{cases} f_A(T_A - t), & 0 < t \leq T_A \\ \alpha(t - T_A)N_0(t - T_A), & t > T_A \end{cases} \quad (21)$$

$$\frac{dN_0}{dt} = -[\alpha(t) + \beta(t)]N_0(t) - 2\begin{cases} f_A(T_A - t), & 0 < t \leq T_A \\ \alpha(t - T_A)N_0(t - T_A), & t > T_A \end{cases} \quad (22)$$

$$\frac{dN_M}{dt} = \beta(t)N_0(t) - 2\begin{cases} f_M(T_M - t), & 0 < t \leq T_M \\ \beta(t - T_M)N_0(t - T_M), & t > T_M \end{cases} \quad (23)$$

$$\frac{dN_B}{dt} = \beta(t)N_0(t) - \lambda N_B(t). \quad (24)$$

Similar equations are obtained for abnormal cells in the active, resting, and blood compartments with a missing maturation compartment because abnormal cells do not possess maturation properties as normal cells. The leukemic state is represented by a similar set of parametric functions but with primes on them such as α', β', and λ'. Interaction between normal and abnormal cells is represented through the control parameters α, β, α', and β' in the forms

$$\alpha(t) = \alpha_0 + \alpha_1 \log\left[\frac{\overline{N}}{N(t) + N'(t)}\right] \quad (25)$$

$$\beta(t) = \alpha_0 + \beta_1 \log\left[\frac{\overline{N}}{N(t) + N'(t)}\right] \quad (26)$$

$$\alpha'(t) = \alpha'_0 + \alpha'_1 \log\left[\frac{\overline{N'}}{N(t) + N'(t)}\right] \quad (27)$$

$$\beta'(t) = \alpha'_0 + \beta'_1 \log\left[\frac{\overline{N'}}{N(t) + N'(t)}\right] \quad (28)$$

where α_0, α_1, β_0, β_1, α'_0, α'_1, β'_0, and β'_1 are positive constants, $\overline{N}$ is the equilibrium level of normal cells, $\overline{N'}$ is the equilibrium level of abnormal cells, $N(t)$ represents the total number of normal cells at time t, and $N'(t)$ is the total population of abnormal cells at time t. Existence of the equilibrium levels of normal and abnormal cells may be due to a situation in which some form of homeostatic co-existence may be present between such cells until the disease takes hold leading to either detection and treatment or death. Rubinow and Lebowitz (1976a) used the cell cycle model given by Eqs. (21)–(24) but with Eq. (23) suppressed along with Eqs. (25)–(28) to capture some essential dynamical features of the acute leukemic state in which the simulations showed that the main effect of the interaction of the abnormal population, which was said to possess an aberrant set of kinetic parameters, with the normal population is to destabilize the normal homeostatic control mechanisms of this latter population. This results in a situation where there are no normal cells but an abundance of leukemic cells. A case of aberrance of the control parameters in Eqs. (25) to (28) may be manifested in a situation in which overproduction of abnormal cells occur such that the normal cell population operates at a level where $N(t) + N'(t) > \overline{N}$ but $N(t) + N'(t) < \overline{N'}$. In this case, $\alpha < \alpha_0$ and $\beta < \alpha_0$ meaning less cells in the normal state would be produced or sent to mature.

On the other hand, $\alpha' > \alpha'_0$ and $\beta' > \alpha'_0$ meaning more abnormal cells would be produced and this does not augur well for the system.

We note that the model proposed by Rubinow and Lebowitz (1976a) was the first attempt to address the problem of modeling a disseminated cancer such as acute myeloblastic leukemia by adopting a cell cycle approach that involved an appropriate inclusion of time delays in the development of normal and abnormal cells. Strengths of the model laid in the inclusion of the Gompertz growth law for normal and abnormal cells and the inclusion of time delays. The Gompertz growth curve, even though empirical, has been shown to fit appreciably well to tumor data (Laird, 1965; Skipper and Perry, 1970; Afenya and Calderón, 2000). However, the model has its limitations like any other model. Most notably, even though the model expresses the dampening of normal cell growth through the control parameters α and β, it does not address the important phenomenon of contact inhibition arising from leukemic crowding and occupation of spaces available for normal cell growth. Nonetheless, the inclusion of such phenomenon in the model could have resulted in the further coupling of the model equations and generated a level of complexity that would have made it difficult to arrive at a major conclusion of that investigative work: the leukemic state is not uncontrolled growth but controlled growth with an aberrant set of parameters. In any case, this conclusion needs to be studied further in the face of current clinical findings and results regarding cancer. The Rubinow-Lebowitz model also brings time delays into focus. The current cell population level of a cellular organism may be based on its past history some time units ago. Therefore, studying such past history which presupposes the existence of a time delay in growth may be important and hold the key to correctly predicting the evolutionary dynamics of the organism. Delays constitute an essential feature in biomedical systems. Replacing an instantaneous interaction in a model by a delayed one, for example, can give a qualitative change in the behavior of the variables that describe a system. The most familiar effect is a change in the stability of a steady state that could drastically affect the system.

The work of Rubinow and Lebowitz was followed by that of Djulbegovic and Svetina in 1985, close to a decade after. Djulbegovic and Svetina introduced modifications to the cell cycle model of Burns and Tannock (1970) and Smith and Martin (1973) by describing the behavior of a single cell population in two phases; A and B. Phase A was the resting phase of the cell cycle and Phase B was the proliferating phase. By drawing parallels and similarities between their work and that of Rubinow and Lebowitz, they analyzed a single cell population model and extended their analysis to cover two co-existing cell populations in the active and resting phases of the cell cycle; a normal cell population and a leukemic cell population. They related the model parameters to cell growth, cell-cell interaction, and cell differentiation. The leukemic cell population differed from the normal population by having different values of model parameters. Djulbegovic and Svetina showed that the alteration of the growth and cell-cell interaction parameters of the model could be eliminated as possible causes of the leukemic state leading to the modification of the differentiation process as the only possible cause of leukemic development. Their models do not address the characteristics of the cell development process but include time delays. They also do not include the Gompertzian growth of the cells. It would be instructive to see whether the same kind of conclusions would be arrived at if such a growth law was included and further coupling due to leukemic crowding was introduced into their model.

More than a decade after the work of Djulbegovic and Svetina (1985) and two decades after that of Rubinow and Lebowitz (1976a) modeling activity on acute leukemia was virtually non-existent until the work of Afenya (1996) and Afenya and Calderón (1996) appeared. Employing the viewpoint of Clarkson (1972) as adopted in earlier works (Rubinow and Lebowitz, 1976a; Djulbegovic and Svetina, 1985), they proposed a simple non-cell cycle model to investigate acute leukemia by having single co-existing populations of normal and abnormal cells with the abnormal cells interacting and inhibiting the normal cells through a mass-action type crowding term. The basic model is as follows:

$$\frac{dN_l}{dt} = g\left(\log \frac{A_l}{N_l}\right) N_l - fN_l, \quad g > 0, \quad f > 0 \qquad (29)$$

$$\frac{dN}{dt} = a\left(\log \frac{A_n}{N}\right) N - bN - cNN_l, \quad a > 0, \quad b > 0, \quad c > 0 \qquad (30)$$

$$N(0) = N_0, \quad N_l(0) = N_{l0}. \qquad (31)$$

Here, A_l and A_n are the respective asymptotic bounds of the malignant and normal cell populations. Parameter a is the intrinsic growth rate of the normal cells and b is their death rate. Parameter c is a measure of the degree of inhibition of the normal cells with respect to the abnormal cells. The parameter g is the intrinsic growth rate of the malignant cells and f is their death rate. An inspection of model system (29)–(31) shows that the death terms in Eqs. (29) and (30) are explicitly separated from and not included in the Gompertz components of the equations. Even though inclusion of the death terms is plausible, separation of such terms may endow the model system with a reasonable level of definitiveness given that it already combines and does not show a number of aspects of the cell growth structure such as cells in the active, resting, marrow, and peripheral blood states.

The analyses and simulations of their models (Afenya, 1996; Afenya and Calderón, 1996) show that among a number of factors supporting malignant development, leukemic inhibition, no matter how small in magnitude or strength, profoundly affects the growth of normal cells. Leukemic inhibition plays the role of altering and driving down the normal cell steady state to a point where normal cell recovery cannot be guaranteed. Model predictions indicate that the inhibitive process is due to the space-occupying effects of a rapidly proliferating and accumulating malignant cell population in addition to nutritional and other factors that lead to the displacement and mortality of the normal clones.

The simple models (Afenya, 1996; Afenya and Calderón, 1996) capture certain essential characteristics of normal and abnormal cell behavior in the malignant state. They explicitly account for and show how malignant inhibition due to cell-cell interaction and leukemic crowding could adversely affect the continued survival of normal neutrophils. Support for such leukemic inhibition is found in the work of Irene *et al.* (1979) where it is pointed out that malignant cells produce chalones that inhibit the growth of normal cells. The work of Clarkson and his co-workers (1967) also shows that certain possibilities could ensure and promote leukemic development. Those possibilities include: (a) competition for essential nutrients between the normal and leukemic cells which could lead to competitive exclusion and extinction of the normal cell population and (b) contact inhibition and production of growth inhibitors by leukemic cells. The problem, though, is how such an inhibitive parameter that is described by c in model Eq. (30) can be measured either *in vivo* or *in vitro*. Measurement of such a parameter may go a long way to express the degree of coupling or interaction existing between normal and malignant cells and reveal a quantitative level of suppression exercised by malignant cells over normal cells and this could have an impact on treatment. The models (Afenya, 1996; Afenya and Calderón, 1996) also include the notion of Gompertzian growth of normal and malignant cells but do not include cell cycle descriptions since the cells are lumped together into single co-existing cell populations. Such lumping constitutes a limitation on the models from a cell cycle oriented viewpoint but they provide some important insights that were built upon in the work of Afenya and Bentil (1998).

One important prediction arising from the models of Afenya (1996) and Afenya and Calderón (1996) was that leukemic inhibition is due to the space-occupying effects of malignant cells. Such effects lead to the occupation by malignant cells of subendosteal sites where hematopoiesis is preferentially resident, thus hampering normal cell growth. In investigating the space-occupying effects, Afenya and Bentil (1998) made a number of assumptions about normal and abnormal cells. These assumptions yielded a space-time model given by:

$$L_t = gL\left(\log \frac{A_l}{L}\right) - fL + d_l \nabla^2 L, \quad g > 0, \quad f > 0 \tag{32}$$

$$N_t = aN\left(\log \frac{A_n}{N}\right) - bN - cNL + d_n \nabla^2 N, \quad a > 0, \quad b > 0, \quad c > 0 \tag{33}$$

with initial conditions

$$N(\mathbf{s}, 0) = N_0, \quad L(\mathbf{s}, 0) = L_0, \tag{34}$$

where the parameters A_l and A_n are the respective asymptotic bounds on the abnormal and normal cell populations. Since malignant activity is taking place in the body, which can be described as a closed environment (Murray, 2003), zero flux boundary conditions of the following form are imposed:

$$(\mathbf{n} \cdot \nabla)\mathbf{W} = 0. \tag{35}$$

Vector **W** is such that $\mathbf{W} = (L, N)$ where $L = L(\mathbf{s}, t)$, $N = N(\mathbf{s}, t)$. Vector **s** represents the spatial dimensions such as the Cartesian coordinate system that we can in this case refer to as the space of distribution, t represents time, and **n** is the unit outward normal to the space of distribution. The parameters a, b, c, g, f are as described in relation to model system (29)–(31). The diffusive coefficients for the abnormal and normal cell populations are given by d_l and d_n, respectively.

Model system (32)–(35) is an expanded form of model system (29)–(31) that includes spatial dimensions with essentially Gompertzian and cell cycle independent kinetics. The Laplacian terms denoted by $\nabla^2 L$ and $\nabla^2 N$ capture the diffusive properties of the abnormal and normal cells in arbitrary spatial dimensions. Analysis of the model was carried out in a one-dimensional geometrical setting in which the operator ∇^2 was set equal to the second order partial derivative $\partial^2/\partial x^2$ of the normal and abnormal cell population with x representing the real number line. In validating the analysis that could naturally be extended to higher spatial dimensions, simulations of the model were considered in two-and three-dimensional space and in radial geometries. Since the bone marrow is the main organ that supports hematopoiesis, the underlying geometries could be conveniently idealized to be stationed within this organ. By so doing, cells could be considered as either distributing over a long slender bone containing the marrow in a one-dimensional fashion or over a rectangular or radial cross-section of the marrow in a two- or three-dimensional setting.

Analyses and simulations of this simple spatio-temporal description of cell evolution in leukemia led to the following instructive predictions and insights:

(1) Existence of two realistic sets of steady states; one in which there is a co-existence of normal and leukemic cells and the other in which there are leukemic cells and no normal cells.
(2) The steady state that represents a situation of co-existence between normal and abnormal cells breaks down in the face of small space-time perturbations while the steady state with only leukemic cells and no normal cells persists. The persistence of this particular steady state indicates that there is an ultimate situation in which malignant cells replace normal cells.

(3) The breakdown in the steady state of co-existence may occur when there is a large leukemic diffusive coefficient compared to a small normal diffusive coefficient. It may also occur when the size of the space of distribution is relatively large. This may mean that in the malignant state normal cell survival cannot occur unless this state is itself disrupted by an external agent, in this case, a drug (Spinolo, 1994). These predictions and the ensuing simulations in Afenya and Bentil (1998) show, in what follows, that

(4) There may be accumulations of cells at certain sites and depletions at other sites and through such processes malignant cells may occupy sites of the normal cells as the propagation of spatial heterogeneities occur. As a result, it may be suggested that the positions occupied by the leukemic cells as they expand may be very fertile areas that are rich in nutrients needed for hematopoiesis. This is because those positions used to be occupied by the displaced normal cells. Thus, abnormal cell numbers may increase very rapidly.

(5) The rapid increase in leukemic cell numbers may also be tied to the phenomenon of contact inhibition through which normal cell production is disrupted and malignant cell growth is stimulated possibly over a wide region of space.

(6) Upon introduction of leukemic cells, existing normal cell colonies go through a process of shrinkage as their positions are invaded by emerging colonies of abnormal cells. The resulting malignant dominance may cause damage to and disturb the colony-forming capabilities of the normal cells.

The model predictions (Afenya and Bentil, 1998) suggested that through certain diffusive processes and mechanisms, the normal cells are displaced from their positions by colonies of abnormal cells, over a wide region of space, and are driven to extinction. Essentially, the malignant cell colonies display a tendency to invade the spaces "designated" for normal cell growth. Also, over a region of space, the rapid increase in the malignant population over a period of time may result in a high abnormal cell density. This could lead to a migration of the malignant cells, possibly through a diffusive process, to regions of low cell density and nutrient availability. This could account for the reasons why other organs of the body become clogged with masses of abnormal cells, as is noted in the work of Henderson (1986). Figures 2(a) and (b) demonstrate some of the simulation results obtained from model system (32)–(35). These figures have been selected as candidates among all the figures that describe the interesting simulations in Afenya and Bentil (1998). They constitute visualization plots of Figure 5 in Afenya and Bentil (1998) that depict trends followed by normal and abnormal cells in numerical experiments designed to replicate the *in vivo* situation. Figure 2(a) represents contour lines for leukemic cells that are "projected" onto the surface plot for normal cells at steady state. Similarly, contour lines representing normal cells are projected onto the surface plot of leukemic cells in Figure 2(b).

In Figure 2(a) the apexes of the inverted cone-like structures are inhabited by shrinking normal cell colonies and their middle and top portions are inhabited by expanding colonies of leukemic cells. The middle and top portions of the various hills in Figure 2(b) are occupied by colonies of leukemia cells and the very bottom portions are occupied by the displaced and shrinking normal cell colonies. These superimposed surface plots and contours suggest the persistence of the steady state in which only leukemic cells are present. They also reveal the spatial heterogeneities that persists when leukemic cells are present, with aggregates of these abnormal cells being the main occupants of the space of distribution that is represented by the coordinate configurations involving the x- and y-axes as measurement axes placed in the bone marrow or elsewhere with the leukemia (L) or normal (N) population serving as the vertical axis in a right-handed three-dimensional coordinate system. It would be instructive to learn what happens in simulations of an expanded model that includes more cell characteristics.

We recall that in their observations, Rubinow and Lebowitz (1976a) noted that the rise in leukemic cell numbers, anthropomorphically speaking, "fools" the normal cells into "thinking" that a high steady state population has been reached by seeing the leukemic population as normal. Thus, normal cell production is shut off apparently to compensate for an over-abundance of cells that are actually not normal resulting in leukemic domination. Since the normal cells by themselves cannot recognize the abnormal ones, it may rather be appropriate to suggest that the curtailment of normal cell production may be directly linked to the biological mechanisms that include the

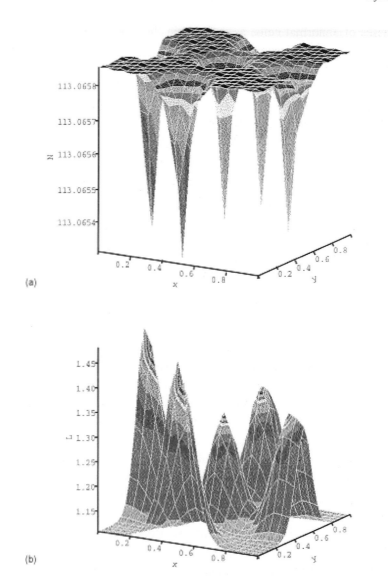

Fig. 2. Visualization plots with axes not matched to scales. Contour lines representing abnormal cells are drawn on the surface plot for normal cells and *vice versa*. The grey and dark areas represent different contour levels. In (a), the tips of the inverted cone-like structures are inhabited by shrinking colonies of normal cells. Their middle and top portions are covered by emergent malignant cell colonies. In (b), the middle and top parts of the various hills are occupied by leukemic cell colonies and their bottom portions are occupied by displaced and shrinking normal cell colonies. The vertical axes in (a) and (b) respectively represent normal and abnormal cell populations multiplied by 10^{10} and the x- and y-axes are measured in non-dimensional units.

phenomenon of leukemic contact inhibition that is explored in the models that have been discussed here (Afenya, 1996; Afenya and Calderón, 1996; Afenya and Bentil, 1998).

We believe that the simple nature of model system (32)–(35) engendered the insightful predictions enumerated above. Afenya (2001a) and Afenya and Calderón (2003) have tried to link model system (32)–(35) to clinical work with a reasonable measure of success. They traced commonalities between model predictions and insights on one hand and findings from work with real time patient data on the other hand (Afenya, 2001a; Afenya and Calderón, 2003) in their quest to validate model system (32)–(35). Specifically, in validating the predictions generated by model systems (29)–(31) and (32)–(35), Afenya (2001a) employs real time patient data supplied by Dr. Wilson Hartz, Jr., formerly the Attending Internist and Hematologist at Sherman Hospital in Elgin, Illinois and Dr. Yoshihito Yawata of Kawasaki Medical School in Japan.

In studies of the data of Dr. Yoshihito Yawata of Japan, agreements were found between the model prediction of the persistence of a steady state that has only abnormal cells and no normal cells and the data of Dr. Yawata that shows a complete packing of the bone marrow by malignant cells (Afenya, 2001a). In trying to find agreements between the model simulations and the data of Dr. Wilson Hartz, Jr., the choice of the abnormal cell type was clearly the leukemic blasts. However, the data involved counts on various normal cell types of two patients with leukemia: promyelocytes, myelocytes, platelets, basophils, metamyelocytes, segmented neutrophils, and band neutrophils, to mention a few. To determine the normal cell type that could serve as a representative of the normal cells, Afenya (2001a) observed behavior stemming from the simulations and used probabilistic and statistical methods that led to the choice of segmented neutrophils (SEGS), compared to other cell types, as the best candidates (with p-value $= 0.0000611$ and p-value $= 0.0042$ for the two respective patients considered) that facilitated the agreements found between the model simulations and data. Figure 3(a) shows the model simulation of normal cell behavior in the malignant state juxtaposed to real time data on the SEGS in Figure 3(b). In considering the model prediction shown in Figure 3(a) about the normal cells, an irregularly oscillatory pattern of decrease was observed in segmented neutrophil (SEGS) numbers (which have a normal

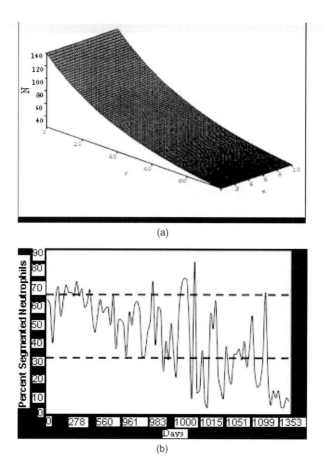

Fig. 3. Decrease in normal cells towards low levels in model simulations and real time patient data when abnormal cells are present. In the simulations in (a) the normal cells decrease monotonically towards low levels while the same pattern of decrease takes place in a patient in (b) but the decrease happens in an irregularly oscillatory fashion. On the average the simulations capture the essential pattern of decrease in the normal cell population. The vertical axis in (a) describes the normal cell population multiplied by 10^{10} and t represents the dimensionless time axis. The variable x represents the spatial axis of measurement that could be made small. In (b), the vertical axis represents percentage of segmented neutrophils and the horizontal axis represents time in days.

range of 31%–66% described by the two dashed horizontal lines in the figure) towards a low level, as is shown in Figure 3(b). It is important to note that the SEGS decreased towards the lower bound of their normal range around Day 988, and went below this lower bound between Days 1091 and

1122, and beyond Day 1190 till the end of the profile. These days are in correspondence with the times (or time ranges) when the blasts displayed abnormal behavior (not shown here — see Afenya (2001a)). Thus, even though these neutrophils oscillated irregularly within their normal range for some period of time, they essentially display a decreasing trend by going below the lower bound of their normal range for a relatively substantial period of time that spans up to the end of the profile. This decreasing trend becomes particularly pronounced towards the end of the profile. On average, the model prediction in Figure 3(a) is in agreement with what could be found in Figure 3(b). A smoothing of the curve in Figure 3(b) tended to look like what is observed in Figure 3(a) (Afenya, 2001a) that essentially captures the pattern of decrease towards lower levels over time when abnormal cells are present. Thus the performance of the model in mimicking behavior observed from the data gives it a reasonable level of validity and makes it relevant when considering candidate models for describing treatment.

It would be important and instructive to expand the models to account for cell cycle behavior in the resting and active phases of cell development and in the process investigate the role played by possible convection and advection mechanisms in the disease state. Such studies may shed further light on malignant development in cancer in general. Even though our special emphasis is on models of acute leukemia, there are a few models of chronic leukemia, specifically chronic myeloid leukemia (CML), that are worth mentioning since chronic leukemia eventually develops into acute leukemia. The models proposed by Wheldon (1975), Wheldon *et al.* (1974), and Mackey and Glass (1977) with regards to studies of CML merit attention. These models have been based on peripheral control of granulopoiesis so they include peripheral feedback loops. They have generated useful insights into oscillatory and chaotic behavior in CML that may aid in the study of other cancers including solid tumors.

2.2. Cell Kinetics

In the models that we have so far considered, the Gompertz growth law was the key driving mechanism governing the cell kinetics. Gompertzian growth of cell populations is to a large extent empirical. The Gompertz function has been found to fit well to solid tumor growth data (Laird,

1965). The literature also shows a number of attempts at justifying Gompertzian growth of solid tumor cells (Calderón and Kwembe, 1991; Frenzen and Murray, 1986; Gyllenberg and Webb, 1989). However, justification for Gompertzian growth of disseminated cancer cells is virtually non-existent. Quite understandably, assumptions made about such growth of leukemia cells for example have been based on the growth of solid tumors. Against this background, we discuss the work of Afenya and Calderón (2000) in which they employ empirical and theoretical arguments to arrive at Gompertzian characterization of the kinetics of disseminated cancer cells.

In the empirical considerations, Afenya and Calderón (2000) discussed the prevailing difficulties in obtaining data on untreated human leukemia since immediately it is detected, its treatment begins. They argued that it would have been ideal to have available for study, large data sets on untreated human leukemia. However, the way in which the disease evolves makes this impossible and this constitutes a problem in itself. In the face of limited definitive data, they opted to work with the data of Skipper and Perry (1970) since they found it to be the most appropriate in addressing the subject matter. Using the method of least squares it was found that the Gompertz growth law was the best fitting curve to the data of Skipper and Perry (1970). The Gompertz curve that fits the data is described by the function:

$$L(t) = 10^{12} \exp[-44.55 e^{-0.0306 t}] \tag{36}$$

where $L(t)$ represents the population of leukemia cells at some time t, which is measured in days. Figure 4 shows a fit of the curve described by Eq. (36) to the data points, with the ordinate representing the leukemic cell population measured on a $\log_{10}$ scale, and the abscissa representing time in days. The Gompertz fit was compared to other growth curves such as the logistic, exponential, and cubic curves as is shown in Figure 5. To determine goodness-of-fit, minimization was carried out on the sum of squared errors for logarithms of the leukemic population sizes given by

$$\Omega = \sum_{i=1}^{p} (\ln L_i - \ln L(t_i))^2 \tag{37}$$

where the quantity p represents the number of data points, L_i represents the predicted populations of leukemic cells at different instances of time,

Mathematical Models of Cancer and Their Relevant Insights 197

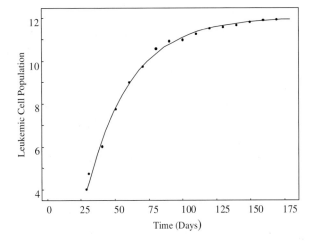

Fig. 4. Fit of the Gompertz curve to leukemia data. The data points are represented by blackened dots. The ordinate represents the leukemia cell population on a $\log_{10}$ scale and the abscissa represents time in days.

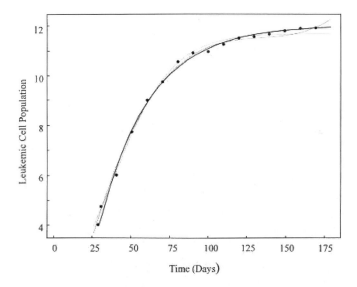

Fig. 5. Comparison of the Gompertz curve to the fit of the logistic, exponential, and cubic curves to leukemia data. The data points are represented by blackened dots. The Gompertz curve is represented by a dark line, the logistic by a thin line, and the cubic by tiny dots. The best-fitting curve is the Gompertz curve followed closely by the logistic. The ordinate represents the leukemia cell population on a $\log_{10}$ scale and the abscissa represents time in days.

and $L(t_i)$ represents the observed population values at those same time instances. Minimization of Eq. (37) for the various curves considered in Figure 5 yielded the Gompertz curve to be the best-fitting curve. This was followed closely by the logistic curve after which came the cubic curve. The poorest fit came from the exponential curve (not shown in the figure).

The empirical considerations above sought to validate the theoretical considerations that we now conduct an appraisal of. In the theoretical studies, it was observed that the bone marrow is a richly cellular and highly vascularized connective tissue (Naeim, 1992; Tavassoli *et al.*, 1983). It is the structure that supports the sophisticated physiological activity of hematopoiesis. It is one of the largest organs in the human body and has a total volume of 30–50 ml per kg body weight with about half of this volume consisting of hemopoietically inactive fatty marrow (Wickramasinghe, 1975). Leukemic stem cells also have their origins in the bone marrow. It is said, though, to be surprisingly uniform (Schrier, 1988). Based on the organizational layout and function of the marrow, the following observations were made (Naeim, 1992; Tavassoli *et al.*, 1983; Wickramasinghe, 1975; Schrier, 1988, 1994):

(a) the bone which encompasses the marrow is tubular with its cross-section showing a cylindrical structure;
(b) there are a great number of small vessels and sinuses in the periphery, in the vicinity of the bone; and
(c) as a consequence, the intensity of hemopoiesis is maximal at the periphery, leaving the central part of the bone (the interior of the marrow) with relatively little or no hemopoeitic activity.

Stemming out of these, a simplified model for malignant transformation dynamics in a cross-section of the marrow was proposed based on the following assumptions:

(1) Cell growth is greater near the boundary than in the center if it is proportional to the blood vessel density or vascularization.
(2) The great number of small vessels and sinuses at the periphery form on the average, aggregates of slender cylindrical rods with small diameters.
(3) The spread of malignancy within the highly vascularized areas that are near the boundary of the marrow results in the propagation of

malignant stem cells outwardly from centers of the blood vessels towards areas of low cell density. This propagation may be proportional to the concentration of metabolite within the bone marrow.

(4) The circulation of nutrients is carried through a diffusive process from the vessels and spread outwardly with respect to their centers.

(5) The maximum reach of the concentration of metabolite, which diffuses outwardly in relation to the small vessels and the amount consumed, occurs at a finite radius R that can be measured from the center of a representative blood vessel.

(6) A scaled representative of hemopoietic activity occurs in tissues that surround a small blood vessel. On the basis of scaling, a representative blood vessel may be considered to have a geometrically small diameter in relation to the greatly vascularized area surrounding it.

(7) The rate of change of the mass of the malignant population of cells located in the highly vascularized areas at time t is proportional to the malignant mass m and the concentration C of metabolite at time t.

Using these observations and assumptions as a background, the proposed model considered a situation in which there is consumption and diffusion of metabolite (at a steady state) in the vicinity of the small vessels, the endosteum, and the hemopoietic marrow. Afenya and Calderón (2000) then employed various mathematical techniques and derivations that led to the sandwiching of the malignant growth in between two Gompertzian expressions thus yielding the equation

$$\frac{dm}{dt} = hm(t)\ln\left[\frac{m_{\max}}{m(t)}\right] \qquad (38)$$

which is a differential equation of Gompertz type, where $m(t)$ represents the malignant mass, $m_{\max}$ is the absolute maximum level to which the malignant mass could get around blood vessels, and h is a positive constant describing the rate of growth of the malignant mass. Afenya and Calderón (2003) have also employed entropy considerations to arrive at Gompertz kinetics of disseminated cancer cells. Since the empirical and theoretical assertions seek to reinforce each other, it is justifiable from an analytical and empirical standpoint, to say that the law that governs the growth of disseminated cancer cells is essentially Gompertzian. Nevertheless, it would be important to carry out more investigations of available data as a way of placing this assertion on firm grounds.

3. CANCER TREATMENT MODELS

As we noted earlier, advances have been made in the treatment of cancer but challenges still remain in the face of guaranteeing prolonged remissions and preventing frequent relapses with the continuing attendant problems such as multidrug resistance. Investigations of various therapeutic strategies have been proceeding steadily alongside debates that are going on about treatment uncertainties (Rowe, 2001). In a pointed remark by Rowe (2001), he says that "it is unknown how many cycles of high dose cytarabine are necessary as the data on this are scanty and it is perhaps incredible how little thinking goes into giving someone four courses rather than two or three courses of high-dose cytarabine with no data whatsoever that this makes a difference. Similarly, in the optimal dose of high-dose cytarabine — anywhere from 1.5 to 3 g/m^2, the optimal duration and the total number of doses have not been established." These sentiments point to the fact that more investigations on all fronts including from the mathematical modeling angle are needed as the fight against cancer continues.

Mathematical modeling of cancer chemotherapy has been going on for more than three decades now and has contributed to ideas of therapy scheduling and dealing with drug resistance. However, as Swierniak *et al.* (1996) admittedly point out, with minor exceptions, practical applications of modeling results have not been encouraging. It is heartening to note, though, that the recent testing of the Norton-Simon hypothesis (1986) by Citron and his co-workers (2003) has opened up a new front in multidisciplinary approaches to cancer chemotherapy that hold good promise for the future. As the search for the best treatment strategies continue, it is becoming clear that mathematical models could be used to address questions related to chemotherapy scheduling with the aim of achieving maximum reductions in the cancer load while at the same time doing least harm to normal cells. The comments of Rowe (2001) above suggests that cancer chemotherapy needs to be looked at as a formal optimization problem, and one branch of applied mathematics that is undoubtedly amenable to addressing such problems is optimal control theory.

Various optimal control models of cancer chemotherapy have been proposed in recent years (Afenya, 2001b; Costa *et al.*, 1995a, b; Martin and Teo, 1994; Matveev and Savkin, 1994, 1995; Swan, 1990; Swan and Vincent, 1977; Swierniak *et al.*, 1996; Zietz and Nicolini, 1979), most of which are in relation to treatment of solid tumors or cancer treatment in general. The models have sought to provide insights arising from cell-cycle specific treatment in some cases and cell-cycle non-specific approaches to therapy in other cases. With regards to our focus on disseminated cancers in this paper we will discuss and review a couple of models that employ optimal control approaches to treatment of such cancers.

3.1. Optimal Control Models

The first application of optimal control theory to a human tumor that was of a disseminated type was done by Swan and Vincent (1977). In their work, they used assumptions about Gompertzian growth of immunoglobulin G (IgG) multiple myeloma cells and developed a single differential equation of drug action on such cells. The equation describing the drug action is as follows:

$$\frac{dL}{dt} = \alpha L \ln \frac{\theta}{L} - \frac{k_1 v L}{k_2 + v}, \quad L(0) = L_0 \qquad (39)$$

where the parameter α represents the growth rate of the tumor, θ represents the greatest size of the tumor, L denotes the number of tumor cells at time t, L_0 is the size of the tumor at the start of treatment, $v(t)$ denotes the drug concentration at the site of action, and k_1 and k_2 are positive constants in the saturation type loss term in the equation. By making suitable changes of variables, the following optimal control model was established:

$$\text{Minimize} \quad J(u) = \int_0^T u(\tau)d\tau, \quad 0 \leq u \leq u_{\max} \qquad (40)$$

$$\text{subject to} \quad \frac{dx}{d\tau} = \frac{pu}{1+u} - x, \quad x(0) = x_0 = C \qquad (41)$$

with a prescribed final state $x(T)$, where $\tau = \alpha t$, $u = v/k_2$, $x = \ln(\theta/L)$, $p = k_1/\alpha$, and $C = \ln(\theta/L_0)$. In an attempt to mimick treatment with the cycle-non-specific drugs melphelan, cyclophosphamide, and prednisone (MCP), it was assumed that the effect of the three drugs could be represented by a single fictitious drug with concentration $u(\tau)$. Thus u is regarded here as a control variable. The integral of u over a treatment interval of dimensionless length T is a measure of toxicity of the anticancer drug concentration in the plasma. This also measures the cumulative drug dose over the entire treatment period as is represented by the performance functional $J(u)$. As Eqs. (40) and (41) show, the aim was to determine the positive continuous time optimal controller u that minimized the toxicity of treatment subject to the growth kinetics of the tumor. Solution of model system (40)–(41) required the construction of the Hamiltonian $H(\tau) = u + \lambda[pu(1+u)^{-1} - x]$ where $H(\tau, u(\tau), x(\tau))$ is written as $H(\tau)$, $\lambda(\tau)$ is the costate variable given by $d\lambda/d\tau = -\partial H/\partial x$, and $\lambda(0)$ is obtained from $H(0) = 0$ on an optimal trajectory. As long as $p > x_0$, the solution was obtained as

$$u(\tau) = \left[1 - (x_0/p)^{1/2}\right]^{-1} e^{\tau/2} - 1 \qquad (42)$$

$$x(\tau) = p\left\{1 - \left[1 - (x_0/p)^{1/2}\right]e^{-\tau/2}\right\}^2 \qquad (43)$$

$$u = (x/p)^{1/2}\left[1 - (x/p)^{1/2}\right]^{-1} \qquad (44)$$

and this satisfied $H(\tau) = 0$ for $\tau \geq 0$.

The solutions show that as τ increases x tends to a constant value which was found to be in accordance with clinically observed features of the chemotherapy of multiple myeloma, where treatment reduces the tumor population to a plateau level. The model also produced a feedback relation in which given the state x, the control could be determined from Eq. (44). However, from Eq. (42), the control becomes unbounded as $\tau \to \infty$ suggesting an increase in the drug concentration over a large treatment time interval but the model simulations (Swan and Vincent, 1977) showed that computed values of u remained below the preassigned value of $u_{\max}$ as in Eq. (40) so the control performed well in reducing the tumor burden. The optimal control solution that suggest the use of a continuous drug delivery method also showed that the accumulated amount of drug was about 1/40 of the accumulated amount with a standard, periodic, discrete-dosage clinical program. The simple nature of the model generated closed form solutions that may be difficult to obtain with increasing complexity. The main conclusion arising from this model after comparisons of the modes of treatment was that the drug should be administered at low doses by employing a continuous drug infusion therapy system instead of the standard discrete-dosage (intermittent) application system. Since the introduction of this model, lots of investigative activities have taken place with some studies advocating high dose treatment of multiple myeloma in association with hematopoietic stem cell rescue and support (Abedi and Elfenbein, 2003; Child *et al.*, 2003; Donato *et al.*, 2004). In light of recent studies, it would be instructive to expand this model to include various treatment characteristics. Such model expansion could also address possible cell cycle kinetics and behavior in an environment of interaction between normal and malignant cells.

Within the context of our special emphasis here on acute leukemia, we note that Afenya (2001b) recently proposed an optimal control model to study the treatment of this disease in the case where there was recombinant hematopoietic growth factor support. The model was obtained by establishing certain assumptions about cell cycle specific and cell cycle nonspecific treatment and superimposing a chemotherapeutic treatment regimen on model system (29)–(31) to get the following:

$$\text{Minimize} \quad J = \int_0^R q \, dt \qquad (45)$$

subject to

$$\frac{dL}{dt} = g\left(\log \frac{L_A}{L}\right) L - fL - ku(t)L \qquad (46)$$

$$\frac{dN}{dt} = a\left(\log \frac{N_A}{N}\right) N - bN - cNL - hu(t)N + G(N(t)) \qquad (47)$$

$$N(0) = N_0, \quad L(0) = L_0 \qquad (48)$$

$$\text{with} \quad u_{\min} \leq u(t) \leq u_{\max} \qquad (49)$$

$$\text{and} \quad \begin{bmatrix} L(R) \\ N(R) \end{bmatrix} = \begin{bmatrix} \hat{L} \\ \hat{N} \end{bmatrix}. \qquad (50)$$

The quantity R in this case denotes the final time of treatment, $[0, R]$ denotes the chemotherapeutic time interval and the parameter q is a constant of proportionality that could be a patient-dependent parameter to be specified by the oncologist based on the prognosis of the patient. Equation (45) shows that the cost of treatment is proportional to the treatment time interval. Quantities L_0 and N_0 are the respective populations of leukemic and normal cells at detection. Quantities $\hat{L}$ and $\hat{N}$ denote the respective final populations of malignant and normal cells at the end of treatment. It was expected that $\hat{L}$ will be small or negligible and $\hat{N}$ will be large enough to guarantee and facilitate normal cell regrowth after treatment is discontinued. From the assumptions, $G(N(t)) = rN$, where r is a constant that may be defined as the recovery rate per unit time of normal cells due to the infusion of recombinant hematopoeitic growth factors. The quantity $u_{\min}$ indicates that a minimum drug level could be set by the oncologist in order to achieve reasonable cell kill. It could just as well be set to zero by the oncologist to depict the point at which no drug is yet applied. Quantity $u_{\max}$ means the drug is to be set at a maximum tolerable concentration by the oncologist or set at an appreciably high dosage if the patient's body could withstand the drug insult.

By letting $x = \log \frac{L_A}{L}$, $y = \log \frac{N_A}{N}$, $\tau = gt$, $\alpha = \frac{a}{g}$, $\beta = \frac{b-r}{g}$, $\gamma = \frac{cL_A}{g}$, $\mu = \frac{f}{g}$, $\eta = \frac{k}{g}$, $\kappa = \frac{q}{g}$, $T = gR$, and $\nu = \frac{h}{g}$ with a choice in which $q = g$, $u_{\min} = 0$, and $u_{\max} = 1$, model system (45)–(50) becomes transformed into the following model:

$$\text{Minimize} \quad J = \int_0^T d\tau \tag{51}$$

subject to

$$\frac{dx}{d\tau} = -x + \mu + \eta u(\tau) \tag{52}$$

$$\frac{dy}{d\tau} = \gamma e^{-x} - \alpha y + \beta + \nu u(\tau) \tag{53}$$

$$x(0) = x_0, \quad y(0) = y_0 \tag{54}$$

$$\text{with} \quad 0 \leq u(\tau) \leq 1 \tag{55}$$

$$\text{and} \quad \begin{bmatrix} x(T) \\ y(T) \end{bmatrix} = \begin{bmatrix} \hat{x} \\ \hat{y} \end{bmatrix} \tag{56}$$

with the expectation that $\hat{x}$ is large, $\hat{y}$ is reasonably small, and T is free. The case $u(\tau) = 0$ is equivalent to no drugs applied and the case $u(\tau) = 1$ is equivalent to the application of high concentrations of drugs.

To solve this problem, a Hamiltonian of the following form was obtained

$$H(\tau, x, y, u) = 1 + \lambda_x[-x + \mu + \eta u(\tau)] + \lambda_y[-\alpha y + \beta + \gamma e^{-x} + \nu u(\tau)]. \quad (57)$$

The costate equations are:

$$\dot{\lambda}_x = \lambda_x + \lambda_y \gamma e^{-x} \quad (58)$$

$$\dot{\lambda}_y = \alpha \lambda_y. \quad (59)$$

The Pontryagin minimum principle (Kirk, 1970; Martin and Teo, 1994), was applied to get

$$(\eta \lambda_x^* + \nu \lambda_y^*) u^*(\tau) \leq (\eta \lambda_x^* + \nu \lambda_y^*) u(\tau), \quad (60)$$

where $u^*(\tau)$ is the control and is a measure of the toxic effects of the drugs that needed to be determined to ensure that a high malignant cell kill was achieved. By letting $S(\tau) = \eta \lambda_x + \nu \lambda_y$, $u^*(\tau)$ was obtained as

$$u^*(\tau) = \begin{cases} 0 & \text{if } S(\tau) > 0 \\ \in (0, 1) & \text{if } S(\tau) = 0 \\ 1 & \text{if } S(\tau) < 0 \end{cases}. \quad (61)$$

By proving a proposition (Afenya, 2001a) to the effect that no time interval during the period of chemotherapy existed over which $S(\tau) \equiv 0$, Eq. (61) became

$$u^*(\tau) = \begin{cases} 1 & \forall \ \tau \in [0, \tau_s) \text{ and} \\ 0 & \forall \ \tau \in [\tau_s, T] \end{cases} \quad (62)$$

where τ_s is the switching time from the calculations. The result in Eq. (62) means that when $u = 1$, heavy doses of drugs are applied and when $u = 0$ a rest period is reached. Essentially, the control obtained is bang-bang. Simulations of the model yielded Figures 6, 7, and 8. In Figure 6, high doses of drugs coupled with the infusion of growth factors were applied and resulted in the depletion of the normal and abnormal cell populations. However, the

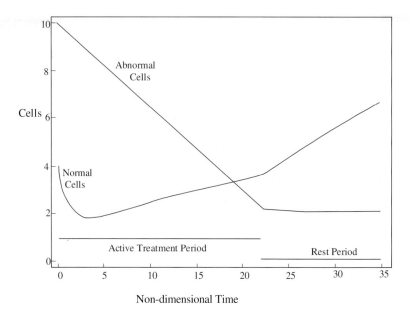

Fig. 6. Intensive treatment with infusions of growth factors. The normal and abnormal cells go through a process of decline during the period of active therapy. However, normal cell recovery begins during this period leading to a domination of the abnormal population during this active treatment period and lasting through the rest period. In this simulation scheme growth factor support was started 2.5 non-dimensional time units after treatment began. The abscissa is measured in non-dimensional time units and the ordinate represents the cell populations measured on a $\log_{10}$ scale.

normal cells went through a process of recovery that dominated the growth of the leukemic population during the period of rest. The model predicted a treatment time of approximately 35 non-dimensional time units where, based on the data employed, one non-dimensional time unit is equivalent to 0.772 days. This treatment time included 22.42 non-dimensional time units (the computed switching time) of intensive high-dose chemotherapy followed by the rest period. Based on the model-predicted treatment time, simulations were ran of a case in which treatment was carried out without the infusion of growth factors. This treatment cycle included 23 non-dimensional time units of intensive therapy followed by a rest period lasting till the 35th unit of time and is depicted in Figure 7. Figure 8 shows simulations beyond the treatment period in the case where growth factors are used during treatment. In this situation normal cells display hemopoietic

Mathematical Models of Cancer and Their Relevant Insights

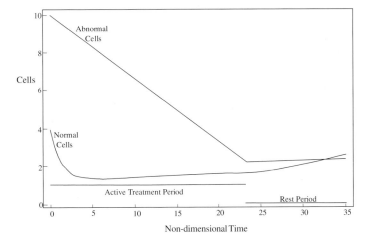

Fig. 7. Intensive treatment without infusions of recombinant hematopoietic growth factors. The normal and abnormal cells go through a process of decline during the period of active treatment but even though the normal population begins to recover, it remains dominated by the abnormal population. The eventual domination gained by the normal cells over the abnormal population only takes place towards the end of the rest period and involves a relatively small advantageous differential. The abscissa is measured in non-dimensional time units and the ordinate represents the cell populations measured on a $\log_{10}$ scale.

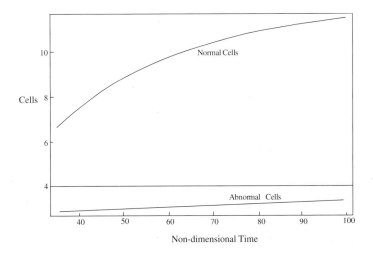

Fig. 8. Growth of the normal and abnormal cell populations beyond the treatment cycle that involved the use of recombinant hemopoietic growth factors. The normal cells evolve towards their normal population level while the abnormal cell population remains below a certain threshold. The abscissa is measured in non-dimensional time units and the ordinate represents the cell populations measured on a $\log_{10}$ scale.

recovery by dominating the leukemic cell population that remains below a certain threshold for some period of time. Here, the normal cell population evolves towards the normal level of around 10^{12} cells.

Afenya (2001b) compared the simulation results and noted that the simulations shown in Figures 6 and 8, cases involving growth factors, indicated that normal cell recovery and domination of the malignant population begins at an appreciable level of chemotherapeutic drug concentration during the interval of active treatment. The process of normal cell domination then continues throughout the rest period and beyond this stated period. It could be observed from Figure 6 that a sizeable differential depicting the advantage gained by the normal cells over the abnormal cells occurs by the end of the rest period. By contrasting and comparing the simulations in Figure 6 to those shown in Figure 7, where no growth factors were utilized, it could be observed from the simulations depicted in Figure 7 that the normal cells recover at a slower pace during the period of active treatment and do not gain dominance over the malignant cells during this period, a direct opposite of what can be seen in Figure 6. The dominance that they gain towards the end of the rest period shows a differential that is relatively small in size. Also, the desired targets for the normal and abnormal cells are not reached in Figure 7, as is the case in Figure 6, within the same treatment time interval of 35 non-dimensional time units. It is worth noting from the simulations in Figures 6 and 7 that the leukemic population displays an inability to regrow aggressively during the rest periods in the two different situations. This possibly stems from the damage and subsequent reductions in size that it sustained due to the heavy drug doses applied during the periods of active treatment thus preventing it from regaining strength right after those periods ended.

In relating the model to clinical studies, some areas of agreement between the model predictions and observations from the clinical data of Yamasaki and his co-workers (1991) were found that included: (a) the creation of a differential that is advantageous to the normal population by the time treatment is discontinued, when growth factors are used during treatment; (b) propensity of the abnormal cell population to regain or maintain dominance in the absence of growth factor infusions during treatment; (c) the tendency of the abnormal cell population to remain at a reasonably low level and not regrow aggressively even beyond the period of active treatment; (d) rise of the normal cell population towards its normal level possibly because of the weakening of the leukemic inhibition, as a result of the treatment and the growth factor stimulation; and (e) an advantageous treatment outcome arising out of using the strategy of intensive treatment alongside growth factor infusions. It was noted from Figure 6 that the model predicts about 17.3 days (22.42 non-dimensional time units) of intensive treatment, which is less than the 25 and 31 days of real time treatment observed in the work of Yamasaki *et al.* (1991). Thus, by assuming all other things being equal and considering that the same amounts of drugs were administered at similar intensity levels, a direct comparison of the model simulations with the clinical data showed that the model shortened the treatment time by 7.7 days in one case and by 13.7 days in another case. The implication here is that the same kind of outcome obtained after 25 or 31 days of treatment could be achieved within the model-predicted optimal time of 17.3 days, and this could go a long way to lessen the costs of hospitalization.

Other model predictions and insights (Afenya, 2001b) that were not necessarily reflected in the observations from the clinical data (Yamasaki *et al.*, 1991) but were found to have useful implications for clinical studies included: (a) *a priori* knowledge of the lengths of the active treatment time, and the rest period over which the abnormal cells may not regrow aggressively, granted that certain patient parameters such as growth and death rates of the cell populations are known; (b) determination of the time to discontinue treatment even before treatment begins; (c) knowledge of the time interval of neutropenia; (d) a threshold below which the abnormal population could be kept for some period of time to guarantee regrowth of the normal population possibly before another course of therapy, if needed, is administered; (e) an interval of time beyond the rest period over which to wait before carrying out any treatment of minimal residual disease; (f) specification of the final populations of normal and abnormal cells to be reached by the end of the rest period to ensure normal cell recovery; and (g) knowledge of the cost of treatment.

The modeling approach adopted here started with imposing a chemotherapeutic treatment regimen on a model (Afenya, 1996; Afenya and Calderón, 1996) that had been shown to capture characteristics of normal and abnormal cell growth during malignant development. The usefulness of such an approach lies in the fact that the strength, robustness, and general kinetics exhibited by the model over which the regimen is being imposed is known and may need no further justification. It is interesting to note that Matveev and Savkin (2001) also obtained similar conclusions about the optimal control even though they used a different performance index in their work with this same model. Nevertheless, model system (45)–(50) can still be improved. It would be of interest to see how the analysis and simulations play out in relation to clinical studies when other cell cycle properties and other normal and abnormal cell behaviors along with different types of performance functionals are introduced into the model. It would also be instructive to account for the occurrence of time delays in this model.

4. PARAMETER ESTIMATION

There are challenges associated with estimation of appropriate model parameters (Mary, 1981a, 1981b) in this area and this is partly or largely due to the scanty or almost nonexistent data on untreated human tumors. This is understandable because of the manner in which detection and treatment of cancer takes place. A physician may not know that a patient has a disease such as acute leukemia until the patient shows up at the clinic to complain about certain disease symptoms and immediately detection takes place, treatment must follow in order to save lives. In such circumstances, measurements that are made may only come from the short time period between detection and treatment and the time following treatment and items that are measured may not mostly match with parameters needed for simulating models. Also, laboratory experiments performed on animals sometimes focus on specific questions that may not necessarily have any relation to parameter estimation issues of relevance to models. In any case, the developing trend in which clinicians such as Citron *et al.* (2003) are testing predictions based entirely on mathematical models hold the promise that such problems could be overcome in the near future as the interactions and collaborations between biomedical and biomathematical researchers increase and mathematical models become fully-fledged resources that are readily available to the biological and medical community.

Within the framework of the challenges mentioned above, model parameter estimation in this area has been based on inferences and manipulations drawn out of relevant existing work in the literature. Parameters have also been directly extracted from the literature, where appropriate. In certain circumstances, model parameters have been obtained as a result of the existence of appropriate data that engender rare opportunities in fitting growth functions as is highlighted in Section 2.2. The parameters used in this chapter that depict the simulations have come directly from some of the work (Afenya, 1996, 2001a, 2001b; Afenya and Calderón, 1996; Afenya and Bentil, 1998) we have appraised here.

5. CONCLUDING REMARKS

We have attempted to bring together the work that has been done in the area of modeling cancers of the disseminated type in this article. As has been noted, not too much modeling activity has been going on with regards to these cancers even though active biomedical research is proceeding very steadily on disseminated cancers and is raising questions that could be addressed through mathematical modeling. The increasing quantitative nature and scope of biomedical research also defines a natural role for mathematical studies of biomedical processes and mechanisms. In our discussions, our comments focused on leukemia as a representative of disseminated cancers since most of the work on such cancers has this disease as a focus. We did place special emphasis on acute leukemia, being a particularly dangerous type of cancer.

Some of the models that we considered were developed based on the cell cycle and some used non-cell cycle approaches. Nevertheless, there are similarities and dissimilarities in the conclusions arising from the models, even though they stem from different modeling circumstances. For example, Rubinow and Lebowitz (1976) developed their model by assuming that normal and abnormal cells existed in resting and active phases of the cell cycle and through analysis of the general characteristics of their model system arrived at the conclusion that normal cells get "fooled" into recognizing abnormal cells as normal and thus shut off their production leading to a situation where there are no normal cells but only abnormal cells, since these cells increase their production. In their studies, Afenya (1996), Afenya and Calderón (1996), and Afenya and Bentil (1998) used a non-cell cycle approach and did not conclude that normal cells get "tricked" by abnormal cells. However, they arrived at the conclusion that through leukemic inhibition normal cells are driven from subendosteal sites where hematopoiesis is preferentially resident leading to an ultimate situation where there are no normal cells but only malignant cells. Djulbegovic and Svetina (1985) used a cell cycle approach in their modeling but focused more on finding the cause responsible for leukemic development. They did not consider the general characteristics of the cell systems. As we have mentioned, it would be appropriate to introduce modifications into these models in order to capture more essential characteristics of normal and abnormal cell behavior.

In discussing model validation with the use of available data, we considered work done by Afenya (2001a) in finding agreements between predictions based on model analysis and simulations and information obtained from the data of Dr. Yawata of Japan and Dr. Hartz, formerly of Sherman Hospital in Elgin, Illinois. We noted that the simulations tended to capture certain characteristics of normal cell behavior when abnormal cells are present. The ability of the models to mimic cell behavior as observed from and validated by data make them relevant when considering candidate models that could be employed in treatment optimization. We also touched on the importance of cell kinetics in the development of normal and abnormal cells in disseminated cancer because the growth kinetics of human cancer may provide a more rational basis for future improvements in treatment. It is important to remark that all the viewpoints that we considered led to Gompertzian growth kinetics of disseminated cancer cells. The compelling reasons and justification for this type of growth lies in the histological structure, function, and geometry of the bone marrow, from where hematopoietic cells originate, which leads naturally to such growth kinetics.

In this discourse we also discussed and appraised some mathematical models of cancer treatment. We pointed out the amenability of the theory of optimal control to determining chemotherapy scheduling and underscored the need for cancer treatment to be viewed as a formal optimization problem. In the process we considered a treatment model of multiple myeloma (Swan and Vincent, 1977) the predictions of which agreed with clinical approaches at the time of its introduction. It was suggested that an expansion of that model may yield additional interesting insights within the context of current clinical studies. A treatment model proposed by Afenya (2001b) that deals with growth factor support during therapy was also considered and the relevance of the predictions from that model to clinical and patient data was traced that prompted recommendations about how the model could be improved. Some treatment models have predicted low dose treatment while others have predicted high dose or dose dense treatment, and these predictions have mirrored the treatment strategies being adopted in the clinic. These different strategies show that no general consensus as yet exists on which strategy is the best among all available strategies and this suggests that more studies that involve model improvements and re-assessments are needed in the fight against cancer.

Throughout our discussions and comments we have attempted to highlight the important insights from the mathematical models that are of relevance to clinical studies. We believe that such an approach will put mathematical models directly in the service of biomedicine as efforts are made to improve techniques and processes that aid in the detection, treatment, and management of cancer. Admittedly, it is difficult to comment on everything related to the subject matter of this chapter. Therefore, we have cited various articles from the literature that may be of relevance. It is our hope that this chapter will further stimulate interdisciplinary work that employs mathematical modeling in this challenging area of research.

References

1. Abedi M and Elfenbein GJ. Novel approaches in the treatment of multiple myeloma. *Med. Health R. I.* 2003; **86**(8): 231–235.
2. Afenya EK and Bentil DE. Models of acute myeloblastic leukemia and its chemotherapy. In: *Computational Medicine, Public Health, and Biotechnology Part I*. Witten M (ed.) World Scientific, New Jersey, 1995, p. 397.
3. Afenya EK and Bentil DE. Some perspectives on modeling leukemia. *Math. Biosci.* 1998; **150**: 113–130.
4. Afenya EK and Calderón CP. A brief look at normal cell decline and inhibition in acute leukemia. *J. Can. Det. Prev.* 1996; **20**(3): 171–179.

5. Afenya EK and Calderón CP. A remark on leukemogenesis. *Int. J. Math. Stat. Sci.* 1999; **8**(2): 199–205.
6. Afenya EK and Calderón CP. Diverse ideas on the growth kinetics of disseminated cancer cells. *Bull. Math. Biol.* 2000; **62**: 527–542.
7. Afenya EK and Calderón CP. Modeling disseminated cancers — a review of mathematical models. *Comm. Theor. Biol.* 2003; **8**(2): 225–253.
8. Afenya EK. Acute leukemia and chemotherapy: a modeling viewpoint. *Math. Biosci.* 1996; **138**: 79–100.
9. Afenya EK. Cancer treatment strategies and mathematical modeling. In: *Mathematical Models in Medical and Health Sciences.* Horn MA, Simonett G and Webb GF (eds.) Vanderbilt University, Nashville, 1998, pp. 1–8.
10. Afenya EK. Use of real time leukaemia data to validate model predictions based on analyses and computer simulations. *Cell Prolif.* 2001a; **34**: 331–345.
11. Afenya EK. Recovery of normal hemopoiesis in disseminated cancer therapy — a model. *Math. Biosci.* 2001b; **172**: 15–32.
12. American Cancer Society Homepage, 2007.
13. Archimbaud E, Leblond V, Fenaux P, Dombret H, Cordonnier C, Dreyfus F, Cony-Makhoul P, Tilly H, Troussard X, Auzanneau G, Thomas X, French M and Marie JP. Timed sequential chemotherapy for advanced acute myeloid leukemia. *Hematol. Cell Ther.* 1996; **38**(2): 161–167.
14. Aye MT, Niho Y, Till JE and McCulloch EA. Studies of leukemic cell populations in culture. *Blood* 1974; **44**: 205–219.
15. Bajzer Z. Gompertzian growth as a self-similar and allometric process. *Growth Dev. Aging* 1999; **63**: 3–11.
16. Bajzer Z, Vuk-Pavlovic S and Huzak M. Mathematical modeling of tumor growth kinetics. In: *A Survey of Models for Tumor-Immune System Dynamics.* Adam JA and Bellomo N (eds.) Birkhauser, Boston, 1997, pp. 89–133.
17. Bassukas ID. Comparative Gompertzian analysis of alterations of tumor growth patterns. *Cancer Res.* 1994; **54**: 4385–4392.
18. Bernstein SH. Growth factors in the management of adult acute leukemia. *Hematol. Oncol. Clin. North Am.* 1993; **7**(1): 255–274.
19. Burns FJ and Tannock IF. On the existence of a G_0-phase in the cell cycle. *Cell Tissue Kinet.* 1970; **3**: 321–334.
20. Calderón CP and Kwembe TA. Modeling tumor growth. *Math. Biosci.* 1991; **103**: 97–114.

21. Campana D, Neale GAM, Coustan-Smith E and Pui CH. Detection of minimal residual disease in acute lymphoblastic leukemia: the St. Jude experience. *Leukemia* 2001; **15**: 278–279.
22. Casey AE. The experimental alteration of malignancy with an homologous mammalian tumor material. *Am. J. Cancer* 1934; **21**: 760–775.
23. Child JA, Morgan GJ, Davies FE, Owen RG, Bell SE, Hawkins K, Brown J, Drayson MT and Selby PJ. High-dose chemotherapy with hematopoietic stem-cell rescue for multiple myeloma. *N. Engl. J. Med* 2003; **348**(19): 1875–1883.
24. Citron ML, Berry DA, Cirrincione C, Hudis C, Winer EP, Gradishar WJ, Davidson NE, Martino S, Livingston R, Ingle JN, Perez EA, Carpenter J, Hurd D, Holland JF, Smith BL, Sartor CI, Leung EH, Abrams J, Schilsky RL, Muss HB and Norton L. Randomized trial of dose-dense versus conventionally scheduled and sequential versus concurrent combination chemotherapy as postoperative adjuvant treatment of node-positive primary breast cancer: First report of Intergroup Trial C9741/Cancer and Leukemia Group B Trial 9741. *J. Clin. Oncol.* 2003; **21**(8): 1431–1439.
25. Clarkson BD, Dowling MD Jr, Gee TS and Burchenal JH. Treatment of acute myeloblastic leukemia, *Bibl. Haematol.* 1973; **39**: 1098–1114.
26. Clarkson BD, Ohkita T, Ota K and Fried J. Studies of cellular proliferation in human leukemia. I. Estimation of growth rates of leukemic and normal hematopoietic cells in two adults with acute leukemia given single injections of tritiated thymidine, *J. Clin. Invest.* 1967; **46**(4): 506–529.
27. Clarkson BD, Ota K, Ohkita T and O'Connor A. Cellular proliferation in acute leukemia, *Proc. Am. Assoc. Cancer Res.* 1964; **5**: 43.
28. Clarkson BD. Acute myelocytic leukemia in adults. *Cancer* 1972; **30**: 1572–1582.
29. Costa MIS, Boldrini JL and Bassanezi RC. Chemotherapeutic treatments involving drug resistance and level of normal cells as a criterion of toxicity. *Math. Biosci.* 1995a; **125**: 211–228.
30. Costa MIS, Boldrini JL and Bassanezi RC. Drug kinetics and drug resistance in optimal chemotherapy. *Math. Biosci.* 1995b; **125**: 191–209.
31. Cronkite EP. Kinetics of leukemic cell proliferation. *Semin. Hematol.* 1967; **4**(4): 415–423.
32. Dancey JT, Deubelbeiss KA, Harker LA and Finch CA. Neutrophil kinetics in man. *J. Clin. Invest.* 1976; **58**: 705–715.
33. DeVita VT, Principles of cancer management: chemotherapy. In: *Cancer: Principles and Practice of Oncology*, 5th ed. DeVita VT, Hellman S and Rosenberg SA (eds.) Lippincott-Raven, 1997, pp. 333–347.

34. Djulbegovic B and Svetina S. Mathematical model of acute myeloblastic leukaemia: an investigation of the relevant kinetic parameters. *Cell Tissue Kinet.* 1985; **18**: 307–319.
35. Dombret H. Granulocytic colony-stimulating factors in the management of patients with acute myeloid leukemia. *Hematol. Cell Ther.* 1996; **38**(3): 231–240.
36. Donato ML, Aleman A, Champlin RE, Weber D, Alexanian R, Ippoliti CM, de Lima M, Anagnostopoulos A and Giralt S. High-dose topotecan, melpphalan and cyclophosphamide (TMC) with stem cell support: a new regimen for the treatment of multiple myeloma. *Leuk. Lymph.* 2004; **45**(4): 755–759.
37. Drewinko B, Alexanian R, Boyer H, Barlogie B and Rubinow SI. The growth fraction of human myeloma cells. *Blood* 1981; **67**: 333–338.
38. Estey E. Hematopoietic growth factors in the treatment of acute leukemia. *Curr. Opin. Oncol.* 1998; **10**: 23–30.
39. Estey EH, Thall PF, Pierce S, Cortes J, Beran M, Kantarjian H, Keating MJ, Andreeff M and Freireich E. Randomized phase II study of fludarabine + cytosine arabinoside + idarubicin ± all-*trans* retinoic acid ± granulocyte colony–stimulating factor in poor prognosis newly diagnosed acute myeloid leukemia and myelodysplastic syndrome. *Blood* 1999; **93**(8): 2478–2484.
40. Fortin P and Mackey MC. Periodic chronic myelogenous leukemia: spectral analysis of blood cell counts and aetiological implications. *Br. J. Haematol.* 1999; **104**: 336–345.
41. Frenzen CL and Murray JD. A cell kinetics justification for Gompertz equation. *SIAM J. Appl. Math.* 1986; **46**: 614–629.
42. Fukuhara T, Miyake T, Maekawa I, Kurosawa M, Suzuki S, Noto S, Mori A, Chiba K, Toyoshima T, Hirano T, Morioka M, Tsutsumi Y, Okabe M and Kakinoki Y. Treatment with low-dose cytosine arabinoside followed by administration of macrophage colony-stimulating factor prolongs the survival of patients with RAEB, RAEB-T, or leukemic phase myelodysplastic syndrome: a pilot study. *Int. J. Hematol.* 2000; **71**(4): 366–371.
43. Gianni M, Terao M, Zanotta S, Barbui T, Rambaldi A and Garattini E. Retinoic acid and granulocyte colony-stimulating factor synergistically induce leukocyte alkaline phosphatase in acute promyelocytic leukemia cells. *Blood* 1994; **83**(7): 1909–1921.
44. Godwin JE, Kopecky KJ, Head DR, Willman CL, Leith CP, Hynes HE, Balcerzak SP and Appelbaum FR. A double-blind placebo controlled trial of G-CSF in elderly patients with previously untreated acute myeloid leukemia: a Southwest Oncology Group Study. *Blood* 1998; **91**(10): 3607–3615.

45. Grignani F. Chronic myelogenous leukemia. *Crit. Rev. Oncol. Hematol.* 1985; **4**: 31–66.
46. Gyllenberg M and Webb GF. Quiescence as an explanation of Gompertzian tumor growth. *Growth Dev. Aging* 1989; **53**: 25–33.
47. Haas R, Ho AD, Del Valle F, Fischer JTh, Ehrhardt R, Dobner H, Witt B, Huberts H, Kaplan E and Hunstein W. Idarubicin/cytosine arabinoside and mitoxantrone/etoposide for the treatment of *de novo* acute myelogenous leukemia. *Semin. Oncol.* 1993; **20**(6): 20–26.
48. Haberman R. *Elementary Applied Partial Differential Equations*: *With Fourier Series and Boundary Value Problems*, 3rd ed. Prentice Hall, NJ, 1997.
49. Hamblin TJ. Disappointments in treating acute leukemia in the elderly. *N. Engl. J. Med.* 1995; **332**(25): 1712–1713.
50. Hanazono Y, Miyazono K, Piao YF, Taketazu F, Chiba S, Miyagawa K, Hirai H, Sakamoto S, Miura Y, Yazaki Y and Takaku F. Treatment of acute nonlymphocytic leukemia by combination of recombinant human granulocyte colony-stimulating factor and cytotoxic agents: a report of six cases. *Int. J. Hematol.* 1992; **55**: 243–248 (1992).
51. Haurie C, Dale DC and Mackey MC. Occurrence of periodic oscillations in the differential blood counts of congenital, idiopathic and cyclical neutropenic patients before and during treatment with G-CSF. *Exp. Hematol.* 1999; **27**: 401–409.
52. Haurie C, Person R, Dale DC and Mackey MC. Hematopoietic dynamics in grey collies. *Exp. Hematol.* 1999; **27**: 1139–1148.
53. Hearn T, Haurie C and Mackey MC. Cyclical neutropenia and the peripheral control of white blood cell production. *J. Theor. Biol.* 1998; **192**: 167–181.
54. Henderson ES. Acute leukemia in adults. In: *Hematologic Malignancies*. Hoogstraten B (ed.) Springer Verlag, Berlin, 1986, pp. 17–30.
55. Herzig RH. High-dose ara-C in older adults with acute leukemia. *Leukemia* 1996; **10**(1): 10–11.
56. Hofer EP, Brucher S, Mehr K and Tibken B. An approach to a biomathematical mode of lymphocytopoiesis. *Stem Cells* 1995; **13**(1): 290–300.
57. Hoffbrand AV and Petit JE. *Essential Haematology*. Blackwell Science, Oxford, 1984.
58. Hokanson JA, Brown BW, Thompson JR, Drewinko B and Alexanian R. Tumor growth patterns in multiple myeloma. *Cancer* 1977; **39**: 1077–1084.
59. Hokanson JA, Brown BW, Thompson JR, Jansson B and Drewinko B. Mathematical model for human myeloma relating growth kinetics and drug resistance. *Cell Tissue Kinet.* 1986; **19**: 1–10.

60. Hsu HC, Chiu CF, Tan TD, Chau WK, Tseng CS and Ho CH. Post-remission intensive consolidation with high-dose cytarabine-based chemotherapy and granulocyte colony-stimulating factor in adults with acute myelogenous leukemia: a preliminary report. *Chin. Med. J.* (*Taipei*) 1995; **56**: 305–311.
61. Huhmann IM, Watzke HH, Geissler K, Gisslinger H, Jager U, Knobl P, Pabinger I, Korninger L, Mannhalter C, Schwarzinger I, Kalhs P, Hass OA and Lechner K. FLAG (fludarabine, cytosine arabinoside, G-CSF) for refractory and relapsed acute myeloid leukemia. *Ann. Hematol.* 1996; **73**: 265–271.
62. Irene TM, Boll K, Sterry K and Maurer HR. Evidence for a rat granulocyte chalone effect on the proliferation of normal human bone marrow and of myeloid leukemias. *Acta Haematol.* 1979; **61**: 130–137.
63. Iwao N, Yoshida M, Hatake K, Hoshino Y, Hagiwara S, Tomizuka H, Shimizu R, Suzuki T, Furukawa Y, Komatsu N, Muroi K, Miwa A, Sakamoto S and Miura Y. Combination chemotherapy of carboplatin and cytosine arabinoside for high-risk leukemia: a pilot study. *Leuk. Res.* 1995; **19**(12): 899–903.
64. Kalaycio M, Pohlman B, Elson P, Lichtin A, Hussein M and Tripp B. Chemotherapy for acute myelogenous leukemia in the elderly with cytarabine, mitoxantrone, and granulocyte-microphage colony-stimulating factor. *Am. J. Clin. Oncol.* 2001; **24**(1): 58–63.
65. Kazarinoff ND and van DerDriessche P. Control of oscillations in hematopoiesis. *Science* 1979; **203**: 1348–1349.
66. Keating M, Estey E, O'Brien S, Kantarjian H, Robertson LE and Plunket W. Clinical experience with fludarabine in leukaemia. *Drugs* 1994; **47**(6): 39–49.
67. Kendal WS. Gompertzian growth as a consequence of tumor heterogeneity. *Math. Biosci.* 1985; **73**: 103–107.
68. Kirk DE. *Optimal Control Theory: An Introduction*. Prentice-Hall, Englewood Cliffs, NJ, 1970.
69. Kitano K, Kobayashi H, Maeyama H, Miyabayashi H and Furuta S. Treatment of acute monoblastic leukaemia by combination of recombinant human macrophage colony-stimulating factor and low dose ara-C. *Br. J. Haematol.* 1993; **85**: 176–178.
70. Kozusko F and Bajzer Z. Combining Gompertzian growth and cell population dynamics. *Math. Biosci.* 2003; **185**: 153–167.
71. Laird AK. Dynamics of tumor growth: comparison of growth rates and extrapolation of growth curve to one cell. *Br. J. Can.* 1965; **19**: 278–291.
72. Lasota A and Mackey MC. Cell division and the stability of cellular replication. *J. Math. Biol.* 1999; **38**: 241–261.

73. Laver J, Shearer P, Krance R, Hurwitz CA, Srivastava DK, Weinstein HJ and Mirro J. A pilot study of continuous infusion ara-C in combination with rhG-CSF in relapsed childhood acute myeloid leukemia. *Leuk. Lymph.* 1996; **26**: 589–593.
74. Löwenberg B, Putten WLJV, Touw IP, Delwel R and Santini V. Autonomous proliferation of leukemic cells *in vitro* as a determinant of prognosis in adult acute myeloid leukemia. *N. Engl. J. Med.* 1993; **328**: 614–619.
75. Mackey MC and Glass L. Oscillation and chaos in physiological control systems. *Science* 1977; **197**: 287–289.
76. Mackey MC. Cell kinetic status of haematopoietic stem cells. *Cell Prolif.* 2001; **34**: 71–83.
77. Mackey MC. Unified hypothesis for the origin of aplastic anemia and periodic hematopoiesis. *Blood* 1978; **51**(5): 941–955.
78. Martin R and Teo KL. *Optimal Control of Drug Administration in Cancer Chemotherapy*. World Scientific, River Edge, NJ, 1994.
79. Mary JY. Inconsistencies of normal human granulopoiesis data. In: *Biomathematics and Cell Kinetics*. Rotenberg M (ed.) Elsevier/North-Holland Biomedical Press, Amsterdam, 1981a, pp. 39–64.
80. Mary JY. Reference data on normal human granulopoiesis. In: *Biomathematics and Cell Kinetics*. Rotenberg M (ed.) Elsevier/North-Holland Biomedical Press, New York, 1981b, pp. 65–86.
81. Maslak PG, Weiss MA, Berman E, Yao TJ, Tyson D, Golde DW and Scheinberg DA. Granulocyte colony-stimulating factor following chemotherapy in elderly patients with newly diagnosed acute myelogenous leukemia. *Leukemia* 1996; **10**: 32–39.
82. Matveev AS and Savkin AV. Optimal control regimens: influence of tumours on normal cells and several toxicity constraints. *IMA J. Math. Appl. Med. Biol.* 2001; **18**: 25–40.
83. Mauer AM and Fisher V. Characteristics of cell proliferation in four patients with untreated acute leukemia. *Blood* 1966; **28**: 428–445.
84. McCulloch EA and Till JE. Blast cells in acute myeloblastic leukemia: a model. *Blood Cells* 1981; **7**: 63–77.
85. McCulloch EA, Howatson AF, Buick RN, Minden M and Izaguirre CA. Acute myeloblastic leukemia considered as a clonal hemopathy. *Blood Cells* 1979; **5**: 261–282.
86. McCulloch EA. Stem cell renewal and determination during clonal expansion in normal and leukaemic haemopoiesis. *Cell Prolif.* 1993; **26**(5): 399–425.
87. McCulloch EA. Stem cells in normal and leukemic hemopoiesis. *Blood* 1983; **62**(1): 1–13.

88. McCulloch EA. The blast cells of acute myeloblastic leukemia. *Clin. Haematol.* 1984; **13**: 503–519.
89. Metcalf D. The nature of leukemia: neoplasm or disorder of hemopoietic regulation. *Med. J. Aust.* 1971; **2**: 739–746.
90. Morley A. Quantifying leukemia. *N. Engl. J. Med.* 1998; **339**(9): 627–629.
91. Murray JD. *Mathematical Biology II.* Springer-Verlag, New York, 2003.
92. Murray JM. An example of the effects of drug resistance on the optimal schedule for a single drug in cancer chemotherapy. *IMA J. Math. Appl. Med. Biol.* 1995; **12**: 55–69.
93. Murray JM. Optimal drug regimens in cancer chemotherapy for single drugs which block progression through the cell cycle. *Math. Biosci.* 1994; **123**: 183–213.
94. Naeim F. *Pathology of Bone Marrow.* Igaku-Shoin Medical Publishers, Inc., New York, 1992.
95. Nel JS and Falkson CI. Granulocyte-macrophage-colony stimulating factor support in patients with acute non-lymphatic leukemia. *Oncology* 1996; **53**: 482–487.
96. Norton L and Simon R. The Norton-Simon hypothesis revisited. *Can. Treat. Res.* 1986; **70**: 163–169.
97. Norton L. A Gompertzian model of human breast cancer growth. *Cancer Res.* 1988; **48**: 7067–7071.
98. Norton L. The critical concepts and emerging role of taxanes in adjuvant therapy. *Oncologist* 2001; **6**: 30–35.
99. Parrado A, Rodriguez-Fernandez JM, Casares S, Noguerol P, Plaza E, Parody R, Espigado I, de Blas JM and Garcia-Solis D. Generation of LAK cells *in vitro* in patients with acute leukemia. *Leukemia* 1993; **7**(9): 1344–1348.
100. Perry S, Moxley JH, Weiss GH and Zelen M. Studies of leukocyte kinetics by liquid scintillation counting in normal individuals and in patients with chronic myelogenous leukemia. *J. Clin. Invest.* 1966; **45**: 1388–1399.
101. Pfreundschuh M, Trumper L, Kloess M, Schmits R, Feller AC, Rube C, Rudolph C, Reiser M, Hossfeld DK, Eimermmacher H, Hasenclever D, Schmitz N and Loeffler M. Two-weekly or 3-weekly chop chemotherapy with or without etoposide for the treatment of elderly patients with aggressive lymphomas: results of the NHL-B2 Trial of the DSHNHL. *Blood* 2004; **104**(3): 634–641.
102. Pfreundschuh M, Trumper L, Kloess M, Schmits R, Feller AC, Rudolph C, Reiser M, Hossfeld DK, Metzner B, Hasenclever D, Schmitz N, Glass B, Rube C and Loeffler M. Two-weekly or 3-weekly chop chemotherapy with or

without etoposide for the treatment of young patients with good-prognosis (normal LDH) aggressive lymphomas: results of the NHL-B1 Trial of the DSHNHL. *Blood* 2004; **104**(3): 626–633.
103. Piantadosi S. A model of growth with first-order birth and death rates. *Comp. Biomed. Res.* 1985; **8**: 220–232.
104. Piccart-Gebhart MJ. Mathematics and oncology: a match for life? *J. Clin. Oncol.* 2003; **21**(8): 1425–1428.
105. Preisler HD, Raza A and Larson RA. Alteration of the proliferative rate of acute myelogenous leukemia cells *in vivo* in patients. *Blood* 1992; **80**(10): 2600–2603.
106. Rohatiner A and Lister TA. Acute myelogenous leukemia in adults. In: *Leukemia*. Henderson ES, Lister TA and Greaves MF (eds.) W. B. Saunders, 1996, pp. 479–508.
107. Rohatiner A and Lister TA. The general management of the patient with leukemia. In: *Leukemia*, 6th ed. Henderson ES, Lister TA and Greaves MF (eds.) W. B. Saunders, Philadelphia, 1996, p. 247.
108. Rose S. A proposal for a new direction to treat cancer. *J. Theor. Biol.* 1998; **195**: 111–128.
109. Rowe JM. Uncertainties in the standard care of acute myelogenous leukemia. *Leukemia.* 2001; **15**: 677–679.
110. Rubinow SI and Lebowitz JL. A mathematical model of neutrophil production and control in normal man. *J. Math. Biol.* 1975; **1**: 187–225.
111. Rubinow SI and Lebowitz JL. A mathematical model of the acute myeloblastic leukemic state in man. *Biophys. J.* 1976a; **16**: 897–910.
112. Rubinow SI and Lebowitz JL. A mathematical model of the chemotherapeutic treatment of acute myeloblastic leukemic state in man. *Biophys. J.* 1976b; **16**: 1257–1271.
113. Rubinow SI and Lebowitz JL. Model of cell kinetics with applications to the acute myeloblastic state in man. *Biosystems* 1977; **8**: 265–266.
114. Sanz MA, Coco FL, Martin G, Avvisati G, Rayon C, Barbui T, Diaz-Mediavilla J, Fioritoni G, Gonzalez JD, Liso V, Esteve J, Ferrara F, Bolufer P, Bernasconi C, Gonzalez M, Rodeghiero F, Colomer D, Petti MC, Ribera JM and Mandelli F. Definition of relapse risk and role of nonanthracycline drugs for consolidation in patients with acute promyelocytic leukemia: a joint study of the PETHEMA and GIMEMA cooperative groups. *Blood* 2000; **96**(4): 1247–1253.
115. Scherbaum O and Rasch G. Cell size distribution and single cell growth in *Tetrahymena pyriformis* GL. *Acta Pathol. Microbiol. Scand.* 1957; **41**: 161–182.

116. Schiller G, Emmanoulides C, Iastrebner MC, M.M and Naeim F. High-dose cytarabine and recombinant human granulocyte colony-stimulating factor for the treatment of resistant acute myelogenous leukemia. *Leuk. Lymph.* 1996; **20**: 427–434.
117. Schrier SL. Hematopoiesis and red blood cell function. *Sci. Am. Med.* 1988; **Section 5, Subsection I**: 2–8.
118. Schrier SL. The leukemias and the myeloproliferative disorders. *Sci. Am. Med.* 1995; **VIII**: 1–24.
119. Staff Editor. Twelve major cancers. *Sci. Am.* 1996; **275**(3): 126–132.
120. Simeoni M, Magni P, Cammia C, De Nicolao G, Croci V, Presenti E, Germani M, Poggesi I and Rocchetti M. Predictive pharmacokinetic-pharmacodynamic modeling of tumor growth kinetics in xenograft models after administration of anticancer agents. *Can. Res.* 2004; **64**(3): 1094–1101.
121. Simpson-Herren L and Lloyd HH. Kinetic parameters and growth curves for experimental tumor systems. *Cancer Chemother. Rep.* 1970; **54**: 143–174.
122. Skipper HE and Perry S. Kinetics of normal and leukemic leukocyte populations and relevance to chemotherapy. *Cancer Res.* 1970; **30**: 1883–1897.
123. Skipper HE. Kinetics of mammary tumor cell growth and implications for therapy. *Cancer* 1971; **28**: 1479–1499.
124. Smith JA and Martin L. Do cells cycle? *Proc. Natl. Acad. Sci. USA* 1973; **70**: 1263–1267.
125. Spinolo JA. Acute myelogenous leukemia. In: *Current Therapy in Cancer.* Foley JF, Vose JM and Armitage JO (eds.) W.B. Saunders Publishing Company, Philadelphia, 1994, pp. 322–326.
126. Steel GG. *Growth Kinetic of Tumours.* Oxford University Press, Oxford, 1977.
127. Stockdale FE. Cancer growth and chemotherapy. *Sci. Am. Oncol.* 1987; **12**: 1–11.
128. Stone RM, Berg DT, George SL, Dodge RK, Paciucci PA, Schulman P, Lee EJ, Moore JO, Powell BL and Schiffer CA. Granulocyte-macrophage colony-stimulating factor after initial chemotherapy for elderly patients with primary acute myelogenous leukemia. *N. Engl. J. Med.* 1995; **332**(25): 1671–1677.
129. Sullivan PW and Salmon SE. Kinetics of tumor growth and regression in IgG multiple myeloma. *J. Clin. Invest.* 1972; **51**: 1697–1708.
130. Swan GW. Role of optimal control theory in cancer chemotherapy. *Math. Biosci.* 1990; **101**: 237–284.
131. Swan GW and Vincent TL. Optimal control analysis in the chemotherapy of IgG multiple myeloma. *Bull. Math. Biol.* 1977; **39**: 317–337.

132. Swierniak A, Polanski A and Kimmel M. Optimal control problems arising in cell-cycle-specific cancer chemotherapy. *Cell Prolif.* 1996; **29**: 117–139.
133. Tanaka M. Recombinant GM-CSF modulates the metabolism of cytosine arabinoside in leukemic cells in bone marrow. *Leuk. Res.* 1993; **17**(7): 585–592.
134. Tavassoli M and Yoffey JM. *Bone Marrow: Structure and Function.* Alan R. Liss, Inc., New York, 1983.
135. Visani G, Tosi P, Zinzani PL, Manfroi S, Ottaviani E, Testoni N, Clavio M, Cenacchi A, Gamberi B, Carrara P, Gobbi M and Tura S. FLAG (fludarabine + high-dose cytarabine + G-CSF): an effective and tolerable protocol for the treatment of "poor risk" acute myeloid leukemias. *Leukemia* 1994; **8**(11): 1842–1846.
136. von Foerster H. Some remarks on changing populations. In: *The Kinetics of Cellular Proliferation.* Stohlman F Jr. (ed.) Grune and Stratton, New York, 1959.
137. Walters RS, Kantarjian HM, Keating MJ, Plunket WK, Estey EH, Andersson B, Beran M, McCredie KB and Freireich EJ. Mitoxantrone and high-dose cytosine arabinoside in refractory acute myelogenous leukemia *Cancer* 1988; **62**: 677–682.
138. Wheldon TE, Kirk J and Finlay HM. Cyclical granulopoiesis in chronic granulocytic leukemia: a simulation study. *Blood* 1974; **43**: 379–387.
139. Wheldon TE. Mathematical models of oscillatory blood cell production. *Math. Biosci.* 1975; **24**: 289–305.
140. Wickramasinghe SN. *Human Bone Marrow.* Blackwell Scientific Publications, London, 1975.
141. Winkler U, Barth S, Schnell R, Diehl V and Engert A. The emerging role of immunotoxins in leukemia and lymphoma. *Ann. Oncol.* 1997; **8**(1): 139–146.
142. Yamasaki Y, Izumi Y, Sawada H and Fujita K. Probable *in vivo* induction of differentiation by recombinant human granulocyte colony stimulating factor (rhG-CSF) in acute promyelocytic leukemia (APL). *Br. J. Haematol.* 1991; **78**: 579–580.
143. Zietz S and Nicolini C. Mathematical approaches to optimization of cancer chemotherapy. *Bull. Math. Biol.* 1979; **41**: 305–324.

Chapter 8
MAJOR EPIGENETIC HYPOTHESES OF CARCINOGENESIS REVISITED

King-Thom Chung

Although cancer has been considered a disease of deoxyribonucleic acid (DNA), the fundamental dogma that cancer is the result of cumulative mutations that alter specific locations in a cell's DNA is facing challenges. Several major hypotheses of carcinogenesis associated with epigenetic factors are reviewed. Warburg proposed that cancer is caused by irreversible damage to the aerobic respiration capacity of normal cells; Pauling suggested that a deficiency of vitamin C was a cause of carcinogenesis; and Szent-Györgyi proposed a bioelectrical conductivity hypothesis. Deficiency of micronutrients such as folic acid, vitamin B_6, B_{12}, choline, methionine, iron, and zinc are factors that cause cancer. A decrease of cellular nicotinamide adenine dinucleotides concentration and niacin deficiency are also suggested to be involved in the evolution of cancer formation. Dysfunctional gap junctional intercellular communication (GJIC) caused by environmental carcinogens was proposed to be a cause of cancer. The alterations in DNA methylation is also considered to be a cause of carcinogenesis. Chromosomal abnormalities and viral infections also cause cancers. Each hypothesis, which has survived long scrutiny and many trials, led to a valuable study of a particular facet in the complicated processes related to both normal and cancerous cells. It is important to study how these various hypotheses are interrelated and attempt to integrate all different facets of cancer cells into a plausible mechanism of carcinogenesis. It would also be beneficial for future research efforts to study the synergistic effects of both epigenetic factors and genetic alterations on normal cells.

Keywords: Environmental carcinogens, bioelectrical conductivity, anaerobic respiration, carcinogenesis, oncogene, deficiency of vitamins and micronutrients, biosynthesis of NAD, hypermethylation/hypomethylation of DNA, gap junction intercellular communications, viral infections, chromosomal abnormalities, and epigenetic carcinogenesis.

1. INTRODUCTION

Cancer is one of our most feared diseases. Lung cancer is the number one killer among all cancers, followed by cancers of the colon and rectum, bladder, breast, uterus, stomach, prostate, and mouth. These are followed by sarcoma, lymphoma, myeloma, and leukemia (American Cancer Society, 2005).

"We must understand cancer at the basic level before we can hope for a cure." (Miller, 1990). Unfortunately, there are many forms of tumors, both benign and cancerous that confuse research scientists and diagnosticians. Normal human tissue consists of two types of cells, the differentiated cells and the stem cells. A stem cell can divide giving rise to new stem cells and differentiated cells. A differentiated cell does not divide and is an end cell that performs specific functions of the tissues. A tumor develops from a single normal stem cell that has undergone carcinogenic processes. Carcinogenesis is not a single step but rather is generally recognized as a multistep and sequential random process with each step involved in the interplay of genetic factors, epigenetic factors within stem cells, and interactions between cells. There are three major phases in carcinogenesis: initiation, promotion, and progression (Tan, 1991) (Figure 1). Initiation is characterized by an irreversible alteration in the genetic expression of a normal stem cell that produces a heritable change in the cell's phenotype. The genetic alteration may include either a gene mutation or another genetic change such as chromosomal variations. An agent or chemical that induces such change is usually called an initiator. The second phase is promotion, which is reversible. This is a stage that facilitates clonal expansion of initiated stem cells. Chemicals that promote this expansion are called promoters. Examples of promoters are phorbol esters and benzoyl peroxide. The mechanism of promotion may act either directly or indirectly by

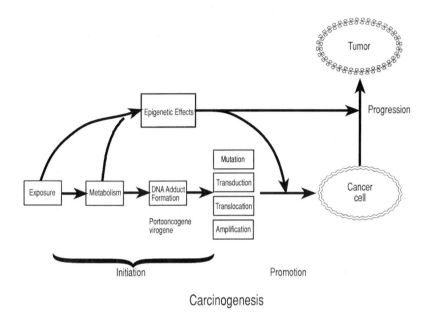

Fig. 1. Three major phases of carcinogenesis.

involving epigenetic factors. Some genetic changes such as gene amplification and chromosomal changes may also play a role in tumor promotion. However, a promotor alone cannot cause cancer. Promotion is effective only following initiation. The third phase is progression, which is involved in the development of metastatic tumor cells, formation of clumps of tumor cells of various sizes, migration of clumps through the circulation system, arrest of clumps, and development of metastatic foci at distant sites. Ample evidence has indicated that genetic mutations or changes are required for progression of tumor cells. This process is also called metastasis. When tumors metastasize, it often means that the disease has reached a terminal stage with the minimum chance of recovery (Tan, 1991).

Many factors have been reported to be involved in the causation of cancer. For example, smoking related carcinogens increase the risk to lung cancer and related epithelial cancers (Bartecchi *et al.*, 1994), dietary factors such as salty food increases stomach cancer risk (Forman, 1991), and red meat or food low in fiber content are related to colorectal cancers (Austoker, 1994; Cross and Sinha, 2004; Parkin *et al.*, 1993; Rogers *et al.*, 1993).

Chronic infections such as proliferative cystitis, hyperplasia, and dysplasia (Cohen, 2002), schistosomiasis (Mostafa et al., 1999; Hussain et al., 2003) have been reported to be related to bladder cancer. Hepatitis B and C infections are related to liver cancer (Beasley, 1988; Harris, 1994). *Helicobacter pylori* infection is related to gastric cancer (Asaka et al., 1997). Brenner et al. (2001) reported a significant correlation between lung cancer and pulmonary diseases such as asthma, bronchitis/emphysema, pneumonia, and tuberculosis. An accumulative exposure to estrogen or dihydrotestosterone is associated with endometrial cancer, breast cancer, and prostate cancer (Colditz et al., 1992; Harris et al., 1992; Henderson et al., 1991; Jick et al., 1980). Urological cancer has been observed in painters, hairdressers, and people who use dyes (Golka et al., 2004). In addition, there are viruses, drugs, and radiation that are also associated with cancer of one type or another. Both genetic and environmental factors are involved with the cause of cancer (Ames et al., 1993, 1995; Ames and Gold, 1997; Ames, 1998, 2001; Cooper, 1982; Croce and Klein, 1984; Doll, 1996; Fink, 1984).

The mainstream of cancer research appears to be related to the explosion of the knowledge of molecular genetics. Boveri (1914) proposed that the origin of malignant tumors arose as a result of mutations in somatic cells. Berenblum and Shubik (1947) proposed that DNA alkylation by alkylating agents and subsequent mutations were the mechanisms by which initiation of cells occurred to be followed by "promotion" to carcinogenesis. The work of Brookes and Lawley (1960) on alkylation of DNA by nitrogen mustards led to the activation of more common carcinogens to products that bind to DNA. James and Elizabeth Miller (1966) put forward the concept of conversion of carcinogens to electrophilic agents that bind to DNA. Since then, the formation of DNA adducts with mutagens/carcinogens has been extensively reported (Beland and Kadlubar, 1985, 1990; Beland and Poirier, 1989; Beland et al., 1997).

The development of short term mutagen assay, particularly the Ames Salmonella/microsome assay (Ames, 1973, 1975, 1979; Maron and Ames, 1983; McCann et al., 1975), and the claim that "carcinogens are mutagens," have stimulated the research on the genetic toxicology. Thousands of chemicals (potential carcinogens) have been screened for mutagenicity in bacteria and carcinogenic activities in animals. The genetic mutation theory has been the main focus of the study of chemical carcinogenesis (Ames, 1989;

Beland and Poirier, 1989; DeMarini *et al.*, 1995; Lijinsky, 1989; Vogelstein *et al.*, 1989).

Equally impressive is the development of oncogenes, protooncogenes, suppressor genes, and virogenes concepts of tumor formation (Burch *et al.*, 1988; Goodman *et al.*, 2004; Hausen, 1991; McMahon *et al.*, 1990; Melhem *et al.*, 1990; Moreau-Gachelin *et al.*, 1990; Pimental, 1991; Stanbridge, 1990; Sukumar and Barbacid, 1990; Tan, 1991; Temin, 1972; Todaro and Huebner, 1969; Weinberg, 1988, 1989). The list of cancer-related mutations has grown to more than 200 oncogenes and 50 suppressors (Hesketh, 1997). The rate at which these molecular markers are being identified continues to increase rapidly. Everyone who is doing cancer research is inclined to focus on the genetical/mutational aspects of carcinogenesis. Genetic alterations of genes or multiple mutations of genes of endogenous or exogenous (like retroviruses) origins have been the main focus of research for at least four decades. Gibbs in his recent review paper in *Scientific American* (2003) stated that the standard dogma for cancer formation was that carcinogens such as ultraviolet sunlight and tobacco, directly alter the DNA sequences of cancer-related genes such as *c-fos, BRAF, c-erbb3*, and mutations in tumor suppressor genes such as *APC, p53, and RB* cause growth-inhibiting proteins encoded by genes to disappear, allowing cells to survive and continue dividing when they should not. At the same time, mutations to oncogenes cause oncoproteins to become hyperactive, prompting the cell to grow in situations in which it normally would not. The excess of oncoproteins and lack of tumor suppressor proteins led mutant cells to reproduce excessively. After many rounds of mutation and expansion, one cell in the mass of mutants breaks free of all restrictions on its growth. The colony invades adjacent tissue in the host organ. In the most advanced stages of its evolution, the cancer leaks cells into the blood stream. These metastatic cells form new colonies at distant sites throughout the body, ultimately interfering with life-critical function (Gibb, 2003). However, this standard dogma is facing challenges. Modifications of the standard dogma such as loss or gain of a chromosome (or part of it) containing the genes, changes in the concentration of other proteins that regulate how genes are transcribed into proteins, early instability of the genes and/or chromosome, and an abnormal number of chromosomes including chromosomes with truncations, extension or swapped segments, have been proposed and remained to be proved (Hahn and Weinberg, 2002; Loeb et al., 2003). Many other abnormalities inside the nuclei of cell have to be considered. All cancer theories will need to incorporate the role of epigenetic phenomena (Gibbs, 2003).

In this article, the author intends to review several almost forgotten epigenetic concepts of carcinogenesis. Epigenetics was originally introduced to mean developmental mechanisms that acted after genes. Here, we simply refer to mechanisms of heredity other than genes, i.e. other than DNA sequences. In other words, mutational and oncogenic hypotheses will not be discussed in detail in this article although these factors are intimately related. An integration of each hypothesis into a plausible mechanism of carcinogenesis was suggested to provide guidance for future research.

2. WHY ARE EPIGENETIC FACTORS IMPORTANT?

Numerous factors are involved in carcinogenesis as mentioned above. The screening of environmental mutagens reveals that a wide variety of environmental compounds result in cancer (Ames et al., 1973, 1975; Chung, 1983, 2000; Chung and Cerniglia, 1992; Chung et al., 1997, 2000, 2006; Claxton et al., 1998; Cooper, 1982; McCann et al., 1975; Sobels and Delehanty, 1981; Tannenbaum et al., 1981; Wolff et al., 1974). More notably is that a number of carcinogens such as acetamide, 3-amino-1-H-1, 2, 4-triazole, 4-amino-2, 3-dimethyl-l-phenyl-3-pyrazolin-5-one, auramine, carbon tetrachloride, cycasin, dietrin, 1,2-dimethylhydrazine, ethionine, ethylcarbamate, 1-hydroxy-safrole, natulan, phenobarbital, safrole, thioacetamide, thiourea, are not mutagens (Ames et al., 1973, 1975). Ames and Gold (1990) stated that half of the chemicals (mitogens) tested (both natural and synthetic) are carcinogens in rodents, and a high percentage of these carcinogens are not mutagens. Lijinsky (1989) pointed out that a large number of liver carcinogens including nitrosomethylethylamine, nitrosodiethanolamine, azoxymethane, methapyrilene, diethyhexyl phthalate, clofibrate, and related hypolipidemic agents are not converted into bacterial mutagens by enzyme preparation from the liver of those species or by hepatocytes. Some important carcinogens including nitrosodiethanolamine,

Table 1. Relationships of mutagenicity to carcinogenicity.

Category	Mutagenicity	Carcinogenicity
I. Mutagenic carcinogen	+	+
II. Non-mutagenic carcinogen	−	+
III. Non-carcinogen mutagen	+	−
IV. Neither mutagenic non-carcinogenic	−	−

saccharin, safrole, diethylstilbestrol, methapyrilene, chloroform, and carbon tetrachloride are consistently insensitive in all the short term assays and are not mutagenic or "premutagenic" or "promutagenic." There are non-mutagenic chemical carcinogens such as DDT and phenobarbital. It is also notable that some mutagens such as hydroxylamine and sodium nitrite, chemical reactions of which with DNA have been very well studied, have stubbornly refused to induce tumors in animals; whereas, other potent mutagens such as alkylsulfonate esters are very weak carcinogens. Table 1 shows the relationships between carcinogens and mutagens. Some carcinogens are not mutagenic, and some mutagens do not induce cancer.

The non-mutagenic carcinogens induce tumors without direct effect on DNA sequence. Weinstein (1991) stated that the non-genotoxic agents that act on cellular targets other than DNA and through epigenetic mechanisms play a role in the evolution of tumors. There are potential epigenetic toxicants such as phorbol esters, vomatoxin, lipid polysaccharides, estrogens, TGF-α a, TNF-α, DDT, interleukin 1-α, prostaglandins, PCB's, airborne particles, methylmercury, unsaturated fatty acids, etc. Many of these agents are cancer related. It is clear that genetic mutation(s) alone is not the sole cause of cancer. Other aspects of carcinogenesis involving non-mutagenic alterations are also important.

Unless a scientist has a grasp of a wide range of carcinogens and basic concepts of all hypotheses about carcinogenesis, he/she is hindered in his/her research. There have been numerous theories of carcinogenesis not directly related to genes or mutations that have been soundly proved. Many of these earlier concepts might be related to the current theories, yet need to be further illustrated. Iverson (1988) emphasizes that "success" is often difficult to measure and that popular paradigms are often copied, while others are neglected or dropped without solid evidence of failure.

This report intends to revisit neglected earlier concepts that have proved sound and speculate on new hypotheses. Those concepts may also help us as a guide to ongoing research. However, this review should be read with an open mind, "... let us remember that are some circumstances a blank mind can be more of an advantage than an intelligent committed one" (Iverson, 1988).

3. THE WARBURG'S HYPOTHESIS

Back in the 1930s, the well known biochemist Otto Warburg proposed that cancer was caused by irreversible damage to the aerobic respiration capacity of normal cells (Warburg, 1956). This hypothesis offers explanations for many biochemical properties of cancer cells, such as the higher concentration and activities of glycolytic enzymes in cancerous cells compared to normal corresponding tissues; deficiency in the mitochondrial electron transport chains in neoplastic tissues; relative inability of the mitochondria to oxidize pyruvate by way of the citric acid cycle, and failure of the normal regulatory mechanism that controls glycolysis and respiration of cancerous cells (Boxer and Devlin, 1961). There were many findings in support of this hypothesis. However, there was also evidence not in accordance with the hypothesis (Bissel et al., 1976). For example, the hypothesis did not consider that a large part of aerobic respiration of cancerous cells is due to fatty acid oxidation. Tumors provided large amounts of lactic acid resulting from glycolysis, but so did many normal tissues (Weinhouse, 1956). The Warburg hypothesis stipulated that all carcinogens should act on the aerobic respiration of normal tissues, but the fact is that not all carcinogens affect aerobic respiration. It is generally believed that as cancers grow, they outgrow their oxygen supply (blood supply) and so suffer an anaerobic crisis; the cells that survive and that continue to grow have been selected for anaerobic metabolism. The anaerobic metabolism of cancer is therefore probably a secondary manifestation or an epiphenomenon, rather then as causal. Nevertheless, the Warburg's ideas are still interesting. Warburg's ideas interpreted the biochemical properties of cancer cells and served as a foundation stone for many biochemical investigations of cancer, which are now known.

4. THE LINUS PAULING HYPOTHESIS: VITAMIN C AND CANCER

Cameron and Rotman (1972) reported that cell proliferation is intimately related to the concentration of vitamin C inside cells. Cameron and Pauling (1973) further showed that there was a correlation between the content of vitamin C in tissues and their anticarcinogenic potential. They indicated that large doses of vitamin C would cause the remission of malignant tumors. In one experiment, they gave 10 g of vitamin C per day to 100 cancer patients who were in the terminal stage of the disease in comparison with a control group of 1000 cancer patients who were also in the terminal state of the disease. The results showed a survival-time of 50 days for the control group, but 210 days for those treated with vitamin C. Analysis of the survival-time curve showed that the survival-time of 90% of the vitamin C-treated group was three-fold that of the control group; whereas, the survival-time of the rest (10% of the vitamin C-treated group) was 20-fold that of the control group (Cameron and Pauling, 1976).

In 1978 they repeated the experiment and similar findings were obtained (Cameron and Pauling, 1978). They found that the level of vitamin C in the plasma and leukocytes of cancer patients was lower than those of normal healthy individuals (Cameron, Pauling and Leibovitz, 1979). Of course, there were many criticisms regarding the methodology of the experiments that Pauling *et al.* conducted. Nevertheless, the results showed clearly that vitamin C possesses an anticancer potential. In numerous cases, vitamin C gave clear supportive treatment to terminal cancer patients, offsetting many of the effects of the disease and prolonging life significantly.

There are many physiological functions of vitamin C, such as stimulation of the formation of collagen, prevention of scurvy, hydroxylation of proline, oxidation of tyrosine and the stimulation of the formation of adrenal steroid hormones. In normal tissues, there is a ground substance called glycosaminoglycan between cells. Cells are normally restrained from proliferating by the highly viscous nature of the intercellular glycosaminoglycan. When cells proliferate, this substance must be digested. Hyaluronidase is the enzyme responsible for the digestion of this ground substance. Hyaluronidase is subject to regulation by a physiological inhibitor (PI) (Figure 2). Studies indicated that PI is a kind of glycosaminoglycan complex

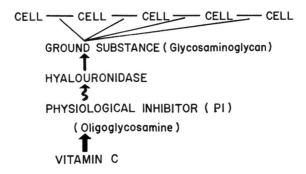

Fig. 2. Vitamin C and cancer.

that requires ascorbic acid for its synthesis and perhaps incorporates residues of ascorbic acid (Cameron and Pauling, 1978). When vitamin C is deficient *in vivo*, the concentration of effective PI in the blood will decrease, and hyaluronidase will digest the ground substance, promoting cell proliferation (Cameron and Rotman, 1972; Cameron, Pauling and Leibovitz, 1979).

The above mechanism explains how a deficiency of vitamin C could in some way promote the development of cancer, i.e. carcinogenesis. But this does not explain the mechanism of supportive effects of treatment of vitamin C. Pauling and his colleagues also showed that many factors involved in host resistance to neoplasia are significantly dependent upon the availability of ascorbate. Therefore, they recommended to the Food and Nutrition Board of the United States that the recommended daily intake (RDI) of vitamin C be increased from 45 mg up to from 250 to 4000 mg (Pauling, 1974). Later, it was found that vitamin C would decrease the incidence of malignant squamous carcinoma in hairless mice irradiated with ultraviolet light (Pauling *et al.*, 1982; Dunham *et al.*, 1982).

There were numerous other reports in addition to those of Pauling and his colleagues in support of the vitamin C anticancer hypothesis. Taper and his group reported that the combined intraperitoneal administration of vitamin C and K produced a distinct chemotherapy-potentiating effect for six different cytotoxic drugs used for cancer therapy (Taper *et al.*, 1987). Another group studied the relationship of diet and pharyngeal cancer and concluded that intake of vitamin C is protective against the disease and that there is a significant increase in risk associated with low intake of this vitamin (Rossing et al., 1989).

In a detailed review, Chen et al. (1989) also pointed out that the consumption of vitamin C-containing foods is associated with lower risk for certain gastric and esophageal cancers. Vitamin C supplementation of diet has been reported to inhibit skin, nerve, lung, and kidney carcinogenesis. Vitamin C has been shown to inhibit tumor cell growth and carcinogen-induced DNA damage (Chen et al., 1989). Ramon et al. (1993) and Mirvish (1994) have reported that vitamin C could delay or prevent the development of gastric cancer, and Kessler et al. (1992) also showed the similar effect on liver cancer. Kono and Hirohata (1996) reported that vitamin C protected against stomach cancer. Wolk et al. (1996) also observed the protective effect of vitamin C against renal cancer in non-smokers.

The mechanism of action by vitamin C is not clearly understood. The role of vitamin C in mesenchymal differentiation has been demonstrated by Franceschi (1992). Animal and *in vitro* studies have shown that vitamin C can effectively inhibit the formation of carcinogenic nitrosamines (Mirvish, 1986, 1993, 1994). Lathia et al. (1988) showed that vitamin C is effective in preventing nitrosation of amines under physiological conditions. Reed et al. (1989) revealed that high doses of vitamin C given to patients at increased risk of developing gastric cancer reduced the intragastric formation of nitrite and nitroso compounds and concluded that intervention with vitamin C therapy might reduce incidence of gastric cancer. Other studies have also demonstrated that vitamin C can reduce the cytotoxic effects of N-hydroxyacetyl-aminofluorene and reduce the covalent binding of 2-acetylaminofluorene (2-AAF) to cellular protein (Holme and Soderland, 1984). Hsieh et al. (1997) demonstrated that vitamin C can reduce the N-acetyltransferase activity of *Klebsiella pneumoniae*, an enzyme involved in the metabolic activation of arylamine carcinogens. Recent work of Piyathilake et al. (2000) indicated that a deficiency of vitamin C was associated with DNA hypomethylation in cancer cells. Hypomethylation and hypermethylation of DNA had been reported to influence carcinogenesis (Baylin et al., 1998; Frisco and Cho, 2002; Jones and Takai, 2001).

However, there were also reports indicating that vitamin C could promote tumor formation in mammals (Schwartz et al., 1993; Shibata et al., 1993) or enhance the chemical mutagenicity in Salmonella (Yamada and

Tomida, 1993). The deficiency of vitamin C is not necessary the sole cause of cancer and a full knowledge of the underlying mechanisms of treatment with vitamin C in relation to carcinogenesis remains to be studied. Vitamin C may also be involved in human aging (Ames, 1998). Much more research is needed along this line and the potential usefulness of vitamin C in prevention and treatment of cancer should not be ignored in the development of research programs.

5. SZENT-GYÖRGYI (BIOELECTRONIC) HYPOTHESIS

The well-known Hungarian biochemist, Albert Szent-Györgyi, a pioneer worker on vitamin C and a Nobel laureate, proposed a bioelectronic hypothesis of cancer (Holden, 1979), in which he attempted to explain the cellular activities of the submolecular level. He assumed that the cancer cell is a α-state cell, which is similar to a fetal cell, and is a non-conductive ("non-stoppable") proliferative cell. The normal cell, on the other hand, is β-state cell, which is conductive but proliferates under rigid regulation. The difference between the α- and β-state cells is that the latter cells possess methylglyoxal in their protein systems. An enzyme, glyoxylase, was believed to digest the methylglyoxal and convert the β-state cell into an α-state (cancer) cell (Figure 3).

According to the Szent-Györgyi explanation, most inanimate systems are built of closed-shell molecules in which electrons lack excitability and mobility. These electrons can be rendered reactive and mobile by taking

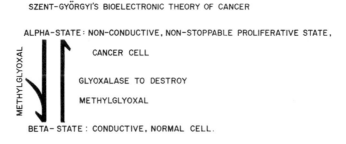

Fig. 3.   Szent-Györgyi's bioelectronic theory of cancer.

some of them out, desaturating the system electronically. Single electron can be taken out of a molecule by transfer to an external acceptor, creating two radicals that form a biradical having no net charge. The living state is such an electronically desaturated state. The universal electron acceptor of the biosphere is oxygen. The theory was that before light and oxygen appeared, a weak electron acceptor could occur through linkage of two C=O groups to glyoxyl and addition of a methyl group. The resulting methylglyoxal, being a weak acceptor, could lead to only a low degree of desaturation and thus to the formation of only simple life forms extent during the dark and anaerobic period — the so called α-state.

The subsequent period, the aerobic period is composed of β-type cells because of the greater occurrence of oxygen, a strong electron acceptor leading to a greater degree of desaturation; therefore, more highly differentiated life forms could develop. When cells are dividing, however, the β-state type cells return partially to the α (proliferative) state. Thus, a cancer cell is trapped in the α state (Szent-Györgyi, 1980).

The process of electron transfer between α and β states depends upon the dielectric constant of the medium and the relative concentration of -SH and methylglyoxal. The structure-building proteins that perform the main biological functions carry with them this chemical mechanism of their desaturation (Szent-Györgyi, 1980). He further stated that central to the mechanism is the amino group of lysine that attaches a methylglyoxal. Through the folding of the side chain of the protein, the CO groups of resulting Schiff bases can come in touch with the NH's of the peptide chain and accept electrons from it, thus desaturating it. Further, ascorbic acid is the catalyst of this charge transfer, which brings protein into the living state (Szent-Györgyi, 1978; Otto, Ladik and Szent-Györgyi, 1979). Purified protein is essentially inanimate matter, according to these hypotheses.

From its formulation, the Szent-Györgyi hypothesis was and remains somewhat a mystery, despite its elegance. Either proof or disproof is not simple or straightforward. Many of the later studies on the regulation of proteins by Szent-Györgyi can not be related to cancer directly. However, we believe Szent-Györgyi, who was a highly imaginative scientist and a Nobel laureate, has made too great a contribution to science to be ignored.

The comments by Dr. Harold Swartz in 1979 are still appropriate. He is "suspicious" of the accuracy of the hypothesis, but believes that Albert

Szent-Györgyi has hereby opened a new direction for further cancer research thought (Holden, 1979). There is little current research directly along this line. However, this type of thinking is still very much attractive.

6. MICRONUTRIENTS AND CANCER

In a recent review, Ames (2001) pointed out that deficiency of micronutrients such as folic acid, vitamin B_6, B_{12}, C, E, niacin, iron and zinc, or an excess amount of iron, could mimic radiation (or xenobiotics) in damaging DNA. The DNA damages include single- and double-strand breaks, and oxidative lesions or both. The population that is deficient for each of these micronutrients seems to have high risks of cancer at most sites. For example, folate deficiency has been reported to be associated with increased risk of colon cancer (Giovannicci *et al.*, 1993, 1995, 1998; Giovannicci, 2002; Mason, 1994; Potter, 2002). Blount *et al.* (1997) demonstrated that folate deficiency caused chromosome breaks in human genes. A level of folate deficiency causing chromosome breaks was present in approximately 10% of the U.S. population and in a much higher percentage of the poor (Ames, 2001). The deficiency of folate causes methylation of uracil to thymine, and subsequent intensive incorporation of uracil into human DNA, which contributes to human carcinogenesis (Chen *et al.*, 1996; Tucker *et al.*, 1996). Uracil in DNA is excised by a repair glycosylase with the formation of a transient single-strand break in the DNA. Two opposing single-strand breaks can cause a double-strand chromosome break. Chromosomal breaks could contribute to the increase risk of cancer and possibly cognitive defect. Folate deficiency causes an increase of an accumulation of homocysteine, which has been associated with a number of diseases (Oakley *et al.*, 1996; Refsum *et al.*, 1998; Rimm *et al.*, 1998). Folate administration will reverse the chromosome breaks and the high uracil level and thus prevent cancer (Blount *et al.*, 1997). Epidemiological and clinical studies indicate that dietary folate intake and blood folate level are inversely associated with colorectal cancer risk. However, recent work of Kim (2004) found that folate possesses dual modulatory effects on carcinogenesis depending on the timing and dose of folate intervention. Folate deficiency has an inhibitory effect; whereas, folate supplementation has a promoting effect on the progression of established neoplasms. In contrast, folate deficiency in normal epithelial tissues appears to predispose them to neoplastic transformation, while the modest levels of folate supplementation suppress the development of tumors in the colon (Kim, 2003). Folate is an important factor in DNA synthesis, stability, and integrity, the repair aberrations of which have been implicated in colorectal carcinogenesis. Folate may also modulate DNA methylation, which is an important epigenetic determinant in gene expression (an inverse relationship), in the maintenance of DNA integrity, chromosal modifications, and in the development of mutations (Kim, 1999).

Choline and methionine deficiency have been shown to be associated with the development of hepatocellular carcinoma in rats (Ghoshal and Farber, 1984, 1993; Mikol et al., 1983). Male rats fed diet deficient in choline and methionine have a reduced content of hepatic choline ($<50\%$ of control), betaine (30% of control), methionine (80% of control), S-adenosylmethionine (SAM) (60% of control), and folate (69% of control) (Selhub et al., 1991). These experimental findings indicated that a diet deficient in choline and methionine caused a stress on the methyl pool and on the folate-dependent remethylation of homocysteine to methionine. Because folate is an essential cofactor in the endogenous formation of methionine from homocysteine, a combined folate-choline-methionine deficiency would cause a more severe methyl donor deficiency than a diet deficient in only choline and methionine (Henning et al., 1989; Tuma et al., 1975). Henning et al. (1996) demonstrated that the carcinogenic effect of methionine-choline deficiency could be enhanced by additional folic acid and niacin deficiency. Animals fed diets deficient in choline or methionine have been shown to have hypomethylated DNA (Tsujiuhi et al., 1999) and changes in specific genes including c-Fos, c-Ha-Ras, and c-Myc (Wainfan and Poirier, 1992).

Vitamin B_{12} deficiency would be expected to cause chromosome breaks by the same mechanisms as folate deficiency. Vitamin B_{12} is required for the methylation of homocysteine to methionine. Homocysteine accumulation is a risk factor for heart disease (Oakley et al., 1996). If either folate or B_{12} is deficient, it will cause uracil to accumulate in DNA. The two deficiencies may act synergistically. In a study of elderly men (Fenech et al., 1997) and young adults (Fenech et al., 1998), increased chromosome breakage was associated with either a deficiency of folate, or B_{12}, or with elevated levels of homocysteine.

Similarly, vitamin B_6 deficiency had been reported to cause prostate cancer (Key et al., 1997) and contribute to heart disease (Rimm et al., 1996). The B_6 deficiency causes a decrease in the enzyme activity of serine hydroxymethyl transferase, which supplies the methylene group for methylenetetrahydrofolate (THF) (Stabler et al., 1997). If the methylene-THF pool is decreased because of B_6 deficiency, then the uracil incorporation, associated with chromosome breaks, would occur. Vitamin B_6 deficiency may also cause the accumulation of tryptophan metabolites in the urine (Birt et al., 1987).

Vitamin E (both α and γ forms of tocopherols) is a fat soluble antioxidant. γ-Tocopherol is also a powerful nucleophile, which can trap electrophilic mutagens that reach membranes. In the membrane, α-tocopherol is an antioxidant, and γ-tocopherol can act as a nucleophile. γ-Tocopherol has been reported to destroy the electrophilic mutagen NO_X by forming nitro-γ-tocopherol, and protecting the lipid, protein, and DNA (Christen et al., 1997; Cooney et al., 1995; Shigenaga et al., 1997). People taking vitamin E supplements (200 IU/day) have been reported to reduce the risk of colon cancer (White et al., 1997). Evidence also suggests that a long-term supplementation with α-tocopherol substantially reduced prostate cancer incidence and mortality in male smokers (Heinonen et al., 1998). Vitamin E deficiency leads to various neuropathologies (Sokol, 1996).

Increased risk of human cancer has been associated with excess iron (Toyokuni, 1996; Yip and Dallman, 1996). Excess iron appears to lead to oxidative DNA damage in rats that is reversed by vitamin E (Zhang et al., 1997). In the presence of excess iron, a series of reactions referred to as a Harber-Weiss cycle (superoxide driven Fenton chemistry) (Halliwell and Gutteridge, 1984), which results in the synthesis of hydroxyl radical (OH·) and other oxidants (Imlay et al., 1988). Products of these reaction have been shown to be genotoxic (causing DNA breaks and/or mutations), which are associated with the initiating events of carcinogenesis (Babbs, 1990; Goodman et al., 2004; Nelson, 1992). On the other hand, iron deficiency also appears to lead to oxidative DNA damage (White et al., 1997). Low iron intake results in anemia, immune dysfunction, and adverse pregnancy outcomes such as prematurity (Yip and Dallman, 1996).

Zinc deficiency has been suggested as a contributor to esophageal cancer in humans, and has been shown to cause esophageal tumors in rats in conjunction with a single low dose of a nitrosamines (Fong et al., 1996). Severe zinc deficiency alone can cause esophageal tumors in rats (Newberne et al., 1997). Zinc deficiency can also reduce the utilization of methyl group from S-adenosyl-methionine (SAM) needed for DNA methylation in rat liver resulting in genomic DNA hypomethylation and histone hypomethylation (Dreosti, 2001; Wallwork and Duerre, 1985). Zinc is involved in over 300 proteins including Cu/Zn superoxide dismutase, in the estrogen receptor, and in synaptic transmission. Over 100 DNA-binding proteins are involved with zinc fingers (Walsh et al., 1994). For example, betaine-homocysterine S-methyltransferase (BHMT) is a zinc metalloenzyme that catalyzes the transfer of methyl group from betaine to homocysteine to produce dimethylglycine (DMG) and methionine (Breksa and Garrow, 2002). Suppressor proteins of $p53$, which prevent mutation by inhibiting cell division and inducing apoptosis in response to DNA lesion, also require zinc for function (Sarkar, 1995). A zinc deficiency in rats has been reported to cause chromosome breakage (Castro et al., 1992), which is involved in cancer formation.

Oxidative stresses that often arise as a result of the imbalance in the human antioxidant states have been implicated in degenerative diseases related to aging and cancer (Aruoma, 1994). Oxidants such as hydrogen peroxide (H_2O_2), ozone (O_3), singlet oxygen and hypochlorous acid (HOCL) and free radicals such as superoxide (O_2^-), hydroxyl ($OH \cdot$), nitric oxide ($NO \cdot$) and peroxyl ($ROO \cdot$) are formed endogenously thorough normal metabolism or from exogenous sources such as cigarette smoke can cause DNA damages (Ames et al., 1993; Ames, 1998; Aruoma, 1994; Beckman and Ames, 1998; Benzie, 1996; Buettner, 1993; Buremmer et al., 1996; Shigenaga et al., 1994). Oxidants can act at several stages in malignant transformation. They induce permanent DNA sequence changes in the form of point mutations (Hsie et al., 1986; Moraes et al., 1990), gene amplification, and rearrangements, which may result in the activation of protooncogenes or the inactivation of tumor suppressor genes (Cerutti, 1991). Oxidants increase the expression of the growth-competence-related protooncogenes c-Fos, c-Jun and c-Myc as well as β-actin (Crawford et al., 1988; Shibanuma et al., 1988). Oxidants can also trigger epigenetic pathways that result in the transitory activation of genes, which participate in the regulation of growth and differentiation (Cerutti, 1991). For example, oxidants have been shown to activate protein kinase (Larsson and Cerutti, 1988) and induce the polyADP ribosylation of chromosomal proteins, both of which participate in the induction of the protooncogene *c-Fos* by oxidants (Cerutti, 1991).

Considerable information is also available on the preventive effect of antioxidants against cancer (Block *et al.*, 1992; Steinmetz and Potter, 1996). Antioxidants may explain much of the protective effect of fruits and vegetables, which were reported to significantly lower the incidence of human cancers if taken in appropriate quantity regularly (Ames *et al.*, 1993; Ames, 1998; Block, 1992; Block *et al.*, 1992; Byer and Guerrero, 1995; Diaz *et al.*, 1997; Diplock,1997; Hercberg *et al.*, 1998; Reddy, 2004). Ferguson *et al.* (2004) recently reviewed the dietary factors in New Zealand that affect epigenetic events and pointed out that dietary factors affecting DNA methylation status might play an important role in colon carcinogenesis.

7. NAD DEFICIENCY AS A FACTOR IN CARCINOGENESIS

The author is particularly interested in the possible role of the niacin cofactor, nicotinamide adenine dinucleotide (NAD; both NAD^+ and NADH) in cancer. The role of NAD as a coenzyme in more than 300 enzymatic reactions of cellular oxidation-reduction metabolism has been long understood (Bernofsky, 1980; Chaykin, 1963; Foster and Moat, 1980). In the past decades, this coenzyme has also been established as an important molecule in signaling pathways. Ziegler (2000) reported that this coenzyme is involved in the synthesis of cyclic ADP-ribose (cADPR) and nicotinic acid adenine dinucleotide phosphate (NAADP). Both cADPR and NAADP have been reported to be potent intracellular calcium-mobilizing agents. In concert with inositol-1,4,5-triphosphate, they participate in cytosolic calcium regulation by releasing calcium from intracellular stores (Ziegler, 2000).

The concentration of this coenzyme (both NAD^+ and NADH) inside of the cell is normally constant. The coenzyme metabolism involves the rapid interconversion of NAD^+ and NADH but does not result in a net withdrawal of NAD from the steady pool. Only immature and proliferating cells have been reported to have lower NAD pools than their normal counterparts

(Jonsson et al., 1988; Robins et al., 1985). However, the NAD concentrations in cancerous cells regardless of the type of cancer are consistently lower than corresponding normal cells (Briggs 1960; Clark and Greenbaum, 1966; Comes and Mustea, 1976; Glock and McLean, 1975; Jedeikin and Weinhouse, 1955; Jedeikin, Thomas and Weinhouse, 1956; Long, 1968; Nodes and Reid, 1964; Schwartz et al., 1974; Smith and King, 1970; Wintzerith et al., 1961) (Table 2). The work of Jacobson et al. (1999) also indicates that there is a decrease of NAD level in the human skin with the occurrence of the malignancy (squamous cell carcinoma).

Then the immediate question is would carcinogens affect the intracellular concentrations of NAD? The answer is yes; a number of carcinogenic azo dyes such as 4-aminoazobenzene (DAB), 4-dimethylamino-3'-methylazobenzene (3'-MeAB), and 4-dimethyl-4'-fluoroazobenzene (4'-FAB), affected the biosynthesis of NAD (Clark and Greenbaum, 1966; Jedeikin et al., 1956). Streptozotocin, a glucose derivative of N-methyl-N-nitrosourea (MNU) has been shown to cause a rapid decrease in the total size of NAD pool of mouse liver, pancreatic islets in mice (Schein et al., 1973), and rats (Ho and Hashim, 1972). Lowering of NAD in mouse liver by several nitrosamines and N-methyl-nitrosourethane has also been described (Schein, 1969). Rankin et al. (1980) also demonstrated that ultimate carcinogens including N-methyl-N'-nitro-N-nitroso-guanidine (MNNG), methylmethanesulfonate, N-acetoxy-2-acetylamino-fluorene-7-bromoethylbenzo(a)-anthracene, and the benzo(a)pyrene derivatives, γ-7,t-8-dihydroxy-9,10-epoxy-7,8,9,10-tetrahydro-benzo(a) pyrene, and benzo(a)pyrene-4,5-epoxide, could cause the reduction of NAD levels in human lymphocytes. There are probably many other carcinogens that also affect the biosynthesis of NAD but have not been extensively investigated.

More interestingly, there appears to be a tendency for NAD concentration in a precancerous cell to decrease with increasing time of treatment of carcinogens (Clark and Greenbaum, 1966; Jedeikin et al., 1956). Some chemical carcinogens and radiation cause the NAD concentration in precancerous cells to decrease (Chang, 1972; Davies et al., 1976; Gunnarson, Berne and Hellerstrom, 1974; Hinz et al., 1973; Ho and Hashim, 1972; Kensler, Suguira and Rhoads, 1940; Schein et al., 1973; Streffer, 1968). The lowering of NAD concentration occurs before the tumor development.

Table 2. Levels of NAD$^+$ and NADH in animal tissues and tumor cells.

Tissue	NAD$^+$ (μg/g)	NADH (μg/g)	NAD$^+$ + NADH (μg/g)	References
Rat				
Rat liver	370 ± 13	204 ± 9	574 ± 17	Glock and McLean (1955)
Rat adrenal	315 ± 136	154 ± 45	469 ± 134	Glock and McLean (1955)
Rat diaphragm	289 ± 7	138 ± 9	427 ± 14	Glock and McLean (1955)
Rat skeleton muscle	278 ± 16	27 ± 5	305 ± 19	Glock and McLean (1957)
Rat cardiac muscle	299 ± 15	184 ± 38	483 ± 23	Glock and McLean (1955)
Rat kidney	223 ± 12	212 ± 54	435 ± 60	Glock and McLean (1955)
Rat mammary gland	227 ± 9	83 ± 3	310 ± 10	Glock and McLean (1955)
Rat brain	133 ± 6	88 ± 36	221 ± 42	Glock and McLean (1955)
Rat spleen	135 ± 12	61 ± 15	196 ± 18	Glock and McLean (1955)
Rat thymus	116 ± 17	35 ± 13	151 ± 20	Glock and McLean (1955)
Rat lung	108	52	160	Glock and McLean (1955)
Rat pancreas	115	78	193	Glock and McLean (1955)
Rat seminal vesicles	128	11	139	Glock and McLean (1955)
Rat ventral prostate	80	17	97	Glock and McLean (1955)
Rat testis	80	71	151	Glock and McLean (1955)
Rat placenta	80	11	101	Glock and McLean (1955)
Rat blood	55 ± 3	36 ± 10	91 ± 8	Glock and McLean (1955)
Mouse liver	517	156	673	Jedeikin and Weinhouse (1955)
Pigeon liver	425	175	660	Jedeikin and Weinhouse (1955)
Rabbit liver	400	156	562	Jedeikin and Weinhouse (1955)

(Continued)

Table 2. (Continued)

Tissue	NAD$^+$ (μg/g)	NADH (μg/g)	NAD$^+$ + NADH (μg/g)	References
Tumors				
Mixed cholangiomas and hepatomas	226 ± 15	83 ± 12	309 ± 19	Glock and McLean (1957)
Hepatomas	255	25	280	Glock and McLean (1957)
Hepatomas	103	37	140	Jedikein and Weinhouse (1955)
Jensen rat sarcoma	118 ± 12	41 ± 9	150 ± 21	Glock and McLean (1957)
Crocker mouse sarcoma	112 ± 22	61 ± 6	174 ± 27	Glock and McLean (1957)
Sarcoma	111 ± 17	22 ± 3	133 ± 17	Glock and McLean (1957)
Carr lymphosarcoma	134 ± 7	37 ± 7	171 ± 9	Glock and McLean (1957)
EL$_4$ mouse leukemia	101 ± 7	33 ± 2	134 ± 7	Glock and McLean (1957)
Walker rat carcinoma	94 ± 9	50 ± 13	145 ± 24	Glock and McLean (1957)
MC/63 mouse carcinoma	131 ± 31	40 ± 3	172 ± 30	Glock and McLean (1957)
Krebs ascites tumour	132 ± 13	36 ± 0.3	168 ± 13	Glock and McLean (1957)
Rhabdomyosarcoma	109	24	133	Jedikein and Weinhouse (1955)
Mouse mammarycarcinoma	105	39	144	Jeideikin and Weinhouse (1955)
Ehrlich ascites	297	0	297	Jedeikin and Weinhouse (1955)

How does the lowering of NAD concentrations happen? The rapid turnover of NAD, an inhibition of biosynthesis, or enhancement of degradation, or both would result in a rapid decrease in the size of the NAD pool. There are two enzymes that are involved in the degradation of NAD, which have been characterized. One is a membrane-bound NAD glycohydrolase (Fukushima *et al.*, 1976). The other is a chromosomal enzyme poly (ADP-ribose) polymerase (PARP) (Hayaishi and Ueda, 1977; Hilz and Stone, 1976). Jacobson *et al.* (1980) demonstrated that treatment of the 3T3 cells with carcinogen MNNG (340 μm) resulted with a seven-fold increase in the activity of PARP, but did not change the activity of NAD glycohydrolase. This suggests that the acute lowering of NAD is caused by an increased synthesis of poly (ADP-ribose). An increased rate of poly (ADP-ribose) synthesis has been observed following the treatment with endonuclease (Miller, 1975), bleomycin (Miller, 1976), X-ray (Benjamin and Gill, 1979), and UV (Berger *et al.*, 1979). Goodwin *et al.* (1978) have also reported that lowering of NAD concentration occurs in L1210 cells following treatment with DNA-damaging agents, γ radiation, and neocarzinostatin. Rankins *et al.* (1980) concluded that lowering of NAD level is a general consequence of DNA damage.

DNA damage also results in the activation of DNA repair mechanisms. Jacobson and Narishimham (1979) reported that NAD-depleted 3T3 cells are incapable of MNNG-induced, unscheduled DNA synthesis, suggesting that DNA repair may not function without the availability of NAD or poly (ADP-ribose) synthesis. Sanford *et al.* (1986) indicated that human susceptibility to cancer resulted from a genetic deficient in DNA repair during the G_2 phase of the cell cycle.

NAD is a substrate for the formation of poly (ADP-ribose), catalyzed by PARP. PARP is a zinc-binding nuclear-dependent enzyme that catalyzes the transfer and polymerization of ADP-ribose onto histone (Mazen *et al.*, 1989; Uchida *et al.*, 1987), onto proteins involved in breaking or joining of DNA strands such as ligase II and topoisomerase I and II (Althaus and Richter, 1987), and onto proteins involved in gene expression and DNA replication (Boulika, 1990; Cardenas-Corona *et al.*, 1987; Tanuma *et al.*, 1983). A number of nuclear proteins including histones H1, H2a, H2b, H3, and H5, and non-histones nuclear lamins, high mobility group (HMG) proteins, and several DNA enzymes such as DNA polymerase α and β, deoxynucleotidyl terminal transferase, Ca^{2+}/Mg^{2+}-dependent endonuclease, DNA topoisomerases, and PARP itself are known to be poly (ADP)-ribosylated by this polymerase (Alderson, 1989).

The poly (ADP)-ribosylation by PARP occurs exclusively in the nucleus and exhibits a cell cycle oscillation with a maximum coinciding with S phase (Alderson, 1991). Poly (ADP)-ribosylation is a post-transcriptional covalent modification of the proteins mentioned above and can cause a substantial alteration of chromatin structures. Alterations of chromatin structure would lead to alteration of gene expression as well as the replication, repair, and recombination processes (Boulika, 1991). Evidence obtained with transgenic mice suggests that the PARP does not play a direct role in DNA break processing. Althaus *et al.* (1999) suggested that PARP is a part of a DNA break signal mechanisms. PARP-associated polymers may recruit signal proteins to sites of DNA breakage and reprogram their functions.

Alvarez-Gonzalez *et al.* (1999) found that enzymatic activities catalyzed by PARP are the subject of a very complex regulatory mechanism. For example, NAD^+ concentration determines the average ADP-ribose polymer size (polymerization reaction), the frequency of DNA strand breaks, and the total number of ADP-ribose chains synthesized (initiation reaction). Chatterjee *et al.* (1999) discovered that the inhibition of PARP would cause a significant increase in sister chromatid exchange (SCE) of the V79 Chinese hamster cells and proposed that PARP is a guardian of the genome that protects against DNA recombination.

The poly (ADP-ribose) is the third naturally occurring species of nucleic acid. It is synthesized by PARP and degraded mainly by poly (ADP-ribose) glycohydrolase (PARG) (Tanuma *et al.*, 1986), and secondarily by a nuclear phosphodiesterase (Boulika, 1991). The half-life of poly (ADP-ribose) is about 30 seconds to 3 minutes (Alvarez-Gonzalez and Althaus, 1989). Ame *et al.* (1999) demonstrated that PARG exists in multiple forms but results from a single gene; and the cellular content of PARG is regulated independently of PARP.

Whitare *et al.* (1995) indicated that the NAD content of cells modulated the expression of tumor suppressor protein p53. Decreased p53 expression correlated with NAD depletion and also occurred in chemically mutated cells carrying greatly reduced PARP activity. The work of Jacobson *et al.* (1999) also showed that NAD depletion in cultured cells from tissues including breast, lung, and skin was accompanied by decreased expression of the tumor suppressor protein p53. Diminished p53 function has been reported to be strongly associated with tumor formation in breast, lung, and skin (Donehowever *et al.*, 1992).

Niacin, a vitamin rich in meat and beans, has been shown to contribute to the maintenance of NAD levels (Jacobson and Jacobson, 1993, 1997). A niacin deficiency can be established by maintaining rats on low niacin diet. Zhang *et al.* (1993) demonstrated that a severe niacin deficiency may increase the susceptibility of DNA to oxidative damages likely due to the low availability of NAD in the liver, blood, and skeletal muscles. Rawling *et al.* (1994) further observed that dietary niacin deficiency lowered tissue NAD and poly (ADP-ribose) concentrations in Fischer-344 rats. Findings of Henning *et al.* (1996) indicated that tissue NAD concentrations were low in animals fed with methyl- and folate-deficient diets, and tumors developed in 100% of these animals. Jacobson *et al.* (1993, 1995, 1996, 1999) in their investigation of the role of niacin in human carcinogenesis indicated that niacin supplement was a potential preventive factor in human carcinogenesis.

It is known that ATP is required by deoxyribonucleotide kinase for biosynthesis of deoxyribonucleotide triphosphate, which is essential for the synthesis of DNA. ATP is also required for polynucleotide ligase activity (Bertazzoni, Mathelet and Carnpagnari,1972; Lindahl and Edelman, 1968; Soderhall and Lindahl, 1973; Tsukada and Ichimura, 1971). The decrease of NAD levels can also lead to a decrease in the ATP levels. The ATP levels are also lower in cancerous cells than their normal counterparts (Comes and Mustea, 1976; Glock and McLean, 1975; Long, 1960; Schwartz *et al.*, 1974). Kirkland (1991) demonstrated that in cultural endothelial cells, the DNA damage could activate PARP to deplete the NAD^+ to the extent that the ATP level drop and cell death might occur. Jacobson *et al.* (1999) illustrated that in human fibroblasts or epithelial cells became severely NAD depleted if grown for four to five population doublings in the absence of nicotinamide. This phenomenon was completely reversible but required 6–24 h in excess nicotinamide. But the nicotinamide depletion has no effect on cell growth rates until NAD content drop to less than 10% of the normal. Tissues at risk for NAD depletion include breast, lung, and skin (Jacobson *et al.*, 1999).

Although it is known that dietary nicotinamide and nicotinic acid serve as precursors of NAD in many human tissues, the mechanism of the conversion of tryptophan to NAD is still unknown in humans. Such a conversion may occur in liver and perhaps in the kidneys (Jacobson *et al.*, 1999). Tryptophan does not appear to be a source of tissue NAD in humans under conditions of restricted niacin intake over a period of a few weeks (Fu *et al.*, 1989). The pathway leading to the biosynthesis of NAD and their control mechanism are complicated (Chung, 1982; Grunicke *et al.*, 1974). In normal human cells, nicotinamide is minimally metabolized through the nicotinamide mononucleotide (NMN), but NMN pathway is the predominant one in Ehrlich ascites tumors (Grunicke *et al.*, 1966). Reztsova *et al.* (1994) studied the rates of quinolinc acid, nicotinic acid, and nicotinamide utilization for the NAD biosynthesis in organs of normal rats and those with lymphosarcoma of Pliss and Walker's carcinosarcoma. It was found that quinolinic acid failed to stimulate the increase in NAD concentration in the liver of Walker's carcinosarcoma-bearing rats. An insignificant rise in NAD concentration was registered in the liver of rats with lymphosarcoma of Pliss following quinolinic acid treatment. The rates of nicotinic acid and nicotinamide utilization in the liver of tumor-bearing animals were lower than in healthy controls. Quinolinic acid utilization was the most intense in the kidney of rats with Walker's carcinosarcoma while nicotinic acid and nicotinamide were mostly utilized in the kidney of rats with lymphosarcoma of Pliss. None of these three precursors caused NAD concentration to increase in Walker's carcinosarcoma cells; whereas, nicotinamide injection was followed by an increase in NAD concentration in Pliss lymphosarcoma cells (Reztsova *et al.*, 1994).

In humans, nicotinamide is postulated to be converted into nicotinic acid by intestinal bacteria. Nicotinic acid is further metabolized to nicotinate mononucleotide (nicotinate MN), which is eventually converted into NAD. The role of intestinal bacteria in this conversion has not been clearly illustrated (Chung, 1982).

NAD is involved in regulating deoxyribonucleic acid (DNA) synthesis (Burzio and Koide, 1970; Lorimer III and Stone, 1977; Morton, 1958; Sugimura, 1973; Yamada *et al.*, 1973). NAD is synthesized exclusively in the cell nucleus and is supplied to the cytoplasm. In the nucleus, NAD serves as a substrate in the formation of poly-ADPR, which is involved in turning off DNA synthesis (Burzio and Koide, 1970; Yamada *et al.*, 1973). When NAD concentration decreases, DNA synthesis will be induced and in turn will lead to increased cell division. Morton (1958) showed that the rate of cell replication is inversely proportional to the concentration of NAD. Thus the lowering of NAD concentration leads to increased cell replication, a phenomenon of cancer cells.

NAD is an electron carrier, and serves as the driving force for electron transport. When the NAD concentration decreases, the electron transport system will be seriously affected. Glycolysis is stimulated in order to replenish the need of NADH. This is in agreement with Otto Warburg's hypothesis, which stated that the cancer cell is due to the damage of aerobic respiration. When NAD concentration is low, the electron conductivity of the cytochrome system will also be low. This is in agreement with the bioelectronic hypothesis of Szent-Györgyi, in which we visualize conceptually that the bioconductor is the electron transport system, rather than the whole cell. If the bioconductor malfunctions due to insufficient concentration of NAD, cancer may result.

Gensler *et al.* (1999) has found that topical nicotinamide has preventive capacity against photocarcinogenesis in mice. They also demonstrated that dietary supplementation with niacin reduced the control incidence of skin cancer in mice after the UV treatment. They further demonstrated that the niacin supplementation would elevate skin NAD content, which was known to modulate the function of DNA strand scission surveillance protein p53 and poly (AP-ribose) polymerase, two proteins critical in cellular responses to UV-induced DNA damage. Spronck and Kirkland (2002) also showed that niacin was required for the maintenance of chromosome stability and may facilitate DNA repair *in vivo* in a tissue that was sensitive to niacin depletion and impaired poly (ADP-ribose) (p-ADPR) synthesis. Niacin supplementation may help to protect the bone marrow cells of cancer patients with compromised nutritional status from side effects of genotoxic chemotherapy drug. In a recent review, Kirkland (2003) stated that niacin participates in a variety of ADP-ribosylation reactions, DNA repair, maintenance of genomic stability, and signaling events for stress such as apoptosis. Disruption of any the processes has the potential to impair genomic stability and deregulate cell division, leading to enhanced cancer risk.

In viewing of the many important roles of NAD in cellular metabolism, it is very likely that NAD metabolism (both synthesis and degradation) is a common target for some carcinogens. The lowering of NAD concentration would not be a consequence of cancer since it occurred before the cancer formation. The lowering of the NAD concentration in cells particularly in the nucleus would trigger a series of biochemical changes including carcinogenic processes. Jacobson *et al.* (1999) strongly stipulated that NAD metabolism is a target for both prevention and treatment of cancer.

8. GAP JUNCTION INTERCELLULAR COMMUNICATION (GJIC) AND CANCER

Another important hypothesis of cancer is that cancer is viewed as a result of a disruption of the homeostatic regulation of a cell's ability to response appropriately to extracellular signals of the body that trigger intracellular signal transducing mechanisms, which modulate gap junctional intercellular communication between the cells within a tissue (Trosko and Ruch, 1998).

Intercellular communication via gap junctions (GJIC) is a fundamental biological process in animal tissues (Hertzberg *et al.*, 1981; Larsen, 1983; Pitts and Finbow, 1986). It might emerge during the evolutionary process from single cell organisms to multicellular organisms. Many new genes appear to accompany these new cellular functions. One of these new genes is the gene coding for a membrane associated protein channel (the gap junction), which allows the passive transfer of ions and small molecular weight molecules between coupled cells. A family of over a dozen of these highly evolutionary-conserved genes (the connexin genes) is coded for the connexin proteins. A hexameric unit of these connexins in one cell (a connexon) couples with a corresponding connexin in a contiguous cell to join the cytoplasm. This serves to synchronize either the metabolic or electrotonic functions of cells within a tissue. Most normal cells within solid tissues, except free-standing cells such as red blood cells, neutrophils, and the stem cells, have functional GJIC (Trosko and Ruch, 1998). GJIC is necessary for normal cellular growth, differentiation, and homeostasis (Schultz, 1985; Neyton and Trautman, 1986).

As Trosko and Ruch (1998) stated: "During the course of evolution from single-celled organism to multicellular organisms, new genes and cellular functions have to accompany that transition. Single-celled organisms survive changes in the environment by adaptively responding to physical and chemical agents such as nutrients and toxins by intracellular signals, which leads to cell proliferation modifications. However, in the multicellular organisms, a delicate orchestration of the regulation of cell proliferation for growth and tissue repair/wound healing and of the differentiation of cells had to occur after fertilization of the egg cell, during embryonic/fetal development, sexual maturation and adulthood/aging of the individual organism. The orchestration of specific cell/tissue/organ and organ system functions is referred to as "homeostasis." Homeostasis in the multicellular organisms is governed via three major communication processes: extracellular-communication via hormones, growth factors, neurotransmitters and cytokines, which trigger intracellular-communication via alterations in second messages (e.g., Ca^{++}, diacyclycerol, pH, ceramides, NO, c-AMP, reactive oxygen species) and activated signal transduction systems to modulate intercellular communication mediated by gap junction channels (Trosko *et al.*, 1993)." Cell adhesion and cell-matrix interactions are considered a subclass of intercellular communication molecules. All of these communication processes are intimately interconnected to maintain its normal development and health. In effect, this communication process must control a cell's ability to proliferate; to differentiate; to apoptosis; if differentiated, to response adaptively (Trosko and Ruch, 1998). One family of highly evolutionary-conserved genes that code for the gap junction proteins or connexins develops at the time multicellularity appears. However, the cancer cell, unlike the normal multicellular cell, does not contact inhibit, does not have growth control, does not terminally differentiate and usually has an abnormal apoptosis response, and does not appear to have functional gap junctional intercellular communication (Loewenstein, 1966; Kanno, 1985; Ruch, 1994; Trosko *et al.*, 1996). As a matter of fact, cancer cells of solid tissues appear to have either dysfunctional homologous or heterologous GJIC. Therefore, it is proposed that reversible disruption of gap junction intercellular communication could play a role during the tumor promotion, and that stable down-regulation of GJIC would lead to the conversion of a premalignant cell to an invasive and metastatic cancer cell

(Trosko and Ruch, 1998). Therefore, one of the molecular mechanisms of carcinogenesis caused by environmental carcinogens might be the interruption of GJIC (Trosko and Chang, 1980; Trosko *et al.*, 1983; Evans *et al.*, 1989; McGuire and Twietmeyer, 1985).

A large number of chemicals that enhance or promote neoplastic transformation does not cause direct DNA damage. These chemicals are generally called tumor promoters. Many of these "non-genotoxic tumor promoters" including pesticides such as DDT, dieldrin, and lindane; pharmaceuticals such as phenobarbital and diazepam; dietary additives such as saccharin and butylated hydroxytoluene; polyhalogenated hydrocarbons such as dioxin; and peroxisome proliferators such as clofibrate, inhibit GJIC in cultured cells and cells within target tissues (Trosko,1987; Klaunig and Ruch, 1990). Most studies indicate that mutagenic (genotoxic) carcinogens do not inhibit GJIC or induce cell proliferation (Budunova and Williams, 1994; Ruch and Klaunig, 1986). However, several classical mutagenic carcinogens have been reported to reduce GJIC (Na *et al.*, 1995). It is suggested that genotoxic carcinogens probably induce neoplastic transformation by mutationally activating proto-oncogenes and inactivating tumor suppressor genes (Trosko *et al.*, 1990). Therefore, chemical carcinogens not necessarily act on solely genetically (mutationally) but could act on epigenetically such as GJIC and cell proliferation.

Furthermore, there are some known oncogenes that were reported to produce products to block the GJIC (Yamasaki, 1990; Trosko *et al.*, 1990). There are also many growth factors such as epidermal growth factor, platelet-derived factor, basic fibroblast growth factor, hepatic growth factor, and transforming growth factor-alpha inhibited GJIC when applied to culture cells (Ruth, 1994). Several oncogenes have been reported to code for growth factors, growth factors receptors, or mitogenic signal transducing elements, and several tumor promoters act as growth factors since they induce cell proliferation. Therefore, oncogenes and tumor promoters have been associated with dysfunctional GJIC. The common mechanism of growth factors, oncogenes, and tumor promoters is to inhibit GJIC and increase cell proliferation, which is a common property of cancer cells (Trosko and Ruch, 1998).

On the other hand, in contrast to the effects of growth factors, oncogenes, and tumor promoters on GJIC, many growth inhibitors and anticancer drugs increase GJIC and connexin expression in target cells (Ruth, 1994). Retinoids, carotenoids, green tea extracts, certain flavonoids, dexamethasone, and cyclic AMP analogues and agonists inhibit neoplastic transformation and/or tumor cell growth in some tissues. These agents were reported to increase gap junction formation and connexin expression (or block the inhibitory effects of tumor promoters on GJIC) (Rogers *et al.*, 1990; Zhang *et al.*, 1992; Sigler and Ruch, 1993; Chaumontet *et al.*, 1994). Certain tumor suppressor gene products were reported to increase GJIC in neoplastic cells. For example, the human chromosome 11 carries one or more tumor suppressor genes (Misra and Srivatsan, 1989; Coleman *et al.*, 1995). Introduction of this chromosome into the neoplastic cells restored normal growth control and reduced tumorgenicity. It is suggested that tumor suppressor gene products inhibit neoplastic transformation by enhancing GJIC in addition to their other known actions on cell cycle genes, signal transduction pathways, and gene expression (Weinberg, 1993). All in all, it is proposed that cancer is the result of dysfunctional gap junctional intercellular communication. The dysfunctional gap junctional intercellular communication concept of cancer seems to be able to relate to other cancer theories. More and more evidences support the theory (Trosko and Ruch, 1998, 2002; Trosko and Chang, 2000, 2001; Ruch and Trosko, 1999; Trosko, 2003; Mesnil, 2004).

9. VIRAL INFECTIONS AND CANCER

Viruses represent one of the main factors that cause normal cells to proliferate and to become malignant. In the beginning of 20th century, Peyton studied a sarcoma that appeared in the breast muscle of a hen, and he eventually discovered that a virus, later named Rous sarcoma virus (RSV), was capable of inducing sarcoma in chicken (Rous, 1911a, b). Some other virus such as myxoma virus causing rabbit tumor was soon reported (Weinberg, 2006).

By 1960, Shope discovered that a DNA virus later called papillomavirus could cause squamous cell carcinoma of the skin (Shope, 1933, 1935; Shope *et al.*, 1958). Later, more than 100 distinct human papillomavirus (HPV) types, all related to the Shope virus, were discovered (Weinberg, 2006).

Temin (1964, 1974a) discovered that the conversion of normal cells into a tumor (cell transformation) could be accomplished by tumor viruses in culture, and the continued actions of proteins produced by the infected virus were required in order to maintain the transformed growth phenotype of the viral infected cells.

The tumor viruses could induce multiple changes in cell phenotypes such as loss of contact inhibition, capable of proliferating indefinitely in culture (immortalized); whereas, the normal cells have a limited proliferative potential in culture and ultimately stop multiplying after a certain predetermined number of cell divisions. The transformed cells would multiply without attachment to the solid substrate provided by the bottom of the Petri dish (anchorage independent); whereas, the normal cells have to have an absolute requirement for tethering to a solid before they would grow. More importantly, the transformed cells acquire the tumor-forming ability — tumorigenicity (Weinberg, 2006).

Studied with cell transformation by DNA tumor viruses — SV40 and polyomavirus indicted that tumor-associated protein (T) antigens found in cancers were induced by viruses. The viral genome of SV40 was found to be integrated into the infected host cells. However, it was found only that the portion of viral genome that contains oncogenic (cancer-causing) information was integrated into the chromosomal DNA of the cancer cells, while the portion that enables the viruses to replicate and construct progeny virus particles were almost always absent or present in only fragmentary form (Weinberg, 2006).

For the RNA viruses such as RSV, Temin (1964, 1972, 1974a, b) found that when these RNA viruses infected cells, they made double stranded DNA (dsDNA) copies of their RNA genomes, and it was these dsDNA versions of the viral genome that became established in the chromosomal DNA of the host cells. Once established in the infected cells, the DNA version of the viral genome was called provirus, which would be replicated along with the chromosomal DNA. The process of making DNA copies of RNA was called reverse transcription, and that was accomplished by the enzyme reverse transcriptase (Baltimore, 1970; Temin, 1970; Temin and Baltimore, 1972). The reverse transcriptases were found to be present in a large group of RNA viruses together called retroviruses. For retroviruses, chromosomal integration is a normal, essential part of the replication cycle; whereas, in the case of DNA tumor viruses such as SV40 and polyomaviruses, chromosomal integration of their genomes is a very rare accident (Weinberg, 2006).

Geneticists working with RSV speculated that there was a gene called *src* gene that had the transformation function that triggered the formation of sarcomas in infected chickens. Bishop and Varmus in 1974 successfully made the DNA probe of the *src* gene (Varmus *et al.*, 1974). Upon using the *src* specific probes, they found that the *src* sequences were present among the DNA sequences of uninfected chicken cells. In other words, *src* gene was found to be a cellular gene. *Src*-related DNA sequences were readily detectable in the genomes of several related bird species and in the DNAs of several mammals. The *src* gene in the genome of a normal organism was termed *c-src* and may carry some role in the life of this organism. However, the viral-transforming gene (*v-src*) closely related to *c-src* was called a potent oncogene because it was capable of transforming normal cells. The origin of the *v-src* gene (oncogene) was proposed to have been envolved from *c-src*. The *v-src* was not naturally present in the genome of the retrovirus ancestral to RSV. This hypothetical viral ancestor was capable of replicating in chicken cells. In fact, such *src*-negative retrovirus — avian leucosis virus (ALV) was common in chickens and was capable of spreading infection from one chicken to another. During the course of infecting a chicken cell, an ancestral virus, similar to this common chicken virus, had acquired sequences from the host-cell genome by some genetic trick. The acquired cellular sequences (e.g., the *src* sequences) were then incorporated into the viral genome, thereby adding a new gene to this virus (like retroviruses). Once present in the genome of RSV, the kidnapped *src* gene could then be altered and exploited by this virus to transform infected cells. The original *c-src* was designated proto-oncogene because it was a precursor of an active oncogene *v-src*. It implies that the genomes of normal vertebrate cells carry a gene that has the potential, under certain circumstances, to induce cell transformation and thus cancer (Weinberg, 2006).

With the similar mechanism as *c-src* proto-oncogene and *v-src* oncogene, a large group of proto-oncogenes were found in the vertebrate genomes. A notably example is *c-myc* (carried by MC29 myelocytomatosis virus). Nevertheless, not every retrovirus acquired random cellular genes could become oncogenic. Only growth-promoting cellular gene acquired by the retrovirus exhibits a cancer-inducing potential.

Mammals were found to harbor retroviruses that were distantly related to ALV and like ALV were capable of acquiring cellular proto-oncogenes and converting them into potent oncogenes. Examples are the feline leukemia virus that acquired the *fes* oncogene in its genome, yielding feline sarcoma virus, and hybrid-rat-mouse leukemia virus, which on a separate occasion acquired two distinct proto-oncogenes. The resulting transforming retroviruses, Harvey and Kristen sarcoma viruses, carry the *H-ras* and *K-ras* oncogenes, respectively in their genome. There were many viral oncogenes and distinct vertebrate proto-oncogenes have been discovered. In each case, a proto-oncogene found in the DNA mammalian or avian species was readily detectable in the genomes of all other vertebrates. There were, for example, chicken, mouse, and the human version of *c-myc*, and these genes seemed to function identically in their respective hosts. Therefore, it became clear that this large repertoire of proto-oncogenes must have been present in the genome of the vertebrates that were the common ancestor of all mammals and birds, and this group of genes, like most others in the vertebrate genome, was inherited by all the modern descendent species (Weinberg, 2006).

So our understanding is that each of the various tumorigenic retroviruses arose when a non-transforming retroviruses, acquired a proto-oncogene from the genome of an infected host cell. In the oncogene-bearing retroviruses, often induced tumors within a short period of time (days or weeks) after they were injected into host animals. However, the non-transforming precursor viruses could also induce cancers, but it would take a much more extended timetable. How did this happen? In the studies of leukemia that ALV-induced in chicken, it was found that the ALV provirus was integrated into the chromosomal DNA immediately adjacent to the *c-myc* proto-oncogene. The close association of the integrated viral genomes and the *c-myc* gene led to a functional link between these two genetic elements. The viral transcriptional promoter, nested within the ALV provirus, disrupted the control mechanisms that normally govern expression of the *c-myc* gene. Instead of being regulated by its own native gene promoter, the cellular *myc* gene was placed directly under viral transcriptional control. As a consequence, rather than being regulated up and down by the host cell, *c-myc* expression was controlled by the foreign usurper that drove its expression unceasingly and at a high rate. In other words, the hybrid viral-cellular gene arose in the chromosomes of leukemic cells that functioned much like the *v-myc* oncogene carried by avian myelocytomatosis virus. This activation of the *c-myc* gene through provirus integration is a low-probability event. Many weeks and many millions of infectious events are required before these malignancies are triggered. This mechanism of proto-oncogene activation is called "Insertional Mutagenesis" (Flint *et al.*, 2000).

There are other types of retrovirus such as human T-cell leukemia virus (HTLV-1) that carry natural oncogene — *tax* gene. The expression of this viral oncogene appears to be an intrinsic and essential component of retroviral replication cycle with host animals, rather than the consequence of rare genetic accidents that yield unusual hybrid genomes such as the genome of RSV (Weinberg, 2006).

The magnitude of virus involved in human cancers is a hotly debated matter. There are several known viruses associated with human cancers. The Epstein-Barr virus (EBV) is a DNA virus associated with Burkitt's lymphoma, a malignant tumor that causes swelling and eventual destruction of the jaw. EBV is also reported to be associated with nasopharyngeal carcinoma and hairy leukoplakia. Human papillomavirus is associated with carcinoma of the uterine cervix. Hepatitis B virus (HBV) has been reported to be a cause of human hepatoma. Kaposi's sarcoma, a cancer of the endothelial cells of the blood vessels or lymphatic system is thought to be associated with herpesvirus. For the RNA virus, human T cell leukemia virus I causes human adult T cell leukemia/lymphoma. Retroviruses are associated with hematopoitic cancers sarcoma and carcinomas.

Some oncologists believe that viruses may play a role in causing half of all types of human cancer. Pfeffer and Voinnet (2006) estimated that 15% of all human cancers are associated with single or multiple viral infections. The discovery of oncogenes in viruses has had a major impact on our understanding of cancer, although there is still much to be learned about cancers in humans. Regulation of the expression of oncogenes is a hot research subject.

O'Shea (2005) reviewed how the DNA tumor viruses subvert the normal cellular checkpoints to achieve successful transformation of the normal cells. The basic mechanisms of DNA viruses involved three signal pathways:

(1) mTOR pathway, (2) DNA damage pathway involving MRN and p53 genes, and (3) growth factor reactors and Rb/p107/p130-HDAC and E2F pathway. Likewise the how hepatitis B virus involved in liver cancer had been extensively studied. HBV has also induced chromosomal instabilities leading to loss of heterozygocity, aneuploidy and polyploidy, which are also a kind of epigenetic mechanism of cancer (Thorgeirsson and Grisham, 2002; Gibbs, 2003). There is much research literature in viral oncology available elsewhere (Moradpour and Wands, 2002; Wands, 2004; Ganem and Prince, 2004). Since viruses can induce tumors, they have to involve mechanisms that uncouple cellular replication from many inter- and extracellular factors that normally control so tightly. Recent studies with DNA viruses have contributed to our understanding of those factors, which are critical tumor target (such as EGFR, PP2A, Rb and p53) and have an important on the development of cancer therapies (O'Shea, 2005). Nevertheless, viral infection is not the sole explanation of the cause of cancer.

10. OTHER EPIGENETIC HYPOTHESES

There are other epigenetic hypotheses of cancer. The epigenetic events are recently considered to be at least as important as gene mutations and genetic changes in cancer initiation and progression (Bangham, 2005; Baylin, 2005). These epigenetic events are often reversible and more frequent than the genetic events. These epigenetic mechanisms address the transcriptional, translational, and/or post-translational level of genes, which affect mainly the activation of oncogene's products or silencing of tumor suppressor gene's proteins. These non-genetic mechanisms involve: (i) aberrant DNA methylation of cytosine at CpG islands of the gene, or aberrant DNA methylation of cytosine in non-promoter region of the gene (Jaffe, 2003; Belinsky, 2004; Baylin, 2005; Flintoft, 2005; Robertson, 2005); (ii) histone acetylation of genes (Lund and Lohuizen, 2004); (iii) loss of imprinting (LOI) of genes (Cui et al., 2003; Bangham, 2005; Robertson, 2005); and (iv) tissue disorganization and gap junction disruption (Ferreira et al., 2001; Jaffe, 2003).

Among the above mechanisms, DNA methylation is considered the main epigenetic process to activate oncogenes and to silence the tumor suppressor genes. Hypomethylation may result in an increased gene transcription. For example, hypomethylation of non-promoter region of DNA and structural elements such as centromeric DNA in the gene might cause enhanced genomic instability (Yamada *et al.*, 2005). Hypermethylation of the CpG islands of the promoter region of the gene to silence tumor suppressor is often an early event to initiate the carcinogenesis cascade. In 75% cases of lung cancer, carcinogenesis is initiated by gene silencing and hence loss of heterozygocity (LOH) of the *RASSE1* and other genes in chromosome 3p (Osaka and Takahashi, 2002; Baylin and Ohm, 2006). In human liver with hepatitis virus infection, in 70% of cases, hepatoma cell carcinoma (HCC) is initiated by reversible event and aberrant methylation of genes as methyltransferase causing elevated level of the *TGF/alpha* gene or elevated level of the *IGF-2* gene (Thorgeirsson and Grisham, 2002). In colon cancer, silencing of the gene *SFRP* (secreted frizzled-related protein) activated the Wnt signal pathways to initiate colon carcinogenesis process (Baylin and Ohm, 2006). Promoter methylation of the *mis-match repair* (*MMR*) gene (*hMLH1* or *hMSH2*) silences *hMLH1* (or *hMSH2*), which further silences *AXIN2* gene impairing the Wnt pathway to start the microsatellite instability pathway in human colon cancer (Jones and Baylin, 2006; Koinuma *et al.*, 2006).

The hypermethylation of the CpG islands of the promoter region of the gene, never occurred in mutants or in chromosomes involving genetic changes. Nevertheless, hypermethylations of the CpG islands of the promoter region of the gene were associated with the allele loss and LOH. Notice that LOH of suppressor genes is the major avenue by means of which most of human cancers are initiated. If one allele is mutated, hypermethylation is often seen as the second inactivating changes (Jone and Baylin, 2002; Baylin and Ohm, 2006). In liver cancer, hypermethylations of promoter often proceed before mutations and genetic changes (Thorgeirsson and Grisham, 2002).

Because 5-methylcytosine is a mutagen, the aberrant methylation of cytosine in the CpG region often leads to gene mutation. For examples, promoter methylation of the O^6-methylguanine-DNA methyltransferase (MGMT) leads to G to A transition mutation. Methylation of cytosine in the coding region increases the mutation rate because of the spontaneous hydrolytic deamination of the methylated cytosine, which caused C to T mutation (Jones and Baylin, 2002). Because methylated CpG islands are preferred binding sites for benzo(a)pyrene diol expoxide and other carcinogens in the tobacco smoke, methylation would enhance carcinogen binding to form DNA adducts and lead to G to T transversion mutation (Jones and Baylin, 2002).

In human colon cancer, Breivik and Gaudernack (1999) showed that either methylating carcinogens or hypermethylation at CpG islands would lead to G/T mismatch, which in turn lead to *MMR* gene deficiency or epigenetic silencing of the *MMR* genes and hence microsatellite instability (MSI). In cancer progression, p53 plays important role in inducing G_1S and G_2 cell cycle arrest and apoptosis. Chen *et al.* (2004, 2005) showed that p53 induce HICI; HICI inhibited SIRTI. Methylating silencing of HICI leads to accumulation of SIRTI, which inhibits and hence impairs p53 function causing defective apoptosis. All in all, DNA methylation (hypermethylation of the promoter region of genes and/or hypomethylation of the other genes) is always associated with the epigenetic process to activate oncogenes and to silence the tumor suppressor genes, or other cellular processes such as apoptosis, which is intimately related with carcinogenesis.

Soto and Sonnenschein (2004) proposed a new cancer paradigm, which they called it "tissue organization field theory (TOFT)." They interpreted that cancer is initiated by a loss of gene (genes) function — an interruption of cell to cell signaling silences critical genes whose function is to repress various genes that guide embryonic development (ED). The critical genes are called *suppressor of embryonic development* (*SED*) genes. Prehn (2005) further interpreted that oncogenes are *ED* genes and tumor suppressor genes are *SED* genes whose inactivation may promote tumor growth by allowing greater expression of *ED* genes. The loss of expression among *SED* genes is usually the result of epigenetically induced disruptions of normal cell to cell signaling. Furthermore, most mutations found in cancer are the result of the lack of repair in silenced or unexpressed *SED* genes. These mutations will eventually hardware or make irreversible the lack of expression in epigenetically silenced *SED* genes within a tumor. A cancer cell will become less and less susceptible to any regression that might involve an epigenetically induced return to a more normal pattern of gene expression and become more difficult to treat. This paradigm is supported by numerous evidences (Prehn, 2005).

Deusberg *et al.* (2000) interpreted cancer is the result of aneuploidy, which is caused by chromosomal truncations, extensions, or swapped segments. Since the carcinogens enter cells and disrupt the spindle apparatus that drags chromosome apart during cell division, the unbalanced spindles improperly separate chromosomes during mitosis, which causes duplication and loss of entire chromosome. Chromosomal abnormalities have been noted in almost every solid cancer studied to date. They argued that chromosomal aneuploidy actually preceded and segregated with chemical carcinogenesis and no gene mutations were required for cancer formation (Gibbs, 2003).

11. CONCLUDING REMARKS AND PERSPECTIVES

It is understood that cancer is caused by multiple reasons. Cancer cells have universal features. Cancerous cells appear not to respond to contact inhibition (Abercrombie, 1979; Borek and Sachs, 1966), fail to terminally differentiate (Markert, 1968; Potter, 1978), appear to be clonally-derived from a stem-like cell (Greaves, 1986; Kondo, 1983; Buick and Pollak, 1984), and continue to genotypically and phenotypically change as the tumor grows (Cairn, 1975; Kerbel *et al.*, 1984). The biological processes of signal transduction and program cell death apoptosis also appear to be altered in cancer cells compared to their normal parental cells (Hill and Treisman, 1995; O'Brian *et al.*, 1993; Thompson, 1995).

From the vast evidence gained over many years, it is obvious that there are multiple factors involved in the formation of cancer cells, and numerous hypotheses have been proposed. It is clear that an individual hypothesis does not need to be expected to clarify the mechanisms of all types of cancer to the exclusion of other hypotheses. The important consideration is how these hypotheses are interrelated. For example, deficiency of nutrients such as folic acid, choline, methionine, and zinc have been related to the status of methylation of DNA that is related to carcinogenesis (Dreosti, 2001; Feruguson *et al.*, 2004; Tsujiuho, 1999). Each epigenetic factor may stimulate the expression of mutations, which give rise to the specific characteristic of cancer cells. Likewise, the mutational event may facilitate the epigenetic manifestation of certain cells. Each hypothesis, which has survived long scrutiny and many trials, may well lead to a valuable detailed study of a particular facet in the complicated biochemical processes related to both normal and cancerous cells. It would be of great interest to find how these individual events are interrelated and concertedly to make a normal cell transformed into an abnormal neoplastic cell.

For instance, the lowering of NAD concentration could result from impairment of biosynthesis from tryptophan caused by carcinogens (especially for non-mutagenic carcinogens), deficiency of tryptophan and/or niacin, or it could result from DNA damage. Since NAD is required in so many biochemical reactions, what kind of biochemical changes might be caused if there is not a sufficient supply of NAD in the long term? Inside of the nucleus, DNA repairing mechanisms are certainly affected, which would facilitate mutations as discussed previously in this paper. In the mitochondria, the electron transport system would be affected and glycolysis might be stimulated, which is somewhat similar to the irreversible damage of the aerobic respiration proposed by Warburg (Figure 4). Deficiency of

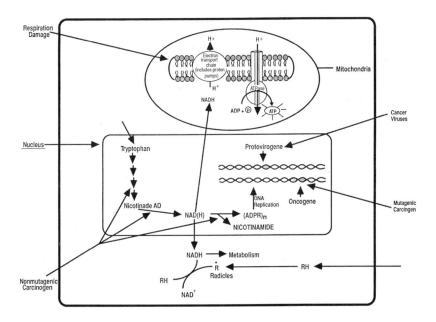

Fig. 4. Proposed action of mutagenic and non-mutagenic carcinogens on carcinogenesis.

one kind of micronutrients including NAD would not be the sole cause of carcinogenesis. However, it may help the manifestation of the properties of transformed cells such as increasing mutations of the oncogenes and suppressor genes. A deficiency of vitamin B_6 in the cells may affect the biosynthesis of NAD resulting in an insufficient supply of NAD in cells as discussed previously. In the meantime, deficiency of B_6 might also cause the accumulation of the intermediate metabolites such as kynurenine, 3-hydroxykynurenine, 8-hydroxyquinoline, kynurenic acid, quinolinic acid, and quinaldic acid in urine. These intermediates are suspected carcinogens that are capable of inducing human bladder carcinogenesis (Birt *et al.*, 1987; Bryan, 1969). Would the deficiency of NAD, vitamin C, and/or other nutrients be related to the GJIC of cells, which is related to many aspects of cancer as reviewed by Trosko and Ruch (1998)? Would the impairment of GJIC be related to the metastasis of cancer cells? These are important questions that need further investigations.

Multiple mutations might have occurred before and after initiation of carcinogenesis. Everyone would agree that cancer is ultimately a disease of the DNA. However, some environmental carcinogens may work at the level of the epigenetic factors (non-mutagenic) such as the metabolism of NAD, vitamins B_6, B_{12}, C, E, folic acid, choline, methionine, and zinc, rather than acting directly on DNA sequences. Chromosomal abnormalities, hypermethylation or hypomethylation of DNA, interruption of cell to cell signaling or gap junction intercellular communications are all important factors in causing cancer. Each epigenetic factor may facilitate the mutations of protooncogenes, expression of oncogenes and/or suppression of the suppressor genes. Epigenetic factors may also be essential for the manifestation of the unique facet of cancer cells such as devoid of aerobic respiration, loss of contact inhibition, uncontrollable replication of DNA, cell proliferations, metastasis, etc.

Mutations alone would not be the only cause of carcinogenesis although mutations (of oncogenes and suppressor genes) might provide the basic mechanisms. Epigenetic factors may work synergistically affecting the genetic mutations and biochemical changes so the cancerous cells result as the outcome of all these carcinogenic processes.

As postulated in Figure 1, epigenetic factors are involved in many stages of cancer development. Both genetic and epigenetic factors interact with each other during the evolution of cancer. Each factor might not be the sole cause of cancer, but is an important part of overall carcinogenetic processes. It is important to keep both genetic and epigenetic factors in mind for any cancer research plan.

To give the research scientists some "handles" on the large number of hypotheses reviewed, it may be helpful to summarize them into two groups: "genetic" and "epigenetic" factors. In visualizing the first, it is clear that proto-oncogenes, oncogenes, or suppressor genes constitute the primary information for expression of carcinogenesis regardless of these genes that are inherited or acquired by viral infections. Would the carcinogens including X-rays and mutagenic chemicals alter cellular proto-oncogenes, oncogenes, or suppressor genes while these genes reside in their normal sites in cellular chromosomes? The results might be a disruption of cellular growth control that destabilized normal cell growth and lead to tumorigenesis. A more detailed investigation on how these epigenetic factors are involved is essential for the illustration of cancer causes. Research on the synergistic effects of genetic factors such as mutagens or oncogenic viruses, epigenetic factors such as how methylation affecting the gene expression and regulation, the expression of GJIC, and the metabolism of folic acid, choline, methionine, vitamins B_6, B_{12}, C, E, NAD, and/or others, should certainly be of important consideration. How all these factors are interrelated how they can be integrated into a plausible mechanism that transforms normal cells into cancer cells are crucial to the understanding of the cause of the carcinogenesis. How we think about the cause of cancer will also determine how we think about the cure of cancer. To understand the whole mechanism might be of vital importance for the design of the complete therapy and prevention of this dreadful disease, which has threatened human health for so long.

ACKNOWLEDGMENTS

The author deeply acknowledged the generous suggestions of the contents of this paper by Dr. Wai-Yuan Tan of the Department of Mathematical Sciences, The University of Memphis, Memphis, TN.

References

1. Abercrombie M. Contact inhibition and malignancy. *Nature* 1979; **281**: 259–262.
2. Alderson T. Ribonucleotide metabolism — fresh approaches to viral and cancer chemotherapy. *Biol. Rev.* 1989; **64**: 159–196.
3. Alderson T. Poly(ADP-ribosylation) processing as a target for the anti-tumor effects of the cell differential agent, hexamethylenebisacetamide, and the N_6-substituted adenosine. *Anti-Cancer Drugs* 1991; **2**: 543–548.
4. Althaus FR and Richter CR. ADP-ribosylation of proteins. *Mol. Biol. Biochem. Biophys.* 1987; **37**: 1–126.
5. Althaus FR, Kleczkowska HE, Malanga M, Muntener CR, Pleschke JM, Ebner M and Auer B. Poly ADP-ribosylation: a DNA break signal mechanism. *Mol. Cell. Biochem.* 1999; **193**: 5–11.
6. Alvarez-Gonzalez R and Althaus FR. Poly(ADP-ribose) catabolism in mammalian cell exposed to DNA-damaging agents. *Mutat. Res.* 1989; **218**: 67–74.
7. Alvarez-Gonzalez R, Watkins TA, Gill PK, Reed JL and Mendoza-Alvarez H. Regulatory mechanisms of poly (ADP-ribose) polymerase. *Mol. Cell. Biochem.* 1999; **193**: 19–22.
8. Ame J-C, Jacobson EL and Jacobson MK. Molecular heterogeneity and regulation of poly(ADP-ribose) glycohydrolase. *Mol. Cell. Biochem.* 1999; **193**: 75–81.
9. American Cancer Society Cancer Facts and Figures. American Cancer Society, Atlanta, GA, 2005.
10. Ames BN, Durston WE, Yamasaki E and Lee FD. Carcinogens are mutagens: a simple test system combining liver homogenates for activation and bacteria for detection. *Proc. Natl. Acad. Sci. USA* 1973; **70**: 2281–2285.
11. Ames BN, McCann J and Yamasaki E. Methods of detecting carcinogens and mutagens with Salmonella/mammalian-microsome mutagenicity test. *Mutat. Res.* 1975; **31**: 347–364.
12. Ames BN. Identifying environmental chemicals causing mutations and cancer. *Science* 1979; **204**: 587–593.
13. Ames BN. Mutagenesis and carcinogenesis: endogenous and exogenous factors. *Environ. Mol. Carcinog.* 1989; **14**(Suppl. 16): 66–77.
14. Ames BN and Gold LS. Too many rodent carcinogens: mitogenesis increases mutagenesis. *Science* 1990; **249**: 970–971.
15. Ames BN, Shigenaga MK and Hagen TM. Oxidants, antioxidants, and the degenerative diseases of aging. *Proc. Natl. Acad. Sci. USA* 1993; **90**: 7915–7922.

16. Ames BN, Gold LS and Willett WC. The causes and prevention of cancer. *Proc. Natl. Acad. Sci. USA* 1995; **92**: 5258–5265.
17. Ames BN and Gold LS. Environmental pollution, pesticides, and the prevention of cancer: misconceptions. *FASEB J.* 1997; **11**: 1041–1052.
18. Ames BN. Micronutrients prevent cancer and delay aging. *Toxicol. Lett.* 1998; **102/103**: 5–18.
19. Ames BN. DNA damage from micronutrient deficiencies is likely to be a major cause of cancer. *Mutat. Res.* 2001; **475**: 7–20.
20. Aruoma OI. Nutrition and health aspects of free radicals and antioxidants. *Food Chem. Toxicol.* 1994; **32**: 671–683.
21. Asaka M, Takeda H, Sugiyama T and Kato M. What role does *Helicobacter pylori* play in gastric cancer? *Gastroenterology* 1997; **113**: S56–60.
22. Austocker J. Diet and cancer. *Br. Med. J.* 1994; **308**: 1610–1614.
23. Babbs C. Free radicals and the etiology of colon cancer. *Free Radic. Biol. Med.* 1990; **8**: 191–200.
24. Baltimore D. RNA-dependent DNA polymerase in virions of RNA tumor viruses. *Nature* 1970; **226**: 1209–1211.
25. Bangham J. A shifting balance. *Nat. Rev. Genet.* 2005; **5**: 251.
26. Bartecchi CE, MacKenzie TD and Schrier RW. The human costs of tobacco use (first of two parts). *N. Engl. J. Med.* 1994; **330**: 907–912.
27. Baylin SB, Herman JG, Graff JR, Vertino PM and Issa JP. Alterations in DNA methylation: a fundamental aspect of neoplasia. *Adv. Cancer Res.* 1998; **72**: 141–196.
28. Baylin SB. DNA methylation and gene silencing in cancer. *Nat. Clin. Pract. Oncol.* 2005; **2**: S4–S11.
29. Baylin SB and Ohm JE. Epigenetic genes silencing in cancer — a mechanism for early oncogenic pathway addiction. *Nat. Rev. Cancer* 2006; **6**: 107–116.
30. Beasley RP. Hepatitis B virus: the major etiology of hepatocellular carcinoma. *Cancer* 1988; **61**: 1942–1956.
31. Beckman KB and Ames BN. The free radical theory of aging matures. *Physiol. Rev.* 1998; **78**: 547–581.
32. Beland FA and Kadlubar FF. Formation and persistence of arylamine DNA adducts *in vivo*. *Environ. Health Prospect.* 1985; **62**: 19–30.
33. Beland FA and Kadlubar FF. Metabolic activation and DNA adducts of aromatic amides and nitroaromatic hydrocarbons. In: *Chemical Carcinogenesis and Mutagenesis I*. Cooper CS and Grover PL (eds.) Springer, Berlin, 1990, pp. 267–325.

34. Beland FA and Poirier MC. DNA adducts and carcinogenesis. In: *The Pathobiology of Neoplasia*. Sirica AE (ed.) Plenum Publishing Corp., New York, 1989, pp. 57–80.
35. Beland FA, Melchior WB, Mourato LLG, Amela Santos M and Marques MM. Arylamine-DNA arylamine conformation in relation to mutagenesis. *Mutat. Res.* 1997; **376**: 13–19.
36. Belinsky SA. Gene-promoter hypermethylation as a biomarker in lung cancer. *Nat. Rev. Cancer* 2004; **4**: 707–717.
37. Benjamin RC and Gill DM. A connection between poly(ADP-ribose) synthesis and DNA damage. *Fed. Proc. Am. Soc. Exp. Biol.* 1979; **38**: 619.
38. Benzie IFF. Lipid peroxidation: a review of causes, consequences, measurement and dietary influences. *Int. J. Food Sci. Nutr.* 1996; **47**: 233–261.
39. Berenblum I and Shubik P. A new, quantitative approach to the study of the stages of chemical carcinogenesis in the mouse's skin. *Br. J. Cancer* 1947; **1**: 383–391.
40. Berger NA, Sikorski GW, Petzold SJ and Kurohara KK. Association of poly (adenosine diphosphoribose) synthesis with DNA damage and repair in normal human lymphocytes. *J. Clin. Invest.* 1979; **63**: 1164–1171.
41. Bernofsky C. Physiologic aspects of pyridine nucleotide regulation in mammals. *Mol. Cell. Biochem.* 1980; **33**: 135–143.
42. Bertazzoni Y, Mathelet M and Campagnari F. Purification and properties of a polynucleotide ligase from calf thymus glands. *Biochem. Biophys. Acta* 1972; **287**: 404–414.
43. Birt DF, Julius AD, Hasegawa R, St. John M and Cohen SM. Effect of L-tryptophan excess and vitamin B_6 deficiency on rat urinary bladder cancer promotion. *Cancer Res.* 1987; **47**: 1244–1250.
44. Bissell M, Rambeck WA, White RC and Bassham JA. Glycerol phosphate shuffle in virus-transformed cells in culture. *Science* 1976; **191**: 856–858.
45. Block G. The data support a role for antioxidants in reducing cancer risk. *Nutr. Rev.* 1992; **52**: 207–213.
46. Block G, Patterson B and Subar A. Fruits, vegetables and cancer prevention: a review of the epidemiologic evidence. *Nutr. Cancer* 1992; **18**: 1–29.
47. Blount BC, Mack MM, Wehr CM, MacGregor JT, Hiatt RA, Wang G, Wickramasinghe SN, Everson RB and Ames B. Folate deficiency causes uracil misincorporation into human DNA and chromosome breakage: implications for cancer and neuronal damage. *Proc. Natl. Acad. Sci. USA* 1997; **94**: 3290–3295.
48. Borek C and Sach L. The difference in contact inhibition of cell replication between normal cells and cells transformed by different carcinogens. *Proc. Natl. Acad. Sci. USA* 1966; **56**: 1705–1711.

49. Boulikas T. Poly(ADP-ribosylation) histones in chromatin replication. *J. Biol. Chem.* 1990; **265**: 14638–14647.
50. Boulikas T. Relation between carcinogenesis, chromatin structure and poly(ADP-ribosylation) (Review). *Anticancer Res.* 1991; **11**: 489–528.
51. Boveri T. Zur Frage Entshehung maligner tumoren. *Gustav Fischer* 1914; Jena.
52. Boxer GE and Devlin TM. Pathways of intracellular hydrogen transport. *Science* 1961; **134**: 1495–1501.
53. Breivik J and Gaudernack G. Genomic instability, DNA methylation, and natural selection in colorectal carcinogenesis. *Semin. Cancer Biol.* 1999; **9**: 245–254.
54. Breksa AP and Garrow TA. Mutagenesis of zinc-binding motif of betaine-homocysteine methyltransferase reveals that Gly 214 is essential. *Arch. Biochem. Biophys.* 2002; **399**: 73–80.
55. Brenner AV, Wang Z, Kleinerman RA, Wang, L, Zhang S, Metayer C, Lei S, Cui H and Lubin HH. Previous pulmonary diseases and risk of lung cancer in Gansu Province, China. *Int. J. Epidemiol.* 2001; **30**: 118–124.
56. Briggs MH. Vitamin and coenzyme content of the hepatoma induced by butter yellow. *Nature* 1960; **187**: 249–250.
57. Brookes P and Lawley PD. The reaction of mustard gas with nucleic acids in vitro. *Biochem. J.* 1960; **77**: 478–484.
58. Bruemmer B, White E, Vaughan TL and Cheney CL. Nutrient intake in relation to bladder cancer among middle-aged men and women. *Am. J. Epidemiol.* 1996; **144**: 485–495.
59. Bryan GT. Role of tryptophan metabolites in urinary bladder cancer. *Am. Ind. Hyg. Assoc. J.* 1969; **30**: 27–34.
60. Budunova IV and Williams GM. Cell culture assays for chemicals with tumor promoting or inhibiting activity based on the modulation of intercellular communication. *Cell Biol. Toxicol.* 1994; **10**: 71–116.
61. Buettner GR. The pecking order of free radicals and antioxidants: lipid peroxidation, a-α-tocopherol, and ascorbate. *Arch. Biophys. Biochem.* 1993; **300**: 535–543.
62. Buick RN and Pollak MN. Prospectives on clonogenic tumor cells, stem cells, and oncogenes. *Cancer Res.* 1984; **44**: 4909–4918.
63. Burch KB, Liu ET and Larrick JW. *Oncogenes: An Introduction to the Concepts of Cancer Genes.* Springer-Verlag, New York, 1988.
64. Burzio L and Koide SS. A functional role of polyADPR in DNA synthesis. *Biochem. Biophys. Res. Commun.* 1970; **40**: 1013–1029.

65. Byers T and Guerrero N. Epidemiologic evidence for vitamin C and vitamin E in cancer prevention. *Am. J. Clin. Nutr.* 1995; **62**: 1385S–1392S.
66. Cairns J. Mutation selection and the natural history of cancer. *Nature* 1975; **255**: 197–200.
67. Cameron E and Rotman D. Ascorbic acid, cell proliferation and cancer. *Lancet* 1972; **1**: 542.
68. Cameron E and Pauling L. Ascorbic acid and the glycosaminoglycan: an orthomolecular approach to cancer and other disease. *Oncology* 1973; **27**: 181–192.
69. Cameron E and Pauling L. Supplementary ascorbate in the supportive treatment of cancer: prolongation of survival times in terminal human cancer. *Proc. Natl. Acad. Sci. USA* 1976; **73**: 3685–3689.
70. Cameron E and Pauling L. Supplementary ascorbate in the supportive treatment of cancer: re-evaluation of prolongation of survival times in terminal human cancer. *Proc. Natl. Acad. Sci. USA* 1978; **75**: 4538–4542.
71. Cameron E, Pauling L and Leibovitz B. Ascorbic acid and cancer: a review. *Cancer Res.* 1979; **39**: 663–681.
72. Cardenas-Corona ME, Jacobson EL and Jacobson MK. Endogenous polymers of ADP-ribose are associated with the nuclear matrix. *J. Biol. Chem.* 1987; **262**: 14863–14866.
73. Castro CE, Kaspin LC, Chen S-S and Nolker SG. Zinc deficiency increases the frequency of single-strand DNA breaks in rat liver. *Nutr. Res.* 1992; **12**: 721–736.
74. Cerutti PA. Oxidant stress and carcinogenesis. *Eur. J. Clin. Invest.* 1991; **21**: 1–5.
75. Chang AY. On the mechanism for the depression of liver NAD by streptozotocin. *Biochem. Biophys. Acta* 1972; **261**: 77–84.
76. Chatterjee S, Berger SJ and Berger NA. Poly(ADP-ribose) polymerase: a guardian of the genome that facilitates DNA repair by protecting against DNA recombination. *Mol. Cell. Biochem.* 1999; **193**: 23–30.
77. Chaumontet C, Bex V, Gaillard-Sanchez I, Seillan-Heberden C, Suschetet M and Martel P. Apigenin and tangeretin enhance gap junctional intercellular communication in rat liver epithelial cells. *Carcinogenesis* 1994; **15**: 2325–2330.
78. Chaykin S. Nicotinamide coenzyme. *Ann. Rev. Biochem.* 1963; **35**: 149–170.
79. Chen LH, Boissonneault GA and Glauert HP. Vitamin C, vitamin E and cancer (Review). *Anticancer Res.* 1989; **8**: 739–748.
80. Chen J, Giovannucci E, Kelsey, K, Rimm EB, Stampfer MJ, Colditz GA, Spiegelman D, Willett WC and Hunter DJ. A methylenetetrahydrofolate

reductase polymorphism and the risk of colorectal cancer. *Cancer Res.* 1996; **56**: 4862–4864.
81. Chen W, Cooper TK, Zahnow CA, Overholtzer M, Zhao Z, Ladanyi M, Karp JE, Gokgoz N, Wunder JS, Andruli IL, Levin AJ, Mankowski JL and Baylin SB. Epigenetic and genetic loss of HIC1 function accentuates the role of p53 in tumorigenesis. *Cancer Cell* 2004; **6**: 387–398.
82. Chen W, Wang D, Yen R, Luo J, Gu W and Baylin S. Tumor suppressor HIC1 directly regulates SIR1 and modulates p53-dependent apoptotic DNA damage response. *Cell* 2005; **123**: 437–448.
83. Christen S, Woodall AA, Shigenaga MK, Southwell-Keely PT, Duncan MW and Ames BN. α-Tocopherol traps mutagenic electrophiles such as NO_X and complements α-tocopherol: physiological implications. *Proc. Natl. Acad. Sci. USA* 1997; **94**: 3217–3222.
84. Chung K-T. An association of carcinogenesis and decrease of cellular NAD concentration. *Chin. J. Microbiol. Immunol.* 1982; **15**: 265–274.
85. Chung K-T. Significance of azo reduction in the carcinogenesis and mutagenesis of azo dyes. *Mutat. Res.* 1983; **114**: 263–281.
86. Chung K-T and Cerniglia CE. Mutagenicity of azo dyes: structure-activity relationships. *Mutat. Res.* 1992; **277**: 201–220.
87. Chung K-T, Kirkovsky L, Kirkovsky A and Purcell WP. Review of mutagenicity of monocyclic aromatic amines: quantitative structure-activity relationships. *Mutat. Res.* 1997; **387**: 1–16.
88. Chung K-T. Mutagenicity and carcinogenicity of aromatic amines metabolically produced from azo dyes. *Environ. Carcinog. Ecotoxicol. Rev. Part C J. Environ. Sci. Health* 2000; **C18**: 51–74.
89. Chung K-T, Chen S-C, Wong TY, Li YS, Wei CI and Chou MW. Mutagenicity studies of benzidine and its analogues: structure-activity relationships. *Toxicol. Sci.* 2000, **56**: 351–356.
90. Chung K-T, Chen S-C and Claxton LD. Review of the *Salmonella typyimurium* mutagenicity of benzidine, benzidine analogues and benzidine-based dyes. *Mutat. Res.* 2006; **612**: 58–76.
91. Clark JB and Greenbaum AL. The concentration and biosynthesis of nicotinamide nucleotides in the livers of rats treated with carcinogens. *Biochem. J.* 1996; **98**: 546–556.
92. Claxton LD, Houk VS and Thomas TJ. Genotoxicity of industrial wastes and effluents. *Mutat. Res.* 1998; **410**: 237–243.
93. Cohen SM. Comparative pathology of proliferative lesions of the urinary bladder. *Toxicol. Pathol.* 2002; **30**: 663–671.

94. Colditz GA, Stampfer MJ, Willett WC, Hunter DJ, Manson JE, Hennekens CH, Rosner BA and Speizer FE. Type of postmenopausal hormone use and risk of breast cancer: 12-year follow-up from the nurses' study. *Cancer Causes Control* 1992; **3**: 433–439.
95. Coleman WB, McCullough KD, Esch GL, Civalier CJ, Livanos E, Weissman BE, Grisham JW and Smith GJ. Suppression of the tumorigenic phenotype of a rat liver epithelial cell line by the p11.2-p12 region of human chromosome 11. *Mol. Carcinog.* 1995; **13**: 220–232.
96. Comes R and Mustea I. The levels of NAD^+ and NADH in blood of patients with cancer. *Neoplasma* 1976; **23**: 451–455.
97. Cooney RV, Harwood PJ, Franke AA, Narala K, Sundstrom A-K, Berggren P-O and Mordan LJ. Products of β-tocopherol reaction with NO_2 and their formation in rat insulinoma (RINm5F) cells. *Free Radic. Biol. Med.* 1995; **19**: 259–269.
98. Cooper GM. Cellular transforming genes. *Science* 1982; **218**: 801–806.
99. Croce CM and Klein G. Chromosome translocations and human cancer. *Sci. Am.* 1984; **252**: 54–60.
100. Crawford D, Zbinden I, Amstad P and Cerutti P. Oxidant stress induces the protooncogenes *c-fos* and *c-myc* in mouse epidermal cells. *Oncogene* 1988; **3**: 27–32.
101. Cross AJ and Sinha R. Meat-related mutagens/carcinogens in the etiology of colorectal cancer. *Environ. Mol. Mutagen.* 2004; **44**: 44–55.
102. Cui H, Cruz-Correa M, Giardiello FM, Hutcheon DF, Kafonek DR, Brandenburg S, Wu Y, He X, Powe NR and Feinber AP. Loss of IGF imprinting: a potential marker of colorectal cancer risk. *Science* 2003; **299**: 1753–1755.
103. Davies M I, Halldorsson H, Shall S and Skidmore CJ. The action of streptozotocin on mouse leukaemia cells. *Biochem. Soc. Trans.* 1976; **4**: 635–637.
104. DeMarini DM, Shelton ML and Levine JG. Mutation spectra of cigarette smoke condensate in Salmonella: comparison to mutations in smoking-associated tumors. *Carcinogenesis* 1995; **16**: 2535–2542.
105. Diaz NN, Frei B, Vita JA and Keaney Jr JF. Antioxidants and atherosclerotic heart disease. *New Engl. J. Med.* 1997; **337**: 408–416.
106. Diplock AT. Commentary: will the "Good Fairies" please prove to us that vitamin E lessens human degenerative disease? *Free Radic. Res.* 1997; **27**: 511–532.
107. Doll R. Nature and nurture: possibilities for cancer control. *Carcinogenesis* 1996; **17**: 177–184.

108. Donehowever LA, Harvey M, Slagle BL, McArthur MJ, Montgomery Jr CA, Butel JS and Bradley A. Mice deficient for p53 are developmentally normal but susceptible to spontaneous tumors. *Nature* 1992; **356**: 215–221.
109. Dreosti IE. Zinc and the gene. *Mutat. Res.* 2001; **475**: 161–167.
110. Duesberg P, Li R, Rasnick D, Rausch C, Willer A, Kraemer A, Yerganian G and Hehlmann R. Aneuploidy precedes and segregates with chemical carcinogenesis. *Cancer Genet. Cytogenet.* 2000; **119**: 83–93.
111. Dunham WB, Zuckercandle E, Reynolds R, Willoughby R, Marcuson R, Barth R and Pauling L. Effects of intake of L-ascorbic acid on the incidence of dermal neoplasms induced in mice by ultraviolet light. *Proc. Natl. Acad. Sci. USA* 1982; **79**: 7532–7536.
112. Evans MG, El-Fouly MH and Trosko JE. Detection of inhibition of intercellular communication by the Scrape-loading/dye transfer technique: a concentration/response study. *In Vitro Toxicol.* 1988/89; **2**: 101–108.
113. Fenech MF, Dreosti IE and Rinaldi JR. Folate, vitamin B_{12}, homocysteine status and chromosome damage rate in lymphocytes of older men. *Carcinogenesis* 1997; **18**: 1329–1336.
114. Fenech M, Aitken C and Rinaldi JR. Folate, vitamin B_{12}, homocysteine status and DNA damage in young Australian adults. *Carcinogenesis* 1998; **19**: 1163–1171.
115. Ferguson LR, Karunasinghe N and Philpott M. Epigenetic events and protection from colon cancer in New Zealand. *Environ. Mol. Mutagen.* 2004; **44**: 36–43.
116. Ferreira R, Naguibneva I, Pritchard LL, Sit-Si-Ali S and Harel-Bellan A. The Rb/chromatin connection and epigenetic control: opinion. *Gene Oncogene* 2001; **20**: 3128–3133.
117. Flintoft L. Silent transmission. *Nat. Rev. Genet.* 2005; **5**: 720–721.
118. Fink L. Unraveling the molecular biology of cancer. *BioScience* 1984; **34**: 75–77.
119. Fong LY, Li J, Farber J and Magee PN. Cell proliferation and esophageal carcinogenesis in the zinc-deficiency rat. *Carcinogenesis* 1996; **17**: 1841–1848.
120. Forman D. The etiology of gastric cancer. In: *Relevance to Human Cancer of N-Nitroso Compounds, Tobacco Smoke and Mycotoxins*, IARC Scientific Publications, No. 105. O'Neill IK, Chen J and Bartsch H (eds.) IARC, Lyon, France, 1991, pp. 22–32.
121. Foster JW and Moat AG. Nicotinamide adenine dinucleotide biosynthesis and pyridine nucleotide cycle metabolism of microbial system. *Microbiol. Rev.* 1980; **44**: 83–105.

122. Franceschi RT. The role of ascorbic acid in mesenchymal differentiation. *Nutr. Rev.* 1992; **50**: 65–70.
123. Friso S and Choi SW. Gene nutrient interactions and DNA methylation. *J. Nutri.* 2002; **132**: S2382–2387.
124. Fu CS, Swendseid ME, Jacob RA and McKee RW. Biochemical markers for assessment of niacin status in young men: levels of erythrocyte niacin coenzymes and plasma tryptophan. *J. Nutr.* 1989; **119**: 1949–1955.
125. Fukushima M, Okayama H, Takahashi Y and Hayaishi O. Characterization of the NAD^+ glycohydrase associated with rat liver nuclear envelope. *J. Biochem.* 1976; **80**: 167–176.
126. Ganem D and Prince M. Hepatitis B viral infection — natural history and clinical consequences. *New Engl. J. Med.* 2004; **350**: 1118–1129.
127. Gensler HL, Williams T, Huang AC and Jacobson EL. Oral niacin prevents photocarcinogenesis and photoimmunosuppression in mice. *Nutr. Cancer* 1999; **34**: 36–41.
128. Ghoshal AK and Farber E. The induction of liver cancer by dietary deficiency of choline and methionine without added carcinogens. *Carcinogenesis* 1984; **5**: 1367–1370.
129. Ghoshal AK and Farber E. Biology of disease choline deficiency, lipotrope deficiency and the development of liver disease including liver cancer: a new perspective. *Lab. Invest.* 1993; **68**: 255–260.
130. Gibbs WW. Untangling the roots of cancer. *Sci. Am.* 2003; **289**: 57–65.
131. Giovannucci E, Stampfer MJ, Colditz GA, Rimm EB, Trichopoulos D, Rosner BA, Speizer FE and Willett W. Folate, methionine and alcohol intake and risk of colorectal adenoma. *J. Natl. Cancer Inst.* 1993; **85**: 875–884.
132. Giovannucci E, Rimm EB, Ascherio A, Stampfer MJ, Colditz GA and Willet WC. Alcohol, low-methionine-low folate diets and risk of colon cancer in men. *J. Natl. Cancer Inst.* 1995; **87**: 265–273.
133. Giovannucci E, Stampfer MJ, Colditz GA, Hunter DJ, Fuchs C, Rosner BA, Speizer FE and Willet WC. Multivitamin use, folate and colon cancer in women in the nurses' health study. *Ann. Intern. Med.* 1998; **129**: 517–524.
134. Giovannucci E. Epidemiological studies of folate and colorectal neoplasia: a review. *J. Nutr.* 2002; **132**: S2350–S2355.
135. Glock GE and McLean P. Levels of oxidized and reduced diphosphopyridine nucleotide and triphosphopyridine nucleotide in tissues. *Biochem. J.* 1975; **65**: 413–416.
136. Golka K, Weise A, Assennato G and Bolt HM. Occupational exposure and urological cancer. *World J. Urol.* 2004; **21**: 382–391.

137. Goodman JE, Hofseth LG, Hussain SP and Harris CC. Nitric oxide and p53 in cancer-prone chronic inflammation and oxyradical overload disease. *Environ. Mol. Mutagen.* 2004; **44**: 3–9.
138. Goodwin PM, Lewis PJ, Davies MI, Skidmore CJ and Shall S. The effect of gamma radiation and neocarzinostatin on NAD and ATP levels in mouse leukemia cells. *Biochem. Biophys. Acta* 1978; **543**: 576–582.
139. Greaves MF. Differentiation-linked leukemogenesis in lymphocytes. *Science* 1986; **234**: 697–704.
140. Grunicke H, Liersch M, Hinz M, Pushendorf B, Richter E and Holzer H. Die Reductung des Nikotinsäureamids für die synthese des NAD in Ehrlich ascites-tumorzellen. *Biochim. Biophys. Acta* 1966; **121**: 228–240.
141. Grunicke H, Keller HJ, Liersch M and Benaguid A. New aspects of the mechanism and regulation of the pyrimidine nucleotide metabolism. *Adv. Enz. Regul.* 1974; **12**: 397–418.
142. Gunnarson R, Berne C and Hellerstrom C. Cytotoxin effects of streptozotocin and N-nitrosomethylurea on the pancreatic B cells with special regard to the role of nicotinamide-adenine dinucleotide. *Biochem. J.* 1974; **140**: 487–497.
143. Hahn WC and Weinberg RA. Rules of making human tumor cells. *New Engl. J. Med.* 2002; **347**: 1593–1603.
144. Halliwell B and Gutteridge JMC. Oxygen toxicity, oxygen radicals, transition metals and disease. *Biochem. J.* 1984; **219**: 1–14.
145. Harris CC. Solving the viral-chemical puzzle of human liver carcinogenesis. *Cancer Epidemiol. Biomarkers Prev.* 1994; **3**: 1–2.
146. Harris JR, Lippman ME, Veronesi U and Willett W. Breast cancer. *N. Engl. J. Med.* 1992; **327**: 319–328.
147. Hausen HZ. Viruses in human cancers. *Science* 1991; **254**: 1167–1173.
148. Hayaishi O and Ueda K. Poly(ADP-ribose) and ADP-ribosylation of proteins. *Ann. Rev. Biochem.* 1977; **46**: 95–116.
149. Heinonen OP, Albanes D, Virtamo J, Taylor PR, Huttunen JK, Hartman AM, Haapakoski J, Malila N, Rautalahti M, Ripatti S, Maenpaa H, Teerenhovi L, Koss L, Virolainen M and Edwards BK. Prostate cancer and supplementation with alpha-tocopherol and beta-carotene: incidence and mortality in a controlled trial. *J. Natl. Cancer Inst.* 1998; **90**: 440–446.
150. Henderson BE, Ross RK and Pike MC. Toward the primary prevention of cancer. *Science* 1991; **254**: 1131–1138.
151. Henning SM, NcKee RW and Swendseid ME. Hepatic content of S-adenosylmethionine, S-adenosylhomocysteine and glutathione in rats receiving treatments modulating methyl donor availability. *J. Nutr.* 1989; **119**: 1478–1482.

152. Henning SM, Swendseid ME and Coulson WF. Male rats fed methyl- and folate-deficient diets with or without niacin develop hepatic carcinomas associated with decreased tissue NAD concentrations and altered poly(ADP-ribose) polymerase activity. *J. Nutr.* 1996; **127**: 30–36.
153. Hercberg S, Galan P, Preziosi P, Alfarez MJ and Vazquez C. The potential role of antioxidant vitamins in preventing cardiovascular diseases and cancers. *Nutrition* 1998; **14**: 513–520.
154. Hertzberg EL, Lawrence TS and Gilula NB. Gap junctional communication. *Ann. Rev. Physiol.* 1981; **43**: 479–499.
155. Hesketh R. *The Oncogene and Tumor Suppressor Gene Facts Book*, 2nd ed. Academic Press, 1997.
156. Hill CS and Treisman R. Transcriptional regulation by extracellular signals: mechanisms and specificity. *Cell* 1995; **80**: 199–211.
157. Hilz H and Stone P. Poly (ADP-ribose) and ADP-ribosylation of proteins. *Rev. Physiol. Biochem. Pharmacol.* 1976; **76**: 1–58.
158. Hinz M, Katsilambros N, Maier V, Schatz H and Pfeiffer EF. Significance of streptozotocin induced nicotinamide adenine dinucleotide (NAD) degradation in mouse pancreatic islets. *Feb. Lett.* 1973; **30**: 225–228.
159. Ho C and Hashim S. Pyridine nucleotide depletion in pancreatic islets associated with streptozotocin-induced diabetes. *Diabetes* 1972; **21**: 789–793.
160. Holden C. Albert Szent-Gyorgyi, electrons and cancer. *Science* 1979; **203**: 522–524.
161. Holme JA and Soderlund EJ. Modulation of genotoxic and cytotoxic effects of aromatic amines in monolayers of rat hepatocytes. *Cell Biol. Toxicol.* 1984; **1**: 95–110.
162. Hsie A, Recio L, Katz D, Lee C, Magner M and Schenley R. Evidence for reactive oxygen species inducing mutations in mammalian cells. *Proc. Natl. Acad. Sci. USA* 1986; **83**: 9616–9620.
163. Hsieh SE, Ho HH, Yen YS and Chung JG. The effect of vitamin C on N-acetyltransferase activity in *Klebsiella pneumoniae. Food Chem. Toxicol.* 1997; **35**: 1151–1157.
164. Hussain SP, Hofseth LJ and Harris CC. Radical causes of cancer. *Nat. Rev. Cancer* 2003; **3**: 276–285.
165. Imlay JA, Chin SM and Linn S. Toxic damage by hydrogen peroxide through the Fenton reaction *in vivo* and *in vitro. Science* 1988; **240**: 640–642.
166. Iverson H. (ed.). *Theories of Carcinogenesis*. Hemisphere Publishing Corporation (a subsidiary of Harper & Row Publishers, Inc.), Washington, DC, 1988.

167. Jacobson EL and Narishimham G. Absence of carcinogen-induced unscheduled DNA synthesis in NAD-depleted 3T3 cells. *Fed. Proc.* 1979; **38**: 619.
168. Jacobson MK, Levi V, Juarez-Salinas H, Barton RA and Jacobson EL. Effects of carcinogenic N-alkyl-N-nitroso compounds on nicotinamide adenine dinucleotide metabolism. *Cancer Res.* 1980; **40**: 1497–1802.
169. Jacobson EL. Niacin deficiency and cancer in women. *J. Am. Coll. Nutr.* 1993; **12**: 412–416.
170. Jacobson EL and Jacobson MK. A biomarker for the assessment of niacin nutritive as a potential preventive factor in carcinogenesis. *J. Intern. Med.* 1993; **233**: 59–62.
171. Jacobson EL, Dame A, Pyrek JS and Jacobson MK. Evaluating the role of niacin in human carcinogenesis. *Biochemie* 1995; **77**: 394–398.
172. Jacobson EL, Huang AC, Williams T and Gensler HL. Chemical prevention by niacin in a mouse model of UV-induced skin carcinogenesis. *Proc. Am. Assoc. Cancer Res.* 1996; **37**: A279.
173. Jacobson EL and Jacobson MK. Tissue NAD as a biochemical measure of niacin status in humans. *Meth. Enzymol.* 1997; **280**: 221–230.
174. Jacobson EL, Shieh WM and Huang AC. Mapping the role of NAD metabolism in prevention and treatment of carcinogenesis. *Mol. Cell. Biochem.* 1999; **193**: 69–74.
175. Jaffe LF. Epigenetic theory of cancer initiation. In: *Advances in Cancer Research*. Elsevier, Inc., USA, 2003.
176. Jedeikin LA and Weinhouse S. Metabolism of neoplastic tissues. VI. Assay of oxidized and reduced diphosphopyridine nucleotide in normal and neoplastic tissues. *J. Biol. Chem.* 1955; **213**: 271–280.
177. Jedeikin LA, Thomas AJ and Weinhouse S. Metabolism of neoplastic tissue X. Diphosphopyridine levels during azo dyes hepatocarcinogenesis. *Cancer Res.* 1956; **16**: 867–872.
178. Jick H, Walker AM, Watkins RN, D'Ewart DC, Hunter JR, Danford A, Madsen S, Dinan BJ and Rothman KJ. Replacement estrogen and breast cancer. *Am. J. Epidemiol.* 1980; **112**: 586–594.
179. Jones PA and Baylin SB. The fundamental role of epigenetic events in cancer. *Nat. Rev. Genet.* 2002; **3**: 415–428.
180. Jones PA and Takai D. The role of DNA methylation in mammalian epigenetics. *Science* 2001; **293**: 1068–1070.
181. Jonsson G, Menard GL, Jacobson EL, Poirier GG and Jacobson MK. Effect of hyperthermia on poly(adenosine diphosphate ribose) glycohydrolase. *Cancer Res.* 1988; **48**: 4240–4243.

182. Kanno Y. Modulation of cell communication and carcinogenesis. *Jpn. J. Physiol.* 1985; **35**: 693–707.
183. Kerbel RS, Frost P, Liteplo R, Carlow DA and Elliot BE. Possible epigenetic mechanisms of tumor progression: induction high frequency heritable but phenotypically unstable changes in the tumorigenic and metastatic properties of tumor cell populations by 5-azacytidine treatment. *J. Cell Physiol.* 1984; **3**: 87–97.
184. Kensler CJ, Suguira M and Rhoads CP. Co-enzyme I and riboflavin content of livers of rats fed butter yellows. *Science* 1940; **91**: 623.
185. Kessler H, Husemann B and Wagner W. Potential protective effect of vitamin C on carcinogenesis caused by nitrosamine in drinking water: an experimental study on Wistar rats. *Eur. J. Surg. Oncol.* 1992; **18**: 275–281.
186. Key TJ, Silcocks PB, Davey KK, Appleby PN and Bishop DT. A case-control study of diet and prostate cancer. *Br. J. Cancer* 1997; **76**: 678–687.
187. Kim YI. Folate and carcinogenesis: evidence, mechanisms and implications. *J. Nutr. Biochem.* 1999; **10**: 68–88.
188. Kim YI. Role of folate on colon cancer development and progression. *J. Nutr.* 2003; **133**: 3731S–3739S.
189. Kim YI. Will mandatory folic acid fortification prevent or promote cancer. *Am. J. Clin. Nutr.* 2004; **80**: 1123–1128.
190. Kirkland JB. Lipid peroxidation, protein thiol oxidation and DNA damage in hydrogen peroxide-induced injury to endothelial cells: role of activation of poly(ADP-ribose) polymerase. *Biochim. Biophys. Acta* 1991; **1092**: 319–325.
191. Kirkland JB. Niacin and carcinogenesis. *Nutr. Cancer* 2003; **46**: 110–118.
192. Klaunig JE and Ruch RJ. Role of intercellular communication in nongenotoxic carcinogenesis. *Lab. Invest.* 1990; **62**: 135–146.
193. Koinuma K, Yamashita Y, Liu W, Hatanaka H, Kurashina K, Wada T, Tokada S, Kaneda R, Choi YL, Fujiwara SI, Miyakura Y, Nagai H and Mano H. Epigenetic silencing of AXIN2 in colorectal carcinoma with microsatellite instability. *Oncogene* 2006; **25**: 139–146.
194. Kondo S. Carcinogenesis in relation to the stem cell mutation hypothesis. *Differentiation* 1983; **24**: 1–8.
195. Kono S and Hirohata T. Nutrition and stomach cancer. *Cancer Causes Control* 1996; **7**: 41–55.
196. Larsen WJ. Biological implications of gap junction structure, distribution and composition: a review. *Tissue Cell* 1983; **15**: 645–671.

197. Larsson R and Cerutti P. Translocation and enhancement of phosphotransferase activity of protein kinase C following exposure of mouse epidermal cells tom oxidants. *Cancer Res.* 1989; **49**: 5627–5632.
198. Lathia D, Braasch A and Theissen U. Inhibitory effects of vitamin C and E on *in-vitro* formation of N-nitrosamine under physiological conditions. *Front. Gastrointest. Res.* 1988; **14**: 151–156.
199. Lijinsky W. A view of the relation between carcinogenesis and mutagenesis. *Environ. Mol. Carcinog.* 1989; **14**(Suppl. 16): 78–84.
200. Lindahl T and Edelman GM. Polynucleotide ligase from myeloid lymphoid tissues. *Proc. Natl. Acad. Sci. USA* 1968; **61**: 680–687.
201. Loeb LA, Loeb KR and Anderson JP. Multiple mutations and cancer. *Proc. Natl. Acad. Sci. USA* 2003; **100**: 776–781.
202. Long C. *Biochemists' Handbook*. Van Nostrand, Princeton, New Jersey, 1968, p. 782.
203. Lorimer III, WS and Stone PR. Control of histone HI climer poly(ADP-ribose) complex formation by poly(ADP-ribose) glycohydrolase. *Exp. Cell Res.* 1977; **106**: 261–266.
204. Loewenstein WR. Permeability of membrane junctions. *Ann. N. Y. Acad. Sci.* 1966; **137**: 441–472.
205. Lund AH and Lohuizen VM. Epigenetic and cancer. *Genes Dev.* 2004; **18**: 2315–2335.
206. Markert C. Neoplasia: a disease of cell differentiation. *Cancer Res.* 1968; **28**: 1908–1914.
207. Maron DM and Ames BN. Revised methods for the Salmonella mutagenicity test. *Mutat. Res.* 1983; **113**: 173–215.
208. Mason JB. Folate and colonic carcinogenesis: searching for a mechanistic understanding. *J. Nutr. Biochem.* 1994; **5**: 170–175.
209. Mazen A, Menissier-de Murcia J, Molinete M, Simonin F, Gradwohl G, Poirier G and de Murcia G. Poly(ADP-ribose) polymerase: a novel finger protein. *Nucleic Acid Res.* 1989; **17**: 4689–4698.
210. McGuire PG and Twietmeyer TA. Aortic endothelial endothelial junctions in developing hypertension. *Hypertension* 1985; **7**: 483–490.
211. McCann J, Choi E, Yamasaki E and Ames BN. Detection of carcinogens and mutagens in the Salmonella/microsome test: assay of 300 chemicals. *Proc. Natl. Acad. Sci. USA* 1975; **72**: 5135–5139.
212. McMahon G, Davis EF, Huber EJ, Kim Y and Wogan GN. Characterization of c-Ki-ras and N-ras oncogenes in aflatoxin B-1-induced rat liver tumors. *Proc. Natl. Acad. Sci. USA* 1990; **87**: 1104–1108.
213. McNeil PG and Twietmeyer TA. Aortic endothelial junctions in developing hypertension. *Hypertension* 1985; **7**: 483–490.

214. Melhem MF, Kazanecki ME, Rao KN, Kunz HW and Gill III TJ. Genetics and diet: synergism in hepatocarcinogenesis in rats. *J. Am. Coll. Nutr.* 1990; **9**: 168–173.
215. Mesnil M. Gap junctions and cancer: implications and perspectives. *Med. Sci.* 2004; **20**: 197–206.
216. Mikol YB, Hoover KL, Creasia D and Poirier LA. Hepatocarcinogenesis in rats fed methyl-deficient, amino acid-defined diets. *Carcinogenesis* 1983; **4**: 1619–1629.
217. Miller EC and Miller JA. Mechanism of chemical carcinogenesis: nature of proximate carcinogens and interactions with macromolecules. *Pharmacol. Rev.* 1966; **18**: 805–838.
218. Miller EG. Stimulation of nuclear poly (adenosine diphosphate-ribose) polymerase activity from HeLa cells by endonucleases. *Biochem. Biophys. Acta* 1975; **395**: 191–200.
219. Miller EG. Stimulation of poly(adenosine diphosphate-ribose) polymerase activity by bleomycin. *Fed. Proc.* 1976; **36**: 906.
220. Miller JA. Genes that protect against cancer. *BioScience* 1990; **40**: 563–566.
221. Mirvish SS. Effects of vitamin C and E on N-nitroso compound formation, carcinogenesis and cancer. *Cancer* 1986; **58**: 1842.
222. Mirvish SS. Role in cancer etiology of vitamin C inhibition of N-nitroso compound formation. *Am. J. Clin. Nutr.* 1993; **57**: 598–599.
223. Mirvish SS. Experimental evidence for inhibition of N-nitroso compound formation as a factor in the negative correlation between vitamin C consumption and the incidence of certain cancers. *Cancer Res.* 1994; **54**(Suppl.): 1948s–1951s.
224. Misra BC and Srivatsan ES. Localization of HeLa cell tumor suppressor gene to the long arm of chromosome 11. *Am. J. Hum. Genet.* 1989; **45**: 565–577.
225. Moradpour D and Wands JR. Molecular pathogenesis of hepatocellular carcinoma. In: *Hepatology: A Textbook of Liver Disease.* Zakim D and Boyer TD (eds.) WB Saunders, Philadelphia, 2002, pp. 1333–1354.
226. Moraes E, Keyse S and Tyrrell R. Mutagenesis by hydrogen peroxide treatment of mammalian cells: a molecular analysis. *Carcinogenesis* 1990; **11**: 283–293.
227. Moreau-Gachelin F, Ray D, de Both NJ, van der Feitz MJM, Tambourin P and Tavitian A. Spi-1 oncogene activation in Rauscher and Friend murine virus-induced acute erythroleukemias. *Leukemia* 1990; **4**: 20–23.
228. Morton RK. Enzymatic synthesis of coenzyme I in relation to chemical control of cell growth. *Nature* 1958; **181**: 540–542.

229. Mostafa MH, Sheweita SA and O'Connor PJ. Relationship between schistosomiasis and bladder cancer. *Clin. Microbiol. Rev.* 1999; **12**: 97–111.
230. Na MR, Koo SK, Kim DY, Park SD, Rhee SK, Kang KW and Joe CO. In vitro inhibition of gap junctional intercellular communication by chemical carcinogens. *Toxicology* 1995; **98**: 199–206.
231. Nelson RL. Dietary iron and colorectal cancer risk. *Free Radic. Biol. Med.* 1992; **12**: 161–168.
232. Newberne PM, Broitman S and Schrager TF. Esophageal carcinogenesis in the rat: zinc deficiency. DNA methylation and alkyltransferase activity. *Pathobiology* 1997; **65**: 253–263.
233. Neyton J and Trautman A. Physiological modulation of gap junctional permeability. *J. Exp. Biol.* 1986; **124**: 93–114.
234. Nodes JT and Reid E. Azo-dye carcinogenesis: ribonucleotides and ribonucleases. *Br. J. Cancer* 1964; **17**: 745–774.
235. Oakley GP Jr, Adams MJ and Dickinson CM. More folic acid for everyone, now. *J. Nutr.* 1996; **126**: 7512S–755S.
236. O'Brian CA, Ward NE and Ioannides CG. Altered signal transduction in carcinogenesis. *Adv. Mol. Cell Biol.* 1993; **7**: 61–88.
237. Osaka H and Takahashi T. Genetic alterations of multiple tumor suppressors and oncogenes in the carcinogenesis and progression of lung cancer. *Oncogene* 2002; **21**: 7421–7434.
238. O'Shea CC. DNA tumor viruses — the spies who lyse us. *Curr. Opin. Genet. Dev.* 2005; **15**: 18–26.
239. Otto P, Ladik J and Szent-Gyorgyi A. Quantum chemical calculation of model systems for ascorbic acid adducts with Schiff bases of lysine side chains: possibility of internal charge transfer in protein. *Proc. Natl. Acad. Sci. USA* 1979; **76**: 3849–3851.
240. Parkin DM, Pisani P and Ferlay J. Estimate of world incidence of eighteen major cancers in 1985. *Int. J. Cancer* 1993; **54**: 594–606.
241. Pauling L. Are recommended daily allowances for vitamin C adequate? *Proc. Natl. Acad. Sci. USA* 1974; **71**: 4442–4446.
242. Pauling L, Willoughby R, Reynolds R, Blaisdell RE and Lawson L. Incidence of squamous cell carcinoma in hairless mice irradiated with ultraviolet light in relation to intake of ascorbic acid (vitamin C) and of D, L-α-tocopheryl acetate (vitamin E). *Int. J. Vitam. Nutr. Res. (Switzerland)* 1982; **23**: 53–82.
243. Pfeffer S and Voinnet O. Viruses, microRNAs and cancer. *Oncogene* 2006; **25**: 6211–6219.
244. Pimentel E. *Oncogenes*, 2nd ed., Vols. I. & II. CRC Press, Inc., Boca Raton, FL, 1991.

245. Pitts JD and Finbow ME. The gap junction. *J. Cell Sci.* 1986; **4**(Suppl.Y): 239–266.
246. Piyathilake CY, Bell WC, Johanningand GL, Cornwel PE, Heimburger DC and Grizzle WE. The accumulation of ascorbic acid by squamous cell carcinomas of the lung and larynx is associated with global methylation of DNA. *Cancer* 2000; **89**: 171–176.
247. Potter VR. Phenotypic diversity in experimental hepatomas: the concept of partially blocked ontogeny. *Br. J. Cancer* 1978; **38**: 1–23.
248. Potter JD. Methyl supply, methyl metabolizing enzymes and colorectal neoplasia. *J. Nutr.* 2002; **132**: S2410–S2412.
249. Prehn RT. The role of mutation in the new cancer paradigm. *Cancer Cell Int.* 2005; **5**: 9–14.
250. Ramon JM, Serra-Majem L, Cerdo C and Oromi J. Nutrient intake and gastric cancer risk: a case-control study in Spain. *Int. J. Epidemiol.* 1993; **22**: 983–988.
251. Rankin PW, Jacobson MK, Mitchell VR and Busbee DB. Reduction of nicotinamide adenine dinucleotide in human lymphocytes. *Cancer Res.* 1980; **40**: 1803–1807.
252. Rawling JM, Jackson TM, Driscoll ER and Kirkland JB. Dietary niacin deficiency lowers tissue poly(ADP-ribose) and NAD^+ concentrations in Fischer-344 rats. *Biochem. J. Nutr.* 1994; **124**: 1597–1603.
253. Reed PI, Johnston BJ, Walters CL and Hill MJ. Effect of ascorbic acid on the intragastric environment in patients at increased risk of developing gastric cancer. *Gastroenterology* 1989; **96**: A411.
254. Reddy BS. Studies with the azoxymethane-rat preclinical model for assessing colon tumor development and chemoprevention. *Environ. Mol. Mutagen.* 2004; **44**: 26–35.
255. Refsum JM, Ueland PM, NygArd O and Vollset SE. Homocysteine and cardiovascular disease. *Ann. Rev. Med.* 1998; **49**: 31–62.
256. Reztsova VV, Filov VA, Ivin BA, Kon'kov SA and Krylova IM. Basic and salvage pathways of NAD biosynthesis in organs of normal rats, tumor-bearing rats and in tumors. *Vopr. Onkol.* 1994; **40**: 68–71 (in Russian).
257. Rimm E, Willett W, Manson J, Speizer F, Hennekens C and Stampfer M. Folate and vitamin B6 intake and risk of myocardial infarction among US women (Abstract). *Am. J. Epidemiol.* 1996; **143**: S36.
258. Rimm EB, Willett WC, Hu FB, Sampson L, Colditz GA, Manson JE, Hennekens C and Stampfer MJ. Folate and vitamin B6 from diet and supplements in relation to risk of coronary heart disease among women. *J. Am. Med. Assoc.* 1998; **279**: 359–364.

259. Robertson T. DNA methylation and human diseases. *Nat. Rev. Genet.* 2005; **6**: 597–610.
260. Robins HI, Dennis WH, Neville AJ, Shecterle LM, Martin PA, Grossman J, Davis TE, Neville SR, Gillis WK and Rusy BF. A non-toxic system for 41.8°C whole-body hypertherma: results of a phase I study using a radiant heat device. *Cancer Res.* 1985; **45**: 3937–3944.
261. Rogers M, Berestcky JM, Hossain MZ, Guo H, Kadle R, Nicholson BJ and Bertram JS. Retinoid-enhanced gap junctional communication is achieved by increased levels of connexion 43 mRNA and protein. *Mol. Carcinog.* 1990; **3**: 335–343.
262. Rogers AE, Zeisel SH and Groopman J. Diet and carcinogenesis. *Carcinogenesis* 1993; **14**: 2205–2217.
263. Rossing MA, Vaughan TL and McKnight B. Diet and pharyngeal cancer. *Int. J. Cancer* 1989; **44**: 593–597.
264. Rous P. A transmissible avian neoplasm: sarcoma of the common fowel. *J. Exp. Med.* 1911a; **12**: 696–705.
265. Rous P. Transmission of a malignant new growth by means of a cell-free filtrate. *J. Am. Med. Assoc.* 1911b; **56**: 198.
266. Ruch RJ. The role of gap junctional intercellular communication in neoplasia. *Ann. Clin. Lab. Sci.* 1994; **24**: 216–231.
267. Ruch RJ and Klaunig JE. The effects of tumor promoters, genotoxic carcinogens and hepatocytotoxins on mouse hepatocyte intercellular communication. *Cell Biol. Toxicol.* 1986; **2**: 469–483.
268. Ruch RJ and Trosko JE. The role of oval cells and gap junctional intercellular communication in hepatocarcinogenesis. *Anticancer Res.* 1999; **19**: 4831–4838.
269. Sanford KK, Parshad R and Gantt R. Genetic influences on chromosomal radiosensitivity, DNA repair and cancer susceptibility. *ATCC Quart. Newsl.* 1986; **6**(4): 1,2,6.
270. Sarkar B. Metal replacement in DNA-binding zinc finger proteins and its relevance to mutagenicity and carcinogenicity through free radical generation. *Nutrition* 1995; **11**: 646–649.
271. Schein PS. 1-Methyl-nitrosourea and dialkylnitrosamine depression of nicotinamide adenine dinucleotide. *Cancer Res.* 1969; **29**: 1226–1232.
272. Schein PS, Cooney DA, McMenamin MG and Anderson T. Streptozotocin diabetes — further studies on the mechanism of depression of nicotinamide adenine dinucleotide concentrations in mouse pancreatic islets and liver. *Biochem. Pharmacol.* 1973; **22**: 2625–2631.
273. Schultz RM. Roles of cell to cell communication communication and development. *Biol. Reprod.* 1985; **32**: 27–42.

274. Schwartz JP, Passonneau JV, Johnson GS and Pastan I. The effect of growth conditions on NAD^+ and NADH concentrations and the NAD^+:DNAH ratio in normal and transformed fibroblasts. *J. Biol. Chem.* 1974; **249**: 4138–4143.
275. Schwartz J, Shklar G and Trickler D. Vitamin C enhances the development of carcinomas in the hamster buccal pouch experimental model. *Oral Surg. Oral Med. Oral Pathol.* 1993; **76**: 718–722.
276. Selhub J, Seyoum E, Pomfret EA and Zeisel SH. Effects of choline deficiency and methotrexate treatment upon liver folate content and distribution. *Cancer Res.* 1991; **51**: 16–21.
277. Shibanuma M, Kuroki T and Nose K. Induction of DNA replication and expression of protooncogene c-myc and c-fos in quiescent Balb/3T3 cells by xanthine/xanthine oxidase. *Oncogene* 1988; **3**: 17–21.
278. Shibata MA, Hirose M, Kagawa M, Boonyaphiphat P and Ito N. Enhancing effect of concomitant L-ascorbic acid administration on BHA-induced forestomach carcinogenesis in rats. *Carcinogenesis* 1993; **14**: 275–280.
279. Shigenaga MK, Hagen TM and Ames BN. Oxidative damage and mitochondrial decay in aging. *Proc. Natl. Acad. Sci. USA* 1994; **91**: 10771–10778.
280. Shigenaga MK, Lee HH, Blount BC, Christen S, Shigeno ET, Yip H and Ames B. Inflammation and NO_X-induced nitration: assay for 3-nitrotyrosine by HPLC with electrochemical detection. *Proc. Natl. Acad. Sci. USA* 1997; **94**: 3211–3216.
281. Shope RE. Infectious papillomatosis of rabbits: with a note on the histopathology. *J. Exp. Med.* 1933; **58**: 607–624.
282. Shope RE. Serial transmission of virus of infectious papillomatosis rabbits. *Proc. Soc. Exp. Biol. Med.* 1935; **32**: 830–832.
283. Shope RE, Mangold R, McNamara LG and Dumbell KR. An infectious cutaneous fibroma of the Virginia white-tailed deer (*Odocoileus virginianus*). *J. Exp. Med.* 1958; **108**: 797–802.
284. Sigler K and Ruch RJ. Enhancement of gap junctional intercellular communication in tumor promoter-treated cells by components of green tea. *Cancer Lett.* 1993; **69**: 15–19.
285. Smith JA and King JB. Biochemical studies on hormone responsive mammary tumors in BR6 mice. *Cancer Res.* 1970; **30**: 2055–2060.
286. Sobels FH and Delehanty J. The first five years of ICPEMC; The International commission for Protection against Environmental Mutagens and Carcinogens. Environmental Mutagens and Carcinogens. In: *Proceedings of the Third International Conference on Environmental Mutagens*, Tokyo, Mishima and Kyoto, September 21–27, 1981, pp. 81–100.

287. Soderhall S and Lindahl T. Two DNA ligase activities from calf thymus. *Biochem. Biophys. Res. Commun.* 1973; **53**: 910–916.
288. Sokol RJ. Vitamin E. In: *Present Knowledge in Nutrition*, Vol. 91. Zeigler EE and Filer Jr LJ (eds.) ILSI Press, Washington, DC, 1996, pp. 10771–10778.
289. Soto A and Sonnenschein C. The somatic mutation theory of cancer: growing problems with the paradigm? *BioEssays* 2004; **26**: 1097–1107.
290. Spronck JC and Kirkland JB. Niacin deficiency increases spontaneous and etoposide-induced chromosomal instability n rat bone marrow cells *in vivo*. *Mutat. Res.* 2002; **508**(1–2): 83–97.
291. Stabler SP, Sampson DA, Wang LP and Allen RH. Elevations of serum cystathionine and total homocysteine in pyridoxine-, folate-, and cobalamin-deficient rats. *J. Nutr. Biochem.* 1997; **8**: 279–289.
292. Stanbridge EJ. Identifying tumor suppressor genes in human colorectal cancer. *Science* 1990; **247**: 12–13.
293. Steinmetz KA and Potter JD. Vegetables, fruit and cancer prevention: a review. *J. Am. Diet. Assoc.* 1996; **96**: 1027–1039.
294. Streffer C. The biosynthesis of NAD in the liver of irradiated mice. *J. Vitaminol. (Kyoto)* 1968; **14**: 130–134.
295. Sugimura T. Poly (adenosine diphosphate ribose). *Prog. Nucleic Acid Res. Mol. Biol.* 1973; **13**: 127–151.
296. Sukumar S and Barbacid M. Specific patterns of oncogene activation in transplacentally induced tumors. *Proc. Natl. Acad. Sci. USA* 1990; **87**: 718–722.
297. Szent-Gyorgyi A. The living state and cancer. *Ciba Found. Symp.* 1978; **67**: 3–18.
298. Szent-Gyorgyi A. The living state and cancer. *Physiol. Chem. Phys.* 1980; **12**: 99–110.
299. Tanuma SI, Johnson LD and Johnson GS. ADP-ribosylation of chromosomal proteins and mouse mammary tumor virus gene expression: glucocoricoids rapidly decrease endogenous ADP-ribosylation of nonhistone high mobility group 14 and 17 proteins. *J. Biol. Chem.* 1983; **258**: 15371–15375.
300. Tanuma SI, Kawashima K and Endo H. Purification and properties of an (ADP-ribose)$_n$ glycohydrolase from guinea pig liver nuclei. *J. Biol. Chem.* 1986; **261**: 965–969.
301. Tan WY. *Stochastic Models of Carcinogenesis*. Marcer-Dekker, Inc., New York, 1991.
302. Tannenbaum SR, Correa P, Newberne PM and Fox JG. Endogenous formation of N-nitroso compounds and gastric cancer. In: *Environmental*

Mutagens and Carcinogens. Sugimura T, Kondo S and Takebe H (eds.) University of Tokyo Press, Alan R. Liss, Inc., New York, 1981, pp. 565–570.
303. Taper HS, De-Gerlache J, Lans M and Roberfroid M. Non-toxic potentiation of cancer chemotherapy by combined C and K-3 vitamin pretreatment. *Int. J. Cancer* 1987; **40**: 575–579.
304. Temin HM. Nature of provirus of Rous sarcoma. *Nat. Cancer Inst. Monography* 1964; **17**: 557–570.
305. Temin HM. RNA-dependent DNA polymerase in virions of Rous sarcoma virus. *Nature* 1970; **226**: 1211–1213.
306. Temin HM. The RNA tumor viruses: background and foreground. *Proc. Natl. Acad. Sci. USA* 1972; **69**: 1016–1020.
307. Temin HM. Introduction to virus-caused cancers. *Cancer* 1974a; **34**: 1347–1352.
308. Temin HM. On the origin of RNA tumor viruses. *Ann. Rev. Genet.* 1974b; **8**: 155–177.
309. Temin HM and Batimore D. RNA-directed DNA synthesis and RNA tumor viruses. *Adv. Virus Res.* 1972; **17**: 129–186.
310. Thorgeirsson SS and Grisham JW. Molecular pathogenesis of human hepatocellular carcinoma. *Nat. Genet.* 2002; **31**: 339–346.
311. Todaro GJ and Huebner RJ. Oncogenes of RNA tumor virus as determinants of cancer. *Proc. Natl. Acad. Sci. USA* 1969; **64**: 1087–1094.
312. Todaro GJ and Huebner R. The viral oncogene hypothesis, new evidence. *Proc. Natl. Acad. Sci. USA* 1972; **69**: 1009–1015.
313. Thompson CB. Apoptosis in the pathogenesis and treatment of disease. *Science* 1995; **267**: 1456–1462.
314. Toyokuni S. Iron-induced carcinogenesis: the role of redox regulation. *Free Radic. Biol. Med.* 1996; **20**: 553–566.
315. Trosko JE and Chang CC. An integrative hypothesis linking cancer, diabetes and atherosclerosis: the role of mutations and epigenetic change. *Med. Hypotheses* 1980; **6**: 455–468.
316. Trosko JE, Jone L and Chang CC. Oncogenes, inhibited intercellular communication and tumor promotion. *Princess Takamatsu Symp.* 1983; **14**: 101–113.
317. Trosko JE. Mechanisms of tumor promotion: possible role of inhibited intercellular communication. *Eur. J. Clin. Oncol.* 1987; **23**: 19–29.
318. Trosko JE, Chang CC, Madhukar BV and Klaunig JE. Chemical, oncogene and growth factor inhibition gap junctional intercellular communication: an integrative hypothesis of carcinogenesis. *Pathobiology* 1990; **58**: 265–278.

319. Trosko JE, Madhukar BV and Chang CC. Endogenous and exogenous modulation of gap junctional intercellular communication: toxicological and pharmacological implications. *Life Sci.* 1993; **53**: 1–19.
320. Trosko JE, Chang CC, Madhukar BV and Dupont E. Intercellular communication: a paradigm for the interpretation of the initiation/promotion/progression model of carcinogenesis. In: *Chemical Induction of Cancer: Modulation and Combination Effects.* Arcos JC (ed.) Boston, 1996, pp. 205–225.
321. Trosko JE and Inoue T. Oxidative stress, signal transduction and intercellular communication in radiation carcinognesis. *Stem Cells* 1997; **15**(Suppl. 2): 59–67.
322. Trosko JE and Ruch RJ. Cell-cell communication in carcinogenesis. *Front. Biosci.* 1998; **15**: D208–236.
323. Trosko JE and Chang CC. Modulation of cell-cell communication in the cause and chemoprevention/chemotherapy of cancer. *Biofactors* 2000; **12**: 259–263.
324. Trosko JE and Chang CC. Role of stem cells and gap junction intercellular communication in human carcinogenesis. *Radiat. Res.* 2001; **115**: 175–180.
325. Trosko JE and Ruch RJ. Gap junctions as targets for cancer chemoprevention and chemotherapy. *Curr. Drug Targets* 2002; **3**: 465–482.
326. Trosko JE. The role of stem cells and gap junctional intercellular communication in carcinogenesis. *J. Biochem. Mol. Biol.* 2003; **36**: 43–48.
327. Tsukada K and Ichimura M. Polynucleotide ligase from rat liver partial hepatectomy. *Biochem. Biophys. Res. Commun.* 1971; **421**: 1156–1161.
328. Tucker KL, Mahnken B, Wilson PW, Jacques P and Selhub J. Folic acid fortification of the food supply: potential benefits and risks for the elderly population. *J. Am. Med. Assoc.* 1996; **276**: 1879–1885.
329. Tuma DJ, Barak AJ and Sorrell MF. Interaction of methotrexate with lipotropic factors in rat liver. *Biochem. Pharmacol.* 1975; **24**: 1327–1331.
330. Uchida K, Morita T, Sato T, Ogura T, Yamashita R, Noguchi S, Suzuki H, Nyunoya H, Miwa M and Sugimura T. Nucleotide sequence of a full-length cDNA for human fibroblast poly(ADP-ribose) polymerase. *Biochem. Biophys. Res. Commun.* 1987; **148**: 617–622.
331. Varmus HE, Heasly S and Bishop JM. Use of DNA-DNA annealing to detect new virus specific DNA sequences in chicken embryo fibroblasts after infection by avian sarcoma virus. *J. Virol.* 1974; **14**: 895–903.
332. Vogelstein B, Fearson CR, Kern SE, Hamilton SR, Preisinger AC, Nakamura Y and White R. Allelotype of colorectal carcinomas. *Science* 1989; **244**: 207–211.

333. Wallwork JC and Duerre JA. Effect of zinc deficiency on methionine metabolism, methylation reactions and protein synthesis in isolated perfused rat liver. *J. Nutr.* 1985; **115**: 252–262.
334. Walsh CT, Sandstead HH, Prasad AS, Newberne PM and Fraker PJ. Zinc: health effects and research priorities for the 1990s. *Environ. Health Perspect.* 1994; **102**: 5–46.
335. Wands JR. Prevention of hepatocellular carcinoma. *New Engl. J. Med.* 2004; **351**: 1567–1570.
336. Warburg O. On the origin of cancer cells. *Science* 1956; **123**: 309–314.
337. Weinberg RA. Finding the antioncogene. *Sci. Am.* 1988; **259**: 44–51.
338. Weinberg RA. Oncogenes, antioncogenes and molecular bases of multistep carcinogenesis. *Cancer Res.* 1989; **49**: 3713–3721.
339. Weinberg R. Tumor suppressor genes. *Neuron* 1993; **11**: 191–196.
340. Weinberg R. Tumor viruses. In: *The Biology of Cancer*. Garland Sciences, New York, 2006, Chapter 3, pp. 57–90.
341. Weinhouse S. On respiratory impairment in cancer cells. *Science* 1956; **124**: 267–268.
342. Weinstein IB. Nonmutagenic mechanism in carcinogenesis: role of protein kinase C in signal transduction and growth control. *Environ. Health Perspect.* 1991; **93**: 175–179.
343. Whitacre CM, Hashimoto H, Hashimoto S, Tsai M-L, Chatterjee S, Berger SJ and Berger NA. Involvement of NAD-poly(ADP-ribose) metabolism in p53 regulation and its consequence. *Cancer Res.* 1995; **55**: 3697–3701.
344. White E, Shannon JS and Patterson RE. Relationship between vitamin and calcium supplement use and colon cancer. *Cancer Epidemiol. Biomarkers Prev.* 1997; **6**: 769–774.
345. Wintzerith M, Klein N, Mandel L and Mandel P. Comparison of pyridine nucleotides in the liver and in an ascites hepatoma. *Nature* 1961; **91**: 467–469.
346. Wolff AW and Oehme FW. Carcinogenic chemicals in food as an environmental issue. *J. Am. Vet. Med. Assoc.* 1974; **164**: 623–629.
347. Wolk A, Lindblad P and Adami H-O. Nutrition and renal cell cancer. *Cancer Causes Control* 1996; **7**: 5–18.
348. Yamada M, Nagao M, Hidaka T and Sugimura T. Effect of poly (ADP-ribose) formation on DNA synthesis and DNA fragmentation in nuclei of rat liver and ascites hepatoma AH-130 cells. *Biochem. Biophys. Res. Commun.* 1973; **54**: 1567–1572.
349. Yamada J and Tomita Y. Enhancing effect of ascorbic acid on the mutagenicity of Trp-P-1 by indigocarmine in the Salmonella microsome. *Biosci. Biotechnol. Biochem.* 1993; **58**: 2197–2200.

350. Yamada Y, *et al.* Opposing effects of DNA hypomethylation on intestinal and liver carcinogenesis. *Proc. Natl. Acad. Sci. USA* 2005; **102**: 13580–13585.
351. Yamasaki H. Gap junctional intercellular communication and carcinogenesis. *Carcinogenesis* 1990; **11**: 1051–1058.
352. Yip R and Dallman PR. Iron. In: *Present Knowledge in Nutrition*. Ziegler EE and Filer LJ Jr (eds.) ILSI Press, Washington, DC, 1996, pp. 277–292.
353. Zhang JZ, Henning SM and Swendseid ME. Poly (ADP-ribose) polymerase activity and DNA strand breaks are affected in tissues of niacin-deficient rats. *J. Nutr.* 1993; **123**: 1349–1355.
354. Zhang D, Okada S, Yu YY, Zheng P, Yamaguchi R and Kasai H. Vitamin E inhibits apoptosis, DNA modification and cancer incidence induced by iron-mediated peroxidation in Wistar rat kidney. *Cancer Res.* 1997; **57**: 2410–2414.
355. Zhang L-X, Cooney RV and Bertram JS. Carotenoids up-regulate connexin-43 gene expression independent of their provitamin A or antioxidant properties. *Cancer Res.* 1992; **52**: 5707–5712.
356. Ziegler M. New functions of a long-known molecule. Emerging roles of NAD in cellular signaling. *Eur. J. Biochem.* 2000; **267**: 1550–1564.

Chapter 9

INDUCTION AND REPAIR OF DNA DAMAGE FORMED BY ENERGETIC ELECTRONS AND LIGHT IONS

Robert D. Stewart and Vladimir A. Semenenko

In this chapter, radiation quantities and units and the induction and repair of radiation-induced DNA damage are briefly reviewed. Results of the authors' theoretical studies of DNA damage induction and outcomes from the excision repair of clustered DNA lesions are reported for selected types of low- and high-LET (linear energy transfer) radiation. Models for the conversion of clusters into small-scale mutations and aberrations are presented and used to illustrate trends in mutagenesis as a function of dose and particle LET.

Keywords: Ionizing radiation, absorbed dose, linear energy transfer, DNA damage, double-strand break, excision repair, non-homologous end joining, homologous recombination, point mutation, chromosome aberration.

1. DOSIMETRIC QUANTITIES AND UNITS

1.1. Absorbed Dose

A convenient concept for the quantification of radiation effects in living and non-living systems is the *absorbed dose*, D. Absorbed dose is defined as the average amount of energy imparted to matter of mass m in the limit as the mass approaches zero (ICRU, 1971). In the International System of Units (SI), absorbed dose is specified in terms of the *gray* (Gy), which equals an imparted energy of one joule (J) per kilogram (kg). An alternative unit for absorbed dose is the *rad*. A rad equals 100 ergs per gram, so 1 rad = 0.01 Gy. Table 1 lists the range of doses typically needed to produce selected biological effects. For comparison, the exposure limits recommended by the

Table 1. Exposure limits and doses needed to produce selected biological effects.

Exposure source or biological effect	Absorbed dose (Gy)
Nearly instantaneous death of exposed individual	>500
Fractionated radiation therapy (local tumor control)	25–175
Gastrointestinal syndrome (death within 3–10 days)	>10
Eventual death of 50% of exposed individuals (LD_{50} dose)	~4
Cancer mortality in 0.1% of exposed individuals	~0.1
Detectable levels of chromosome aberrations	0.05 – 0.25
Occupational exposure limit	<0.05
Annual dose from environmental sources (in the US)	<0.005
Public (non-occupational) exposure limit	<0.001

National Council on Radiation Protection and Measurements (NCRP, 1987) are also summarized in Table 1.

1.2. Linear Energy Transfer (LET)

The biological effects of radiation, per unit absorbed dose, tend to increase as the local concentration of ionizations and excitations along the radiation path increases. Most of this ionization and excitation is due to the action of charged particles. Charged particles may constitute the primary radiation or they may arise through interactions of neutral particles with matter (e.g., photoelectric effect or Compton scattering of photons, neutron capture).

The concept of LET (ICRU, 1971), which is defined as the average energy lost by a charged particle (dE) per unit distance traveled through a medium (dl), is ultimately motivated by the idea that the biological effects of radiation are more closely related to the spatial pattern of ionization and excitation within biological targets than to the intrinsic properties of the radiation. That is, if two different types of radiation produce similar spatial patterns of ionization and excitation, the biological consequences of these radiations will also be similar. To better differentiate among small and larger scale energy deposition patterns, the concept of LET is sometimes subdivided into *restricted* and *unrestricted* LET. The restricted LET, L_Δ, is defined as the average energy loss due to collisions which transfer energy less than some specified value Δ (ICRU, 1971), i.e., $L_\Delta = (dE/dl)_\Delta$. Although the restricted LET is defined in terms of an energy cutoff (Δ) instead of a particle range, energy losses less than Δ are considered "local" because the distance a charged particle can penetrate through matter decreases rapidly as the kinetic energy of that particle decreases. For charged particles in a condensed medium such as water or tissue, cutoff energies on the order of 100 eV correspond to penetration distances of the same order of magnitude as a single atom (i.e., < 1 nm). The unrestricted LET, L_∞, corresponds to the limiting case when all energy transfers are considered local. In the literature, LET usually means unrestricted LET rather than restricted LET (i.e., LET = L_∞).

Figure 1 shows the unrestricted LET for electrons, protons and α particles in water (ICRU, 1984; ICRU, 1993). Energetic electrons are nearly always considered a "low-LET" radiation whereas protons and α particles are often considered a "high-LET" radiation. A more pragmatic definition is to consider all radiations with a LET less than about 1 keV/μm a low-LET radiation. All other radiations are high-LET radiations. With this definition,

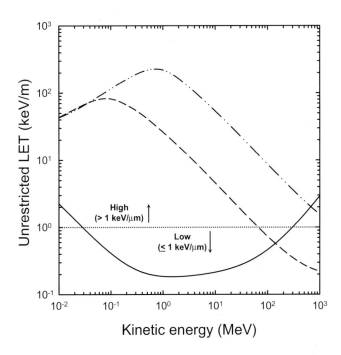

Fig. 1. Unrestricted LET for electrons (solid line), protons (dashed line) and α particles (dash-dot-dot line) in water.

electrons with kinetic energies greater than about 30 keV and less than 300 MeV are low-LET radiations as are protons with kinetic energies greater than about 70 MeV. Any α-particle with a kinetic energy less than 1 GeV is considered a high-LET radiation.

1.3. Microdosimetry

For a region of matter, the expected number of times energy is deposited in a small target, such as a cell or a cell nucleus, is usually well approximated by a Poisson distribution (ICRU, 1983), i.e.,

$$p(n|\bar{n}) = e^{-\bar{n}}\frac{\bar{n}^n}{n!}. \tag{1}$$

Here, $\bar{n}$ is the expected number of times a target is hit. A hit, also called a *radiation event*, occurs when a source particle, or one or more of the secondary particles it produces, imparts energy to matter in a target of interest. The average number of times a target is hit equals the absorbed dose divided by the frequency-mean specific energy per event, i.e., $\bar{n} \equiv D/\bar{z}_F$ (ICRU, 1983). The initial amount of ionization and excitation produced in a target per hit is proportional to the average energy deposited in the target per hit, i.e., proportional to $m\bar{z}_F$ where m is the mass of the target. The frequency-mean specific energy per event, $\bar{z}_F$, may thus be interpreted as a measure of the amount (severity) of initial chemical change (damage) produced in a target per hit.

For a spherical target composed of water, the frequency-mean specific energy per event can be estimated using (ICRU, 1983)

$$\bar{z}_F(\text{Gy}) = 0.204\frac{\bar{y}_F}{d^2}, \tag{2}$$

where $\bar{y}_F$ is the frequency-mean lineal energy (keV/μm) and d is the diameter of the target (μm). For energetic ions traversing micrometer-sized targets, the frequency-mean lineal energy is approximately equal to L_∞ (ICRU, 1983). For photons and electrons, Monte Carlo radiation transport methods (e.g., Bolch and Kim, 1994; Stewart *et al.*, 2002) or measured data (ICRU, 1983) are needed to accurately estimate $\bar{y}_F$ and $\bar{z}_F$. Table 2 lists the frequency-mean specific energy for spherical water targets 5 μm in diameter (representing an average cell nucleus) irradiated with selected types of ionizing radiation. Estimate of the frequency-mean specific energy for 10 keV electrons is from (Stewart et al., 2002). Estimates for 100 keV and 1 MeV electrons are from (Bolch and Kim, 1994). For protons and α particles, the frequency-mean specific energy was calculated using Eq. (2). Total stopping power values (ICRU, 1984; ICRU, 1993) were used to approximate LET and are also shown in Table 2. The total stopping power is defined as the sum of the collision (average rate of energy loss per unit path length due to Coulomb collisions that result in the ionization and excitation of atoms) and nuclear (average rate of energy loss per unit path length due to the transfer of energy to recoiling atoms in elastic collisions) stopping powers.

Data are listed in Table 2 for three electron energies representative of low-LET radiation. Protons with energies up to 250 MeV are often used in radiation therapy for the treatment of cancer (Amaldi and Kraft, 2005). The α-particle energies of 6.29 MeV and 5.49 MeV correspond to the types of radiation emitted by ^{220}Rn and ^{222}Rn, respectively, the major contributors to internal dose from natural radiation background (Thorne, 2003). Data for 1 MeV protons and 3.5 MeV α particles are shown because these types of radiation are sometimes used in radiobiological experiments.

From Eq. (1), it follows that the probability a target is hit one or more times is $1 - \exp(-D/\bar{z}_F)$. As the results in Figure 2 illustrate, the distribution of damage (hits) among a collection of irradiated cells is substantially different for low and high doses of radiation. For doses of radiation larger than $10\bar{z}_F$, the nucleus of more than 99.99% of irradiated cells is hit one

Table 2. LET and frequency-mean specific energy in a 5 μm spherical target for selected types of ionizing radiation.

Radiation Type	L_∞ (keV/μm)	$\bar{z}_F$ (Gy)
1 MeV electrons	0.19	0.0012
250 MeV protons	0.39	0.0032
100 keV electrons	0.41	0.0032
10 keV electrons	2.3	0.0065
1 MeV protons	26	0.21
6.29 MeV α particles	75	0.61
5.49 MeV α particles	83	0.68
3.5 MeV α particles	113	0.92

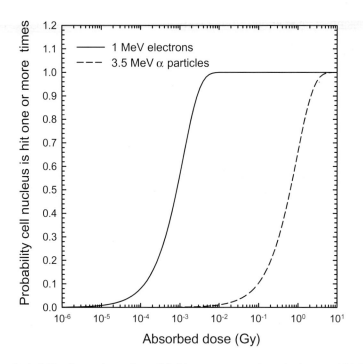

Fig. 2. Probability the nucleus of a cell is hit one or more times by low- and high-LET radiation.

or more times. For lower doses of radiation, some cells are hit one or more times while others do not sustain any damage. The nucleus of less than 1% of the cells is hit for doses less than $0.01\bar{z}_F$.

For the same dose of radiation, the probability that the nucleus of a particular cell sustains one or more hits tends to *increase* as the particle LET *decreases*. A 0.1 Gy dose of 1 MeV electrons will deposit energy in the nucleus of nearly all of the irradiated cells whereas 0.1 Gy of 3.5 MeV α particles will only deposit energy in the nucleus of about 9.6% of the cells. However, the severity of the initial amount of damage tends to *increase* with *increasing* LET, i.e., $\bar{z}_F$ increases with increasing LET. The biological consequences of ionizing radiation ultimately depend on both the average number of times a target is hit, which is proportional to D, and the severity of the initial damage per hit, which is proportional to $\bar{z}_F$.

2. INDUCTION OF DNA DAMAGE

2.1. Classification of DNA Damage

Exposure to ionizing radiation produces many different types of DNA damage, including DNA lesions such as damaged bases and strand breaks (Ward, 1988). Although similar types of DNA lesions are produced by endogenous processes and various physical and chemical agents (Friedberg et al., 1995), the relative yield and spatial arrangement of damaged nucleotides can be quite different. Groups of several damaged nucleotides within one or two helical turns of the DNA, often referred to as multiply damaged sites (Ward, 1988) or clustered damages (Goodhead, 1994), are known to be a hallmark of ionizing radiation. Theoretical considerations suggest that, in addition to isolated lesions, low-LET radiation can create clusters with as many as ten lesions (Semenenko and Stewart, 2004). High-LET radiation is capable of producing damage of even greater complexity, i.e., up to 25 lesions per cluster (Semenenko and Stewart, 2004).

All configurations of DNA damage can be classified into one of three mutually exclusive categories: double-strand breaks (DSBs), single-strand breaks (SSBs) and base damage (Bd). To provide additional information about the nature of the initial damage formed by radiation, SSBs and DSBs are sometimes subdivided into various types of simple and complex damage by the number of lesions forming the cluster (Semenenko and Stewart, 2004, 2006). Idealized schematics showing examples of various types of damage are presented in Figure 3. Other damage classification schemes, such as the one proposed by Charlton and Humm (1988) and extended by Nikjoo et al. (1999), can be useful when the contribution of complex forms of damage to the overall yields of DSBs and SSBs is of interest.

2.2. Mechanisms

Radiation damages the DNA through direct and indirect mechanisms (Ward, 1988). In the direct effect, radiation directly transfers energy to the DNA molecule. The indirect effect arises when the deposition of energy near the DNA forms reactive species that diffuse and interact with the DNA. The primary chemical species believed responsible for the indirect effect is

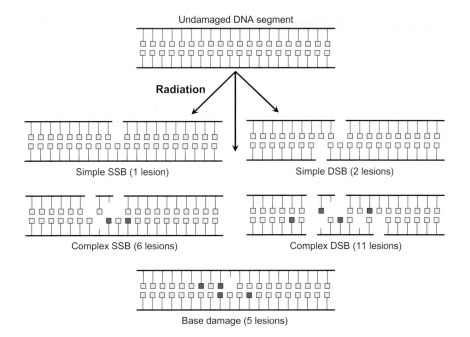

Fig. 3. Examples of the types of DNA damage formed by ionizing radiation. Light squares denote undamaged bases. Damage to individual nucleotides may comprise missing or damaged (dark squares) bases and strand breaks (accompanied by base loss).

hydroxyl radical (·OH). First, radiation ionizes a water molecule:

$$H_2O \rightarrow H_2O^+ + e^-. \tag{3}$$

The H_2O^+ is a positively charged ion that interacts with a nearby water molecule to form an ·OH radical through the reaction:

$$H_2O^+ + H_2O \rightarrow H_3O^+ + \cdot OH. \tag{4}$$

Hydroxyl radicals have nine electrons, and one of them is unpaired and highly reactive. The interactions of ·OH radicals with the DNA can produce base damage and strand breaks, i.e., the same types of lesions as the direct mechanism. The average diffusion distance of a hydroxyl radical in a cellular milieu is about 6 nm (Roots and Okada, 1975) or about three times the diameter of the DNA double helix, which implies that the chance an ·OH radical will damage the DNA decreases rapidly with distance beyond about 6 nm.

For electrons, protons and α particles, damage clusters usually involve fewer than 20 base pairs (bp) of damaged and undamaged DNA (Semenenko and Stewart, 2004). A linear segment of DNA composed of 20 bp is approximately 6.8 nm long and 2 nm in diameter, which implies that clusters are formed when radiation imparts energy in targets with dimensions smaller than about 10 nm. For clusters forming DSBs, at least two to five ionizations must occur in sites of diameters 1–4 nm (Brenner and Ward, 1992). For atomic and molecular-sized targets such as these, the discrete nature of the interactions of radiation with matter becomes important. The energy deposits formed by ionizing radiation are sometimes categorized as spurs (6 – 100 eV), blobs (100 – 500 eV) and short tracks (500 – 5000 eV). Spurs occupy roughly spherical regions of matter with a diameter less than 5 nm, and blobs are about 5–10 nm in diameter (Pimblott et al., 1996). Short tracks occupy spheroid-shaped regions of matter with dimensions on the order of 25 to 75 nm (Pimblott et al., 1996). For 1 MeV electrons having a stopping power of 0.19 keV/μm, the most probable energy loss in a single event in water is 22.5 eV, the average energy loss is 38 eV (Pimblott et al., 1990), and the average distance between energy loss events is about 200 nm. Because the average distance between energy deposits is larger than the dimensions of the target for cluster formation (i.e., 10 nm), low-LET radiation that passes near a short DNA segment usually only creates one spur, blob or short track sufficiently close to cause damage. Because the probability that multiple energy deposits are formed close to a small DNA segment increases with increasing LET, the number of lesions per cluster will also tend to increase with increasing LET.

The occurrence of a 10 eV energy deposit in a 10 nm site composed of water corresponds to a frequency-mean specific energy of about 3000 Gy. For larger energy deposits (blobs or short tracks) or multiple energy deposits (high-LET radiation), the mean specific energy per event will be even larger. This implies that, for a 10 nm site, the frequency-mean specific energy per event will be larger than about 3000 Gy for low- and high-LET radiation. For doses smaller than $0.1\bar{z}_F = 300$ Gy, a single radiation event ($p(1|0.1) = 0.09$; refer to Eq. (1)) will occur at least 20 times as often as two or more events ($p(2|0.1) = 0.0045$). This observation suggests that, for doses in the biologically-relevant range, clusters of DNA lesions are mainly formed by energy deposits associated with one event (track) rather than two or more events.

A one-event mechanism implies that the initial yield of DNA damage (i.e., the yield of damage before repair starts) is proportional to absorbed dose whereas a two-event mechanism implies that the initial damage yield is proportional to the square of the absorbed dose, i.e., $p(2|\bar{n}) = e^{-\bar{n}} \bar{n}^2/2 = \bar{n}^2/2 +$ higher order terms. Experiments show that the induction of many types of DNA damage is proportional to absorbed dose up to hundreds or even thousands of Gy (Frankenberg-Schwager, 1990). For human cells with a DNA content of about 6 Gbp per cell, the initial number of DSBs per cell increases linearly with increasing dose up to at least 100 Gy (Rothkamm and Löbrich, 2003). In yeast cells, the initial DSB yield per cell increases linearly with dose up to at least 2400 Gy (Frankenberg-Schwager et al., 1979). The production of base damage is also linear with absorbed dose up to at least a few hundreds of grays (Ward, 1995). These observations all imply that the single-hit mechanism is responsible for the induction of DNA damage by radiation.

Prise et al. (1998) compiled large amounts of data on DSB induction by ionizing radiation in various eukaryotic cells. DSB yields measured using up-to-date techniques, such as pulsed-field gel electrophoresis (PFGE), and expressed per unit genome length are similar among yeast and mammalian cells (all estimates fall in the range from 4.2 to 6.9 DSBs Gy^{-1} Gbp^{-1}), despite order-of-magnitude differences in genome sizes. This observation suggests that the induction of DSBs and other forms of clustered damage is, as a first approximation, proportional to the genome size. To produce equal numbers of DNA clusters per cell, yeast cells need to be exposed to 250 times as high a dose as do human cells. Because damage anywhere within the DNA has the potential to kill or cause mutations, cells with a small genome are much more resistant to ionizing radiation than are cells with large genomes.

2.3. Initial Yield and Characteristics

Although a variety of experimental techniques are available to quantify the overall yields of SSBs, DSBs and some types of base damage, measurements do not provide information about the exact location of lesions within the DNA (Ward, 1998). Computational approaches (Friedland et al., 1999; Holley and Chatterjee, 1996; Nikjoo et al., 2001) are needed to obtain maps of damage to individual nucleotides. Detailed simulations of the initial physical and chemical processes resulting in DNA damage ("track-structure simulations") have been shown to give approximate numerical agreement with measured damage yields (e.g., see Campa *et al.*, 2005; Friedland *et al.*, 2003, 2005; Rydberg *et al.*, 2002).

One disadvantage of track-structure calculations is that Monte Carlo simulations of the cascade of physical and chemical events resulting in DNA damage span time scales from about 1 ps to 1 μs and are thus very computationally intensive. The fast Monte Carlo damage simulation (MCDS) algorithm (Semenenko and Stewart, 2004, 2006) provides a computationally efficient alternative to detailed track-structure simulations.[a] Overall SSB and DSB yields and yields of selected types of simple and complex SSBs and DSBs compare favorably to the ones predicted by detailed track-structure calculations (Semenenko and Stewart, 2006). The MCDS algorithm can generate damage configurations for electrons with kinetic energies $\geq$80 eV, protons with energies $\geq$105 keV and α particles with energies $\geq$2 MeV (Semenenko and Stewart, 2006).

Table 3 provides a summary of DNA damage yields predicted by the MCDS algorithm for selected types of low- and high-LET radiation. 3.5 MeV α particles produce about 2.9 times more DSBs Gy^{-1} $cell^{-1}$ than 1 MeV electrons. However, 3.5 MeV α particles produce 1.7 and 3.7 times fewer SSBs and Bd Gy^{-1} $cell^{-1}$, respectively, than 1 MeV electrons. The initial yield of SSBs and base damages decreases with increasing LET because of competition between the formation of DSB and non-DSB

Table 3. DNA damage yields (Gy^{-1} $cell^{-1}$) for selected types of ionizing radiation in an average human cell containing 6 Gbp of DNA.

Radiation type	DSBs	SSBs	Bd	Total
1 MeV electrons	49.8	1136	2567	3753
250 MeV protons	50.1	1135	2562	3747
100 keV electrons	50.1	1135	2559	3744
10 keV electrons	53.3	1123	2483	3659
1 MeV protons	95.6	944	1558	2598
6.29 MeV α particles	127	776	981	1884
5.49 MeV α particles	131	750	908	1789
3.5 MeV α particles	144	664	703	1511

[a]The MCDS computer program is available at http://rh.healthsciences.purdue.edu/mcds/

categories of damage. That is, a cluster can either be a DSB or a non-DSB cluster. The total number of lesions formed anywhere within the DNA, per unit absorbed dose, is approximately the same for low- and high-LET radiation (Semenenko and Stewart, 2006). However, as particle LET increases, the degree of lesion clustering increases, which implies that the same number of lesions are organized into fewer clusters. The total number of clusters formed by 3.5 MeV α particles is about 2.5 times lower than the total cluster yield from 1 MeV electrons. As the degree of lesion clustering increases, the chance that at least one strand break is formed on the opposing strand of the DNA increases (i.e., a DSB is created). Because a cluster must either be a DSB or a non-DSB cluster (SSB or site of base damage), the yield of non-DSB clusters tends to decrease with increasing particle LET, as shown in Table 3.

Figure 4 shows relative yields of SSBs and DSBs produced by electrons, protons and α particles. Also shown in Figure 4 are the relative yields of

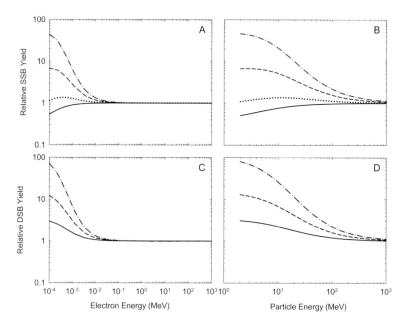

Fig. 4. Yields of SSBs (A, B) and DSBs (C, D) for electrons (A, C) and α particles (B, D) relative to the corresponding values for 1 MeV electrons. Solid line — relative yield of total breaks, dotted line — relative yield of breaks composed of more than 1 lesion, dashed line — more than 3 lesions, dash-dot line — more than 5 lesions.

complex SSBs and DSBs. 1 MeV electrons ($L_\infty = 0.19\,\text{keV}/\mu\text{m}$) are used as the reference radiation. For 1 MeV electrons, approximately 50% of the DSBs are simple DSBs (2 lesions per cluster), 30% of the DSBs are composed of 3 lesions, 13% are composed of 4 lesions, 5% are composed of 5 lesions and less than 2% are very complex DSBs (≥ 6 lesions). About 60% of the SSBs are composed of 1 lesion, 27% are composed of 2 lesions, 9% are composed of 3 lesions and about 4% are composed of 4 or more lesions. In the limit of very high kinetic energies (low LET), the relative SSB and DSB yields are nearly the same as the damage yields for 1 MeV electrons. As the kinetic energy of the electron decreases, particle LET increases (see Figure 1) and the relative yield of complex SSBs and DSBs increases. The same trends in relative damage yields occur for α particles. For the energy range most relevant to radiation protection (i.e., $\leq 10\,\text{MeV}$), α particles produce over 20-fold more very complex DSBs (> 5 lesions) than 1 MeV electrons.

3. REPAIR OF BASE DAMAGE AND SINGLE-STRAND BREAKS

3.1. Mechanisms

Many different types of oxidative DNA damage, ranging from modified bases to AP (apurinic/apyrimidinic) sites to strand breaks, are repaired by base excision repair (BER) (Wallace, 1998). Two modes of the BER process have been observed in both prokaryotes (Dianov and Lindahl, 1994) and eukaryotes (Frosina *et al.*, 1996; Klungland and Lindahl, 1997; Matsumoto *et al.*, 1994). The excision and replacement of a single nucleotide, termed short-patch BER, occurs in 75–90% of cases (Nilsen and Krokan, 2001). The other mode, long-patch BER, results in the removal of fragments up to 13 nucleotides long (Memisoglu and Samson, 2000), although replacement of only 2 nucleotides is most frequent (Nilsen and Krokan, 2001).

The BER process is accomplished through a series of steps that include (1) excision of a damaged base, (2) cleavage of the DNA backbone with the removal of the phosphodeoxyribose group, (3) gap-filling synthesis by a DNA polymerase, and (4) sealing of the gap by a DNA ligase. BER is a highly conserved repair pathway among all eukaryotes, i.e., from yeast to humans (Memisoglu and Samson, 2000). Significant homologies have also been demonstrated among bacterial and human proteins that participate in BER (Wallace, 1998).

3.2. Excision Repair Outcomes and Kinetics

Several lines of evidence suggest that clustered damage sites can be converted into DSBs as a result of unsuccessful excision repair of radiation-induced DNA damage. An increase in DSB yields has been observed in Chinese hamster cells during post-irradiation incubation (Ahnström and Bryant, 1982; Dugle *et al.*, 1976). Normal bacterial cells are more radiosensitive than mutants lacking DNA glycosylases (Blaisdell and Wallace, 2001). This result is attributed to aborted BER of closely spaced lesions within clustered damage sites that leads to the formation of potentially lethal DSBs. Formation of secondary DSBs has been observed in DSB repair-deficient Chinese hamster ovary xrs5 cells (Gulston *et al.*, 2004) and human lymphoblastoid TK6 cells (Yang *et al.*, 2004) after exposure to ionizing radiation. We refer to these DSBs as enzymatic DSBs to distinguish them from the prompt DSBs that are formed directly by ionizing radiation.

The misrepair of clustered damage also has the potential to cause local changes in the DNA sequence, such as base substitutions, small deletions and insertions. Such events are classified as point mutations in mutagenesis studies (Grosovsky *et al.*, 1988). Weinfeld *et al.* (2001) hypothesized that, when a cluster is repaired, a polymerase may need to insert a nucleotide opposite a lesion present in the other strand, which may have mutagenic consequences. Ward *et al.* (1994) suggested that aberrant repair of clustered damage may result in enzymatic DSBs which, if they occur within an exon of the gene, may be converted into point mutations.

The BER processing of radiation-induced DNA damage other than DSBs can be simulated using the Monte Carlo excision repair (MCER) model (Semenenko *et al.*, 2005; Semenenko and Stewart, 2005).[b] Three possible outcomes of the excision repair process are considered within the MCER model: (1) correct restitution of damage, (2) production of single-base substitutions through a synthesis opposite either damaged or undamaged template, and (3) formation of enzymatic DSBs. Table 4 lists

---

[b]The MCER computer program is available at http://rh.healthsciences.purdue.edu/mcer/.

Table 4. Short-patch BER outcomes for selected types of ionizing radiation.

Radiation type	Correct repair (%)		Enzymatic DSBs (%)
	SSBs	Base damage	
1 MeV electrons	96.2	99.4	2.59
250 MeV protons	96.1	99.4	2.60
100 keV electrons	96.1	99.4	2.61
10 keV electrons	95.8	99.3	2.79
1 MeV protons	91.2	98.3	5.47
6.29 MeV α particles	86.1	97.1	8.06
5.49 MeV α particles	85.2	96.9	8.50
3.5 MeV α particles	82.2	96.1	9.93

probabilities of excision repair outcomes calculated with the MCER model for selected types of radiation. The reported values correspond to processing of all damage though the predominant short-patch BER pathway. In the MCER model, enzymatic DSBs are only formed through the aborted repair of clusters containing at least one strand break (i.e., base damage does not contribute to the formation of enzymatic DSBs).

In the MCER model, excision repair is hypothesized to occur through repetitive cycles of lesion removal until either all nucleotides within the repair patch are replaced or an enzymatic DSB is formed. Because long-patch BER involves the removal of patches several nucleotides in length, more than one damaged nucleotide may sometimes be removed during a repair cycle. The time required to repair a cluster is expected to increase as the number of repair cycles increases. The MCER model thus provides data to estimate pathway-specific relative repair rates for different classes of SSBs and base damage.

As a first approximation, the expected rate of change in the number of the ith type of cluster (SSB or base damage), denoted $U_i(t)$, can be adequately modeled by the differential equation

$$\frac{dU_i(t)}{dt} = \sigma_i \dot{D}(t) - \lambda_i U_i(t). \tag{5}$$

Here, σ_i is the initial yield of the ith type of cluster (Gy^{-1} cell^{-1}) and λ_i is the probability per unit time the ith type of cluster is repaired. The rate constant, λ_i, is related to the half-time for cluster repair, τ_i, by $\lambda_i = \ln(2)/\tau_i$. Let κ_i be the average number of cycles needed to repair the ith type of cluster, and let

τ_0 denote the half-time for the repair of a one-cycle cluster. The half-time for repair of the ith type of cluster may now be expressed as $\tau_i = \kappa_i \tau_0$. Because the MCDS program provides estimates of σ_i and combined MCDS/MCER simulations provide estimates of κ_i, the only truly adjustable parameter in the repair model is half-time for the repair of a one-cycle cluster (τ_0). For SSBs and most types of base damage, repair half-times are usually less than 30 minutes.

For constant dose rates, $dU_i(t)/dt \to 0$ and the expected number of unrepaired clusters per cell approaches the asymptotic value $\sigma_i \dot{D}/\lambda_i$. For the dose rates found in the terrestrial environment $\sigma_i \dot{D}/\lambda_i$ is $\sim 10^{-5}$ cell^{-1} ($\sigma_i < 3800$ Gy^{-1} cell^{-1}, $\tau_0 \sim 5$ minutes, $\dot{D} \sim 1$ mGy per year). Except for exposure to very low doses or dose rates, the background level of damage may be neglected in most radiobiological experiments, i.e., $U_i(0) = 0$. Alternatively, for an acute dose of radiation, $\dot{D}(t) = 0$ for $t > 0$, $U_i(0) = \sigma_i D$, and the integration of Eq. (5) gives

$$U_i(t) = D\sigma_i e^{-\lambda_i t}. \tag{6}$$

The total number of unrepaired clusters at time t is the sum of Eq. (6) over all types of clusters, i.e.,

$$U(t) = D \sum_i \sigma_i e^{-\lambda_i t}. \tag{7}$$

Figure 5 shows the multi-exponential repair kinetics predicted by Eq. (7) for the spectrum of damage configurations produced by 1 MeV electrons and 3.5 MeV α particles. Also shown in Figure 5 are the mono-exponential repair kinetics that arise when $\tau_i \to \tau_{\text{avg}}$. Here, $\tau_{\text{avg}} = \kappa_{\text{avg}} \tau_0$ and

$$\kappa_{\text{avg}} \equiv \frac{\sum \kappa_i \sigma_i}{\sum \sigma_i}. \tag{8}$$

As illustrated in Figure 5, the model predicts that the overall lifetime of the clusters processed by BER tends to increase (the repair rate tends to decrease) as the particle LET increases. The damage formed by 3.5 MeV α particles is repaired with a 2.3 times (3.30/1.42) greater half-time than the damage produced by 1 MeV electrons (assuming mono-exponential kinetics). The estimate of $\tau_{\text{avg}} = 7$ min ($\tau_0 = 5$ min) for low-LET radiation (1 MeV electrons) is consistent with estimates directly derived from measured data (Frankenberg-Schwager, 1990).

Induction and Repair of DNA Damage

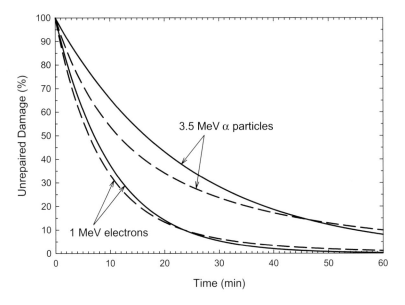

Fig. 5. Base excision repair of SSBs and base damage produced by 1 MeV electrons and 3.5 MeV α particles. Solid lines: mono-exponential kinetics with $\tau_{avg} = \kappa_{avg}\tau_0$, where $\tau_0 = 5$ min, κ_{avg} is 1.42 for 1 MeV electrons and 3.30 for 3.5 MeV α particles. Dashed lines: the multi-exponential kinetics that arise when each type of cluster is considered separately, i.e., Eq. (7) with $\tau_i = \kappa_i\tau_0$.

3.3. Point Mutations Arising from Base Damage and Single-Strand Breaks

The outcome from the attempted repair of a cluster (i.e., a "repair event") is either correct repair, repair with mutation or formation of an enzymatic DSB. After an acute dose of radiation, the expected number of repair events at time t must equal the initial number of clusters minus the number of unrepaired clusters at time t, i.e., $D\sigma_i - U_i(t)$. In the limit when $t \to \infty$, $U_i(t) \to 0$ and the expected number of repair events equals $D\sigma_i$. Because correct repair, misrepair and aborted repair are mutually exclusive events, the expected number of point mutations arising from the ith type of cluster, denoted m_i, is

$$m_i = D\sigma_i(1 - a_i - q_i). \tag{9}$$

Here, a_i is the probability a cluster is correctly repaired and q_i is the probability the cluster is converted into an enzymatic DSB and the excision repair

Table 5. Point mutations (Gy^{-1} $cell^{-1}$) that arise from SSBs and base damage. Initial damage yield (σ_i) was computed using the MCDS model (Table 3), and the probability of correct repair (a_i) and aborted repair (q_i) were computed using the MCER model (Table 4).

Radiation type	L_∞ (keV/μm)	Base damage	SSBs	Total
1 MeV electrons	0.19	16.3	14.2	30.5
250 MeV protons	0.39	16.4	14.3	30.7
100 keV electrons	0.41	16.4	14.3	30.7
10 keV electrons	2.3	17.3	15.3	32.6
1 MeV protons	26	26.1	31.5	57.6
6.29 MeV α particles	75	28.4	45.1	73.5
5.49 MeV α particles	83	28.4	47.0	75.4
3.5 MeV α particles	113	27.4	52.3	79.7

process aborts. Table 5 lists the expected number of point mutations, per unit dose, arising from the misrepair of base damage and SSBs. Electrons with kinetic energies from 10 keV to 1 MeV produce nearly same number of point mutations ($\sim$30 mutations Gy^{-1} $cell^{-1}$). 3.5 MeV α particles produce about 2.6 times as many point mutations as 1 MeV electrons. For electrons, base damage and SSBs contribute about equally to the point mutation yield even though the initial yield of base damage is about twice as high as the initial SSB yield (see Table 3). Base damage and SSBs contribute about equally to the point mutation yield because the excision repair of a SSB is more error prone than the excision repair of base damage (see Table 4). However for 3.5 MeV α particles, 34% of the point mutations are due to base damage and 66% are due to SSBs. The contribution of SSBs to the overall mutation yield increases with increasing LET because the ratio of SSBs to base damage increases with increasing LET.

4. REPAIR OF DOUBLE-STRAND BREAKS

4.1. Mechanisms

The DSB is considered the most critical form of DNA damage induced by ionizing radiation, although most of the over 40 DSBs Gy^{-1} $cell^{-1}$ formed in a typical mammalian cell by low-LET radiation (electrons and X-rays) are correctly rejoined. Correct rejoining does not necessarily mean correct repair. *Correct repair* means that the original DNA base sequence is restored, whereas *correct rejoining* means that break ends associated with the same DSB are rejoined to each other (refer to Figure 3). Unrejoined and misrejoined DSBs produce large-scale rearrangements of the genetic material (Hlatky *et al.*, 2002; Sachs *et al.*, 1997), including reciprocal and non-reciprocal exchanges and acentric fragments. Exchanges are formed when break ends associated with pairs of DSBs are misrejoined to each other, i.e., as described by the breakage and reunion theory of aberration formation (Hlatky *et al.*, 2002; Savage, 1998). Exchanges and other chromosome and chromatid aberrations have been implicated in phenotypic alterations, cell death, genomic instability and neoplasia (Bedford, 1991; Brown and Attardi, 2005; Lengauer *et al.*, 1998; van Gent *et al.*, 2001).

The two main pathways eukaryotic cells use to rejoin DSBs are homologous recombination (HR) and non-homologous end joining (NHEJ) (Jackson, 2002; Jeggo, 2002). The NHEJ and HR pathways are highly conserved among all eukaryotes. The NHEJ is an important repair mechanism in all phases of the cell cycle whereas HR is only important in late S and G_2 phases of the cell cycle (Rothkamm *et al.*, 2003). The dominant repair mechanism in simple eukaryotes, such as the yeasts *S. cerevisiae* and *S. pombe* (Jackson, 2002), is HR.

In the HR pathway, a damaged chromosome is aligned with an undamaged chromosome that has sequence homology. The genetic information stored in the undamaged chromosome is then used as a template for the repair of the damaged chromosome. Because the HR pathway has an undamaged template to guide the repair process, DSBs may be repaired without any loss of genetic information (i.e., correct repair). In contrast, the NHEJ pathway does not require a template with sequence homology. Instead, a series of proteins bind to the break ends (some of these proteins have exo- and endo-nuclease activity that is responsible for the removal of damaged nucleotides) and directly rejoin (ligate) the break ends. In CHO-K1 cells, the rejoining of DSBs formed by the I-*Sce*I endonuclease produces deletions of 1 to 20 bp (Liang *et al.*, 1998). Rare deletions up to 299 bp as well as the aberrant insertion of 45 to 205 bp also occur (Liang *et al.*, 1998). Ionizing radiation can create DSBs composed of more than 10 or even 20 lesions (Semenenko and Stewart, 2004), and the NHEJ pathway is unlikely to correctly repair complex DSBs such as those formed by ionizing radiation (Jeggo, 2002). Experiments also suggest that the complexity of a DSB impacts on the rate of DSB rejoining. The results of an *in vitro* DSB repair assay using plasmids and HeLa cell extracts suggests that simple DSBs are rejoined as much as 15 times faster than complex DSBs (Pastwa et al., 2003). Fast and slow rejoining DSBs have also been inferred from the analysis of cell survival data (Guerrero et al., 2002; Stewart, 2001). Some DSBs may also remain unrejoined for many days or may not ever be rejoined (Rothkamm and Löbrich, 2003). The latter situation may arise as a mechanism for the targeted elimination of damaged cells.

4.2. Repair Kinetics, Chromosome Aberrations and Small-Scale Mutations

In the repair-misrepair (RMR) model (Tobias, 1985) and lethal and potentially lethal (LPL) model (Curtis, 1986), DSBs are rejoined through first- and second-order processes. The first-order process represents the correct rejoining of break ends associated with one DSB, and the second-order process results in the formation of exchange-type aberrations through the pairwise interaction of break ends associated with two different DSBs. As a first approximation, the conversion of DSBs into exchanges can be described by a system of coupled, first-order non-linear differential equations, i.e.,

$$\frac{dL(t)}{dt} = \Sigma \dot{D}(t) - (\lambda + \eta \varepsilon) L(t) - \eta L^2(t), \tag{10}$$

$$\frac{dY(t)}{dt} = \eta \varepsilon L(t) + \eta L^2(t). \tag{11}$$

Here, $dL(t)/dt$ and dY/dt are the per-cell average rates of change in the expected number of DSBs and exchanges, respectively; $\dot{D}(t)$ is the instantaneous absorbed dose rate at time t (Gy h^{-1}); Σ is the initial number of DSBs formed by ionizing radiation (Gy^{-1} cell^{-1}); $\varepsilon = \bar{z}_F \Sigma$ is the expected number of DSBs formed per track (radiation event). The rate of first-order damage repair is characterized by the rate constant λ (h^{-1}) or, alternatively, the effective half-time for DSB rejoining $\tau \equiv \ln(2)/\lambda$. The second-order DSB interaction rate is governed by rate constant η (h^{-1}). The terms $\Sigma \dot{D}(t)$, $\lambda L(t)$, $\eta \varepsilon L(t)$ and $\eta L^2(t)$ describe production of DSBs, first-order

Induction and Repair of DNA Damage

removal of DSBs, intra-track DSB interaction and inter-track DSB interaction, respectively.

For low doses and fully developed endpoints (exchanges are scored when all DSBs are rejoined), the above system of equations can be approximated by (see Appendix)

$$Y(D) = \alpha D + \beta G D^2, \tag{12}$$

where

$$\alpha = \rho \bar{z}_F \Sigma^2, \quad \beta = \frac{1}{2}\rho \Sigma^2, \tag{13}$$

$$G = \frac{2}{D^2} \int_{-\infty}^{\infty} dt\, \dot{D}(t) \int_{-\infty}^{t} dt'\, e^{-\lambda(t-t')} \dot{D}(t') \tag{14}$$

and $\rho \equiv \eta/\lambda = \tau\eta/\ln(2)$. The biophysical interpretation of Eq. (14), the generalized Lea-Catcheside dose-protraction factor (Sachs *et al.*, 1997), is that a DSB is created at time t' and, if not rejoined, may interact in pairwise fashion with a second DSB produced at time t (Sachs and Brenner, 1998). For long irradiation times (low dose rates), the protraction factor approaches zero, and $Y(D) \to \alpha D$. In the limit of very short irradiation times (high dose rate), $G \to 1$ and $Y(D) = \alpha D + \beta D^2$. For other irradiation schemes, G is between 0 and 1. For a single dose of radiation delivered at constant dose rate during time interval T, G can be computed using

$$G(\lambda, T) = \frac{2}{(\lambda T)^2} \left[e^{-\lambda T} + \lambda T - 1 \right]. \tag{15}$$

The protraction factor for two acute doses, D_1 and D_2, separated by a time interval T is

$$G(D_1, D_2, \lambda, T) = \frac{D_1^2 + D_2^2 + 2D_1 D_2 e^{-\lambda T}}{(D_1 + D_2)^2}. \tag{16}$$

For low-LET radiation, the half-time for DSB rejoining is about 1 h, and the pairwise interaction rate, η, is about $10^{-4}\,\text{h}^{-1}$ (Stewart, 2001; Guerrero *et al.*, 2002), which implies that $\rho \sim 10^{-4}$. Table 6 lists the trends in α and β predicted by Eq. (13) with $\rho = 10^{-4}$. The initial DSB yield (Σ parameter) was computed using the MCDS algorithm, and data from Table 2 were used for $\bar{z}_F$. For constant ρ, the model predicts that α and β increase

Table 6. Predicted linear-quadratic (LQ) model parameters for the formation of exchanges through pairwise interaction of DSBs formed directly by ionizing radiation ($\rho = 10^{-4}$).

Radiation type	$\bar{z}_F$ (Gy)	Σ (Gy^{-1})	α (Gy^{-1})	β (Gy^{-2})
1 MeV electrons	0.0012	49.8	0.000298	0.124
250 MeV protons	0.0032	50.1	0.000803	0.126
100 keV electrons	0.0032	50.1	0.000803	0.126
10 keV electrons	0.0065	53.3	0.00185	0.142
1 MeV protons	0.21	95.6	0.192	0.46
6.29 MeV α particles	0.61	127	0.98	0.81
5.49 MeV α particles	0.68	131	1.17	0.86
3.5 MeV α particles	0.92	144	1.91	1.04

with increasing LET. The trend in α predicted by Eq. (13) is consistent with estimates of α derived from cell survival data (Furusawa et al., 2000) for particles with LET below about 200 keV/μm. For LET values above 200 keV/μm, estimates of α derived from cell survival data decrease with increasing LET, which may be due to slower DSB rejoining kinetics (larger τ and/or smaller η) associated with the formation of complex DSBs. For CHO cells, half-times for DSB rejoining are two to five times longer for 80 keV/μm carbon ions than for 200 kilovolt peak X-rays (Hirayama et al., 2005).

After an acute dose of radiation, the expected number of DSBs initially formed in the nuclear DNA of a cell is ΣD. Because DSBs are either rejoined correctly or incorrectly, the expected number of correctly rejoined DSBs is

$$D\Sigma - 2Y(D) = D\Sigma - 2(\alpha D + \beta G D^2). \qquad (17)$$

The factor of 2 arises in Eq. (17) because two DSBs are required to form each exchange. For low dose rates ($G \cong 0$) and low doses ($\alpha D \gg \beta G D^2$), Eq. (17) reduces to $D\Sigma - 2\alpha D$. Because the number of correctly rejoined DSBs cannot be negative, Eq. (17) is only valid for absorbed doses less than $(\Sigma - 2\alpha)/2G\beta$. As illustrated in Figure 6, the expected number of correctly rejoined DSBs decreases with increasing dose because of the increasing importance of second-order repair. For 1 MeV electrons with $\rho = 10^{-4}$ ($\alpha = 2.98 \times 10^{-4}$ Gy^{-1}, $\beta = 0.124$ Gy^{-2}), Eq. (17) predicts that 50% of the initial DSBs will be rejoined after an acute dose of 100.4 Gy (filled square in Figure 6). For comparison, Rothkamm and Löbrich (1999) found

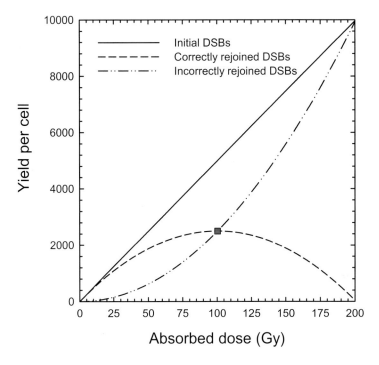

Fig. 6. Expected number of correctly and incorrectly rejoined DSBs after an acute dose of low-LET radiation. Filled square indicates the absorbed dose (100.4 Gy) at which 50% of the initial DSBs are correctly rejoined ($\rho = 10^{-4}$, $\Sigma = 49.8\,\mathrm{Gy}^{-1}$, $\bar{z}_F = 0.0012\,\mathrm{Gy}$).

that approximately 50% of the DSBs formed within the human hypoxanthine guanine phosphoribosyl transferase (HPRT) gene are rejoined correctly after 80 Gy, which corresponds to $\rho = 1.25 \times 10^{-4}$.

Enzymatic DSBs formed through the excision repair of clustered lesions (refer to Sections 3.1 and 3.2) may also contribute to the formation of exchanges and small-scale mutations. Table 7 lists estimates of α and β for the special case when enzymatic DSBs are just as likely to interact in pairwise fashion as DSBs formed directly by radiation. If enzymatic DSBs are involved in the formation of exchanges, estimates of α and β will more than double compared to the values reported in Table 6. Even if most or all enzymatic DSBs are correctly rejoined, the misrepair of enzymatic DSBs may produce substantial numbers of small-scale deletions and insertions. The combined MCDS/MCER model predicts that enzymatic DSBs will

Table 7. Predicted linear-quadratic (LQ) model parameters for the formation of exchanges through pairwise interaction of direct and enzymatic DSBs ($\rho = 10^{-4}$).

Radiation type	$\bar{z}_F$ (Gy)	Σ (Gy^{-1})	α (Gy^{-1})	β (Gy^{-2})
1 MeV electrons	0.0012	79.2	0.000753	0.314
250 MeV protons	0.0032	79.7	0.00203	0.318
100 keV electrons	0.0032	79.7	0.00203	0.318
10 keV electrons	0.0065	84.6	0.00465	0.358
1 MeV protons	0.21	147.3	0.456	1.08
6.29 MeV α particles	0.61	189.6	2.19	1.80
5.49 MeV α particles	0.68	189.6	2.58	1.90
3.5 MeV α particles	0.92	194.7	4.05	2.20

form about as many small-scale mutations as the DSBs formed directly by ionizing radiation.

ACKNOWLEDGMENTS

This work is supported by the Office of Science (BER), US Department of Energy, Grant Nos. DE-FG02-03ER63541 and DE-FG02-03ER63665.

References

1. Ahnström G and Bryant PE. DNA double-strand breaks generated by the repair of X-ray damage in Chinese hamster cells. *Int. J. Radiat. Biol. Relat. Stud. Phys. Chem. Med.* 1982; **41**: 671–676.
2. Amaldi U and Kraft G. Radiotherapy with beams of carbon ions. *Rep. Prog. Phys.* 2005; **68**: 1861–1882.
3. Bedford JS. Sublethal damage, potentially lethal damage, and chromosomal aberrations in mammalian cells exposed to ionizing radiations. *Int. J. Radiat. Oncol. Biol. Phys.* 1991; **21**: 1457–1469.
4. Blaisdell JO and Wallace SS. Abortive base-excision repair of radiation-induced clustered DNA lesions in *Escherichia coli*. *Proc. Natl. Acad. Sci. USA* 2001; **98**: 7426–7430.
5. Bolch WE and Kim E-H. Calculation of electron single event distributions for use in internal beta microdosimetry. *Radiat. Prot. Dosimetry* 1994; **52**: 77–80.

6. Brenner DJ and Ward JF. Constraints on energy deposition and target size of multiply damaged sites associated with DNA double-strand breaks. *Int. J. Radiat. Biol.* 1992; **61**: 737–748.
7. Brown JM and Attardi LD. The role of apoptosis in cancer development and treatment response. *Nat. Rev. Cancer* 2005; **5**: 231–237.
8. Campa A, Ballarini F, Belli M, Cherubini R, Dini V, Esposito G, Friedland W, Gerardi S, Molinelli S, Ottolenghi A, Paretzke H, Simone G and Tabocchini MA. DNA DSB induced in human cells by charged particles and gamma rays: experimental results and theoretical approaches. *Int. J. Radiat. Biol.* 2005; **81**: 841–854.
9. Charlton DE and Humm JL. A method of calculating initial DNA strand breakage following the decay of incorporated ^{125}I. *Int. J. Radiat. Biol.* 1988; **53**: 353–365.
10. Curtis SB. Lethal and potentially lethal lesions induced by radiation–a unified repair model. *Radiat. Res.* 1986; **106**: 252–270. Erratum in: *Radiat. Res.* 1989; **119**: 584.
11. Dianov G and Lindahl T. Reconstitution of the DNA base excision-repair pathway. *Curr. Biol.* 1994; **4**: 1069–1076.
12. Dugle DL, Gillespie CJ and Chapman JD. DNA strand breaks, repair, and survival in X-irradiated mammalian cells. *Proc. Natl. Acad. Sci. USA* 1976; **73**: 809–812.
13. Frankenberg-Schwager M, Frankenberg D, Blöcher D and Adamczyk C. The influence of oxygen on the survival and yield of DNA double-strand breaks in irradiated yeast cells. *Int. J. Radiat. Biol. Relat. Stud. Phys. Chem. Med.* 1979; **36**: 261–270.
14. Frankenberg-Schwager M. Induction, repair and biological relevance of radiation-induced DNA lesions in eukaryotic cells. *Radiat. Environ. Biophys.* 1990; **29**: 273–292.
15. Friedberg EC, Walker GC and Siede W. *DNA Repair and Mutagenesis*. ASM Press, Washington, DC, 1995, pp. 1–58.
16. Friedland W, Jacob P, Paretzke HG, Merzagora M and Ottolenghi A. Simulation of DNA fragment distributions after irradiation with photons. *Radiat. Environ. Biophys.* 1999; **38**: 39–47.
17. Friedland W, Jacob P, Bernhardt P, Paretzke HG and Dingfelder M. Simulation of DNA damage after proton irradiation. *Radiat. Res.* 2003; **159**: 401–410.
18. Friedland W, Dingfelder M, Jacob P and Paretzke HG. Calculated DNA double-strand break and fragmentation yields after irradiation with He ions. *Radiat. Phys. Chem.* 2005; **72**: 279–286.

19. Frosina G, Fortini P, Rossi O, Carrozzino F, Raspaglio G, Cox LS, Lane DP, Abbondandolo A and Dogliotti E. Two pathways for base excision repair in mammalian cells. *J. Biol. Chem.* 1996; **271**: 9573–9578.
20. Furusawa Y, Fukutsu K, Aoki M, Itsukaichi H, Eguchi-Kasai K, Ohara H, Yatagai F, Kanai T and Ando K. Inactivation of aerobic and hypoxic cells from three different cell lines by accelerated ^{3}He-, ^{12}C- and ^{20}Ne-ion beams. *Radiat. Res.* 2000; **154**: 485–496.
21. Goodhead DT. Initial events in the cellular effects of ionizing radiations: clustered damage in DNA. *Int. J. Radiat. Biol.* 1994; **65**: 7–17.
22. Grosovsky AJ, de Boer JG, de Jong PJ, Drobetsky EA and Glickman BW. Base substitutions, frameshifts, and small deletions constitute ionizing radiation-induced point mutations in mammalian cells. *Proc. Natl. Acad. Sci. USA* 1988; **85**: 185–188.
23. Guerrero M, Stewart RD, Wang JZ and Li XA. Equivalence of the linear-quadratic and two-lesion kinetic models. *Phys. Med. Biol.* 2002; **47**: 3197–3209.
24. Gulston M, de Lara C, Jenner T, Davis E and O'Neill P. Processing of clustered DNA damage generates additional double-strand breaks in mammalian cells post-irradiation. *Nucleic Acids Res.* 2004; **32**: 1602–1609.
25. Hirayama R, Furusawa Y, Fukawa T and Ando K. Repair kinetics of DNA-DSB induced by X-rays or carbon ions under oxic and hypoxic conditions. *J. Radiat. Res.* 2005; **46**: 325–332.
26. Hlatky L, Sachs RK, Vazquez M and Cornforth MN. Radiation-induced chromosome aberrations: insights gained from biophysical modeling. *Bioessays* 2002; **24**: 714–723.
27. Holley WR and Chatterjee A. Clusters of DNA damage induced by ionizing radiation: formation of short DNA fragments. I. Theoretical modeling. *Radiat. Res.* 1996; **145**: 188–199.
28. ICRU. *Radiation Quantities and Units*. Report 19, International Commission on Radiation Units and Measurements, Bethesda, MD, 1971.
29. ICRU. *Microdosimetry*. Report 36, International Commission on Radiation Units and Measurements, Bethesda, MD, 1983.
30. ICRU. *Stopping Powers for Electrons and Positions*. Report 37, International Commission on Radiation Units and Measurements, Bethesda, MD, 1984.
31. ICRU. *Stopping Power and Ranges for Protons and Alpha Particles*. Report 49, International Commission on Radiation Units and Measurements, Bethesda, MD, 1993.

32. Jackson SP. Sensing and repairing DNA double-strand breaks. *Carcinogenesis* 2002; **23**: 687–696.
33. Jeggo PA. The fidelity of repair of radiation damage. *Radiat. Prot. Dosimetry* 2002; **99**: 117–122.
34. Klungland A and Lindahl T. Second pathway for completion of human DNA base excision-repair: reconstitution with purified proteins and requirement for DNase IV (FEN1). *EMBO J.* 1997; **16**: 3341–3348.
35. Lengauer C, Kinzler KW and Vogelstein B. Genetic instabilities in human cancers. *Nature* 1998; **396**: 643–649.
36. Liang F, Han M, Romanienko PJ and Jasin M. Homology-directed repair is a major double-strand break repair pathway in mammalian cells. *Proc. Natl. Acad. Sci. USA* 1998; **95**: 5172–5177.
37. Matsumoto Y, Kim K and Bogenhagen DF. Proliferating cell nuclear antigen-dependent abasic site repair in *Xenopus laevis* oocytes: an alternative pathway of base excision DNA repair. *Mol. Cell. Biol.* 1994; **14**: 6187–6197.
38. Memisoglu A and Samson L. Base excision repair in yeast and mammals. *Mutat. Res.* 2000; **451**: 39–51.
39. NCRP. *Radiation Exposure of the US Population from Consumer Products and Miscellaneous Sources.* Report 95, National Council on Radiation Protection and Measurements, Washington, DC, 1987.
40. Nikjoo H, O'Neill P, Terrissol M and Goodhead DT. Quantitative modelling of DNA damage using Monte Carlo track structure method. *Radiat. Environ. Biophys.* 1999; **38**: 31–38.
41. Nikjoo H, O'Neill P, Wilson WE and Goodhead DT. Computational approach for determining the spectrum of DNA damage induced by ionizing radiation. *Radiat. Res.* 2001; **156**: 577–583.
42. Nilsen H and Krokan HE. Base excision repair in a network of defence and tolerance. *Carcinogenesis* 2001; **22**: 987–998.
43. Pastwa E, Neumann RD, Mezhevaya K and Winters TA. Repair of radiation-induced DNA double-strand breaks is dependent upon radiation quality and the structural complexity of double-strand breaks. *Radiat. Res.* 2003; **159**: 251–261.
44. Pimblott SM, LaVerne JA, Mozumder A and Green NJB. Structure of electrons tracks in water. 1. Distribution of energy deposition events, *J. Phys. Chem.* 1990; **94**: 488–495.
45. Pimblott SM, LaVerne JA and Mozumder A. Monte Carlo simulation of range and energy deposition by electrons in gaseous and liquid water. *J. Phys. Chem.* 1996; **100**: 8595–8606.

46. Prise KM, Ahnström G, Belli M, Carlsson J, Frankenberg D, Kiefer J, Löbrich M, Michael BD, Nygren J, Simone G and Stenerlöw B. A review of dsb induction data for varying quality radiations. *Int J. Radiat. Biol.* 1998; **74**: 173–184.
47. Rothkamm K and Löbrich M. Misrejoining of DNA double-strand breaks in primary and transformed human and rodent cells: a comparison between the HPRT region and other genomic locations. *Mutat. Res.* 1999; **433**: 193–205.
48. Rothkamm K, Kruger I, Thompson LH and Löbrich M. Pathways of DNA double-strand break repair during the mammalian cell cycle. *Mol. Cell Biol.* 2003; **23**: 5706–5715.
49. Rothkamm K and Löbrich M. Evidence for a lack of DNA double-strand break repair in human cells exposed to very low X-ray doses. *Proc. Natl. Acad. Sci. USA* 2003; **100**: 5057–5062.
50. Roots R and Okada S. Estimation of life times and diffusion distances of radicals involved in X-ray-induced DNA strand breaks of killing of mammalian cells. *Radiat. Res.* 1975; **64**: 306–320.
51. Rydberg B, Heilbronn L, Holley WR, Löbrich M, Zeitlin C, Chatterjee A and Cooper PK. Spatial distribution and yield of DNA double-strand breaks induced by 3–7 MeV helium ions in human fibroblasts. *Radiat. Res.* 2002; **158**: 32–42.
52. Sachs RK, Hahnfeld P and Brenner DJ. The link between low-LET dose-response relations and the underlying kinetics of damage production/repair/misrepair. *Int. J. Radiat. Biol.* 1997; **72**: 351–374.
53. Sachs RK and Brenner DJ. The mechanistic basis of the linear-quadratic formalism. *Med. Phys.* 1998; **25**: 2071–2073.
54. Savage JR. A brief survey of aberration origin theories. *Mutat. Res.* 1998; **404**: 139–147.
55. Semenenko VA and Stewart RD. A fast Monte Carlo algorithm to simulate the spectrum of DNA damages formed by ionizing radiation. *Radiat. Res.* 2004; **161**: 451–457.
56. Semenenko VA, Stewart RD and Ackerman EJ. Monte Carlo simulation of base and nucleotide excision repair of clustered DNA damage sites. I. Model properties and predicted trends. *Radiat. Res.* 2005; **164**: 180–193.
57. Semenenko VA and Stewart RD. Monte Carlo simulation of base and nucleotide excision repair of clustered DNA damage sites. II. Comparisons of model predictions to measured data. *Radiat. Res.* 2005; **164**: 194–201.
58. Semenenko VA and Stewart RD. Fast Monte Carlo simulation of DNA damage formed by electrons and light ions. *Phys. Med. Biol.* 2006; **51**: 1693–1706.

59. Stewart RD. Two-lesion kinetic model of double-strand break rejoining and cell killing. *Radiat. Res.* 2001; **156**: 365–378.
60. Stewart RD, Wilson WE, McDonald JC and Strom DJ. Microdosimetric properties of ionizing electrons in water: a test of the PENELOPE code system. *Phys. Med. Biol.* 2002; **47**: 79–88.
61. Thorne MC. Background radiation: natural and man-made. *J. Radiol. Prot.* 2003; **23**: 29–42.
62. Tobias CA. The repair-misrepair model in radiobiology: comparison to other models. *Radiat. Res.* 1985; **8**(Suppl.): S77–S95.
63. van Gent DC, Hoeijmakers JH and Kanaar R. Chromosomal stability and the DNA double-stranded break connection. *Nat. Rev. Genet.* 2001; **2**: 196–206.
64. Wallace SS. Enzymatic processing of radiation-induced free radical damage in DNA. *Radiat. Res.* 1998; **150**(Suppl.): S60–S79.
65. Ward JF. DNA damage produced by ionizing radiation in mammalian cells: identities, mechanisms of formation, and reparability. *Prog. Nucleic Acid Res. Mol. Biol.* 1988; **35**: 95–125.
66. Ward JF, Jones GDD and Milligan JR. Biological consequences of non-homogenous energy deposition by ionising radiation. *Radiat. Prot. Dosim.* 1994; **52**: 271–276.
67. Ward JF. Radiation mutagenesis: the initial DNA lesions responsible. *Radiat. Res.* 1995; **142**: 362–368; Erratum in: *Radiat. Res.* 1995; **143**: 355.
68. Ward JF. Nature of lesions formed by ionizing radiation. In: *DNA Damage and Repair: DNA Repair in Higher Eukaryotes*, Vol. 2. Nickoloff JA and Hoekstra MF (eds.) Humana Press Inc., Totowa, NJ, 1998, pp. 65–84.
69. Weinfeld M, Rasouli-Nia A, Chaudhry MA and Britten RA. Response of base excision repair enzymes to complex DNA lesions. *Radiat. Res.* 2001; **156**: 584–589.
70. Yang N, Galick H and Wallace SS. Attempted base excision repair of ionizing radiation damage in human lymphoblastoid cells produces lethal and mutagenic double strand breaks. *DNA Repair* 2004; **3**: 1323–1334.

APPENDIX

Here, we derive an approximate solution for differential Eqs. (10) and (11) in the limit of low doses. The system of equations reduces to the linear-quadratic formalism (Eq. (12)) providing that (1) the majority of

DSBs are removed by correct restitution rather than by pairwise DSB interaction ($\eta \ll \lambda$) and (2) the number of exchanges is determined after all DSBs have been rejoined, i.e., as $t \to \infty$. The integration of Eq. (10) with respect to time gives

$$\int_{-\infty}^{\infty} \frac{dL(t)}{dt} dt = \int_{-\infty}^{\infty} dL(t) = L(\infty) - L(-\infty) = 0 - 0 = 0. \quad (A.1)$$

Here, we have assumed that all DSBs are eventually repaired or misrepaired so that $L(\infty) \to 0$ for any exposure condition provided $\lim_{t \to \infty} \dot{D} = 0$. Equation (A.1) is an expression of the conservation of damage principle:

$$\Sigma \int_{-\infty}^{\infty} \dot{D}(t) dt - (\lambda + \eta \varepsilon) \int_{-\infty}^{\infty} L(t) dt - \eta \int_{-\infty}^{\infty} [L(t)]^2 dt = 0. \quad (A.2)$$

Note that $\int_{-\infty}^{\infty} \dot{D}(t) dt = D$.

Suppose the binary misrepair interaction rate, η, is small compared to the rate of linear misrepair, λ, and that $L(t) = L_0(t) + \eta f(t)$, where $L_0(t)$ is the solution to Eq. (A.2) for $\eta = 0$. This expression can be thought of as the first two terms in a Taylor series expansion of $L(t)$ in powers of η. Now, Eq. (A.2) becomes

$$\Sigma D - (\lambda + \eta \varepsilon) \int_{-\infty}^{\infty} [L_0(t) + \eta f(t)] dt$$

$$- \eta \int_{-\infty}^{\infty} [L_0(t) + \eta f(t)]^2 dt = 0. \quad (A.3)$$

To order zero in η, Eq. (A.3) reduces to

$$\Sigma D - \lambda \int_{-\infty}^{\infty} L_0(t) dt = 0. \quad (A.4)$$

To order zero in η, the solution of Eq. (A.3) is

$$\int_{-\infty}^{\infty} L_0(t) dt = \frac{\Sigma D}{\lambda}. \quad (A.5)$$

Induction and Repair of DNA Damage

Integrating Eq. (11) and keeping only zeroth and first-order terms in η yields

$$Y(\infty) = \int_{-\infty}^{\infty} \frac{dY(t)}{dt} dt = \eta\varepsilon \int_{-\infty}^{\infty} L(t)dt + \eta \int_{-\infty}^{\infty} [L(t)]^2 dt$$

$$= \eta\varepsilon \int_{-\infty}^{\infty} [L_0(t) + \eta f(t)] dt + \eta \int_{-\infty}^{\infty} [L_0(t) + \eta f(t)]^2 dt$$

$$= \eta\varepsilon \int_{-\infty}^{\infty} L_0(t)dt + \eta \int_{-\infty}^{\infty} [L_0(t)]^2 dt. \quad (A.6)$$

Equation (A.6) is written in terms of the integrals of the $\eta = 0$ solutions. It has been shown (Sachs et al., 1997) that

$$\int_{-\infty}^{\infty} [L_0(t)]^2 dt = \frac{\Sigma^2}{2\lambda} GD^2. \quad (A.7)$$

Substituting right-hand sides of Eqs. (A.5) and (A.7) into Eq. (A.6), one obtains

$$Y(\infty) = \eta\varepsilon \frac{\Sigma D}{\lambda} + \eta \frac{\Sigma^2}{2\lambda} GD^2. \quad (A.8)$$

Equation (A.8) has a form of Eq. (12) where α and β are defined as

$$\alpha = \eta\varepsilon \frac{\Sigma}{\lambda} = \varepsilon\rho\Sigma, \quad \beta = \eta \frac{\Sigma^2}{2\lambda} = \frac{1}{2}\rho\Sigma^2, \quad (A.9)$$

where $\rho = \eta/\lambda$. Because $\varepsilon = \bar{z}_F \Sigma$, Eq. (A.9) may also be re-written as

$$\alpha = \rho\bar{z}_F \Sigma^2, \quad \beta = \frac{1}{2}\rho\Sigma^2. \quad (A.10)$$

Chapter 10
RADIATION-INDUCED BYSTANDER EFFECTS

Linda C. DeVeaux

Recent results indicate that cells not directly targeted with radiation exhibit responses directly attributed to their exposure to the irradiated cells or some product of those cells. Many responses, called "bystander effects," have been measured in unirradiated cells, ranging from transient gene expression changes to permanent, heritable changes to the genetic material. In this review, current literature regarding the transmission and nature of the molecule responsible for the bystander effect and the variables surrounding both production of and receipt of these signals is presented. Finally, consideration of the risks posed and the evolutionary implications of this phenomenon are discussed.

Keywords: Radiation, bystander cells, reactive oxygen species, genomic instability, gap-junction communication, medium transfer, signaling.

1. BYSTANDER EFFECTS

1.1. Introduction

Radiation effects exhibited by cells have long been thought to be the result of deposition of the energy within the nucleus, where the interaction results in damage to the DNA. However, recent research has required a reconsideration of the events occurring at both the molecular and cellular level. The use of light-ion microbeams has facilitated such research by delivering tightly focused beams for irradiation of individual cells in a culture, as well as even smaller targets such as individual nuclei or extranuclear compartments (Hei *et al.*, 1997; Shao *et al.*, 2004; Wu *et al.*, 1999). Using such microbeams, it has been demonstrated that the nucleus, long thought to be the major target of radiation damage, need not be hit in order for a radiation-induced response to be invoked in a given cell (Wu *et al.*, 1999). Just as surprisingly, individual cells need not receive dose in order to show evidence of radiation effects. These untargeted "bystander" cells display a range of effects that clearly depend on their exposure to cells that have been targeted with radiation, or a signaling factor produced by targeted cells. These "bystander effects" manifest as radiation-dependent changes in unirradiated cells as a result of intercellular communication from directly irradiated cells. Bystander effects have been demonstrated in many systems at very low doses of radiation, where damage from direct ionizations is limited. There appear to be two distinct mechanisms through which signals eliciting bystander effects are communicated to responding cells. One mechanism requires that the irradiated and unirradiated cells be in direct contact and linked through gap junctions. In a distinct process, signaling molecules are secreted into the medium by the irradiated cells, and subsequently taken up by unirradiated cells. Clastogenic, or chromosome-damaging, factors in the blood of irradiated individuals may represent a circulating form of bystander signaling that affects cells in tissues far from the site of origin.

Current radiation protection standards are based on a log-linear cellular response to damage. The estimated risk of increased cancer due to exposure to low doses of radiation is interpolated from much higher dose exposures. This Linear-No Threshold model assumes that there is no exposure level below which there is no increased risk. Bystander effects may have significant consequences for an understanding of low-dose and therapeutic radiation risk in humans, because this log-linear relationship may not be true at low doses. Effects at low doses are fundamentally different from those seen at high doses, and are not predicted by the standard radiobiological paradigm. Instead of responding as a collection of individual cells, tissues respond as a whole to perturbations in the system. Many variables affect the overall response, and thus the dose-response relationship, including radiation type, the tissue involved, and the endpoint measured. The purpose of this chapter is to provide an overview of current research and theories surrounding radiation-induced bystander effects, and to direct the interested reader to primary literature and review articles on particular topics.

1.2. Definition of Bystander Effects

The basic concept of bystander effects is that radiation may have a larger target than the nucleus, since cellular responses to the initial damage can modulate and expand the impact. Mothersill *et al.* (2004) define bystander effects as "responses detected in cells that were not hit directly by radiation but were influenced in some way by the hit received by another cell." This broad definition allows the inclusion of effects that are manifested in cells that were present at the time of irradiation, but were not within the radiation field, as well as in cells that were not even physically present during irradiation. In this case, radiation-dependent effects can be seen after exposure to cell-derived factors present in medium taken from irradiated cells. This definition may also allow inclusion of delayed effects, such as radiation-induced genomic instability, where radiation-dependent effects are not detectable until many generations after the damage has been sustained and do not appear in all clonal derivatives of the target cell. The cells that display the damage, although derived from the directly-irradiated cells, are separated temporally from the initiating event (for reviews, see Barcellos-Hoff and Brooks, 2001; Goldberg and Lehnert, 2002; Little, 2003; Morgan, 2003; Morgan *et al.*, 2002). These delayed effects of radiation and the immediate responses of bystander cells may be propagated by the same or overlapping signaling pathways (Morgan, 2003), and may be different manifestations of the same event.

The radiation-induced responses documented in bystander cells are varied and, in some cases, appear to be contradictory. Virtually any change that has been observed after direct deposition of energy within a cell has also been observed in bystander cells. There are many variables, however, that must be considered when describing the "bystander effect" itself. It is likely that there is not one single effect, but rather a related group of responses. While it is likely that these outcomes are part of a global response to cellular damage, that has not yet been clearly established; however, many of the effectors of inflammation response and intracellular signaling pathways are involved in the bystander signaling process (McBride *et al.*, 2004). Within a population of cells, those receiving dose must produce a signal that is received by unirradiated cells. Both aspects of this process are affected by the genetic status of both the directly targeted and the recipient cell (Mothersill and Seymour, 2001). Since radiation quality affects the type of damage in directly targeted cells, this may result in markedly different responses being elicited in these cells that ultimately affect the responses in the non-targeted cells. Depending on the response that is actually measured in the bystander cells, the overall effect may be considered to be detrimental or advantageous — what is deleterious at the cellular level may be beneficial to the tissue, or the whole organism. Several recent review articles address this issue (Barcellos-Hoff and Brooks, 2001; Brooks, 2004; Mothersill and Seymour, 2004, 2005).

1.3. History of Bystander Effects

There have been a number of reports dating back to the 1950s that blood from individuals exposed to radiation, either accidentally or therapeutically, can cause chromosomal aberrations in unirradiated blood or cells in culture. These observations suggest that some soluble, circulating signal is causing the genetic changes. Several recent review articles detail the history of these "clastogenic factors" (Brooks, 2004; Goldberg and Lehnert, 2002; Little, 2003; Morgan, 2003; Mothersill and Seymour, 2001). The relevance of these effects in terms of whole organism risks is not clear. Interestingly, the blood from patients with any of a number of genetic diseases that confer chromosomal instability (Bloom's syndrome, Fanconi's anemia, ataxia telangiectasia) also contain this clastogenic ability, in which the generation of reactive oxygen species is involved (Emerit, 1994; Mothersill and Seymour, 2001). Mammary tumors were more prevalent in unirradiated rats animals injected with blood from irradiated rats or sheep, demonstrating an organismal level of response to these factors (Souto, 1962). However, other experiments using partial tissue irradiation of lungs (reviewed in Brooks (2004)) suggest that the scope of these *in vivo* effects may be limited to the tissue irradiated. Further support for this comes from experiments using internally-deposited radioactive materials, which deliver at a very low dose rate. In these cases, tumor development was limited to the tissue of deposition, suggesting that any factor, if produced, was not present in sufficient concentration to elicit a response far from the tissue of origin (Brooks, 2004).

An increased interest in these cellular signaling responses to ionizing radiation came with the report in 1992 of far more cells showing radiation-induced changes than could be accounted for by the number of cells actually receiving nuclear irradiation (Nagasawa and Little, 1992). The number of publications since then reporting similar radiation-induced effects in unirradiated cells far exceeds the scope of this chapter. Most of these recent reports deal with *in vitro* systems, where individual cells or cultures of cells are irradiated, and effects are measured in cultured cells; demonstrations of bystander effects *in vivo* are fewer (reviewed in Morgan (2003)). Recent results from this laboratory demonstrated radiation-dependent mutations in yeast cells not present during irradiation, blurring the distinction between *in vivo/in vitro* systems (DeVeaux et al., 2006). This may represent a fundamental response to generalized stress that is the evolutionary precursor to the effects seen in higher cells.

Use of charged particle and ultrasoft X-ray microbeams (reviewed in Prise et al. (2003)) has allowed precise deposition of discrete numbers of particles in individual cells. This has allowed the demonstration that only one cell needs to receive dose to initiate an effect equivalent to 10–20 cells responding to the bystander signal. The deposition of a single α-particle elicits a response in neighboring cells, and the response is the same whether the target cell receives only one or multiple hits. In other words, the signal appears to be either "on" or "off". In microbeam studies, the spatial distribution of the cells showing a response is not focused around the targeted cell; instead responses are seen throughout. However, since many of the responses require days to be visualized, often the short-term dissemination of the signal itself is not apparent. Recently, Hu et al. (2006) examined the time-dependence of bystander signaling using γ-H2AX immunofluorescent staining to measure immediate production of double-strand breaks in both directly irradiated and bystander cells. These lesions were visible in the unirradiated cells as early as two minutes post-irradiation. Breaks were detected earliest in cells closest to the irradiated cells, although there was no difference in the levels of double-strand breaks in any of the bystander cells by thirty minutes post-irradiation, regardless of distance from the irradiation area, which contained the directly-irradiated cells.

Some common characteristics of the various bystander effects reported can be summarized. Responses are induced after low doses of radiation to the directly targeted cells, and the amount of signal does not increase with dose. In situations where few cells in a population have received dose, bystander effects can dramatically influence total population response; however, as more cells receive direct irradiation, the impact of these secondary effects on the total population response may diminish. This illustrates a key point in overall consideration of bystander effects: the endpoint that is measured in the bystander cells as the result of the irradiation of the targeted cells may or may not be the only change, or even the most significant change. Many laboratories measure a particular endpoint that has been optimized for the system being studied. These changes represent a "snapshot" of the overall changes that have occurred in the cells, but the processive changes that occur from time of exposure to a bystander signal are difficult to document. In addition, it is often difficult to detect events, particularly subtle changes, which one is not looking for. With this in mind, a picture of the progressive changes occurring in bystander cells is slowly emerging.

1.4. Bystander Endpoints

The changes that have been investigated in bystander cells are those that have been seen as effects of direct deposition of energy within a cell. Since direct irradiation produces a multitude of short-term as well as long-term changes, many different endpoints have been investigated in bystander cells. Given the presumed integrating nature of the transmitted signal, it is perhaps surprising that such a wide assortment of endpoints have been observed. This variety should suggest that the signal that is transmitted from irradiated cells to unirradiated cells has a wide target, or initiates a cascade of events. Nagasawa and Little reported increases in sister-chromatid exchange in cells not receiving radiation (Nagasawa and Little, 1992), which was supported by work in other laboratories (Deshpande et al., 1996). Other inherited responses observed, in addition to this genome-altering event, include an altered mutation frequency, micronucleus formation, genomic instability, transformation from the normal to malignant state, and apoptosis (Azzam et al., 2001, 2002; Banerjee et al., 2005; Deshpande et al., 1996; Hei et al., 1997; Huo et al., 2001; Kashino et al., 2004; Lorimore et al., 1998; Lyng et al., 2000; Maguire et al., 2005; Mothersill et al., 2001; Mothersill and Seymour, 1997; Nagasawa and Little, 1999; Sawant et al., 2001; Watson et al., 2000; Zhou et al., 2000, 2002). Several lines of evidence indicate that DNA damage occurs in the bystander cells, including phosphorylation of γ-H2AX, indicative of double-strand break occurrence, and induction of the RPA protein, involved in general DNA repair (Balajee *et al.*, 2004; Hu *et al.*, 2006). More transient responses have also been measured, and include changes in expression and phosphorylation state of the tumor suppressor protein p53 and other proteins involved in cell signaling pathways, cell-cycle arrest, altered response to subsequent challenging doses of radiation (either an adaptive response or increased sensitivity), and altered cell proliferation capabilities (Azzam *et al.*, 1998, 2000, 2001, 2002; Iyer and Lehnert, 2002; Iyer *et al.*, 2000; Matsumoto *et al.*, 2001). Immediate changes detected include initial events in the apoptotic cascade, such as increased cytoplasmic Ca^{2+} levels and mitochondrial membrane permeability as well as changes in mitochondrial mass and distribution, which in turn affect levels of apoptotic proteins (Lyng *et al.*, 2000, 2002; Maguire *et al.*, 2005).

The mutations induced in bystander cells have been analyzed, and compared to the mutations in the directly targeted cells that presumably distributed the signal responsible for the bystander changes (Huo *et al.*, 2001; Zhou *et al.*, 2000). Mutations in bystander cells were predominantly point mutations, rather than the small deletions characteristic of damage induced by direct irradiation by high LET alpha particles. These point mutations are reminiscent of the damage induced by indirect means, such as reactive oxygen species produced in cells directly targeted with radiation. This suggests that different mechanisms are responsible for the resulting mutations in directly targeted and bystander cells, and that after the initial event the damage creates a distinct mutagenic factor that does not predominate in the initially irradiated cell.

Genomic instability may be considered a temporal bystander effect when manifested in progeny of cells that have received dose. This inherited trait results in a "hypermutable" phenotype, which is often associated with transformed cell lines and is characteristic of cancer cells *in vivo* (Morgan *et al.*, 2002). Radiation-induced genomic instability has been demonstrated in descendants not only of directly-irradiated cells, but also in descendants of bystander cells (Lorimore *et al.*, 1998). This bystander genomic instability has been observed both *in vitro* and *in vivo* (Watson *et al.*, 2000). The mechanism of these delayed responses is not well understood, but has been shown to be unrelated to increased double-strand breaks in the genomic

DNA (Morgan et al., 2002). Interestingly, Nagar et al. reported a cytotoxic signal from cells displaying genomic instability that is distinct from bystander signaling (Nagar and Morgan, 2005; Nagar et al., 2003). This "death-inducing factor" was only produced by genomically-unstable clones and not by irradiated cells of the same type. The relationship of this transmissible cytotoxic factor, which also generates genomic instability in its survivors, to radiation-induced bystander signaling remains to be determined.

1.5. Transmission of Signal

Bystander effects have been documented *in vitro* in both transformed and untransformed mammalian cells from various tissues, and appear to be transmitted by two mechanisms. These two modes of transmission are not necessarily independent, but may have signaling components in common. Responses to low fluences of alpha particles have been shown to be mediated by intercellular communication through gap junctions, aqueous channels that form between adjacent cells in tissues of multicellular organisms that allow small molecules to pass (Little, 2003). A bystander effect in these cases has been shown to be inhibited by chemicals that eliminate this communication between neighboring cells (Azzam et al., 1998, 2001). Experiments using cells that do not produce gap junctions due to null mutations in the connexin proteins have demonstrated that a functional gap junction is necessary for the response to be seen (Azzam et al., 2001). Transmission of the signal appears to involve the p53 pathway (Azzam et al., 2001; Iyer and Lehnert, 2000). Considerable evidence supports the involvement of reactive oxygen species, which have been shown to be involved in p53-dependent signaling (Azzam et al., 2002; Johnson et al., 1996; Little et al., 2002). Bystander-mediated increased sister chromatid exchange was inhibited by superoxide dismutase, implicating reactive oxygen species in transmission of the soluble intercellular signal (Iyer et al., 2000). Evidence for the involvement of nitric oxide in transmission of this signal through the medium was demonstrated by inhibition of p53 accumulation in the presence of a nitric oxide scavenger (Matsumoto et al., 2001). Extracellular signaling through such cytokines as transforming growth factor β, and direct membrane interactions suggest that communication other than through gap-junctions may be involved or necessary (Barcellos-Hoff and Brooks, 2001; Nagasawa et al., 2002).

Bystander effects have also been observed in situations where gap-junction communication clearly was not involved. This non-gap junction-mediated response has been observed in sparsely distributed cells after irradiation with charged-particle microbeams, where the low cell densities precluded cell-cell contact (Sawant et al., 2001). Other methodologies utilized physical separation of the irradiated and unirradiated cells, although both were present at the time of irradiation. In some cases, the α-particles used for irradiation deposited their energy in the first layer of cells, and thus were unable to penetrate to the second layer of bystander cells (Suzuki et al., 2004; Zhou et al., 2002). In other experiments, unirradiated cells were shielded from direct irradiation. In either method, although the bystander cells were physically present during the irradiation, transmission of the radiation-induced damage signal was necessarily through the medium. However, studies involving co-culturing techniques indicate that bystander cells need not even be present at the time of irradiation in order to receive signal. Demonstration of the separation of the signal from cell contact was clearly made by induction of response in cells that were not even present at the time of irradiation, but were exposed to medium from irradiated cells (Iyer et al., 2000; Matsumoto et al., 2001; Mothersill and Seymour, 1998). Culture medium from cells that have received irradiation applied to non-irradiated cells elicits an effect that is clearly dependent on the initial irradiation to the target cells. Since the affected cells were not present during the irradiation and had no contact with the irradiated cells, this effect must be the result of an unidentified signaling factor released into the medium by the irradiated cells. Since these early observations, the release of a signal (or signals) and the effect on unirradiated cells has been definitively established (reviewed in Morgan (2003)).

1.6. Identification of Signal

The nature of the signal molecule(s) involved in transmitting the various bystander effects from irradiated cells to unirradiated cells has yet to be determined. It seems likely that a signal transmitted through gap junctions would have different characteristics from a signal that is secreted into the medium. Signals transmitted through gap junctions would not be exposed to the extracellular fluid and thus need not be hardy enough to withstand the conditions experienced there. Gap junction-dependent signals are limited by the size of the junction opening, which is determined by cell type and particular connexin proteins that form the channel. This direct transmission from cytoplasm to cytoplasm through gap junctions bypasses the requirement for secretion from the emitting cell, recognition (receptors) in the recipient cell membrane, and transduction of the signal across the membrane. Because the signal molecule never leaves the confines of the cell, identification of such molecules relies on indirect methods. Decrease in a response after inhibition of gap junction communication, for example, suggests a maximum size for the molecule responsible. Other indirect methods include measurement of intracellular levels of reactive oxygen species (ROS), and varying the genetic makeup of the transmitting and/or receiving cells (Azzam *et al.*, 1998; Little *et al.*, 2002). There is also evidence of membrane signaling involvement in transmission of signals (Nagasawa *et al.*, 2002). These types of experiments, however, do not distinguish between mechanisms necessary to transmit a signal from one that is sufficient. There may be multiple signals that must all be produced in order to elicit a response, and disruption of any one destroys the transmission of the signal.

The signaling components responsible for gap-junction dependent and independent mechanisms may have overlapping components and not be mutually exclusive. Iyer *et al.* (2000) investigated the involvement of soluble factors produced in response to α-irradiation, based on earlier work demonstrating an increase in production of both intracellular and extracellular reactive oxygen species (Narayanan *et al.*, 1997). The source of the extracellular ROS may be from interaction of the α-particle with extracellular matrix components, or it may stem from increased intracellular levels. Another signaling molecule shown to be induced by irradiation is nitric oxide, as evidenced by increases in the enzyme responsible for its production (Leach *et al.*, 2002). The importance of the increased production of this molecule after irradiation lies in its ability to diffuse across cell membranes and thus interact with cells some distance from the point of origin. In addition, the longevity of this signal molecule's production is consistent with many reports, and the lack of a single receptor precludes any generalized response predictions (Shao *et al.*, 2004). NO is implicated in regulation of the activity of the cytokine TGF-β1, which may play a role in the tissue response to individual cell irradiation (Barcellos-Hoff and Brooks, 2001; Brooks, 2005; Vodovotz *et al.*, 1999).

Signals that are secreted from irradiated cells by gap-junction independent mechanisms provide a means for directly determining the nature of the molecule(s) involved. Medium transfer protocols (Maguire *et al.*, 2005; Mothersill and Seymour, 1997, 1998, 2001; Mothersill *et al.*, 2001, 2004), where medium from irradiated cells can be removed and applied to unirradiated cells, provide a means for isolating the factor(s) using the medium as a reagent. The major problem in this approach, as put forth at the Bystander Effects meeting held at McMaster University in October 2004, is the presence of a reproducible and robust assay for the bystander effect in question (Various authors, 2005). Early work by Mothersill and Seymour (1998) indicated that the effect carried in the irradiated cell culture medium (ICCM, or conditioned medium) was resistant to freezing, but was heat-labile, suggesting a protein component. In recent work by Springer *et al.* (2005), utilizing mass spectrometry to analyze the proteins shed by cells, no significant differences were seen between irradiated and unirradiated cells. These results highlight the necessity for a robust assay to allow fractionation of the medium and identification of those fractions that contain signaling activity. In this way, sufficient concentrations of the molecule(s) could be obtained.

The event that triggers the bystander signal production in directly irradiated cells may also be complex. It has been assumed that DNA damage as a result of ionizations is responsible for initiating the cascade. Ward (2002) reviewed the possible DNA lesions that could be implicated in the generation of the signal produced in the targeted cells. Given the extremely low doses under which bystander cells exhibit responses, it seemed unlikely that double-strand breaks (DSB) themselves are the primary event. A recent report using γ-HSAX immunofluorescent staining to visualize DSB events, however, showed that these lesions appeared not only in the directly irradiated cells, but also in the unirradiated bystander cells (Hu *et al.*, 2006). The recent report of cytoplasmic irradiation induction of bystander signaling brings into question whether or not DNA lesions are the primary event, or could be a result of the signal already in progress (Shao, Folkard *et al.*, 2004).

1.7. Status of Sending and Receiving Cells

There is evidence of cell type-specificity in the ability to both produce and respond to the bystander signal. The differences in the cell lines must determine the ability or lack thereof to participate in these processes. Medium from irradiated epithelial cells affected unirradiated fibroblasts, but the reverse was not true, indicating that tissue-specific gene expression status, as well as genetic background, are important (Mothersill and Seymour, 1997). By definition, cell lines derived from different individuals are not isogenic. The role of individual genes can only be precisely determined by comparing cell lines that differ at only one locus. The production of and response to a bystander signal are separate and separable abilities, and the genetic factors determining each must be examined as such (Mothersill *et al.*, 2000). Several studies have investigated the role of genes whose products are integral to DNA repair. Mothersill, Seymour and coworkers utilized a reporter cell (one that consistently displays a bystander response) to assay different cell lines for production of a signal (Mothersill *et al.*, 2001). Using this assay, they have shown that medium from irradiated cell lines containing various DNA repair defects elicited larger bystander responses (lower survival) than closely matched repair-proficient lines (Mothersill *et al.*, 2004). Medium exchange between deficient and proficient lines also demonstrated the importance of the genetic status of the recipient (bystander) cells, as demonstrated in the following examples. Cells defective in the non-homologous end-joining (NHEJ) repair pathway showed an increase in sensitivity to the bystander signal, displaying mutation rates much higher than wild type cells (Nagasawa *et al.*, 2003). The mutations displayed in the repair-deficient bystander cells consisted mainly of deletions, whereas mutations in wild-type bystander cells were primarily point mutations (Huo *et al.*, 2001; Little *et al.*, 2003). These results suggest that oxidative lesions generated in the bystander cells were not effectively repaired in the NHEJ mutants. This is supported by the finding that cells deficient in homologous recombinational repair had a bystander response comparable to wild type cells (Nagasawa *et al.*, 2005). Recipient cells that were defective in base excision repair, however, showed wild type levels of bystander effects in medium transfer experiments, indicating that signals generating double-strand breaks, and not base damage, were transferred through the medium (Kashino *et al.*, 2004).

What is almost certainly as important as the genetic status of the sending and receiving cells is the epigenetic status, including the environment in which the cell resides. Even within a tissue, where cells may be in similar differentiated states, the location, stage in cell cycle, oxygen status, etc., and thus gene expression will differ from cell to cell. This is akin to the relationship that children in a family have to one another — although each presumably arises from the same genetic pool, and is raised in the same macroenvironment (house, family), their relationship within the family hierarchy is different, and thus each one experiences the environment differently. Similarly, cells within a tissue have relationships to the other cells in that tissue, and each experiences a different microenvironment, which may influence either production of or receipt of a bystander signal. It is difficult to reproduce this relationship with cells in culture, so the linking of the *in vitro* work to *in vivo* experiments will be particularly helpful in determining the global effect of bystander signaling on organismal well-being.

1.8. Dependence on Radiation Type

Much of the literature concerning bystander signaling has focused on low fluences of α particles (high linear energy transfer, or LET) where few cells in a population actually received dose. The DNA damage incurred by interaction with a single high LET particle along its track is significantly different from low LET interactions, and consists of complex lesions in a localized region. Bystander effects have clearly been demonstrated after irradiation with more sparsely ionizing radiation. What is not clear is if the two types of radiation invoke the same cellular response. Even demonstrations of similar outcomes in the bystander cells would not define the initial lesion responsible in the irradiated cells, or the integration of signaling pathways that probably occurs. Recent work using microbeam irradiation has demonstrated that nuclear traversal is not necessary in order for a cell to produce a bystander signal (Shao *et al.*, 2004), which suggests a signaling pathway that is unlinked to direct DNA damage. UV-induced apoptosis in unirradiated cells exposed to medium of irradiated cells has also been reported (Banerjee *et al.*, 2005). Thus, the questions of radiation quality, type and dose delivery (time frame, for example) and the outcome in either the producing or responding cells are largely unexplored.

The effect of the energy deposition on the cell is also dependent on the oxygen state of the cell, which influences not only direct damage to the DNA, but clearly invokes signaling cascades within the cell (Brooks, 2005). The energy/REDOX status has also been shown to be important for a response to bystander signaling through the medium (Mothersill *et al.*, 2000).

1.9. Bystander Effects and Cancer Risk

Recognized or not, bystander effects have been part of the measured response of cells to low level exposure to radiation. The demonstration of an effect in cells that were outside a radiation field has implications for risks at these low doses, where a log-linear response to dose has been assumed. Bystander effects, if leading to cancer induction, show that there is no threshold below which the risk for changes to the cell is zero, as even single cells receiving one particle can induce a response in other cells. The question remains, then, if there is evidence that bystander effects do affect (either positively or negatively) induction of cancer either in the bystander cells from the same tissue, or at sites removed from the initial target.

Much of the uncertainty surrounding the global effects of bystander signaling are based on the scale of the response measured. The responses exhibited by individual cells may or may not be detrimental to the organism as a whole, and thus any particular effect may be viewed as either positive or negative, depending on the level or organization that is being considered. A signal that causes an individual cell to undergo apoptosis may appear to be detrimental at that level, but to the tissue or the organism, the death of that cell may be advantageous (Barcellos-Hoff and Brooks, 2001; Mothersill and Seymour, 2004, 2005). Other outcomes of bystander signaling may increase the ability of individual cells or cells organized in tissues to contend with such insults, and thus be advantageous to the survival of those cells. Such an adaptive response conferred through bystander mechanisms has been reported, although there are conflicting results surrounding the degree of protection observed relative to the endpoint measured (Iyer and Lehnert, 2002; Mitchell *et al.*, 2004).

1.10. Evolutionary Considerations of Bystander Effects

At some point during evolution, the bystander signaling mechanism must have become established in cells, either because it provides a useful function, or indirectly as a byproduct of a necessary cellular mechanism. The widespread existence of this response to radiation in cells of varying origin suggests that this phenomenon is a normal cellular response to stress, which is present in most (if not all) living systems. We recently identified that a stress-induced bystander signal is present in the fission yeast *Schizosaccharomyces pombe* (DeVeaux et al., 2006). Physical stresses — heat, UV and electron beam radiation — as well as the chemical stress of exposure to the DNA-damaging antibiotic bleomycin, induced mutations in cells not present at the time of the stress. These effects were strikingly similar to the bystander effects reported in mammalian cells. That multiple stresses elicit this intercellular response is also very reminiscent of the intracellular adaptive response seen in *S. cerevisiae*, where priming doses with heat or radiation resulted in increased cross-resistance to subsequent challenges by either heat or ionizing radiation (Mitchel and Morrison, 1982). Until fairly recently, microbes were thought to be isolated organisms, where the growth of each individual had no impact on others until nutrients were depleted and toxic wastes built up. However, the existence of complex biofilms has proven that each microbial cell is not an island, and that interaction and communication is more typical of actual environmental conditions than is growth in isolated laboratory cultures. Quorum-sensing, where individual bacteria emit small molecules that are sensed by other members of the community, allows adjustment of growth to compensate for population size (Fuqua et al., 2001). Biofilms are often composed of cells from multiple species. Even individual cells from the same species within biofilms display vastly different gene expression profiles, suggesting that a type of differentiation occurs within these communities, reminiscent of tissue structure in higher organisms. If unicellular members of a microbial community are exposed to injury in the form of stress, it would be advantageous for the damaged members to communicate a signal to allow the remaining cells to adjust to the change in conditions.

UV-induced bystander effects have also been reported in sponges, where the gases nitric oxide and ethylene were shown to be important in transmission (Muller et al., 2006). These demonstrations of stress-induced communication in non-vertebrate cells are fundamentally important for delineating the basic processes involved in bystander signaling mechanisms. By reducing the complexity of the model systems used for study, the role of genetic and environmental factors can more easily be dissected. Gene expression analyses, already underway in some systems, will allow identification of the cellular responses at the transcriptional level (Chaudhry, 2006).

2. SUMMARY

Because of the complexity inherent in living systems, responses to stresses, such as radiation, are not log-linear with dose applied. This is exemplified by bystander effects, where low doses of radiation induce a signal to be produced by stressed cells, which is communicated to other unaffected cells. The response elicited in bystander cells may be either positive or negative, depending on the endpoint measured. An endpoint that appears to be detrimental at the cellular level may be beneficial at the organismal level. Bystander effects appear to be part of the normal cellular response to stress, and are present in cells of diverse origin and complexity.

ACKNOWLEDGMENTS

The work of LCD described in this chapter was supported by United States Department of Defense contract DAD13-03-C-0054 and by U.S. Department of Energy grant DE-AC07-99ID13727. The author would like to thank Antone L. Brooks and Jeffrey Perkel for helpful comments, and Christine Sestero and Jonathon R. Smith for assistance in researching the literature.

References

1. Azzam EI, de Toledo SM, Gooding T and Little JB. Intercellular communication is involved in the bystander regulation of gene expression in human cells exposed to very low fluences of alpha particles. *Radiat. Res.* 1998; **150**: 497–504.
2. Azzam EI, de Toledo SM and Little JB. Direct evidence for the participation of gap junction-mediated intercellular communication in the transmission of damage signals from alpha-particle irradiated to nonirradiated cells. *Proc. Natl. Acad. Sci. USA* 2001; **98**: 473–478.
3. Azzam EI, De Toledo SM, Spitz DR and Little JB. Oxidative metabolism modulates signal transduction and micronucleus formation in bystander cells from alpha-particle-irradiated normal human fibroblast cultures. *Cancer Res.* 2002; **62**: 5436–5442.

4. Azzam EI, de Toledo SM, Waker AJ and Little JB. High and low fluences of alpha-particles induce a G1 checkpoint in human diploid fibroblasts. *Cancer Res.* 2000; **60**: 2623–2631.
5. Balajee AS, Ponnaiya B, Baskar R and Geard CR. Induction of replication protein a in bystander cells. *Radiat. Res.* 2004; **162**: 677–686.
6. Banerjee G, Gupta N, Kapoor A and Raman G. UV induced bystander signaling leading to apoptosis. *Cancer Lett.* 2005; **223**: 275–284.
7. Barcellos-Hoff MH and Brooks AL. Extracellular signaling through the microenvironment: a hypothesis relating carcinogenesis, bystander effects, and genomic instability. *Radiat. Res.* 2001; **156**: 618–627.
8. Brooks AL. Evidence for "bystander effects" *in vivo*. *Hum. Exp. Toxicol.* 2004; **23**: 67–70.
9. Brooks AL. Paradigm shifts in radiation biology: their impact on intervention for radiation-induced disease. *Radiat. Res.* 2005; **164**: 454–461.
10. Chaudhry MA. Bystander effect: biological endpoints and microarray analysis. *Mutat. Res.* 2006; **597**: 98–112.
11. Deshpande A, Goodwin EH, Bailey SM, Marrone BL and Lehnert BE. Alpha-particle-induced sister chromatid exchange in normal human lung fibroblasts: evidence for an extranuclear target. *Radiat. Res.* 1996; **145**: 260–267.
12. DeVeaux LC, Durtschi LS, Case JG and Wells DP. Bystander effects in unicellular organisms. *Mutat. Res.* 2006; **597**: 78–86.
13. Emerit I. Reactive oxygen species, chromosome mutation, and cancer: possible role of clastogenic factors in carcinogenesis. *Free. Radic. Biol. Med.* 1994; **16**: 99–109.
14. Fuqua C, Parsek MR and Greenberg EP. Regulation of gene expression by cell-to-cell communication: acyl-homoserine lactone quorum sensing. *Annu. Rev. Genet.* 2001; **35**: 439–468.
15. Goldberg Z and Lehnert BE. Radiation-induced effects in unirradiated cells: a review and implications in cancer. *Int. J. Oncol.* 2002; **21**: 337–349.
16. Hei TK, Wu LJ, Liu SX, Vannais D, Waldren CA and Randers-Pehrson G. Mutagenic effects of a single and an exact number of alpha particles in mammalian cells. *Proc. Natl. Acad. Sci. USA* 1997; **94**: 3765–3770.
17. Hu B, Wu L, Han W, Zhang L, Chen S, Xu A, Hei TK and Yu Z. The time and spatial effects of bystander response in mammalian cells induced by low dose radiation. *Carcinogenesis* 2006; **27**: 245–251.
18. Huo L, Nagasawa H and Little JB. HPRT mutants induced in bystander cells by very low fluences of alpha particles result primarily from point mutations. *Radiat. Res.* 2001; **156**: 521–525.

19. Iyer R and Lehnert BE. Alpha-particle-induced increases in the radioresistance of normal human bystander cells. *Radiat. Res.* 2002; **157**: 3–7.
20. Iyer R and Lehnert BE. Effects of ionizing radiation in targeted and nontargeted cells. *Arch. Biochem. Biophys.* 2000; **376**: 14–25.
21. Iyer R and Lehnert BE. Low dose, low-LET ionizing radiation-induced radioadaptation and associated early responses in unirradiated cells. *Mutat. Res.* 2002; **503**: 1–9.
22. Iyer R, Lehnert BE and Svensson R. Factors underlying the cell growth-related bystander responses to alpha particles. *Cancer Res.* 2000; **60**: 1290–1298.
23. Johnson TM, Yu ZX, Ferrans VJ, Lowenstein RA and Finkel T. Reactive oxygen species are downstream mediators of p53-dependent apoptosis. *Proc. Natl. Acad. Sci. USA* 1996; **93**: 11848–11852.
24. Kashino G, Prise KM, Schettino G, Folkard M, Vojnovic B, Michael BD, Suzuki K, Kodama S and Watanabe M. Evidence for induction of DNA double strand breaks in the bystander response to targeted soft X-rays in CHO cells. *Mutat. Res.* 2004; **556**: 209–215.
25. Leach JK, Black SM, Schmidt-Ullrich RK and Mikkelsen RB. Activation of constitutive nitric-oxide synthase activity is an early signaling event induced by ionizing radiation. *J. Biol. Chem.* 2002; **277**: 15400–15406.
26. Little JB. Genomic instability and bystander effects: a historical perspective. *Oncogene* 2003; **22**: 6978–6987.
27. Little JB, Azzam EI, de Toledo SM and Nagasawa H. Bystander effects: intercellular transmission of radiation damage signals. *Radiat. Prot. Dosimetry* 2002; **99**: 159–162.
28. Little JB, Nagasawa H, Li GC and Chen DJ. Involvement of the nonhomologous end joining DNA repair pathway in the bystander effect for chromosomal aberrations. *Radiat. Res.* 2003; **159**: 262–267.
29. Lorimore SA, Kadhim MA, Pocock DA, Papworth D, Stevens DL, Goodhead DT and Wright EG. Chromosomal instability in the descendants of unirradiated surviving cells after alpha-particle irradiation. *Proc. Natl. Acad. Sci. USA* 1998; **95**: 5730–5733.
30. Lyng FM, Seymour CB and Mothersill C. Early events in the apoptotic cascade initiated in cells treated with medium from the progeny of irradiated cells. *Radiat. Prot. Dosimetry* 2002; **99**: 169–172.
31. Lyng FM, Seymour CB and Mothersill C. Initiation of apoptosis in cells exposed to medium from the progeny of irradiated cells: a possible mechanism for bystander-induced genomic instability? *Radiat. Res.* 2002; **157**: 365–370.

32. Lyng FM, Seymour CB and Mothersill C. Production of a signal by irradiated cells which leads to a response in unirradiated cells characteristic of initiation of apoptosis. *Br. J. Cancer* 2000; **83**: 1223–1230.
33. Maguire P, Mothersill C, Seymour C and Lyng FM. Medium from irradiated cells induces dose-dependent mitochondrial changes and BCL2 responses in unirradiated human keratinocytes. *Radiat. Res.* 2005; **163**: 384–390.
34. Matsumoto H, Hayashi S, Hatashita M, Ohnishi K, Shioura H, Ohtsubo T, Kitai R, Ohnishi T and Kano E. Induction of radioresistance by a nitric oxide-mediated bystander effect. *Radiat. Res.* 2001; **155**: 387–396.
35. McBride WH, Chiang CS, Olson JL, Wang CC, Hong JH, Pajonk F, Dougherty GJ, Iwamoto KS, Pervan M and Liao YP. A sense of danger from radiation. *Radiat. Res.* 2004; **162**: 1–19.
36. Mitchel RE and Morrison DP. Heat-shock induction of ionizing radiation resistance in *Saccharomyces cerevisiae*, and correlation with stationary growth phase. *Radiat. Res.* 1982; **90**: 284–291.
37. Mitchel RE and Morrison DP. Heat-shock induction of ionizing radiation resistance in *Saccharomyces cerevisiae*. Transient changes in growth cycle distribution and recombinational ability. *Radiat. Res.* 1982; **92**: 182–187.
38. Mitchell SA, Marino SA, Brenner DJ and Hall EJ. Bystander effect and adaptive response in C3H 10T(1/2) cells. *Int. J. Radiat. Biol.* 2004; **80**: 465–472.
39. Morgan WF. Is there a common mechanism underlying genomic instability, bystander effects and other nontargeted effects of exposure to ionizing radiation? *Oncogene* 2003; **22**: 7094–7099.
40. Morgan WF. Non-targeted and delayed effects of exposure to ionizing radiation: I. Radiation-induced genomic instability and bystander effects *in vitro*. *Radiat. Res.* 2003; **159**: 567–580.
41. Morgan WF. Non-targeted and delayed effects of exposure to ionizing radiation: II. Radiation-induced genomic instability and bystander effects *in vivo*, clastogenic factors and transgenerational effects. *Radiat. Res.* 2003; **159**: 581–596.
42. Morgan WF, Hartmann A, Limoli CL, Nagar S and Ponnaiya B. Bystander effects in radiation-induced genomic instability. *Mutat. Res.* 2002; **504**: 91–100.
43. Mothersill C, Rea D, Wright EG, Lorimore SA, Murphy D, Seymour CB and O'Malley K. Individual variation in the production of a "bystander signal" following irradiation of primary cultures of normal human urothelium. *Carcinogenesis* 2001; **22**: 1465–1471.

44. Mothersill C and Seymour C. Medium from irradiated human epithelial cells but not human fibroblasts reduces the clonogenic survival of unirradiated cells. *Int. J. Radiat. Biol.* 1997; **71**: 421–427.
45. Mothersill C and Seymour C. Radiation-induced bystander effects: are they good, bad or both? *Med. Confl. Surviv.* 2005; **21**: 101–110.
46. Mothersill C and Seymour C. Radiation-induced bystander effects: past history and future directions. *Radiat. Res.* 2001; **155**: 759–767.
47. Mothersill C and Seymour CB. Cell-cell contact during gamma irradiation is not required to induce a bystander effect in normal human keratinocytes: evidence for release during irradiation of a signal controlling survival into the medium. *Radiat. Res.* 1998; **149**: 256–262.
48. Mothersill C and Seymour CB. Radiation-induced bystander effects–implications for cancer. *Nat. Rev. Cancer* 2004; **4**: 158–164.
49. Mothersill C, Seymour RJ and Seymour CB. Bystander effects in repair-deficient cell lines. *Radiat. Res.* 2004; **161**: 256–263.
50. Mothersill C, Stamato TD, Perez ML, Cummins R, Mooney R and Seymour CB. Involvement of energy metabolism in the production of "bystander effects" by radiation. *Br. J. Cancer* 2000; **82**: 1740–1746.
51. Muller WE, Ushijima H, Batel R, Krasko A, Borejko A, Muller IM and Schroder HC. Novel mechanism for the radiation-induced bystander effect: nitric oxide and ethylene determine the response in sponge cells. *Mutat. Res.* 2006; **597**: 62–72.
52. Nagar S and Morgan WF. The death-inducing effect and genomic instability. *Radiat. Res.* 2005; **163**: 316–323.
53. Nagar S, Smith LE and Morgan WF. Characterization of a novel epigenetic effect of ionizing radiation: the death-inducing effect. *Cancer Res.* 2003; **63**: 324–328.
54. Nagasawa H, Cremesti A, Kolesnick R, Fuks Z and Little JB. Involvement of membrane signaling in the bystander effect in irradiated cells. *Cancer Res.* 2002; **62**: 2531–2534.
55. Nagasawa H, Huo L and Little JB. Increased bystander mutagenic effect in DNA double-strand break repair-deficient mammalian cells. *Int. J. Radiat. Biol.* 2003; **79**: 35–41.
56. Nagasawa H and Little JB. Induction of sister chromatid exchanges by extremely low doses of alpha-particles. *Cancer Res.* 1992; **52**: 6394–6396.
57. Nagasawa H and Little JB. Unexpected sensitivity to the induction of mutations by very low doses of alpha-particle radiation: evidence for a bystander effect. *Radiat. Res.* 1999; **152**: 552–557.

58. Nagasawa H, Peng Y, Wilson PF, Lio YC, Chen DJ, Bedford JS and Little JB. Role of homologous recombination in the alpha-particle-induced bystander effect for sister chromatid exchanges and chromosomal aberrations. *Radiat. Res.* 2005; **164**: 141–147.
59. Narayanan PK, Goodwin EH and Lehnert BE. Alpha particles initiate biological production of superoxide anions and hydrogen peroxide in human cells. *Cancer Res.* 1997; **57**: 3963–3971.
60. Prise KM, Folkard M and Michael BD. A review of the bystander effect and its implications for low-dose exposure. *Radiat. Prot. Dosimetry* 2003; **104**: 347–355.
61. Sawant SG, Randers-Pehrson G, Geard CR, Brenner DJ and Hall EJ. The bystander effect in radiation oncogenesis: I. Transformation in C3H 10T1/2 cells *in vitro* can be initiated in the unirradiated neighbors of irradiated cells. *Radiat. Res.* 2001; **155**: 397–401.
62. Shao C, Aoki M and Furusawa Y. Bystander effect in lymphoma cells vicinal to irradiated neoplastic epithelial cells: nitric oxide is involved. *J. Radiat. Res. (Tokyo)* 2004; **45**: 97–103.
63. Shao C, Folkard M, Michael BD and Prise KM. Targeted cytoplasmic irradiation induces bystander responses. *Proc. Natl. Acad. Sci. USA* 2004; **101**: 13495–13500.
64. Souto J. Tumour development in the rat induced by blood of irradiated animals. *Nature* 1962; **195**: 1317–1318.
65. Springer DL, Ahram M, Adkins JN, Kathmann LE and Miller JH. Characterization of medium conditioned by irradiated cells using proteome-wide, high-throughput mass spectrometry. *Radiat. Res.* 2005; **164**: 651–654.
66. Suzuki M, Zhou H, Geard CR and Hei TK. Effect of medium on chromatin damage in bystander mammalian cells. *Radiat. Res.* 2004; **162**: 264–269.
67. Various authors. Radiation-induced genomic instability and bystander effects; implications for evolutionary biology. *Radiat. Res.* 2005; **163**: 473–476.
68. Vodovotz Y, Chesler L, Chong H, Kim SJ, Simpson JT, DeGraff W, Cox GW, Roberts AB, Wink DA and Barcellos-Hoff MH. Regulation of transforming growth factor beta1 by nitric oxide. *Cancer Res.* 1999; **59**: 2142–2149.
69. Ward JF. The radiation-induced lesions which trigger the bystander effect. *Mutat. Res.* 2002; **499**: 151–154.
70. Watson GE, Lorimore SA, Macdonald DA and Wright EG. Chromosomal instability in unirradiated cells induced *in vivo* by a bystander effect of ionizing radiation. *Cancer Res.* 2000; **60**: 5608–5611.

71. Wu LJ, Randers-Pehrson G, Xu A, Waldren CA, Geard CR, Yu Z and Hei TK. Targeted cytoplasmic irradiation with alpha particles induces mutations in mammalian cells. *Proc. Natl. Acad. Sci. USA* 1999; **96**: 4959–4964.
72. Zhou H, Randers-Pehrson G, Waldren CA, Vannais D, Hall EJ and Hei TK. Induction of a bystander mutagenic effect of alpha particles in mammalian cells. *Proc. Natl. Acad. Sci. USA* 2000; **97**: 2099–2104.
73. Zhou H, Suzuki M, Geard CR and Hei TK. Effects of irradiated medium with or without cells on bystander cell responses. *Mutat. Res.* 2002; **499**: 135–141.

Chapter 11

A STOCHASTIC MODEL OF HUMAN COLON CANCER INVOLVING MULTIPLE PATHWAYS

Wai Y. Tan, Li J. Zhang, Chao W. Chen and J. M. Zhu

Based on recent biological studies, in this chapter we have developed a stochastic model for human colon cancer involving five different pathways. These pathways are: the sporadic LOH pathway (about 70–75%), the familial LOH pathway (about 10–15%), the FAP pathway (about 1%), the sporadic MSI pathway (about 10%) and the HNPCC pathway (about 5%). For this model, we have combined the data augmentation method (equivalent to the EM algorithm in sampling theory framework) with the genetic algorithm (GA algorithm) and the state space model to estimate the genetic parameters of these pathways. We use the Bayesian approach to estimate the parameters through the posterior modes of the parameters by combining the genetic algorithm with the mean numbers of state variables. We have applied this model to fit and analyze the SEER data of human colon cancers from NCI/NIH. Our results indicate that the model not only provides a logical avenue to incorporate biological information but also fits the data much better than other models including the four-stage single pathway model. This model not only would provide more insights into human colon cancer but also would provide useful guidance for its prevention and control and for prediction of future cancer cases.

Keywords: FAP, generalized Bayesian procedure, GA algorithm, HNPCC, LOH, MSI, multiple pathways, state space model.

1. INTRODUCTION

In the past 15 years, molecular biologists and geneticists have revealed the basic molecular and genetic mechanisms for human colon cancer. These mechanisms have been linked to two avenues: the chromosomal instability (CIN) involving chromosomal aberrations and loss of heterozygocity (LOH), and the micro-satellite instability (MIS) involving mis-match repair genes and the creation of mutator phenotype (Chapelle, 2004; Fodde et al., 2001; Green and Kaplan 2003; Hawkins and Ward, 2001; Sparks et al., 1998; Ward et al., 2001). Most pathways of the CIN avenue (also referred to as LOH pathways) involve inactivation through genetic and/or epigenetic mechanisms, or loss, or mutation of the suppressor *APC* gene in chromosome 5q (about 75% of all human colon cancers) whereas almost all pathways of the MSI avenue involve mutation or epigenetic inactivation of the *Mis-Match Repair* (*MMR*) genes (about 15% of all colon cancers). This leads to multiple pathways for the generation of human colon cancer tumors with each pathway following a stochastic multi-stage model with intermediate transformed cells subjecting to stochastic proliferation (birth) and differentiation (death). The goal of this chapter is to develop a stochastic model for human colon cancer to incorporate these biological information and pathways. Specifically, we will proceed to develop a stochastic multi-stage model involving five pathways — the sporadic LOH pathway, the FLOH pathway (familial LOH pathway), the FAP pathway (familial adenomatous polyposis pathway), the sporadic MSI pathway and the HNPCC pathway (hereditary non-polyposis colon cancer pathway), as these pathways will cover almost all of human colon cancer cases.

In Section 2, we will describe the basic cancer biology for human colon cancer. Because each pathway in human colon cancer is an extended multi-event stochastic model of carcinogenesis, in Section 3, we will derive the basic theory for this type of models. In Section 4, we will develop a statistical model of carcinogenesis for human colon cancer involving multiple pathways. In Section 5, we will develop a state space model, a generalized Bayesian method and the Gibbs sampling procedures to estimate the parameters via data augmentation (equivalent to the EM-algorithm in sampling theory) and the genetic algorithm by combining with the means of state variables of the state space model. In Section 6, we will illustrate the model and the methods by fitting the NCI/NIH SEER data sets of human colon cancer. Finally in Section 7, we will discuss the usefulness of the models and the approach and some possible extensions.

2. A BRIEF SUMMARY OF COLON CANCER BIOLOGY

As discussed in the introduction, genetic studies have indicated that there are two major avenues by means of which human colon cancer is derived: the chromosomal instability and the micro-satellite instability. The first avenue is associated with the LOH pathways and the latter associated with the micro-satellite pathways. This leads to five different pathways: the sporadic LOH (about 70%), the familial LOH (about 10–15%), the FAP (about 1%), the sporadic MSI (about 10–15%) and the HNPCC (about 4–5%). For sporadic pathways, the individuals at birth are normal individuals and do not carry any mutated or inactivated suppressor genes. For FAP, the individual has inherited a mutated *APC* gene in chromosome 5 at birth. For HNPCC, the individuals has inherited a mutated mis-match repair gene, mostly *hMLH1* or *hMSH2*. For the familial colon cancer, the individuals have inherited a low penetrating mutated gene such as the *APC* allele *APCI1307K* at birth. Potter (1999) has proposed two more pathways, but the frequency of these pathways are quite small and are negligible.

2.1. The LOH Pathway of Human Colon Cancer (The APC-β – Catenin – Tcf – myc Pathway)

The LOH pathway involves loss or inactivation of the suppressor genes — the *APC* gene in chromosome 5q, the *Smad-4* gene in chromosome 18q and the *p53* gene in chromosome 17p; see **Remark 1**. This pathway accounts for about 70–75% of all colon cancers. It has been referred to as the LOH pathway because it is characterized by aneuploidy /or loss of chromosome segments (chromosomal instability). The FAP colon cancer (about 1%) is a special case of this pathway in which the patient has already inherited a mutated *APC* gene at birth. The FLOH pathway is a special case of this pathway in which the individuals have inherited a low-penetrating gene such as the *APC* allele *APCI1307K* gene at birth.

Morphological studies have indicated that inactivation, or loss or mutation of *APC* creates dysplastic aberrant cript foci (ACF) which grow into dysplastic adenomas. These adenomas grow to a maximum size of about 10 mm^3; further growth and malignancy require the abrogation of differentiation and apoptosis which are facilitated by the inactivation, or mutation

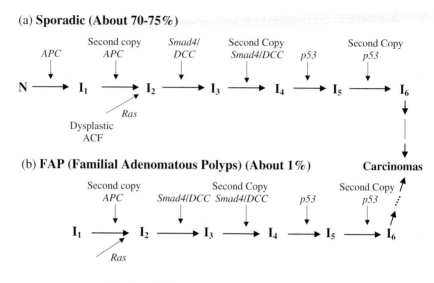

Fig. 1. LOH (loss of heterozygosity) pathway.

or loss of *Smad-4* gene in 18q and the *p53* gene in 17p. The mutation or activation of the oncogene *H-ras* in chromosome 11p and/or mutation and/or activation of the oncogene *src* in chromosome 20q would speed up these transitions by promoting the proliferation rates of the respective intermediate initiated cells (Jessup et al., 2002). This pathway is represented schematically by Figure 1.

The model in Figure 1 is a six-stage model. However, because of the haplo-insufficiency of the *Smad4* gene (Alberici et al., 2006) and the haplo-insufficiency of the *p53* gene (Lynch and Milner, 2006), one may reduce this six-stage model into a four-stage model by combining the third stage and the fourth stage into one stage and by combining the fifth stage and the sixth stage into one stage. This may help explain why for single pathway models, the four-stage model fits the human colon cancer better than other single pathway multi-stage models (Luebeck and Moolgavkar, 2002). Recent biological studies by Green and Kaplan (Green and Kaplan, 2003) and others have also shown that the inactivation or deletion or mutation of one copy of the *APC* gene in chromosome 5 can cause defects in microtubule plus-end attachment during mitosis dominantly, leading to aneuploidy and chromosome instability. This would speed up the mutation or inactivation of the second copy of the *APC* gene and increase fitness of the *APC*-carrying cells in the microevolution process of cancer progression. This could also help explain why the *APC* LOH pathway is more frequent than other pathways.

From the above documentation and discussion, biologically one may assume a four-stage model for the sporadic LOH pathway. It follows that one may assume a three-stage model for the FAP pathway whereas the FLOH is a three-stage model with low penetration of the inherited gene. In what follows, we denote by I_j the jth stage cells in the LOH pathways. Then for sporadic LOH, the model is $N \to I_1 \to I_2 \to I_3 \to I_4 \to$ *cancer tumor*; for FAP, the model is $I_1 \to I_2 \to I_3 \to I_4 \to$ *cancer tumor* and for FLOH, $I_1^{(*)} \to I_2 \to I_3 \to I_4 \to$ *cancer tumor*, where $I_1^{(*)}$ denotes first-stage intermediate cell with a low-penetrating *APC* allele. To account for the low penetration of the inherited gene one may practically model FLOH by a four-stage model with the constraint that the proliferation rate and the death rate are zero for the first stage cells; then the first mutation rate is the penetrating rate of the inherited gene.

Remark 1: As observed by Sparks *et al.* (1998), instead of the *APC* gene, this pathway can also be initiated by mutation of the oncogene β-*catenin*; however, the proportion of human colon cancer due to mutation of β-*catenin* is very small (less than 10%) as compared to the *APC* gene, due presumably to the contribution of the *APC* on chromosome instability (Green and Kaplan, 2003); also, because epigenetic changes are more frequent than genetic changes, it may be due to some epigenetic mechanisms (Breivik and Gaudernack, 1999; Jones and Baylin, 2002; Baylin and Ohm, 2006) to silence the *APC* gene as epigenetic silencing of tumor suppressor genes may lead to LOH of chromosomal segment involving these genes (Jones and Baylin, 2002; Baylin and Ohm, 2006).

2.2. The MSI (Micro-Satellite Instability) Pathway of Human Colon Cancer

This pathway accounts for about 15% of all colon cancers and appears mostly in the right colon. It has been referred to as the MSI pathway or the mutator phenotype pathway because it is initiated by the mutations or epigenetic methylation of the *mis-match repair* genes (the *MMR* genes, mostly *hMLH1* in chromosome 3p21 and *hMSH2* in chromosome 2p16) creating a mutator phenotype to significantly increase the mutation rate of

(a) Sporadic (About 10-15%)

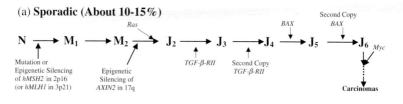

(b) HNPCC (Hereditary Non-Polyposis Colon Cancer) (≤ 5%)

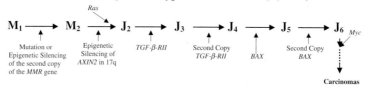

Fig. 2. MSI (microsatellite instability) pathway.

many critical genes 100 to 1000 times. Normally these critical genes are *TGF-β RII*, *BAX* (The X protein of *bcl-2* gene), *IGF2R*, or *CDX-2*. The *MMR* genes are *hMLH1*, *hMSH2*, *hPMS1*, *hPMS2*, *hMSH6* and *hMSH3*; mostly *hMLH1* (50%) and *hMSH2* (40%). The HNPCC (about 5% or less) is a special case of this pathway in which the individuals have inherited a mutant *MMR* gene (mostly *hMLH1* and *hMSH2*). This pathway is represented schematically by Figure 2.

As in the LOH pathway, assuming haploinsufficiency of tumor suppressor genes, one may approximate this pathway by a five-stage model. (Our analysis of the SEER data suggests a five-stage model for this pathway.) Morphologically, mutation or methylation silencing of the *MMR* gene *hMLH1* or *hMSH2* generates hyperplastic polyp which leads to the generation of serrated adenomas. These adenomas develop into carcinomas after the inactivation, or loss or mutations of the *TGF-β RII* gene and the *BAX* gene, thus abrogating differentiation and apoptosis. (*BAX* is an anti-apoptosis gene.) In what follows, we denote by J_i the ith stage cells in the MSI pathways. Then for sporadic MSI, the model is $N \to J_1 \to J_2 \to J_3 \to J_4 \to J_5 \to$ cancer tumor; for HNPCC, the model is $J_1 \to J_2 \to J_3 \to J_4 \to J_5 \to$ cancer tumor.

Recent biological studies (Baylin and Ohm, 2006; Koinuma *et al.*, 2006) have shown that both the CIS and the MSI pathways involve the Wnt signalling pathway, the TGF-β signalling and the p53-Bax signalling but different genes in different pathways are affected in these signalling processes. (In the CIS pathway, the affected gene is the *APC* gene in the Wnt signalling, the *Smad4* in the TGF-β signalling and *p53* gene in the p53-Bax signalling; on the other hand, in the MSI pathway, the affected gene is the *Axin 2* gene in the Wnt signalling, the *TGF-β-Receptor II* in the TGF-β signalling and the *Bax* gene in the p53-Bax signalling.) Because point mutation of genes are in general very small compared to epigenetic changes, one may speculate that colon cancer may actually be initiated by some epigenetic mechanisms (Breivik and Gaudernack, 1999; Jones and Baylin, 2002; Baylin and Ohm, 2006). In fact, Breivik and Gaudernack (1999) showed that in human colon cancer, either methylating carcinogens or hypermethylation at C_pG islands would lead to G/T mismatch which in turn leads to *MMR* gene deficiency or epigenetic silencing of the *MMR* genes and hence MSI (microsatellite instability); alternatively, either hypo-methylation, or bulky-adduct forming (BAF) carcinogens such as alkylating agents, UV radiation and oxygen species promote chromosomal rearrangement via activation of mitotic check points (MCP), thus promoting CIS (chromosomal instability). A recent review by Baylin and Ohm (2006) have demonstrated that epigenetic events may lead to LOH and mutations of many genes which may further underline the importance of epigenetic mechanisms in cancer initiation and progression.

Based on the above biological studies, in this chapter we thus postulate that the incidence data of human colon cancer are described and generated by a multi-stage model involving five pathways as defined above. In this chapter, because of haploid-insufficiency of the tumor suppressor genes {*Smad-4, p53, Axin, Bax, TG F-β-ReceptorII*}, the number of stages for the FAP pathway, the HNPCC pathway, the sporadic MSI pathway (family LOH pathway) and the sporadic LOH pathway are assumed as $\{k_1 = 3, k_2 = 4, k_3 = 5, k_5 = 4\}$ respectively. Since the family LOH pathway (FLOH) is a three-stage pathway with low penetration for the inherited gene, as illustrated in Section 2, we will model this by a four-stage model under the constraint that the proliferation rate and death rate of the first stage cell are zero; in this case $k_4 = 4$ and the first mutation rate is the penetrating rate of the inherited gene.

Because in the multiple-pathways model of human colon cancer, each pathway follows a single pathway multi-stage model, in the next section we thus describe and characterize the most general single pathway model of carcinogenesis.

3. THE STOCHASTIC MULTI-STAGE MODEL OF CARCINOGENESIS

The most general k-stage model for a single pathway is the extended k-stage ($k \geq 2$) multi-event model proposed by Tan and Chen (1998). This is an extension of the multi-event model first proposed by Chu (1985) and studied by Tan (1991) and Little (1995). It views carcinogenesis as the end point of k ($k \geq 2$) discrete, heritable and irreversible events (mutations or genetic changes or epigenetic changes) with intermediate cells subjected to stochastic proliferation and differentiation. It takes into account cancer progression by assuming that malignant cancer tumors develop from primary I_k cells through clonal expansion, where a primary I_k cell is a k-stage cell generated directly from a I_{k-1} cell. As an example we consider the sporadic LOH pathway of human colon cancer as described above (in this example, $k = 4$).

Let N denote normal stem cells, T the cancer tumors and I_j the jth stage intermediate cells arising from the $(j-1)$th stage intermediate cells ($j = 1, \ldots, k$) by mutation or some genetic changes or epigenetic changes. Then the model assumes $N \to I_1 \to I_2 \to \cdots \to I_k$ with the N cells and the I_j cells subject to stochastic proliferation (birth) and differentiation (death). It takes into account cancer progression by following Yang and Chen (1991) to postulate that cancer tumors develop from primary I_k cells by clonal expansion (i.e., stochastic birth-death process).

Given the above extended multi-event model, the next question is how to develop analytical mathematical results. Because the classical Markov theory as described in Tan (1991) is too complicated to be of much use, to analyze the above model we thus propose an alternative approach through stochastic differential equations. It is shown by Tan and Chen (1998) through probability generating function method that the stochastic differential equation method is equivalent to the classical Markov theory method but is more powerful.

3.1. Stochastic Equations of State Variables

Let $\{I_0(t) = N(t), I_j(t), j = 1, \ldots, k\}$ and $T(t)$ denote the number of normal stem cells, the number of the jth stage initiated cells ($j = 1, \ldots, k$) and the number of detectable cancer tumors at time t, respectively. For $i = 0, 1, \ldots, k - 1$, let $b_i(t), d_i(t)$ and $\alpha_i(t)$ denote the birth rate, the death rate and the mutation rate of the I_i cells at time t respectively. Put $\gamma_i(t) = b_i(t) - d_i(t)$ and let $\{B_i(t), D_i(t), M_i(t)\}$ denote the numbers of birth and death of I_i cells and the number of mutations from $I_i \to I_{i+1}$ during $[t, t + \Delta t)$ respectively. Then given $I_i(t)$ I_i cells at time t, to the order of $o(\Delta t)$, the conditional probability distribution of $\{B_i(t), D_i(t), M_i(t)\}$ given $I_i(t)$ is multinomial with total number $I_i(t)$ and with probabilities $\{b_i(t)\Delta t, d_i(t)\Delta t, \mu_i(t)\Delta t\}$. To the order of $o(\Delta t)$, the conditional expected values of $\{B_i(t), D_i(t), M_i(t)\}$ given $I_i(t)$ are given respectively by $\{I_i(t)b_i(t)\Delta t, I_i(t)d_i(t)\Delta t, I_i(t)\mu_i(t)\Delta t\}$. From this, one derives the following stochastic equations for the state variables (see **Remark 2**):

$$dN(t) = N(t + \Delta t) - N(t) = B_0(t) - D_0(t) = N(t)\gamma_0(t)\Delta t + e_0(t)\Delta t, \tag{1}$$

$$dI_j(t) = I_j(t + \Delta t) - I_j(t) = M_{j-1}(t) + B_i(t) - D_i(t)$$
$$= \{I_{j-1}(t)\alpha_{j-1}(t) + I_j(t)\gamma_j(t)\}\Delta t + e_j(t)\Delta t, \quad j = 1, \ldots, k - 1. \tag{2}$$

In Eqs. (1) – (2), the random noises $\{e_j(t)\Delta t, j = 1, \ldots, k - 1\}$ are derived by subtracting the conditional mean values from the random transition variables. It can easily be shown that these random noises have expected value zero and are uncorrelated with the state variables $\{N(t), I_i(t), i = 1, \ldots, k - 1\}$. Further, because the size of the organ in adult normal individuals is practically stable, one may practically assume $\gamma_0(t) = 0$.

Remark 2: Because genetic changes and epigenetic changes occur during cell division, to the order of $o(\Delta t)$, the probability is $\alpha_i(t)\Delta t$ that one I_i cell at time t would give rise to one I_i cell and one I_{i+1} cell at time $t + \Delta t$ by genetic changes or epigenetic changes. It follows that the mutation process of $I_i \to I_{i+1}$ would not affect the population size of I_i cells but only increase the size of the I_{i+1} population.

3.2. The Expected Number of $I_j(t)$

Let $u_j(t) = EI_j(t)$ be the expected number of $I_j(t)$, $j = 0, \ldots, k-1$. By taking expectation over both sides of Eqs. (1) – (2), we have the following system differential equations for these expected numbers:

$$\frac{du_0(t)}{dt} = u_0(t)\gamma_0(t), \tag{3}$$

$$\frac{du_j(t)}{dt} = u_{j-1}(t)\alpha_{j-1}(t) + u_j(t)\gamma_j(t), \quad j = 1, \ldots, k-1. \tag{4}$$

Solving the above system of equations, we obtain:

$$u_0(t) = u_0(0) e^{\int_0^t \gamma_0(x)dx}, \tag{5}$$

$$u_j(t) = u_j(0) e^{\int_0^t \gamma_j(x)dx} + \int_0^t u_{j-1}(y)\alpha_{j-1}(y) e^{\int_y^t \gamma_j(x)dx}, \tag{6}$$

for $j = 1, \ldots, k-1$.

Assume that $\{I_j(0) = T(0) = 0, j = 1, \ldots, k-1\}$. If $\{\gamma_j(t) = \gamma_j, \alpha_j(t) = \alpha_j \text{ for } (j = 0, 1, \ldots, k-1)\}$, then the above solution of Eqs. (5) – (6) reduces to:

$$u_0(t) = u_0(0) e^{\int_0^t \gamma_0(x)dx}, \tag{7}$$

$$u_j(t) = EI_j(t) = I_0(0) \left\{ \prod_{r=0}^{j-1} \alpha_r \right\} \sum_{u=0}^{j} A_j(u) e^{\gamma_u t}, \tag{8}$$

for $j = 1, \ldots, k-1$

where $A_j(u) = \{\prod_{\substack{v=0 \\ v \neq u}}^{j} (\gamma_u - \gamma_v)\}^{-1}$.

In particular, we have:

$$u_{k-1}(t) = EI_{k-1}(t) = \{\alpha_{k-1}\}^{-1} \sum_{u=0}^{k-1} \theta_k(u) e^{\gamma_u t}, \tag{9}$$

where $\theta_k(u) = I_0(0) \{\prod_{r=0}^{k-1} \alpha_r\} A_{k-1}(u)$.

3.3. The Probability Distribution of the Number of Detectable Tumors

Because, as shown by Yang and Chen (1991), malignant cancer tumors arise from primary cancer tumor cells by clonal expansion with birth rate $b_T(s, t)$ and $d_T(s, t)$, where s is the time that the primary cancer cell (i.e., the I_k cell) arises from the I_{k-1} cell by mutation, one may readily model the development of malignant cancer tumors from cancer tumor cells by using stochastic birth and death process of cancer tumor cells. Let $T(t)$ be the number of detectable malignant tumor per individual by time t and let $P_T(s, t)$ be the probability that a primary cancer cell at time s develops into a detectable cancer tumor at time t. (Explicit formula for $P_T(s, t)$ has been given in Tan and Chen (1998) and in Tan (2002, Chapter 8).) Then, given $\{I_{k-1}(s), s \leq t\}$, it has been shown in Tan (2002) that

$$T(t)|\{I_{k-1}(s), s \leq t\} \sim \text{Poisson}(\lambda(t)), \tag{10}$$

where

$$\lambda(t) = \int_{t_0}^{t} I_{k-1}(s)\alpha_{k-1}(s) P_T(s, t)ds.$$

From Eq. (10), the probability that cancer tumors develop during $[t_{j-1}, t_j)$ is

$$Q_T(j) = E\{e^{-\int_0^{t_{j-1}} I_{k-1}(s)\alpha_{k-1}(s) P_T(s,t_{j-1})ds} - e^{-\int_0^{t_j} I_{k-1}(s)\alpha_{k-1}(s) P_T(s,t_j)ds}\}$$
$$= E\{e^{-\int_0^{t_{j-1}} I_{k-1}(s)\alpha_{k-1}(s) P_T(s,t_{j-1})ds}(1 - e^{-R(t_{j-1},t_j)})\}, \tag{11}$$

where $R(t_{j-1}, t_j) = \int_0^{t_j} I_{k-1}(s)\alpha_{k-1}(s) P_T(s, t_j)ds - \int_0^{t_{j-1}} I_{k-1}(s)\alpha_{k-1}(s) P_T(s, t_{j-1})ds$.

Let the time unit $\Delta t \sim 1$ corresponding to three or six months or longer. Then, for human colon cancers, one may practically assume $P_T(s, t) \approx 1$ for $t - s \geq 1$. Thus, combining with the observation that $\alpha_{k-1}(s) = \alpha_{k-1}$ is usually very small ($\leq 10^{-6}$), $Q_T(j)$ is closely approximated by:

$$Q_T(j) \approx E\{e^{-\sum_{i=0}^{t_{j-1}} I_{k-1}(i)\alpha_{k-1}}(1 - e^{-R(j)})\},$$

where $R(j) = \sum_{i=t_{j-1}+1}^{t_j-1} I_{k-1}(i)\alpha_{k-1}$.

4. A STATISTICAL MODEL AND THE PROBABILITY DISTRIBUTION OF CANCER INCIDENCE DATA

The NCI/NIH SEER project (Ries *et al.*, 2001) gives number of new cancer cases for each of the age groups and the total number of normal people at risk for each of the age groups respectively. Thus, in the SEER project from NCI/NIH, the colon cancer data are given by $\{(Y_j, n_j), j = 1, \ldots, k\}$, where Y_j is the observed number of new colon cancer cases in the jth age group $[t_{j-1}, t_j)$ with n_j people at risk since birth for this age group. (In the SEER data, $k = 18$ and $t_j - t_{j-1}$ equals to five years.) In this type of data, the n_j's are fixed but the Y_j's are random variables. Because for human beings cancer development is restricted to his/her life time, for a slight better fit, we restrict human cancer to human's life time by changing n_j into $m_j = n_j p$, where p is the probability that humans develop cancer during his/her life time. In this paper, p is estimated by $\hat{p} = \sum_{j=1}^{18} \hat{p}_j$, where $\hat{p} = \frac{Y_j}{n_j}$. It is easy to see that $E\hat{p} = p$ and the variance of $\hat{p}$ is $\text{Var}(\hat{p}) = \sum_{j=1}^{18} \frac{1}{n_j} p_j(1 - p_j)$. Since in the SEER data $n_j > 10^7$ and $0 < p_j(1 - p_j) < 1$, $\text{Var}(\hat{p}) < \frac{18}{10^7}$ which is almost zero; thus one may practical assume $\hat{p} = p$.

To derive the probability distribution of Y_j given m_j, let $\{i = 1, i = 2, \ldots, i = 5\}$ correspond to the pathways (FAP, HNPCC, MSI, FLOH, LOH) respectively. Let ω_i denote the probability for entertaining the ith pathway and let $Q_i(j)$ denote the conditional probability that the individual will develop colon cancer during the jth age group given that this person develops colon cancer by following the ith pathway during his/her life time.

We assume that each individual develops colon cancer tumor by the same mechanism independently of one another. Then for each individual, the probability that this individual would develop colon cancer tumor during the jth age group $[s_{j-1}, s_j)$ is given by $Q_j = \sum_{i=1}^{5} \omega_i Q_i(j)$. Then the probability distribution of Y_j given that m_j people are at risk for colon cancer during his/her life time in this age group is:

$$P\{Y_j | m_j\} = \binom{m_j}{Y_j} Q_j^{Y_j} [1 - Q_j]^{m_j - Y_j}.$$

Put $Y = (Y_j, j = 1, \ldots, k)$. Then the joint density of Y given $\{m_j, j = 1, \ldots, k\}$ is:

$$P\{Y|m_j, j = 1, \ldots, k\} = \prod_{j=1}^{k} P\{Y_j|m_j\}.$$

Because m_j is very large and Q_j is usually very small, $P\{Y_j|m_j\}$ is closely approximated by a Poisson distribution with mean $\lambda_j = m_j Q_j = \sum_{i=1}^{5} \lambda_i(j)$, where $\lambda_i(j) = m_j \omega_i Q_i(j)$. That is,

$$P(Y_j|m_j) \sim e^{-\lambda_j} \frac{1}{Y_j!} \lambda_j^{Y_j}. \tag{12}$$

In the above distribution, because each pathway follows a multi-stage multi-event model as given in Section 3, $Q_i(j)$ is similar to the $Q_T(j)$ given by Eq. (11) in Section (3.3) but with parameters from the ith pathway. Assuming that the ith pathway is a k_i stage multi-event model, then

$$Q_i(j) \approx E\{e^{-\sum_{t=0}^{t_j-1-1} I_{k_i-1}^{(i)}(t)\alpha_{k_i-1}^{(i)}} (1 - e^{-R^{(i)}(j)})\},$$

where $\{I_{k_i-1}^{(i)}(t), \alpha_{k_i-1}^{(i)}\}$ are the $\{I_{k-1}(t), \alpha_{k-1}\}$'s from the ith pathway model and $R^{(i)}(j) = \sum_{t=t_{j-1}}^{t_j-1} I_{k_i-1}^{(i)}(t)\alpha_{k_i-1}^{(i)}$.

Since the $\alpha_{k_i-1}^{(i)}$'s are usually very small and since in the SEER data $\{m_j, m_j - Y_j\}$ are usually very large, as shown in Tan et al. (2004) one may further approximate $\lambda_i(j) = m_j Q_i(j)$ by:

$$\lambda_i(j) \approx \exp\left\{\log m_j - \sum_{t=0}^{t_j-1} EI_{k_i-1}^{(i)}(t)\alpha_{k_i-1}^{(i)} + \log(1 - e^{-ER^{(i)}(j)})\right\},$$

where $ER^{(i)}(j) = \sum_{t=t_{j-1}}^{t_j-1} EI_{k_i-1}^{(i)}(t)\alpha_{k_i-1}^{(i)}$.

The above equation shows that the probability distribution of the Y_j's depends on each pathway of the stochastic model of colon cancer mainly through the expected number of the last stage cells.

4.1. Data Augmentation and the Expanded Model

The probability density given in Eq. (12) involves five pathways. To derive information from each pathway, as in the EM-algorithm, we expand the model involving $\{Y_j, j = 1, \ldots, k\}$ by defining the variables $Z_i(j)$, where $Z_i(j)$ is the number of colon cancer cases in the jth age group derived by the ith pathway. Then the $Z_i(j)$'s are not observable but $\sum_{i=1}^{5} Z_i(j) = Y_j$. Further, the conditional probability distribution of $\underset{\sim}{Z} = \{Z_i(j), i = 1, 2, 3, 4\}$ given Y_j is a four-dimensional multinomial distribution with parameters $\{Y_j, g_i(j), j = 1, \ldots, 4\}$, where for $i = 1, \ldots, 5$,

$$g_i(j) = \frac{\omega_i Q_i(j)}{\sum_{r=1}^{5} \omega_r Q_r(j)} = \frac{m_j \omega_i Q_i(j)}{\sum_{r=1}^{5} m_j \omega_r Q_r(j)}$$

$$= \frac{\omega_i \lambda_i(j)}{\sum_{r=1}^{5} \omega_r \lambda_r(j)}.$$

That is, the conditional density of $\underset{\sim}{Z}_j = \{Z_i(j), i = 1, 2, 3, 4\}$ given Y_j is:

$$f\{\underset{\sim}{Z}_j | Y_j\} = \binom{Y_j}{Z_i(j), i = 1, \ldots, 5} \prod_{i=1}^{5} \{g_i(j)\}^{Z_i(j)},$$

where $Z_5(j) = Y_j - \sum_{i=1}^{4} Z_i(j)$, $\binom{Y_j}{Z_i(j), i=1,\ldots,5} = \frac{Y_j!}{\prod_{i=1}^{5} Z_i(j)!}$ and $g_5(j) = \frac{\omega_5 \lambda_5(j)}{\sum_{r=1}^{5} \omega_r \lambda_r(j)} = 1 - \sum_{i=1}^{4} g_i(j)$.

Put $\mathbf{Z} = \{\underset{\sim}{Z}(j), j = 1, \ldots, k\}$ and let Θ be the collection of all parameters. Then, the conditional density of $\mathbf{Z}$ given $\{\mathbf{Y}, \Theta\}$ is

$$P(\mathbf{Z}|\mathbf{Y}, \Theta) = \prod_{j=1}^{k} P(\underset{\sim}{Z}_j | Y_j). \tag{13}$$

The joint density of $\{Y_j, \underset{\sim}{Z}_j\}$ given m_j is, noting that $\sum_{i=1}^{5} Z_i(j) = Y_j$:

$$f\{Y_j, \underset{\sim}{Z}_j | m_j\} = f(Y_j|m_j) f\{\underset{\sim}{Z}_j | Y_j\}$$

$$= e^{-\lambda_j} \frac{(\lambda_j)^{Y_j}}{Y_j!} \binom{Y_j}{Z_i(j), i=1,\ldots,5} \prod_{i=1}^{5} \{g_i(j)\}^{Z_i(j)}$$

$$= \prod_{i=1}^{5} \frac{1}{Z_i(j)!} e^{-m_j \omega_i Q_i(j)} \{m_j \omega_i Q_i(j)\}^{Z_i(j)}$$

$$= \prod_{i=1}^{5} \frac{1}{Z_i(j)!} e^{-\omega_i \lambda_i(j)} \{\omega_i \lambda_i(j)\}^{Z_i(j)}, \qquad (14)$$

where $\lambda_i(j) = m_j Q_i(j)$.

It follows that the joint density of $\{\mathbf{Y}, \mathbf{Z}\}$ given Θ is:

$$f\{\mathbf{Y}, \mathbf{Z}|\Theta\} = \prod_{j=1}^{k} f\{Y_j, \underset{\sim}{Z}_j | m_j\}. \qquad (15)$$

4.2. The Genetic Parameters

To account for the proportion of the five pathways (FAP, HNPCC, MSI, HNPCC, FLOH, LOH), based on survey in the literature (Chapelle, 2004; Fodde et al., 2001; Green and Kaplan, 2003; Hawkins and Ward, 2001; Sparks et al., 1998) we will assume $\{\omega_1 = 0.01, \omega_2 = 0.05, \omega_3 = 0.10, \omega_4 = 0.14, \omega_5 = 0.70 = 1 - \sum_{i=1}^{4} a_i\}$ as the estimated proportions of these pathways.

For identifying the genetic parameters, let the mutation rates of $I_i \to I_{i+1}$ be α_i, $(i = 0, 2, 3)$ $(I_0 = N)$ in the LOH pathways (LOH, FLOH, FAP). Let $\{\alpha_1, \alpha_1^{(*)} = c_1 \alpha_1, \alpha_1^{(**)} = c_2 \alpha_1, (c_i > 0, i = 1, 2)\}$ be the mutation rate (or inactivation rate) for $I_1 \to I_2$ for LOH, FLOH and FAP respectively and let $\alpha_0^{(*)}$ be the penetration rate of the inherited gene in the FLOH pathway; see Remark 3. Let $\gamma_i(t) = b_i(t) - d_i(t)$ be the proliferation rate of the I_i stage cells ($i = 1, 2, 3$) for the LOH, FAP and FLOH pathways. Then the genetic parameters for the LOH, FAP and FLOH pathways are

$\Theta_1 = (\alpha_i, i = 0, 1, 2, 3, \alpha_0^{(*)}, c_1, c_2, \gamma_i(t), i = 1, 2, 3, t = 1, \ldots, t_M)$. Similarly, let the mutation rates of $J_i \to J_{i+1}$ be β_i, $(i = 0, 2, 3, 4)$ in the MSI and HNPCC pathways, $\{\beta_1, \beta_1^{(*)} = d\beta_1, d > 0\}$ the mutation rate for $J_1 \to J_2$ for MSI and HNPCC pathways respectively and $\eta_i(t)$ the proliferation rate of the J_i stage cells ($i = 1, 2, 3, 4$) for the MSI and HNPCC pathways. Then the genetic parameters for the MSI and HNPCC pathways are $\Theta_2 = (\beta_i, i = 0, 1, 2, 3, 4, d, \eta_i(t), i = 1, 2, 3, 4, t = 1, \ldots, t_M)$. Then the collection of all parameters are $\Theta = (\Theta_1, \Theta_2)$. To fit the model to the data, because the *APC* gene and the *MMR* genes are tumor suppressor genes and because the *MMR* genes affect mainly the mutation rates of relevant genes, one may assume $\gamma_1(t) = \eta_1(t) = \eta_2(t) = 0$; furthermore, because the limitation of proliferation of the I_i ($i = 2, 3$) and J_j ($j = 3, 4$) cells derive mainly from apoptosis, as a close approximation, one may also assume $\{\gamma_2(t) = \gamma_2 e^{-\delta_1 t}, \gamma_3(t) = \gamma_3\}$ and $\{\eta_3(t) = \eta_3 e^{-\delta_2 t}, \eta_4(t) = \eta_4\}$. Then $\Theta_i, i = 1, 2$ reduce respectively to $\Theta_1 = \{\alpha_i . i = 0, 1, 2, 3, c_1, c_2, \gamma_2, \gamma_3, \delta_1\}$ and $\Theta_2 = \{\beta_i, i = 0, 1, 2, 3, 4, d, \eta_3, \eta_4, \delta_2\}$. From this, one would need to estimate 18 unknown parameters. In the SEER data, there are 18 independent observations (the Y_j's) and there are prior information and information from the stochastic system model for the estimation of these unknown parameters.

Remark 3: As illustrated by Herrero-Jimenez *et al.* (1998), colon stem cells are located at the bottom layer of the crypt of the colon and there are eight transit layers between the stem cell and the terminal layer in the lumen. Cells in the colon are constantly undergo tissue renewal. As cells in the terminal layer undergo apoptosis, cells from the previous layers, composed of transit cells, divide and replace those cells that had been lost; similarly the loss of transit cells are replenished by transit cells from previous layers. Thus, as more transition cells divide to replenish the terminal layer, these cells are eventually replenished by stem cells which undergo asymmetric division. Herrero-Jimenez *et al.* (1998) showed that if one of the *APC* gene in either a transition cell or terminal cell acquired mutation to initiate the cancer process, that mutation would be lost unless the second copy of the *APC* gene has also been lost or inactivated. Thus, if the individual has inherited an *APC* gene, then the chance of the inactivation or mutation of the second copy of *APC* is much greater than one would expect in sporadic LOH. This same principle also apply to MSI.

5. THE STATE SPACE MODEL AND THE GENERALIZED BAYESIAN APPROACH FOR ESTIMATING THE UNKNOWN PARAMETERS

State space model is a stochastic model which consists of two sub-models: The stochastic system model which is the stochastic model of the system and the observation model which is a statistical model based on available observed data from the system. For the state space model of human colon cancer, the stochastic system model is the stochastic multiple pathways model consisting of five pathways with each pathway following an extended multi-event model described in Section 3; the observation model of this state space model is the expanded statistic model based on the observed number of colon cancer cases and the unobserved number of colon cancer cases from each pathway over different age groups as described in Section 4. This state space model takes into account the basic mechanisms of the system and the random variation of the system through its stochastic system model and incorporate all these into the observed data from the system; furthermore, it validates and upgrades the stochastic model through its observation model and the observed data of the system. (Notice that the expanded statistical model in Section 4 depends on the stochastic system model through the expected number of cancer intermediate states from each of the five pathways.) As illustrated in Tan (2002, Chapters 8–9), the state space model has many advantages over both the stochastic model and the statistical model when used alone since it combines information and advantages from both of these models.

By integrating this state space model with the prior distribution of the parameters and the Gibbs sampling procedures, in this section we will develop a generalized Bayesian approach to estimate the unknown parameters.

5.1. The Prior Distribution of the Parameters

For the prior distributions of $\Theta = (\Theta_1, \Theta_2)$, because biological information have suggested some lower bounds and upper bounds for the mutation rates and for the proliferation rates, we assume

$$P(\Theta_1, \Theta_2) \propto c(c > 0) \tag{16}$$

where c is a positive constant if these parameters satisfy some biologically specified constraints; and equal to zero for otherwise. These biological constraints are:

(i) For the mutation rates in the LOH, FLOH and FAP pathways, $1 < N_0\alpha_0 < 1000$ ($N \to I_1$ in LOH), $10^{-4} < \alpha^* < 10^{-2}$ (α^* for the penetration rate of the inherited gene in FLOH), $10^{-6} < \alpha_i < 10^{-4}, i = 1, 2, 3$ for the mutation from $I_i \to I_{i+1}$ in FAP, LOH and FLOH, $10^{-6} < c_1\alpha_1 < 10^{-4}$ for FLOH and $10^{-4} < c_2\alpha_1 < 10^{-2}$ for FAP. For the proliferation rates of FAP, LOH, FLOH, $\gamma_1(t) = 0, \gamma_2(t) = \gamma_2 e^{-\delta_1 t}, 10^{-4} < \gamma_2 < 2*10^{-2}, 10^{-4} < \delta_1 < 5*10^{-2}, \gamma_3(t) = \gamma_3, 10^{-2} < \gamma_3 < 0.5$.

(ii) For the mutation rates in the MSI and HNPCC pathways, $10 < N_0\beta_0 < 1000$ ($N \to I_1$ in MSI), $10^{-5} < d\beta_1 = \beta_1^* < 10^{-3}, 10^{-6} < \beta_1 < 10^{-3}, 10^{-5} < \beta_2 < 10^{-3}$, and $10^{-5} < \beta_i < 10^{-2} (i = 3, 4)$. For the proliferation rates in MSI and HNPCC, $\eta_i(t) = 0, i = 1, 2, \eta_3(t) = \eta_3 e^{-\delta_2 t}, 10^{-3} < \eta_3 < 5*10^{-1}, 10^{-3} < \delta_2 < 5*10^{-2}, \eta_4(t) = \eta_4, 10^{-3}\eta_4 < 0.5$.

We will refer the above prior as a partially informative prior which may be considered as an extension of the traditional non-informative prior given in Box and Tiao (1973).

5.2. The Posterior Distribution of the Parameters Given $\{Y, Z\}$

Let $P(\Theta_1, \Theta_2) = P(\Theta_1)P(\Theta_2)$ be the prior distribution of the parameters $\Theta = \{\Theta_1, \Theta_2\}$. From Eqs. (14) – (16), the joint posterior density of $\Theta = \{\Theta_1, \Theta_2\}$ given $\{Y, Z\}$ is

$$P(\Theta_1, \Theta_2 | Y, Z)$$

$$\propto P(\Theta_1, \Theta_2) \prod_{j=1}^{k} P(Y_j, \underset{\sim}{Z}(j))$$

$$\propto P(\Theta_1, \Theta_2) \prod_{j=1}^{n} \prod_{i=1}^{5} \left(\frac{1}{Z_i(j)!} e^{-\omega_i m_j Q_i(j)} [\omega_i m_j Q_i(j)]^{Z_i(j)} \right). \quad (17)$$

Notice that aside from the prior distribution of $\{\Theta_i, i = 1, 2\}$, the above posterior density is a product of Poisson densities.

5.3. The Multi-Level Gibbs Sampling Procedure for Estimating Parameters

Given the above probability distributions, the multi-level Gibbs sampling procedure for deriving estimates of the unknown parameters are given by:

(a) Step 1: Generating Z Given Y and Θ (The Data-Augmentation Step).

Given Y_j and given the parameter values of Θ, use the conditional distribution $P\{Z|Y, \Theta\}$ given $\{Y, \Theta\}$ in Eq. (13) to generate Z. Denote it by $\hat{Z}$.

Because generating a large sample of $Z_i(j)$ and using the sample mean $\hat{Z}_i(j)$ as the outcome is equivalent to let $\hat{Z}_i(j)$ be the conditional mean value of $Z_i(j)$ given Y_j, to speed up convergence one may take the conditional mean value of $Z_i(j)$ as $\hat{Z}_i(j)$ given data and the parameter values. That is, we take:

$$\hat{Z}_i(j) = \hat{g}_i(j),$$

where $\hat{g}_i(j)$ is derived from $g_i(j)$ by substituting the given parameter values. Numerically this is equivalent to the E-step of the E-M algorithm in the sampling theory framework.

(b) Step 2: Estimation of $\Theta = \{\Theta_1, \Theta_2\}$ Given $\{Y, Z = \hat{Z}\}$.

Given Y and $Z = \hat{Z}$ from Step 1 and using the mean values of the state variables $I_{k_{i-1}}^{(i)}(t)$ from the stochastic system model, apply the GA algorithm to derive the posterior modes of $\Theta = \{\Theta_1, \Theta_2\}$ under the prior distributions given by Section 5.1. These posterior modes are taken as estimates of the Θ_i ($i = 1, 2$). Denote the estimates by ($\hat{\Theta}_i, i = 1, 2$).

(c) Step 3: Recycling Step.

With $\{Z = \hat{Z}, \Theta_i = \hat{\Theta}_i, i = 1, 2\}$ given above, go back to Step (a) and continue until convergence.

The proof of convergence of the above steps can be proved using similar procedure used by Tan (2002, Chapter 3). At convergence, the $\{\hat{\Theta}_i, i = 1, 2\}$ are the generated values from the posterior distribution of $\{\Theta_i, i = 1, 2\}$ given Y independently of Z (for proof, see Tan (2002), Chapter 3). Repeat the above procedures one then generates a random sample of Θ from the posterior distribution of $\Theta = \{\Theta_i, i = 1, 2\}$ given Y; then one uses the sample mean as the estimates of ($\Theta_i, i = 1, 2$) and use the sample variances and covarainces as estimates of the varainces and covariances of these estimates.

In Step 2, we use the genetic algorithm to derive estimates of $\Theta = (\Theta_i, i = 1, 2)$. In the next section, we thus provide a brief description of this algorithm.

5.4. The Genetic Algorithm

The genetic algorithms are a class of search techniques inspired from the biological process of evolution by means of natural selection. The basic principle is that those with the largest fitness will be selected as the generation progresses. For the model in this paper, we will use the multi-stage model defined in Eqs. (1) – (2) in Section 3 and the probability density in Section 3.4 to construct the function as individuals in the population. For these individuals, we will define the fitness by the negative of the sum of deviances from the five pathways given $\{Y, Z, \underset{\sim}{\omega} = (\omega_i, i = 1, \ldots, 5)\}$. From Eq. (15), the deviance from the ith pathway given $\{Y, Z, \underset{\sim}{\omega}\}$ is

$$D_i = \sum_{j=1}^{k} \{-[\omega_i m_j Q_i(j) - Z_i(j)] + \log\{\omega_i m_j Q_i(j)/Z_i(j)\}\}.$$

Given the fitness, the genetic algorithm would choose the parameter values to maximize the fitness according to evolutionary principle as described above. In this paper we will use the genetic algorithm "PIKAIA" to derive estimates for each pathway. This genetic algorithm is in the public domain and can be downloaded from the web (C:PIKAIA Homepage.htm).

6. APPLICATION AND RESULTS

In this section, we will apply the above model to the NCI/NIH colon cancer data from the SEER project. Given in Table 1 are the numbers of people at risk and colon cancer cases in the age groups together with the predicted cases from the model. There are 18 age groups with each group spanning over five years.

To fit the data, we have assumed that $\gamma_0 = 0$ because of the observation that the size of normal stem cells are constants in almost all tissues.

Table 1. Colon cancer data from SEER (overall population, November 2005).

Age group	Number of people at risk	Observed colon cancer cases	Total predicted colon cancer
0 – 4	53294480	2	0
5 – 9	53293961	2	5
10 – 14	54689411	35	30
15 – 19	55510333	84	101
20 – 24	56897771	242	267
25 – 29	59952627	615	622
30 – 34	59741985	1305	1272
35 – 39	55106562	2466	2286
40 – 44	49703250	4306	3878
45 – 49	43521011	7460	6214
50 – 54	38602075	12780	9864
55 – 59	33327540	19532	15051
60 – 64	28839949	27678	22958
65 – 69	25020118	36698	34444
70 – 74	20612388	42850	46179
75 – 79	15855151	43969	50183
80 – 84	10465639	35820	35529
≥ 85	8590708	32499	32314

Also, we assumed the proportions $\{0.7, 0.10, 0.01, 0.15, 0.4\}$ for the five pathways $\{sporadic\ LOH, familial\ LOH, FAP, MSI, HNPCC\}$ respectively. Given in Table 2 are the estimates of the mutation rates, the birth and death rates of the I_j cells. For comparison purposes, the predicted colon cancer cases in the age groups are also presented in Table 3. Given in Figure 3 are the plots for the estimated probabilities to cancer tumor for each pathways.

From these results, we have made the following observations:

(a) As shown by results in Table 1, the predicted number of cancer cases are very close to the observed cases in all pathways. This indicates that the model fits the data well and that one can safely assume that the human colon cancer can be described by a mixture model of five pathways. The AIC (Akaike Information Criterion) and the BIC (Bayesian Information Criterion) from the model are 4851 and 4847 which are smaller than the AIC of 5864 and the BIC value of 5865 from a single pathway four-stage model respectively (Luebeck and Moolgavkar, 2002). This shows that the

Table 2. Estimates of parameters for each pathway.

	Sporadic LOH pathway			
	I_0	I_1	I_2	I_3
Mutation rate	1.72E-06	8.32E-06	7.02E-05	8.18E-04
	±1.86E-06	±1.37E-06	±2.21E-05	±1.63E-05
Proliferation rate	0	0	9.86E-03	3.07E-02
	N/A	N/A	±6.81E-03	±7.69E-03
Growth limiting para.	0	0	5.17E-02	5.17E-02
	N/A	N/A	±3.78E-02	±3.78E-02
	Familial LOH pathway			
	$I_0{}^*$	I_1	I_2	I_3
Mutation rate	9.60E-04	5.80E-09	7.02E-05	8.18E-04
	±5.21E-04	±2.92-10	±2.21E-05	±1.63E-05
Proliferation rate	0	0	9.86E-03	3.07E-02
	N/A	N/A	±6.81E-03	±7.69E-03
Growth limiting para.	0	0	5.17E-02	5.17E-02
	N/A	N/A	± 3.78E-02	±3.78E-02
	FAP pathway			
	I_1	I_2	I_3	
Mutation rate	3.64E-09	7.02E-05	8.18E-04	
	±2.73E-10	±2.21E-05	±1.63E-05	
Proliferation rate	0	9.86E-03	3.07E-02	
	N/A	±6.81E-03	±7.69E-03	
Growth limiting para.	N/A	5.17E-02	5.17E-02	
	N/A	±3.78E-02	±3.78E-02	

(Continued)

Table 2. (Continued)

MSI pathway

	J_0	J_1	J_2	J_3	J_4
Mutation rate	9.39E-06 ±8.40E-07	6.09E-06 ±2.47E-06	5.17E-05 ±2.55E-05	5.51E-04 ±7.25E-05	5.39E-04 ±2.51E-04
Proliferation rate	0	0	0	3.56E-02 ±1.56E-02	4.46E-02 ±1.20E-03
Growth limiting para.	N/A N/A N/A	N/A N/A N/A	N/A N/A N/A	5.70E-02 ±3.01E-02	5.70E-02 ±3.01E-02

HNPCC pathway

	J_1	J_2	J_3	J_4
Mutation rate	4.32E-09 ±1.10E-10	5.17E-05 ±2.55E-05	5.51E-04 ±7.25E-05	5.39E-04 ±2.51E-04
Proliferation rate	0	0	3.56E-02 ±1.56E-02	4.46E-02 ±1.20E-03
Growth limiting para.	N/A N/A N/A	N/A N/A N/A	5.70E-02 ±3.01E-02	5.70E-02 ±3.01E-02

I_0^*: Mutation with low penetration.

Table 3. Predicted number of different pathways.

Age group	Observed cases	Predicted colon cancer				
		LOH	FAM	FAP	MSI	HNPCC
0 – 4	2	0	0	0	0	0
5 – 9	2	4	0	1	0	0
10 – 14	35	22	2	6	0	0
15 – 19	84	77	6	17	0	1
20 – 24	242	206	16	40	1	2
25 – 29	615	486	40	87	3	5
30 – 34	1305	991	86	173	9	13
35 – 39	2466	1767	162	305	22	30
40 – 44	4306	2973	289	498	51	67
45 – 49	7460	4753	495	716	109	141
50 – 54	12780	7595	857	887	236	299
55 – 59	19532	11689	1392	872	489	609
60 – 64	27678	17756	2238	747	1006	1212
65 – 69	36698	26186	3414	519	2036	2288
70 – 74	42850	34178	4423	223	3766	3590
75 – 79	43969	35933	4361	84	5939	3866
80 – 84	35820	24400	2781	40	6500	1807
≥85	32499	19956	4454	36	7597	270

multiple pathway model fits better than the single pathway four-stage model as proposed by Luebeck and Moolgavkar (2002).

(b) From Table 3, it is observed that for the sporadic LOH pathway, the largest number of cancer cases appeared in the age group between 75 and 79 years old while the mean time to cancer onset is 76 years old. For the FAP pathway, the largest incidence of cancer cases appeared in the 50 to 54 years old and the mean time to cancer is 48 years old. It appeared that there are 25 years difference between the FAP and sporadic LOH pathways. For the familial LOH pathway, the largest cases are in the 70–74 years old with the mean time to cancer around 72 years old. That is, the familial LOH pathway appeared to be between the FAP and sporadic LOH pathways.

(c) Comparing with the LOH pathways, the MSI pathways have one more stage than the LOH pathways respectively. For the MSI pathway, the largest cancer cases appeared in the last age group (age ≥ 85) with the mean time to cancer around 86 years old. For the HNPCC pathway, the largest number of

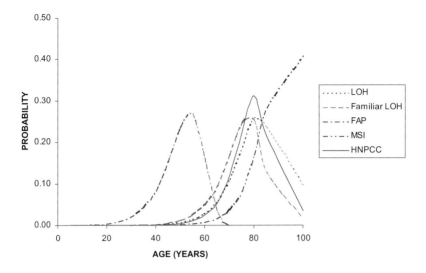

Fig. 3. Probability plot of cancer onset for the five pathways

cancer cases is in the 75–79 years old with the mean time to cancer around 75 years old.

(d) Results in Table 2 showed that the mis-match genes have increased the mutation rates (around 10^{-4} of (J_3, J_4) stage cells in the MSI and HNPCC pathways about ten to 15 times than those (about 10^{-6}) of (I_2, I_3) stage cells in the LOH pathways. As shown in Table 2, these increases have speeded up the time to cancer in MSI and HNPCC pathways by about five to ten years. On the other hand, the mis-match genes do not seem to have increased the proliferation rates (birth rate-death rate) of cells in the MSI and HNPCC pathways as compared to those of the I_i ($i = 2, 3$) cells in the LOH pathways.

(e) From results in Table 2, it is observed that the proliferating rate of I_3 cells is almost double that of I_2 cells in the LOH pathways; on the other hand, it is noted that the proliferating rate of J_4 cells is considerably smaller than that of J_3 cells, due probably to the apoptotic effect of the *Bax* gene in the MSI and HNPCC pathways.

(f) Results in Table 2 showed that the transition from $I_0 \to I_1$ in the LOH pathway had increased the transition rate of $I_1 \to I_2$ in the LOH pathways about 100 times; but this is not the case in the FAP and familial LOH pathways. Also, we observed that in the MSI and HNPCC pathways, the mutation rate of $J_0 \to J_1$ is about the same as that from J_1 to J_2.

7. CONCLUSIONS AND DISCUSSION

To model human colon cancer, to date only single pathway multi-stage models have been proposed (Luebeck and Moolgavkar, 2002). However, recent studies of cancer molecular biology have indicated very clearly that human colon cancer is developed through multiple pathways (Chapelle, 2004; Fodde *et al.*, 2001; Fodde *et al.*, 2001; Green and Kaplan, 2003; Hawkins and Ward, 2001; Sparks *et al.*, 1998; Ward *et al.*, 2001). This indicates that single pathway models are not realistic and hence may lead to incorrect prediction and confusing results in developing preventing and control procedures. To incorporate recent biological results, in this chapter we have developed a stochastic and state space model for human colon cancer involving 5 pathways. Using these models we have developed a generalized Bayesian procedure to estimate the unknown parameters and to predict future cancer cases. We have applied these models and procedure to the NCI SEER data. Our results showed that the proposed multiple pathway model fitted better than the single pathway four-stage model as proposed by Luebeck and Mookgavkar (2002). (The respective AIC and BIC for the multiple pathway model are 4847 and 4851 which are smaller than the AIC (5864) and BIC (5865) of the single pathway four-stage model.) respectively.

Because the number of stages and the mutation rates of intermediate cells in different pathways are different and different drugs may affect different pathways, we believe that it is important to incorporate biological mechanisms into mathematical models. Thus, we believe that our model will be very useful for developing efficient prevention and controlling procedures for human colon cancer and for prediction of future human colon cancer. In this preliminary study, we have not yet compared the multiple pathway model with the single pathway model regarding prediction of future cancer cases and evaluation of treatment protocols for human colon cancer. This will be our future research, we will not go any further here.

ACKNOWLEDGMENTS

The research of this paper is supported by a research grant from NCI/NIH, grant number R15 CA113347-01.

References

1. Alberici P, Jagmohan-Changur S, De Pater E, *et al*. Smad4 haplo-insufficiency in mouse models for intestinal cancer. *Oncogene* 2006; **25**: 1841–1851.
2. Baylin SB and Ohm JE. Epigenetic silencing in cancer — a mechanism for early oncogenic pathway addiction. *Nat. Rev. Cancer* 2006; **6**: 107–116.
3. Box GEP and Tiao GC. *Bayesian Inference in Statistical Analysis*. Addison-Wesley, Reading, MA, 1973.
4. Breivik J and Gaudernack G. Genomic instability, DNA methylation, and natural selection in colorectal carcinogenesis. *Semin. Cancer Biol.* 1999; **9**: 245–254.
5. de la Chapelle A. Genetic predisposition to colorectal cancer. *Nat. Rev. Cancer* 2004; **4**: 769–780.
6. Chu KC. Multi-event model for carcinogenesis: a model for cancer causation and prevention. In: *Carcinogenesis: A Comprehensive Survey Volume 8: Cancer of the Respiratory Tract-Predisposing Factors*. Mass MJ, Ksufman DG, Siegfied JM, Steel VE and Nesnow S. (eds.) Raven Press, New York, 1985, pp. 411–421,
7. Fodde R, Smit R and Clevers H. APC, signal transduction and genetic instability in colorectal cancer. *Nat. Rev. Cancer* 2001; **1**: 55–67.
8. Fodde R, Kuipers J, Rosenberg C, *et al*. Mutations in the APC tumor suppressor gene cause chromosomal instability. *Nat. Cell Biol.* 2001; **3**: 433–438.
9. Lengauer C, Kinzler KW and Vogelstein B. Genetic instability in colorectal cancers. *Nature* 1997; **386**: 623–627.
10. Green RA and Kaplan KB. Chromosomal instability in colorectal tumor cells is associated with defects in microtubule plus-end attachments caused by a dominant mutation in APC. *J. Cell Biol.* 2003; **163**: 949–961.
11. Hawkins NJ and Ward RL. Sporadic colorectal cancers with micro-satellite instability and their possible origin in hyperplastic polyps and serrated adenomas. *J. Natl. Cancer Inst.* 2001; **93**: 1307–1313.

12. Herrero-Jimenez P, Thilly G, Southam PJ, et al. Mutation, cell kinetics, and subpopulations at risk for colon cancer in the United States. Mutat. Res. 1998; **400**: 553–578.
13. Hisamuddin IM and Yang VW. Genetics of colorectal cancer. Med. Gen. Med. 2004; **6**(3): 13.
14. Jackson AL and Loeb LA. The contribution of endogenous sources of DNA damage to the multiple mutations in cancer. Mutat. Res. 2001; **477**: 7–21.
15. Jass JR, Biden KG, Cummings MC, Simms LA, et al. Characterization of a subtype of colorectal cancer combining features of the suppressor and mild mutator pathways. J. Clin. Pathol. 1999; **52**: 455–460.
16. Jessup JM, Gallic GG and Liu B. The molecular biology of colorectal carcinoma: importance of the Wg/Wnt signal transduction pathway. In: *The Molecular Basis of Human Cancer*. Coleman WB and Tsongalis GJ (eds.) Humana Press, Totowa, New Jersey, 2002, Chapter 13, pp. 251–268.
17. Jones PA and Baylin SB. The fundamental role of epigenetic events in cancer. Nat. Rev. Genet. 2002; **3**: 415–428.
18. Koinuma K, Yamashita Y, Liu W, Hatanaka H, Kurashina K, Wada T, et al. Epigenetic silencing of AXIN2 in colorectal carcinoma with microsatellite instability. Oncogene 2006; **25**: 139–146.
19. Laurent-Puig P, Blons H and Cugnenc P-H. Sequence of molecular genetic events in colorectal tumorigenesis. Eur. J. Cancer Prevent. 1999; **8**: S39–S47.
20. Lengauer C, Kinzler KW and Vogelstein B. Genetic instability in colorectal cancers. Nature 1997; **386**: 623–627.
21. Little MP. Are two mutations sufficient to cause cancer? Some generalizations of the two-mutation model of carcinogenesis of Moolgavkar, Venson and Knudson, and of the multistage model of Armitage and Doll. Biometrics 1995; **51**: 1278–1291.
22. Loeb KR and Loeb LA. Significance of multiple mutations in cancer. Carcinogenesis 2000; **21**: 379–385.
23. Lynch CJ and Milner J. Loss of one p53v allele results in four-fold reduction in p53 mRNA and protein: a basis for p53 haplo-insufficiency. Oncogene 2006; **25**: 3463–3470.
24. Peltomaki P. Deficient DNA mismatch repair: a common etiologic factor for colon cancer. Hum. Mol. Genet. 2001; **10**: 735–740.
25. Potter JD. Colorectal cancer: molecules and population. J. Natl. Cancer Inst. 1999; **91**: 916–932.
26. Ries LAG, Eisner MP, Kosary CL, Hankey BF, Miller MA, Clegg L and Edwards BK (eds.). *SEER Cancer Statistic Review, 1973–1998*. National Cancer Institute, Bethesda, MD, 2001.

27. Sparks AB, Morin PJ, Vogelstein B and Kinzler KW. Mutational analysis of the APC/beta-catenin/Tcf pathway in colorectal cancer. *Cancer Res.* 1998; **58**: 1130–1134.
28. Stoler DL, Chen N, Basik M, *et al.* The onset and extent of genomic instability in sporadic colorectal tumor progression. *Proc. Natl. Acad. Sci. USA* 1999; **96**: 15121–15126.
29. Tan WY. *Stochastic Models of Carcinogenesis*. Marcel Dekker, New York, 1991.
30. Tan WY. *Stochastic Models with Applications to Genetics, Cancers, AIDS and Other Biomedical Systems*. World Scientific, Singapore and River Edge, New Jersey, 2002.
31. Tan WY and Chen CW. Stochastic modelling of carcinogenesis: some new insight. *Math. Comp. Model.* 1998; **28**: 49–71.
32. Tan WY and Chen CW. Cancer stochastic models. In: *Encyclopedia of Statistical Sciences*, revised ed. John Wiley and Sons, New York, 2005.
33. Tan WY, Zhang LJ and Chen CW. Stochastic modeling of carcinogenesis: state space models and estimation of parameters. *Disc. Cont. Dyn. Syst. Ser. B* 2004; **4**: 297–322.
34. Ward R, Meagher A, Tomlinson I, O'Connor T, *et al.* Microsatellite instability and the clinicopathological features of sporadic colorectal cancer. *Gut* 2001; **48**: 821–829.
35. Yang GL and Chen CW. A stochastic two-stage carcinogenesis model: a new approach to computing the probability of observing tumor in animal bioassays. *Math. Biosci.* 1991; **104**: 247–258.

Chapter 12

CANCER RISK ASSESSMENT OF ENVIRONMENTAL AGENTS BY STOCHASTIC AND STATE SPACE MODELS OF CARCINOGENESIS

Wai Y. Tan, Chao W. Chen and Li J. Zhang

This article illustrates how to use stochastic models and state space models to assess risk of environmental agents. In this state space model, the stochastic system model is the general stochastic multi-stage model of carcinogenesis whereas the observation model is a statistical model based on cancer incidence data. To analyze the stochastic system model, in this chapter we introduce the stochastic equations for the state variables and derive the probability distributions of these variables. In this chapter we also introduce the genetic algorithm, the multi-level Gibbs sampling procedure and the predicted inference procedure to estimate the unknown genetic parameters and to predict the state variables. As an application, the model and the method are used to illustrate how the arsenic in drinking water induces the bladder cancer in human beings using cancer incidence data given in Morale *et al.* (2000). Our analysis clearly indicated that in the induction of human bladder cancer the arsenic in drinking water was both an initiator and a weak promoter. These results are not possible by purely statistical methods.

Keywords: Cancer risk assessment, dose response curves, extended k-stage multi-event model, genetic algorithm, Gibbs sampling procedures, observation model, predicted inference procedures, stochastic difference equations, stochastic system model, state space model.

1. INTRODUCTION

Human beings are constantly exposed to environmental agents such as arsenic in drinking water. For environmental health and public safety, it is important to assess if these agents can cause cancer and/or other serious diseases.

To answer these questions, statisticians have employed different statistical models and methods to analyze exposure data from human beings; see for example, Morales *et al.* (2000). As illustrated in Morales *et al.* (2000), these methods have ignored information from cancer biology. Furthermore, given that these carcinogens can cause cancer, purely statistical analysis can not provide information about the mechanism on the action of the carcinogens. For example, it is not possible by purely statistical method to classify the carcinogens into initiators or promoters. These latter information are extremely important for developing efficient prevention procedures.

Van Ryzin (1980) has shown that the same data can be fitted equally well by different models; yet different models gave very different results in risk assessment of environmental agents. It follows that to assess risk of environmental agents at low dose levels, it is important to use biologically supported stochastic models of carcinogenesis. This has led EPA to revise guidelines for assessing effects of environmental agents. This new 1996 EPA cancer guidelines (EPA, 1996) call for the use of biologically supported model if sufficient biological information is available to support such a model.

Based on this guideline, in this chapter we will propose new models and methods to analyze human exposure data. This new models and methods are based on most recent cancer biology as proposed in Tan and Chen (1998, 2005). Thus, for assessing cancer risk of environmental agents, we will use the extended stochastic multi-event model of carcinogenesis proposed by Tan and Chen (1998). This model is described in Section 2.

In Section 3, we will introduce the genetic algorithm, the multi-level Gibbs sampling procedures and the predictive inferences procedures to estimate the unknown parameters (mutation rates, birth rates and death rates of initiated cells). In Section 4, to validate the model and to predict the state variables, we will develop state space models combining the extended multi-event model with the statistic model based on the number of cancer cases or number of deaths from cancer cases. In Sections 5 and 6, we will illustrate how to develop confidence intervals for probabilities of developing cancers and how to develop dose-response curves for cancer risk of environmental agents. In Section 7, to illustrate the procedures and model, we will apply the model and the methods to the data in Morale et al. (2000). Finally in Section 8, we give some conclusions.

2. A GENERAL STOCHASTIC MODEL OF CARCINOGENESIS

The most general single-pathway model for carcinogenesis is the extended k-stage ($k \geq 2$) multi-event model to be referred to as the general stochastic k-stage model of carcinogenesis; see Tan and Chen (1998, 2005). This model views carcinogenesis as the end point of $k(k \geq 2)$ discrete, heritable and irreversible events (mutations or genetic or epigenetic changes) with intermediate cells subjected to stochastic proliferation and differentiation and with cancer tumors developed from primary last-stage cells by clonal expansion. This model is an extension of the model proposed by Chu (1985) and Little (1995); see also Tan (1991, Chapter 6).

To illustrate, denote by $N(I_0)$ the normal stem cell, I_j the jth stage initiated cell ($j = 1, \ldots, k$). Then the general stochastic k-stage model of carcinogenesis is characterized by:

(a) $N \rightarrow I_1 \rightarrow I_2 \rightarrow \cdots \rightarrow I_k$.
(b) The I_j cells subject to stochastic proliferation (birth) and differentiation (death) and mutation to generate I_{j+1} cells.
(c) Cancer tumors develop from primary I_k cells by clonal expansion (i.e., stochastic birth-death process), where a primary I_k cell is an I_k cell which arise directly from an I_{k-1} cell.

As an example, we consider the APC-β-catenin-Tcf-myc pathway in human colon cancer (Sparks et al., 1998; Fodde et al., 2001). This pathway accounts for about 80% of all human colon cancer and is characterized by a six-stage model given in Figure 1. In Figure 1, the individual is in the first stage if one copy of the *APC* gene in chromosome 5q has been mutated or deleted or inactivated; the individual is in the second stage if the other copy of the *APC* gene in chromosome 5q has also been mutated or deleted or inactivated. The individual who has sustained the first two genetic changes is in the third stage if one copy of the *Smad4* gene in chromosome 18q

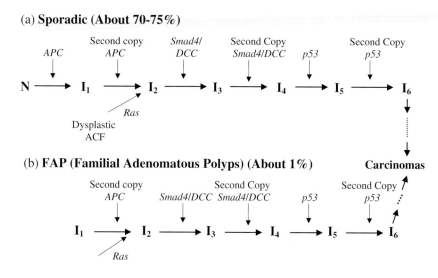

Fig. 1. LOH (loss of heterozygosity) pathway.

has been inactivated or deleted or mutated. The individual who is in the third stage enter the fourth stage if the second copy of the *Smad4* gene in 18q has been inactivated or deleted or mutated. The individual in the fourth stage enters the fifth stage if the first copy of the *p53* gene in 17p has been inactivated or deleted or mutated; the individual in the fifth stage is in the sixth stage if the second copy of the *p53* gene in 17p has been inactivated or deleted or mutated. In this model, the *ras* oncogene in chromosome 12q and the *src* gene in chromosome 20q are considered as promoter genes which promote cell proliferation of the initiated cells when these genes are mutated. Also, because of the haplo-insufficiency of the *Smad4* gene (Alberici *et al.*, 2006) and the haplo-insufficiency of the *p53* gene (Lynch and Milner, 2006), one may reduce this six-stage model into a four-stage model by combining the third stage and the fourth stage into one stage and by combining the fifth stage and the sixth stage into one stage. This may help explain why the four-stage model fits the human colon cancer better than other stage models (Luebeck and Moolgavkar, 2004).

The model in Figure 1 has also been referred to as the CIS (chromosome instability) model or LOH (loss of heterozygocity) model because it is characterized by aneuploidy/or loss of chromosome segments (chromosomal instability). Another avenue by mean of which human colon cancer

(a) **Sporadic (About 10-15%)**

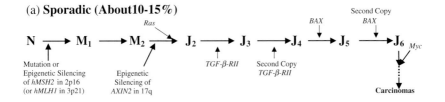

(b) **HNPCC (Hereditary Non-Polyposis Colon Cancer) (≤ 5%)**

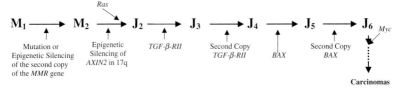

Fig. 2.   MSI-H (high level of MSI) pathway.

is generated is the MSI (microsatellite instabilty) pathway involving the mismatch repair genes *hMLH1, hMSH2, hPMS1, hPMS2, hMSH6* and *hMSH3* (Hawkins and Ward, 2001; Ward *et al.*, 2001). This pathway is described in Figure 2 and accounts for about 15% of all human colon cancer; it has also been referred to as the mutator phenotype pathway because it is initiated by the mutations or methylation of the mis-match repair genes (mostly *hMLH1* (50%) and *hMSH2* (40%)) creating a mutator phenotype to significantly increase the mutation rate 100 to 1000 times of many critical genes as *TGF-β RII, BAX, IGF2R,* or *CDX-2*. Recent biological studies (Koinuma *et al.*, 2006) have shown that both the CIS and the MSI pathways involve the Wnt signalling pathway, the TGF-β signalling and the p53-Bax signalling but different genes in different pathways are affected in these signalling processes. (In the CIS pathway, the affected gene is the *APC* gene in the Wnt signalling, the *Smad4* in the TGF-β signalling and p53 gene in the p53-Bax signalling; on the other hand, in the MSI pathway, the affected gene is the *Axin 2* gene in the Wnt signalling, the *TGF-β-Receptor II* in the TGF-β signalling and the *Bax* gene in the p53-Bax signalling.)

As another important example, we give in Figure 3 the multi-stage model for generating the squamous cell carcinoma of non-small cell lung cancer. This pathway and its molecular biological supports are given in Osaka and Takahashi (2002) and Wistuba *et al.* (2002).

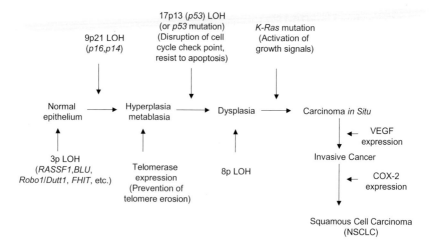

Fig. 3. Histopathology lesions and genetic pathway of squamous cell carcinoma of non-small cell lung cancer (NSCLC).

2.1. The Stochastic Difference Equations for State Variables

Given the model as above, an important issue is how to develop mathematical analysis. Because the traditional approach by Kolmogolov equations are too complicated to yield analytical results, in this chapter we will propose an alternative approach through stochastic equations (Tan, 2002; Tan et al., 2004; Tan and Chen, 1998, 2005). It is shown in Tan and Chen (1998) that this approach is equivalent to the traditional approach but is more powerful and can yield much deeper results which are not possible by the classical approach.

To illustrate the approach for the above model, let $I_i(t)$ denote the number of I_i cells at time t and let $T(t)$ denote the number of cancer tumors at time t.

Define for $j = 0, 1, \ldots, k-1$:

- $B_j(t)$ = Number of new I_j cells generated by stochastic cell proliferation (birth) of I_j cells during $(t, t+1]$,
- $D_j(t)$ = Number of death of I_j cells during $(t, t+1]$,
- $M_j(t)$ = Number of new I_j cells arising from I_{j-1} cells by mutation or some genetic changes during $(t, t+1]$, $j = 1, 2, \ldots, k$.

Let $b_i(t), d_i(t)$ and $\alpha_i(t)$ denote the birth rate, the death rate and the mutation rate of the I_i cells at time t respectively. Then,

$$[B_j(t), D_j(t), M_{j+1}] | I_j(t) \sim ML[I_j(t); b_j(t), d_j(t), \alpha_j(t)].$$

By the conservation law, the stochastic difference equations for the state variables are:

$$\Delta I_0(t) = I_0(t+1) - I_0(t) = B_0(t) - D_0(t)$$
$$= I_0(t)\gamma_0(t) + \epsilon_0(t), \qquad (1)$$

$$\Delta I_j(t) = I_j(t+1) - I_j(t) = M_{j-1}(t) + B_j(t) - D_j(t)$$
$$= \{I_{j-1}(t)\alpha_{j-1}(t) + I_j(t)\gamma_j(t)\} + \epsilon_j(t), \quad j = 1, \ldots, k-1, \qquad (2)$$

where $\gamma_j(t) = b_j(t) - d_j(t)$, $j = 0, 1, \ldots, k-1$.

In the above equations, the random noises $\{e_j(t), j = 1, \ldots, k-1\}$ are derived by subtracting the conditional mean values from the random transition variables. These random noises have expected value zero and are un-correlated with the state variables $\{N(t), I_i(t), i = 1, \ldots, k-1\}$.

2.2. The Probability of Developing Cancer Tumors

It is shown in Tan (2002) that given $\{I_{k-1}(s), s \leq t\}$ cells, the conditional probability distribution of $T(t)$ is Poisson with mean $\lambda(t) = \sum_{i=0}^{t} I_{k-1}(i)\alpha_{k-1}(i)$. Hence the probability that a normal person at time 0 will develop cancer for the first time during the age period $[t_{j-1}, t_j)$ is

$$P_j = E\{e^{-\sum_{i=0}^{t_{j-1}-1} I_{k-1}(i)\alpha_{k-1}(i)} - e^{-\sum_{i=0}^{t_j-1} I_{k-1}(i)\alpha_{k-1}(i)}\}.$$

2.3. Probability Distribution of the State Variables

Denote by $\underset{\sim}{X}(t) = \{I_0(t), I_i(t), i = 1, \ldots, k-1\}, t = 0, 1, \ldots, t_M$. Then the probability density function of $\underset{\sim}{X} = \{\underset{\sim}{X}(0), \underset{\sim}{X}(1), \ldots, \underset{\sim}{X}(t_M)\}$ is

$$P(X) = P(\underset{\sim}{X}(0)) \prod_{t=1}^{t_M} \{P[I_0(t)|I_0(t-1)]$$

$$\times \prod_{i=1}^{k-1} P[I_i(t)|I_{i-1}(t-1), I_i(t-1)]\}$$

and

$$P\{I_0(t+1)|I_0(t)\}$$
$$= \sum_{i=0}^{I_0(t)} \binom{I_0(t)}{i}\binom{I_0(t)-i}{a_0(i;t)}$$
$$\times b_0(t)^i d_0(t)^{a_0(i;t)} (1-b_0(t)-d_0(t))^{I_0(t+1)-2i} \quad (3)$$

$$P[I_r(t+1)|I_{r-1}(t), I_r(t)]$$
$$= \sum_{j=0}^{I_r(t+1)-I_r(t)} g_{r-1}(j,t) \sum_{i=0}^{I_r(t)} \binom{I_r(t)}{i}\binom{I_r(t)-i}{a_r(i,j;t)}$$
$$\times b_r(t)^i d_r(t)^{a_r(i,j;t)} (1-b_r(t)-d_r(t))^{I_r(t+1)-2i},$$
$$r = 1, \ldots, 5, \quad (4)$$

where

$$a_0(i;t) = I_0(t) - I_0(t+1) + i$$
$$a_r(i,j;t) = I_r(t) - I_r(t+1) + i + j, \quad r = 1, \ldots, k-1$$

and for $i = 0, 1, \ldots, k-1$, the $g_i(j, t)$ is the density of a Poisson distribution with mean $\lambda_i(t)$.

3. THE DATA FOR RISK ASSESSMENT OF ENVIRONMENTAL AGENTS

In most cases, the data available for cancer risk assessment of environmental agents are given by:

$$(n_i(j), y_i(j)), \quad (i = 1, \ldots, m, j = 1, \ldots, r),$$

where $n_i(j)$ = the number of person-years (i.e., number of persons at risk who have age in the age group $[t_{j-1}, t_j))$ under exposure to the carcinogen with the ith dose level and $y_i(j)$ = the number of cancer cases (or death from cancer in question) in the age group $[t_{j-1}, t_j)$ under the ith dose level; as an example, see Table 1 in Morale et al. (2000) for cancer risk assessment of arsenic in drinking water in Taiwan. In this data, the $n_i(j)$'s are fixed numbers but the $y_i(j)$ is an outcome of a random variable $Y_i(j)$, where

$$Y_i(j)|n_i(j) \sim \text{Binomial}\{n_i(j), P_i(j)\},$$

with $P_i(j)$ being the P_j under the ith exposure dose level. When $n_i(j)$ is very large and $P_i(j)$ very small, then

$$Y_i(j)|n_i(j) \sim \text{Poisson}\{\lambda_i(j) = n_i(j)P_i(j)\}. \qquad (5)$$

Given the data $\{n_i(j), y_i(j), j = 1, \ldots, r, i = 1, \ldots, m\}$, the likelihood is $L = \prod_{i=1}^{m} L_i$ with $L_i = L_i(\Theta_i|Y_i)$ being the likelihood from the ith dose level:

$$L_i(\Theta_i|Y_i) = \prod_{j=1}^{k} f_i(y_i(j), \lambda_i(j)), \qquad (6)$$

where $f_i(y_i(j), \lambda_i(j))$ is the density of a Poisson variable with mean $\lambda_i(j)$, Θ_i the collection of all unknown parameters (i.e., mutation rates, birth rates, death rates of I_r cells) under the ith dose level and $Y_i = \{y_i(j), j = 1, \ldots, r\}$.

4. STATE SPACE MODELS OF CARCINOGENESIS AND THE PREDICTION OF STATE VARIABLES

To predict the state variables (i.e., the numbers of the intermediate initiated cancer cells) and to validate the model, we will develop state space models for risk assessments of environmental agents.

State space models (Kalman filter models) of stochastic systems are stochastic models consisting of two sub-models: the stochastic system model which is the stochastic model of the system and the observation model which is a statistical model based on some data from the system. Thus, the state space model of the system takes into account the basic mechanisms of the system and the random variation of the system through its stochastic system model and incorporate all these into the observed data from the system; furthermore, it validates and upgrades the stochastic model through its observation model and the observed data of the system. It is advantageous over both the stochastic model and the statistical model when used alone since it combines information and advantages from both of these models. (For more specific advantages of state space models, see Tan (2002, Chapters 8 and 9).)

5. A STATE SPACE MODEL FOR CANCER RISK ASSESSMENT

For the state space model based on data in Table 1, the stochastic system model is given in Section 2 and is represented by a system of stochastic difference equations and the probability distributions of the state variables describing the system for the numbers of initiated cells and cancer tumors; the observation model is represented by a statistical model based on available data in Table 1 from the system as described in Section 3.

5.1. The Stochastic System Model and Probability Distributions

For implementing the Gibbs sampling procedures and EM algorithm, we expand the model by introducing dummy random variables $U(t) = \{M_{i-1}(t), B_i(t), i = 1, \ldots, k-1\}$ and put $\underset{\sim}{U} = \{U(t), t = 0, 1, \ldots, t_M - 1\}$.

Then, the conditional density of $\{X, U\}$ given $\underset{\sim}{X}(0)$ is:

$$P\{X, U | \underset{\sim}{X}(0)\} = \prod_{t=1}^{t_M} P\{\underset{\sim}{X}(t) | \underset{\sim}{X}(t-1), \underset{\sim}{U}(t-1)\}$$
$$\times P\{\underset{\sim}{U}(t-1) | \underset{\sim}{X}(t-1)\}. \quad (7)$$

In the above distribution,

$$P\{\underset{\sim}{U}(t)|\underset{\sim}{X}(t)\} = C_1(t) \prod_{i=1}^{k-1} g_{i-1}\{M_{i-1}(t); t\}[b_i(t)]^{B_i(t)}$$

$$\times [1 - b_i(t)]^{I_i(t) - B_i(t)}, \tag{8}$$

where $C_1(t) = \prod_{i=1}^{k-1} \binom{I_i(t)}{B_i(t)}$; and

$$P\{\underset{\sim}{X}(t+1)|\underset{\sim}{X}(t), \underset{\sim}{U}(t)\} = \prod_{i=1}^{k-1} \binom{I_i(t) - B_i(t)}{\eta_i(t)} \left[\frac{d_i(t)}{1 - b_i(t)}\right]^{\eta_i(t)}$$

$$\times \left[1 - \frac{d_i(t)}{1 - b_i(t)}\right]^{\zeta_i(t)},$$

where

$$\eta_i(t) = I_i(t) - I_i(t+1) + M_{i-1}(t) + B_i(t), \quad i = 1, \ldots, k-1,$$
$$\zeta_i(t) = I_i(t+1) - M_{i-1}(t) - 2B_i(t), \quad i = 1, \ldots, k-1.$$

The above distribution results indicate that given the parameters and given X, one can readily generate U; similarly, given the parameters and given U, one can readily generate X.

5.2. The Observation Model

For the data sets in Section 3, the observation model is then given by the conditional probability distribution of $y_i(j)$ given $n_i(j)$ and given the state variables.

As shown in Tan et al. (2004), given $\underset{\sim}{I}_{k-1}(i, j) = \{I_{k-1}(t; r, i, j), 0 \leq t \leq t_j, r = 1, \ldots, y_i(j)\}$, the density of the conditional distribution of $y_i(j)$ given $\underset{\sim}{I}_{k-1}(i, j)$ is

$$f_Y\{y_i(j)|\underset{\sim}{I}_{k-1}(i,j)\}$$

$$= \frac{1}{y_i(j)!} \exp\{-n_i(j)P_{ij}\} \prod_{r=1}^{y_i(j)} \{n_i(j)e^{-\bar{X}_r(t_{j-1};i,j)}[1-e^{-\bar{Z}_r(j;i)}]\}$$

$$= \frac{1}{y_i(j)!} \exp\{y_i(j)\log n_i(j) - \sum_{r=1}^{y_i(j)}[\bar{X}_r(t_{j-1};i,j)$$

$$- \log(1-e^{-\bar{Z}_r(j;i)})] - n_i(j)P_{ij}\}, \qquad (9)$$

where

$$\bar{X}_r(t;i,j) = \sum_{l=1}^{t} I_{k-1}(l;r,i,j)\alpha_{k-1}(i,l),$$

$$\bar{Z}_r(j;i) = \bar{X}_r(t_j;i,j) - \bar{X}_r(t_{j-1};i,j).$$

It can easily be shown that $Ef_Y\{y_i(j)|\underset{\sim}{I}_{k-1}(i,j)\}$ is the density of the Poisson distribution with mean $n_i(j)P_{ij}$.

For the above state space model, we will use the genetic algorithm, predictive inferences and multi-level Gibbs sampling procedure to estimate the unknown parameters and to predict the state variables. These procedures are given in Section 5.

6. THE GENETIC ALGORITHM AND THE PREDICTED INFERENCE PROCEDURES

6.1. The Genetic Algorithm

The genetic algorithm is an optimization process based on evolution principles involving gene mutation or chromosomal aberrations, mating types (referred to as crossing over in GA), and selection based on fitness. We will use the genetic algorithm "pikaia" in the web which can be downloaded from web using the e-mail address "knapp@hao.ucar.edu" or "paulchar@hao.ucar.edu". This genetic algorithm was written in Fortran and is most convenient to apply. We will use this genetic algorithm in combination with the extended stochastic model of carcinogenesis to derive MLE of $\{\Theta_i, P_i(j), j = 1, \ldots, r, i = 1, \ldots, m\}$ by maximizing the likelihood L.

To apply this genetic algorithm to obtain MLE, we will use the expected numbers $EI_{k-1}(t)$ from an extended k-stage multi-event model and the deviance function to construct the fitness function $-\chi_i$, where χ_i is the deviance function for Poisson distribution defined by:

$$\chi_i = 2 \sum_{i=1}^{r} \sum_{j=1}^{l} \{[\lambda_i(j) - y_i(j)] - y_i(j) \log[\lambda_i(j)/y_i(j)]\}. \quad (10)$$

To derive efficient estimates of parameters and to predict the observed numbers, we will use the following predictive inference procedure to improve on the estimates of the parameters and the prediction.

6.2. The Predictive Inference Procedures

The predictive inference procedure iterates through the following loops:

(a) Given $(Y_i, i = 1, \ldots, m)$, use genetic algorithm to derive estimates $\Theta = (\Theta_i, i = 1, \ldots, m)$ by maximizing L and denote the estimates of $\Theta = (\Theta_i, i = 1, \ldots, m)$ as $\Theta^{(*)} = (\Theta_i^{(*)}, i = 1, \ldots, m)$.

(b) Given $\Theta = \Theta^{(*)}$ from (a), use the extended multi-event model and the Poisson distribution as given above to generate new $y_i(j)$. Denote the newly generated $y_i(j)$ by $\hat{y}_i(j)$.

(c) Using $\{n_i(j), \hat{y}_i(j), j = 1, \ldots, r, i = 1, \ldots, m\}$ as new data, go to step (a) and repeat the process until convergence.

The estimates of $\{\Theta_i, i = 1, \ldots, m\}$ at convergence will be the improved estimates of $\{\Theta_i, i = 1, \ldots, m\}$; the generated $\hat{y}_i(j)$ at convergence is the improved predicted value of the observation $y_i(j)$.

The convergence of the above procedures can be proved by using similar procedures given in Tan (2002, Chapter 3).

We have developed an efficient computer program in Fortran to implement the above predictive procedures. Applying these procedures to some generated data, the above procedures can accurately pick up the correct model and give very close estimates of the unknown parameters in the stochastic multi-event models.

7. DEVELOPING CONFIDENCE INTERVALS FOR PROBABILITIES OF DEVELOPING CANCER BY GENETIC ALGORITHM

Using the predictive procedure as given in Section 4, one can readily derive confidence intervals for the probabilities $\{P_i(j), \sum_{j=1}^{r} P_i(j)\}$ or functions $g(P_i(j), j = 1, \ldots, r)$ of $P_i(j)$. The basic procedure is given as follows:

(1) Given data $\{n_i(j), y_i(j), j = 1, \ldots, r, i = 1, \ldots, m\}$, use the predictive inference procedures in Section 4 to derive estimates of $\{\Theta_i, i = 1, \ldots, m\}$.
(2) Using the estimates in (1) to generate a large sample of $\{Y(n) = (\hat{y}_i(j)(n), j = 1, \ldots, r, i = 1, \ldots, m), n = 1, \ldots, N\}$.
(3) For each n, use the data generated in (2) and the genetic algorithm to derives estimate of the parameters and estimates $\hat{P}_i(j)(n)$ of $P_i(j)$ and hence estimate $\hat{g}_i(n)$ of $g(P_i(j), j = 1, \ldots, r)$.
(4) Plot the $\{\hat{g}_i(n), n = 1, \ldots, N\}$ over n and from the plot obtain the 0.25 percentile $a_i = g_i(0.025)$ and the 0.975 percentile $b_i = g_i(0.975)$. A 0.95% confidence interval for $g(P_j(i), j = 1, \ldots, r)$ is (a_i, b_i).

8. DEVELOPING DOSE-RESPONSE CURVES OF ENVIRONMENTAL AGENTS BY GENETIC ALGORITHM

To derive dose-response curves for the probability of developing cancer by the environmental agent, for each dose level we will use the genetic algorithm to derive estimates of $P_i(j)$ and $P_i = \sum_{j=1}^{r} P_i(j)$. Then we will use $P_i(j)$ and P_i to assess how the dose level $d(i)$ affects $P_i(j)$ and P_i. These are the so-called dose response curves and these type of curves reflect the basic mechanism by means of which the environmental agent affects the process of carcinogenesis. For example, if the agent is an initiator so that it affects the process only through the increase of mutation rates of the cells, then one would expect that $P_i(j) = \exp\{h(d(i))g(t_{j-1}, t_j)\}$, where $h(d(i))$

is a function of $d(i)$ only and $g(t_{j-1}, t_j)$ a function of t_{j-1} and t_j only; on the other hand, if the agent is a complete carcinogen so that it affects both the mutation rate of normal stem cells and the proliferation rate of the initiated cells, then the dose response curve is very complex and may even decrease as dose level increases until some threshold dose, due to apoptosis and/or due to lack of cell immortalization (see Hanahan and Weiberg (2000)). Notice that the data in Table 1 of Morale et al. (2000) revealed that during the dose level interval [0, 300], the observed deaths from cancers decreased as the dose level increases; these observed results implies that the agent might be a complete carcinogen and that the last stage involves genes for apoptosis and/or immortalization.

9. AN APPLICATION AND ILLUSTRATION

In this section we apply the above model and methods to the data in Table 1. This analysis will demonstrate how the arsenic in drinking water would affect human health by inducing bladder cancer. This was not done before

Table 1. Person-years at risk by age, sex and arsenic level with observed number of deaths from bladder cancer.

Arsenic (μg/L)	Age (year)*				Total
	20–30	30–49	50–69	≥70	
Male					
< 100	35,818(0)	34,196(1)	21,040(6)	4,401(10)	95,455(17)
100–299	18,578(0)	16,301(0)	10,223(7)	2,166(2)	47,268(9)
300–599	27,556(0)	25,544(5)	15,747(15)	3,221(12)	72,068(32)
≥ 600	16,609(0)	15,773(4)	8,573(15)	1,224(8)	42,179(27)
Total	98,561(0)	91,814(10)	55,583(43)	11,012(32)	256,970(85)
Female					
< 100	27,901(0)	32,471(3)	21,556(9)	5,047(9)	86,975(21)
100–299	13,381(0)	15,514(0)	11,357(9)	2,960(2)	43,212(11)
300–599	19,831(0)	24,343(0)	16,881(19)	3,848(11)	64,903(30)
≥ 600	12,988(0)	15,540(0)	9,084(21)	1,257(7)	38,869(28)
Total	74,101(0)	87,868(3)	58,878(58)	13,112(29)	233,959(90)

Notes: *–Values in parentheses are number of deaths from bladder.

although statisticians have tried to use statistical models to assess cancer risks using this data. (In the past, biologically supported models were not attempted to analyze this data, not to mention any analysis using biologically supported cancer models.)

Applying the above model and methods, we found that a three-stage stochastic multi-stage model is best to describe the carcinogenesis process by mean of which the bladder cancer develops. Given in Tables 2 and 4 are estimates of mutation rates and cell proliferation rates of different stage initiated cells. Given in Tables 3 and 5 are the estimates of the estimate of probabilities to cancer tumor under different dose levels. Plotted in Figures 4 and 5 are dose response curves for these probabilities.

Table 2. Estimates of parameters for the three-stage homogeneous model (male).

Parameters	Dose			
	50	200	450	800
$\lambda_0(i)$	12.76 ±3.37	14.66 ±3.74	15.79 ±3.28	17.84 ±4.07
$b_1(i)$	$2.32E-02$ ±$1.31E-05$	$2.42E-02$ ±$2.21E-05$	$2.47E-02$ ±$9.18E-06$	$2.51E-02$ ±$8.89E-06$
$d_1(i)$	$5.01E-03$ ±$7.83E-06$	$4.99E-03$ ±$8.33E-06$	$5.00E-03$ ±$3.92E-06$	$5.00E-03$ ±$3.92E-06$
$b_1(i)-d_1(i)$	$1.82E-02$ ±$1.53E-05$	$1.92E-02$ ±$2.36E-05$	$1.97E-02$ ±$9.98E-06$	$2.01E-02$ ±$9.72E-06$
$\alpha_1(i)$	$6.52E-07$ ±$7.91E-08$	$4.07E-07$ ±$9.56E-08$	$6.40E-07$ ±$5.08E-8$	$7.74E-07$ ±$4.72E-8$
$b_2(i)$	$3.88E-04$ ±$3.07E-04$	$8.25E-04$ ±$8.67E-04$	$5.24E-03$ ±$6.00E-04$	$1.24E-02$ ±$8.61E-04$
$d_2(i)$	$8.01E-03$ ±$1.45E-03$	$5.75E-03$ ±$2.36E-03$	$4.64E-03$ ±$5.25E-04$	$4.03E-03$ ±$4.57E-04$
$b_2(i)-d_2(i)$	$-7.63E-03$ ±$1.48E-03$	$-4.93E-03$ ±$2.52E-03$	$6.05E-04$ ±$7.97E-04$	$8.41E-03$ ±$9.74E-04$
$\alpha_2(i)=\alpha_2$	$5.75E-06 \pm 6.42E-09$			

Notes: $\lambda_{00} = 7.86 \pm 0.75$ $\lambda_{01} = 0.12 \pm 1.64\text{E}-02$
$b_{10} = 0.0205 \pm 0.0001$ $b_{11} = 0.0007 \pm 0.0001$
$d_{10} = 0.005 \pm 0.0002$ $d_{11} = 0 \pm 0.0001$
$\alpha_{10} = 3.86\text{E}-07 \pm 3.19E-07$ $\alpha_{11} = 8.48\text{E}-02 \pm 1.43\text{E}-01$
$b_{20} = -0.0172 \pm 0.0097$ $b_{21} = 0.0040 \pm 0.0017$
$d_{20} = 0.0137 \pm 0.0005$ $d_{21} = -0.0015 \pm 0.0001$

Table 3. Estimates of probabilities (male).

Age	Dose			
	50	200	450	800
20–30	$2.79E-06$ $\pm 2.75E-06$	$1.08E-05$ $\pm 7.18E-06$	$1.81E-05$ $\pm 6.05E-06$	$2.41E-05$ $\pm 9.83E-06$
30–50	$5.26E-05$ $\pm 9.55E-06$	$8.59E-05$ $\pm 2.62E-05$	$1.33E-04$ $\pm 2.20E-05$	$2.22E-04$ $\pm 4.36E-05$
50–70	$4.61E-04$ $\pm 4.07E-05$	$3.42E-04$ $\pm 4.89E-05$	$8.07E-04$ $\pm 7.52E-05$	$1.31E-03$ $\pm 1.41E-04$
>70	$2.52E-03$ $\pm 2.66E-04$	$1.15E-03$ $\pm 3.10E-04$	$4.69E-03$ $\pm 2.94E-04$	$7.60E-03$ $\pm 4.40E-04$

Table 4. Estimates of parameters for the three-stage homogeneous model (female).

Parameters	Dose			
	50	200	450	800
$\lambda_0(i)$	64.33 ± 7.44	97.44 ± 9.84	109.92 ± 9.30	119.77 ± 11.32
$b_1(i)$	$1.76E-02$ $\pm 1.07E-05$	$1.92E-02$ $\pm 1.16E-05$	$2.01E-02$ $\pm 5.58E-06$	$2.08E-02$ $\pm 4.96E-06$
$d_1(i)$	$4.99E-03$ $\pm 5.87E-06$	$5.00E-03$ $\pm 6.68E-06$	$5.00E-03$ $\pm 3.05E-06$	$5.00E-03$ $\pm 2.89E-06$
$b_1(i) - d_1(i)$	$1.26E-02$ $\pm 1.22E-05$	$1.42E-02$ $\pm 1.33E-05$	$1.51E-02$ $\pm 6.36E-06$	$1.58E-02$ $\pm 5.74E-06$
$\alpha_1(i)$	$2.59E-07$ $\pm 3.20E-08$	$2.00E-07$ $\pm 3.68E-08$	$2.54E-07$ $\pm 1.68E-08$	$2.81E-07$ $\pm 1.81E-08$
$b_2(i)$	$1.74E-02$ $\pm 1.09E-03$	$1.88E-02$ $\pm 1.71E-03$	$1.91E-02$ $\pm 6.41E-04$	$1.86E-02$ $\pm 8.18E-04$
$d_2(i)$	$4.10E-03$ $\pm 4.97E-04$	$5.11E-03$ $\pm 9.19E-04$	$5.42E-03$ $\pm 3.54E-04$	$5.48E-03$ $\pm 4.22E-04$
$b_2(i) - d_2(i)$	$1.33E-02$ $\pm 1.20E-03$	$1.37E-02$ $\pm 1.94E-03$	$1.37E-02$ $\pm 7.33E-04$	$1.31E-02$ $\pm 9.21E-04$
$\alpha_2(i) = \alpha_2$	$1.36E-06 \pm 5.87E-08$			

Notes:
$\lambda_{00} = 30.00 \pm 5.69$ $\qquad \lambda_{01} = 0.21 \pm 3.18E-02$
$b_{10} = 0.0131 \pm 0.0001$ $\qquad b_{11} = 0.0012 \pm 0.0001$
$d_{10} = 0.005 \pm 0.0001$ $\qquad d_{11} = 0 \pm 0.0001$
$\alpha_{10} = 2.03E-07 \pm 9.18E-08$ $\qquad \alpha_{11} = 3.64E-02 \pm 7.96E-02$
$b_{20} = 0.0157 \pm 0.0015$ $\qquad b_{21} = 0.0005 \pm 0.0003$
$d_{20} = 0.0022 \pm 0.0005$ $\qquad d_{21} = 0.0005 \pm 0.0001$

Table 5. Estimates of probabilities (female).

Age	Dose			
	50	200	450	800
20–30	$3.84E-06$ $\pm 1.16E-06$	$6.61E-06$ $\pm 8.07E-07$	$8.74E-06$ $\pm 5.17E-07$	$1.27E-05$ $\pm 9.25E-07$
30–50	$4.25E-05$ $\pm 1.22E-05$	$6.85E-05$ $\pm 7.93E-06$	$9.03E-05$ $\pm 5.29E-06$	$1.36E-04$ $\pm 1.04E-05$
50–70	$2.05E-04$ $\pm 5.54E-05$	$3.11E-04$ $\pm 3.42E-05$	$4.15E-04$ $\pm 2.43E-05$	$6.50E-04$ $\pm 5.31E-05$
>70	$7.71E-04$ $\pm 1.99E-04$	$1.13E-03$ $\pm 1.20E-04$	$1.56E-03$ $\pm 9.16E-05$	$2.56E-03$ $\pm 2.20E-04$

(a) The estimates of $\lambda_0(d)$ indicates that the arsenic in drinking water is an initiator to initiate the carcinogenesis process. It increase linearly with dose. Comparing results between Table 2 and Table 4, it appeared that the rates in females were at least five to seven times greater than those in males; further these differences increase with dose levels.

(b) The proliferation rates of the first stage (I_1) and second stage (I_2) initiated cells increase as dose level increases, albeit slowly. This indicates that the arsenic in drinking water is also a weak promoter for initiated stem cancer cells.

(c) The proliferation rates of I_2 cells appears to be smaller than those of I_1 cells, indicating that the rapid proliferation after two stages may have invoked the apoptosis process to slow down the proliferation. Notice that for dose levels ≤ 200, the rates in males are even negative.

(d) The estimates of the mutation rates of I_1 and I_2 cells indicate that the arsenic in drinking water would not affect the mutation process in initiated cells.

(e) The dose response curves in Figures 4 and 5 clearly indicate that the arsenic in drinking water is carcinogenic although males in higher age groups the curves may first decrease slightly due to competing death and possibly apoptosis.

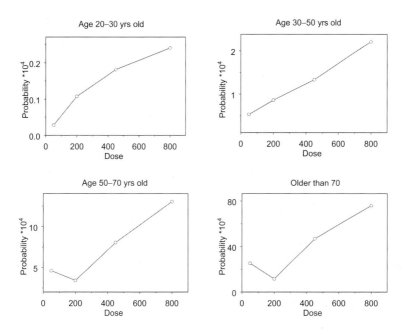

Fig. 4. Plot of the probabilities estimates to cancer under different dose levels (male).

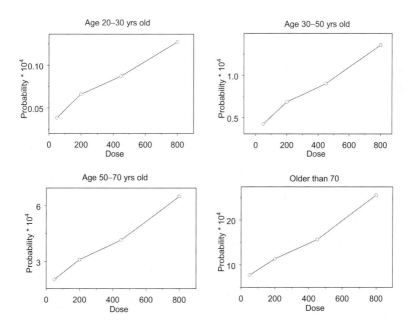

Fig. 5. Plot of the probabilities estimates to cancer under different dose levels (female).

10. CONCLUSIONS

To incorporate biologically-supported models of carcinogenesis into cancer risk assessment, in this chapter we have proposed a state space model approach to assess cancer risk of human beings to exposed environmental agents. In this chapter we have introduced the genetic algorithm, the multi-level Gibbs sampling procedures and the predictive inference procedures to analyze the model and data. We have applied the model and methods to the human cancer incidence data given in Morales *et al.* (2000) to assess cancer risk of bladder cancer for the human population exposed to arsenic in the drinking water. We have shown that the arsenic in drinking water is both an initiator and a weak promoter in inducing bladder cancer in humans. This results are not possible by traditional statistical models and methods. This analysis shows that the proposed model and methods are very promising as a general model and methods for assessing cancer risk of environmental agents. To make this model and methods to be used in a wide range of data including animal carcinogenicity experiments, further researches are definitely needed; this will be topics of our future researches.

ACKNOWLEDGMENTS

The research of this paper is supported by a research grant from NCI/NIH, grant number R15 CA113347-01.

References

1. Alberici P, Jagmohan-Changur S, De Pater E, *et al.* Smad4 haplo-insufficiency in mouse models for intestinal cancer. *Oncogene* 2006; **25**: 1841–1851.
2. Chu KC. Multi-event model for carcinogenesis: a model for cancer causation and prevention. In: *Carcinogenesis: A Comprehensive Survey Volume 8: Cancer of the Respiratory Tract — Predisposing Factors.* Mass MJ, Ksufman DG, Siegfied JM, Steel VE and Nesnow S (eds.) Raven Press, New York, 1985, pp. 411–421.
3. Fodde R, Smit R and Clevers H. APC, signal transduction and genetic instability in colorectal cancer. *Nat. Rev. Cancer* 2001; **1**: 55–67.

4. Hanahan D and Weinberg RA. The hallmarks of cancer. *Cell* 2000; **100**: 57–70.
5. Hawkins NJ and Ward RL. Sporadic colorectal cancers with microsatellite instability and their possible origin in hyperplastic polyps and serrated adenomas. *J. Natl. Cancer Inst.* 2001; **93**: 1307–1313.
6. Koinuma K, Yamashita Y, Liu W, Hatanaka H, Kurashina K, Wada T, *et al.* Epigenetic silencing of AXIN2 in colorectal carcinoma with microsatellite instability. *Oncogene* 2006; **25**: 139–146.
7. Little MP. Are two mutations sufficient to cause cancer? Some generalizations of the two-mutation model of carcinogenesis of Moolgavkar, Venson and Knudson, and of the multistage model of Armitage and Doll. *Biometrics* 1995; **51**: 1278–1291.
8. Luebeck EG and Moolgavkar SH. Multistage carcinogenesis and colorectal cancer incidence in SEER. *Proc. Natl. Acad. Sci. USA* 2002; **99**: 15095–15100.
9. Lynch CJ and Milner J. Loss of one p53v allele results in four-fold reduction in p53 mRNA and protein: a basis for p53 haplo-insufficiency. *Oncogene* 2006; **25**: 3463–3470.
10. Morales KH, Ryan L, Kuo TL, Wu MM and Chen CJ. Risk of internal cancers from arsenic in drinking water. *Environ. Health Perspect.* 2000; **108**: 655–661.
11. Osada H and Takahashi T. Genetic alterations of multiple tumor suppressor genes and oncogenes in the carcinogenesis and progression of lung cancer. *Oncogene* 2002; **21**: 7421–7434.
12. Sparks AB, Morin PJ, Vogelstein B and Kinzler KW. Mutational analysis of the APC/beta-catenin/Tcf pathway in colorectal cancer. *Cancer Res.* 1998; **58**: 1130–1134.
13. Tan WY. *Stochastic Models of Carcinogenesis*. Marcel Dekker, New York, 1991.
14. Tan WY. *Stochastic Models with Applications to Genetics, Cancers, AIDS and Other Biomedical Systems*. World Scientific, Singapore and River Edge, New Jersey, 2002.
15. Tan WY and Chen CW. Stochastic modeling of carcinogenesis: some new insights. *Math. Comp. Model.* 1998; **28**: 49–71.
16. Tan WY and Chen CW. Cancer stochastic models. In: *Encyclopedia of Statistical Sciences*, revised ed. John Wiley and Sons, New York, 2005.
17. Tan WY, Zhang LJ and Chen CW. Stochastic modeling of carcinogenesis: state space models and estimation of parameters. *Disc. Cont. Dyn. Syst. Ser. B* 2004; **4**: 297–322.
18. U.S. EPA. Proposed guidelines for carcinogen risk assessment. *Fed. Regist. Notice* 1996; **61**(79): 17960–18011.
19. Van Ryzin J. Quantitative risk assessment. *J. Occup. Med.* 1980; **22**: 321–326.

20. Ward R, Meagher A, Tomlinson I, O'Connor T, *et al*. Microsatellite instability and the clinicopathological features of sporadic colorectal cancer. *Gut* 2001; **48**: 821–829.
21. Wistuba II, Mao L and Gazdar AF. Smoking molecular damage in brochial epithelium. *Oncogene* 2002; **21**: 7298–7306.

Chapter 13

STOCHASTIC MODELS FOR PRENEOPLASTIC LESIONS AND THEIR APPLICATION FOR CANCER RISK ASSESSMENT

Annette Kopp-Schneider, Iris Burkholder,
Jutta Groos and Lutz Edler

> Two stochastic models, the multistage model and a geometric model, are presented to describe formation and growth of cancer precursor lesions. Both models are applied to data from animal experiments by maximum likelihood methods. The models are used to test biological hypotheses about the process of formation and growth of preneoplastic lesions and to describe the dose-response relationship for model parameters.
>
> *Keywords*: Carcinogenesis model, birth-death process, Poisson process, skin papillomas, foci of altered hepatocytes.

1. INTRODUCTION

Risk assessment is commonly understood as the use of all available information to draw inference about health effects associated with exposure of individuals or populations to hazardous agents. The fundamental goal of most risk assessment analyses is to establish a safe level of human exposure to the agents under study. With this aim in mind, a risk assessment paradigm was developed consisting of five main elements, namely problem formulation, hazard identification, hazard characterization, exposure assessment and risk characterization. For a detailed description of the risk characterization of chemicals in food and diet see e.g., Renwick *et al.* (2003), who addressed amongst others the following issues: Methods of hazard characterization suitable for different risk assessments, characterization of the dose-response relationship through mathematical modelling, use of the mode or mechanism of action and/or the toxicokinetics for dose-response characterization, incorporation of the extent of uncertainty and variability in the resulting output for further refinements of dose-response characterization.

Besides the derivation of health-based guidance values, risk assessors are interested in the estimation of the risks associated with different dose levels of exposure. Depending on quality and size of available data, several approaches to describe the risk have been used. Among these approaches, analysis of tumor incidence data or time to tumor data is one of the building blocks in cancer risk assessment. However, for sufficient precision of cancer risk estimates and evaluation, the sample sizes of these data must be very large and the time period must cover a major proportion of the subject's lifespan. In the event that the sample size in tumor incidence data is small, the statistical power of many risk assessment procedures is usually small which may lead to anticonservative risk assessment (e.g., using NOAEL (No Observed Adverse Effect Level), as point of departure on the basis of tumor incidence rates). Since in many practical problems, increasing the number of subjects is often very difficult, for more accurate and efficient cancer risk assessment, other sources of information are needed. One source of information is based on data on preneoplastic lesions. Preneoplastic lesions were identified as a surrogate endpoint predictive of the course of cancer development because of their tight connection to malignant tumors. In human carcinogenesis precursor lesions have been identified amongst others for colon, prostate, endometrial, skin and liver cancer (e.g., Solakidi *et al.*, 2005) while in experimental carcinogenesis the skin (e.g., Marks and Fürstenberger, 1987), liver (e.g., Bannasch, 1996) and colon (Bird and Good, 2000) served as preferred organs studied for cancer risk assessment. Epidemiological studies with a rigorous quantitative evaluation of precancerous lesions have not been performed yet.

The quantitative evaluation of preneoplastic lesions in animal experimental studies has been found useful for risk assessment. As one example, we refer to experimental skin carcinogenesis where papillomas can be observed over time. In these studies, animals are usually observed weekly over six to 12 months, and the number of papillomas exceeding a prespecified size is recorded. Although difficulties in individual identification of papillomas and the fact that papillomas may grow together caused some limitations on accurate papilloma recording, the quantitative longitudinal evaluation of papilloma counts has proven valuable for the evidence of carcinogenicity as well as for investigation of the mechanisms of carcinogenic action of various agents (Enzmann et al., 1998).

For liver cancer, foci of altered hepatocytes (FAH) have been generated by a multitude of experimental protocols (Bannasch, 1996; Hasegawa and Ito, 1994; Pitot, 1990) and have also been identified in humans (Su et al., 1997). Recently, a protocol employing juvenile rats has been standardized and investigated in a large interlaboratory study (Ittrich et al., 2003) to establish the rat liver foci bioassay for cancer risk assessment. Longitudinal observation of FAH has not been possible so far because histopathological staining of liver sections is needed for their identification. Specifically, liver sections are stained with hematoxylin and eosin (H&E) to demonstrate biochemical changes in hepatocytes, particularly changes in the amount and/or activity of enzymes, and in the content of glycogen. Most notably, staining for the expression of the placental form of the glutathione S-transferase (GSTP) has turned out to be one of the most reliable and easily detectable cytochemical markers for FAH. Staining with GSTP produces high-contrast sections which can be evaluated morphometrically with semi-automated devices. While the classical animal carcinogenicity bioassay requires two years and employs four groups with 50 animals each, an interlaboratory study (Ittrich et al., 2003) had shown that a shorter time period (12 weeks) using four dose and one control group with eight animals per gender each was sufficient to demonstrate the carcinogenicity of a substance. In addition, evaluation of FAH sizes and numbers provides valuable insight into the carcinogenic mechanism which would be much more difficult to obtain from tumor incidence data only. A drawback in the analysis of FAH data originates from stereological limitations. Those are caused by the fact that thin sections of three-dimensional objects are taken and only two-dimensional size information can be measured. It has been shown that model-based analysis can be used to overcome the stereological problems and allows for evaluation of the number and sizes of FAH separately (Kopp-Schneider, 2003; Kopp-Schneider et al., 2006) instead of restricting the analysis to summary measures such as the area fraction of FAH.

In colon carcinogenesis, aberrant crypt foci (ACF) have been suggested to represent possible precursor lesions for colonic adenomas and carcinomas. They were first described by Bird *et al.* (1989) in carcinogen-treated mice, see also Mori *et al.* (2005). ACF similar to those in rodents have been reported in colonic mucosa in humans. They have been used as a suitable endpoint when searching for chemopreventive agents or when assessing the carcinogenic potency of environmental agents.

Over the last 50 years, models have been proposed to describe the process of malignant tumor formation, initiated by the works of Kendall (1960) and Neyman and Scott (1967). They already stated that the process of carcinogenesis is inherently a stochastic process, at least as long as it is not known why certain individuals get cancer under conditions where others are unaffected. Therefore, models of carcinogenesis are formulated in the framework of stochastic processes. Today, a number of approaches exist to describe the formation of malignancies (cf. Kopp-Schneider, 1997). These range from statistical formulas for carcinoma incidence, e.g., the Weibull model, to the description of accumulating damage in an organism (Yakovlev and Tsodikov, 1996), even to complicated computer simulation-based models including relationships between neighboring cells. The most prominent models are cell-kinetic models which have ignored cell-cell interactions, however. One of the strengths of the formulation of the carcinogenic process within the framework of stochastic processes is that maximum likelihood methods can be applied and nested models can be compared by likelihood ratio test.

The appropriateness of models to describe reality can only be assessed when models are applied to data. Application of cell kinetic models to epidemiological data describing the occurrence of malignancies in a heterogeneous population may only give qualitative results and will not be sufficient for drawing conclusions about the actual values of biological parameters. Application of size information of preneoplastic lesions obtained from well-defined experimental conditions, in contrast, can be used to understand and quantitatively describe the process of carcinogenesis. However, it has to be always kept in mind that models represent crude simplifications of reality. As stochastic models are often mathematically more difficult, they need stronger assumptions regarding the interdependency of cells as compared to deterministic models which exclusively describe mean behavior.

In this chapter, the multistage model and a geometric model for preneoplastic lesions are introduced and mathematical derivations of key quantities are presented. Three applications of both models are shown: the first one addresses skin papilloma data whereas the other two are concerned with FAH data. We conclude the chapter with the discussion of the use of the models for risk assessment.

2. MODELING PRENEOPLASTIC LESIONS

It belongs to the paradigm of carcinogenesis modelling to include a premalignant stage which is used to describe incidence and frequency of preneoplastic lesions. Usually it is assumed that preneoplastic lesions are generated according to a Poisson process, an assumption which makes the model mathematically tractable. Carcinogenesis models differ in the way in which preneoplastic lesions change their size and progress to malignancy. When data collection is restricted to the count of detectable preneoplastic lesions, the model does not need to specify the size distribution of preneoplastic lesions but should only provide the probability that a lesion can be detected.

In the following, two different concepts are presented as to how preneoplastic lesions change their size and progress to malignancy. First, a cell-kinetic model termed the multistage model is discussed, and then a geometric, or color-shift, model is introduced.

2.1. The Multistage Model with Clonal Expansion of Intermediate Cells

The most prominent biologically based models for carcinogenesis are the multistage models. These models are formulated at the cellular level, describing the malignant transformation of normal cells as a process involving mutations, cell proliferation/differentiation and occasionally repair. The development of multistage models goes back to Armitage and Doll (1954, 1957) and Kendall (1960). Later Neyman and Scott (1967) applied a multistage model to analyze urethane-induced hyperplastic foci observed in the lungs of mice. Research in multistage models was stimulated by a review article by Whittemore and Keller (1978). Moolgavkar and Venzon (1979) formulated a two-stage model, and have applied it to incidence data from animal experiments and epidemiological studies (see e.g., Moolgavkar, Cross *et al.*, 1990; Moolgavkar, Luebeck *et al.*, 1990;

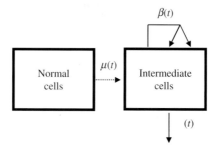

Fig. 1. A cell-kinetic model for formation and growth of intermediate cells. The arrow from normal to intermediate cells is dashed to point out that several steps may be involved in the transformation of a normal to an intermediate cell.

Moolgavkar and Luebeck, 1992). The mathematical and biological foundations of this model are reviewed in the book by Tan (1991).

Mathematically, the general assumption is that the first cell in a clone of intermediate cells is generated through a Poisson process with intensity $\mu(t)$ and that intermediate cells are subject to a linear birth-death process with birth rate $\beta(t)$ and death/differentiation rate $\delta(t)$ (see Figure 1). The formation process is either time-homogeneous or may vary in time.

Virtually all multistage models referenced above, even those which assume that a number of steps are necessary for a normal cell to become intermediate, approximate the formation rate of intermediate cells through a Poisson process. At this point it should be noted that the above-mentioned rate parameters of multistage models are not jointly identifiable and hence cannot be estimated from time-to-tumor data alone (Heidenreich, 1996; Hanin and Yakovlev, 1996; Hanin 2002).

The growth behavior of premalignant cells in a multistage model is described by a birth-death process. Since cells are assumed to evolve independently of each other, the rates of the process are linear in the number of cells. Expressions for the size $Y(t)$ of a cell population subject to linear birth-death process with birt rate $\beta(t)$ and death rate $\delta(t)$ can readily be derived. As shown in Tan (1991, p. 60), for $t > s \geq 0$

$$P(Y(t) = j | Y(s) = 1, Y(t) > 0) = \begin{cases} 1 - \dfrac{1}{g(t,s) + G(t,s)}, & j = 0 \\ \dfrac{g(t,s) + G(t,s)^{j-1}}{(g(t,s) + G(t,s))^{j+1}}, & j \geq 1 \end{cases} \quad (1)$$

where

$$g(t, s) = \exp\left[-\int_s^t (\beta(x) - \delta(x))dx\right]$$

and

$$G(t, s) = \int_s^t \beta(x)g(t, x)dx.$$

The size distribution can be derived explicitly for piecewise constant birth and death rates (see Kopp-Schneider, 1992, 1997).

The fact that the intermediate cell population is subject to a stochastic birth-death process implies that colonies of intermediate cells with a common cell progenitor may die out. The asymptotic extinction probability is given by the ratio of the death rate to the birth rate, or by 1, if the death rate is higher than the birth rate. As a consequence of the stochasticity of the process, every clone of cells has non-zero probability to become extinct, even if the birth rate is much higher than the death rate, as long as the death rate is not zero.

Assuming that the number of normal stem cells as well as the mutation rate of normal stem cell and the birth rate and death rate of intermediate cells are all constants over time, the size distribution of a non-extinct intermediate cell clone has been derived by Dewanji et al. (1989, Eq. (3.9)), see also Geisler and Kopp-Schneider (2000):

$$P(Y(t) = j | Y(t) > 0, Y(s) = 1)$$
$$= \frac{1}{j}\left(\frac{\beta}{\delta}p_0(t, s)\right)^j \frac{1}{(\beta - \delta)(t - s) - \ln[1 - p_0(t, s)]} \quad (2)$$

for $j \geq 1$ with

$$p_0(t, s) = \begin{cases} \dfrac{\delta(1 - e^{(\delta-\beta)(t-s)})}{\beta - \delta e^{(\delta-\beta)(t-s)}} & \text{for } \beta \neq \delta \\ \dfrac{\beta(t - s)}{1 + \beta(t - s)} & \text{for } \beta = \delta \end{cases}.$$

In many cases, the actual size of premalignant cell clones is not observable and a clone can only be counted if its size exceeds a threshold of, say, M cells. This is the case for skin papilloma count data, which contain information about the number of observable papilloma but do not provide actual papilloma sizes. Papillomas are counted when their size exceeds a threshold of, say, M cells. The corresponding probability of a papilloma to be observable can be easily derived from (2).

For the application of cell-kinetic models to FAH data, matters are complicated by the fact that morphometric evaluation of liver sections results in a record of focal transect areas. Two steps are needed before a cell kinetic model can be used to describe FAH formation and growth. First, the cell numbers of intermediate clones have to be translated into FAH volumes. This involves assumptions about the size of a single FAH cell as well as information about the density of the cell packing. The second step is the stereological transformation to link two-dimensional observations to three-dimensional model predictions. For this step, two different approaches could be used:

(i) The data about two-dimensional focal transections could be used to reconstruct the three-dimensional FAH size distribution which in turn would be compared to the model prediction. A number of methods have been proposed for this aim. Unfortunately, all approaches are unstable in the presence of a limited number of observed transections, which means that small changes in the two-dimensional data can result in large changes in three-dimensional FAH size distribution. This is especially true when very small profiles are present.

(ii) The model predictions are transformed from three to two dimensions using the Wicksell transformation (Wicksell, 1925). The observed data are utilized within the maximum likelihood approach to link two-dimensional observations to three-dimensional model predictions. This has become the method of choice for statistical model fit.

The cell kinetic model described above can be extended to include more than one intermediate stage as shown in Figure 2:

The number and size of intermediate cell colonies of the different intermediate types cannot be derived explicitly. Therefore, the formation process for cells of later intermediate types is approximated by a Poisson process with the intensity depending on the expected instead of the actual random

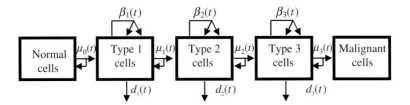

Fig. 2. A multistage model with three types of intermediate cells which are all subject to birth-death processes.

number of cells in the preceding stage, thus transforming the model to a multi-pathway model (Sherman and Portier, 1994).

2.2. A Geometric Model for Colonies of Intermediate Cells

The multistage model describes growth and progression of intermediate lesions based on the behavior of single independent cells. The geometry of the lesion is represented by conversion of cell numbers to lesion volume. In contrast, the color-shift model uses a geometric approach and describes colonies of intermediate cells as entities of spherical shape which change their size by increase of radius (Kopp-Schneider et al., 1998). This model has been used for description of liver focal lesion data. Again, it is assumed that centers of foci are generated according to a Poisson process. Foci start growing from the time point of formation onward, initially starting from either a single cell or a whole cluster of cells. Growth occurs through clonal expansion of foci cells and/or recruitment of neighbor cells.

In essence, the color-shift model allows for several sequential types ("colors") of foci. When focal lesions are generated, it is assumed that they all have the same size, described by radius r_0, and the same phenotype. Each focal lesion may pass through a sequence of phenotypes. The changes in phenotype are irreversible. Details of the original model are summarized in Appendix A and can be found in Kopp-Schneider et al. (1998); generalizations are given in Burkholder and Kopp-Schneider (2002) and in Groos and Kopp-Schneider (2006). A graphical representation is given in Figure 3.

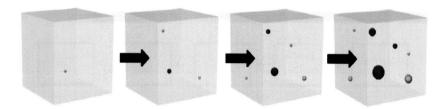

Fig. 3. The color-shift model. FAH of the first type are due to a persistent change in the phenotype of one or several normal hepatocytes, followed by clonal expansion and/or recruitment of neighboring cells. Change in FAH phenotype occurs through alteration of expression in the entire colony (all cells simultaneously) through a process that does not require a mutation. In this case, the entire colony moves onto the next stage ("change color") rather than a single cell in the colony due to mutation.

In the original color-shift-model, foci of different phenotypes all increase in size with the same rate in a deterministic exponential function from time point of formation, τ, onward, i.e., $R(t) = r_0 e^{b(t-\tau)}$

As a first generalization, foci were allowed to change their growth behavior once their phenotype changed (Burkholder and Kopp-Schneider, 2002). As the deterministic growth function was apparently too stiff, modifications of the model were considered in which every focus may grow according to its individual random growth rate which comes from a certain distribution (Groos and Kopp-Schneider, 2006), e.g., b comes from a beta-distribution with the density

$$f(b) = \frac{1}{B(p,q)} \frac{b^{p-1}(a-b)^{q-1}}{a^{p+q-1}} \mathbf{1}[0,a]^{(b)}, \quad p, q, a > 0 \qquad (3)$$

where $\mathbf{1}$ denotes the indicator function and $B(p, q)$ denotes the beta-function

$$B(p,q) = \int_0^1 z^{p-1}(1-z)^{q-1} dz. \qquad (4)$$

2.3. Comparison of Multistage and Color-Shift Model

The two types of models incorporate two types of hypotheses about formation, growth and phenotype change of preneoplastic lesions. The cell kinetic multistage model regards each focus cell as independent from all other cells with the ability to divide or to die or differentiate. It is assumed that cells go through different phenotypes independently. In accordance with the "mutation hypothesis," carcinogenesis is the result of mutations and subsequent clonal expansion of mutated cells. In contrast, the color-shift model allows for several adjacent cells to simultaneously express the phenotype of early focal lesion and for the whole foci to change their phenotype.

Biological studies suggest a gradual metabolic shift within larger cell populations during progression from early to advanced preneoplastic foci of altered hepatocytes. This "field effect hypothesis" is formalized through the color-shift model. This model has so far only been applied to liver focal lesion data, whereas multistage models have been applied to both FAH (e.g., Kopp-Schneider *et al.*, 2006; Moolgavkar *et al.*, 1990; Portier *et al.*, 1996) and papilloma data (e.g., Kopp-Schneider and Portier, 1995). In the competitive application of both models to the same FAH data set with three successive FAH phenotypes the color-shift model has shown a slight advantage over the multistage model (Burkholder and Kopp-Schneider, 2002; Geisler and Kopp-Schneider, 2000).

Comparative validation of both models using statistical methods is hampered by the fact that the two models are not hierarchically nested and, therefore, likelihood ratio tests cannot be used to formally compare model fits. However, computer simulations have shown that the maximum value of the likelihood can be used to identify the appropriate model even if the models are not nested provided that the models have a similar structure (Kopp-Schneider, 2001).

From a risk assessment perspective, parameters in both models can be assumed to depend on the dose of a putative carcinogenic agent. With this tool, biologically based dose-response analyses can be performed, and the action of the agent on the rate of formation and growth of preneoplastic lesions can be examined. Here, the advantage of applying two distinct models to the data is that it can be studied whether the resulting dose-response relationship is robust to changes in the model. One would consider the evidence for a certain dose-response relationship stronger if the results could be replicated with both models. A simultaneous evaluation of one data set with different models was performed by Groos and Kopp-Schneider (2006); a comparative evaluation of a large dose-response data set using two distinct models is currently under investigation.

3. APPLICATION OF CARCINOGENESIS MODELS TO PRENEOPLASTIC LESION DATA

3.1. Mouse Skin Carcinogenesis: Testing Biological Hypotheses about Papilloma and Carcinoma Formation

A two-stage cell kinetic model was applied to data from a mouse skin painting study, assuming that papillomas are the precursors of carcinomas and that papilloma cells are generated from normal cells through a single mutation (Kopp-Schneider and Portier, 1995). The experiment followed a standard initiation-promotion protocol where promotion was stopped at prespecified time points. Animals were initiated with 7,12-dimethylbenz(a)anthracene (DMBA) followed by twice weekly applications of 12-O-tetradecanoylphorbol-13-acetate (TPA). TPA was shown to act as a promoter in the classical experimental setup: given without prior initiation, virtually no papillomas are generated, whereas given with prior initiation, papilloma yield is greatly enhanced. In addition, it has been shown that TPA increases the rate of cell division (Morris and Argyris, 1983). In this experiment, four groups of 60 animals each were used with different lengths of promotion with TPA: 10, 20, 30 and 40 weeks. Animals were followed until death or sacrifice due to carcinoma or date of termination of the study at 60 weeks. Individual weekly papilloma counts were recorded, where a papilloma was assumed detectable when its size exceeded 1 mm in diameter. In addition, time to first carcinoma was recorded by retrospectively determining the time of its first appearance from laboratory records when it was confirmed histologically. Carcinoma grow very rapidly and usually the time span between first observation and histological proof was at most two weeks.

Application of carcinogenesis models to papilloma and carcinoma data aimed at testing hypotheses concerning formation of papillomas and their progression to carcinomas. Using the two-stage model to simultaneously predict papilloma number and carcinoma incidence hypothesizes that all papillomas are precursors of carcinomas and that carcinoma formation is a process involving two rate-limiting steps. The following assumptions were made for application of the two-stage model. (a) Initiation occurred instantaneously at the moment of exposure to DMBA and no intermediate cells were formed after the initiating event. (b) Birth and death rates of papilloma cells are constant for every experimental step, i.e., the rates remain constant before, during and after promotion, and birth rates before and after promotion are identical. The rates were assumed identical over the population of animals. The birth rates before and during promotion were taken from the literature (Morris and Argyris, 1983), and a sensitivity analysis showed that the resulting model fit is in essence unchanged by variation in the choice of birth parameters. A total of five parameters were estimated from the data: the mean number of intermediate cells generated during instantaneous initiation, the death rates during and after promotion (the interval before the start of promotion was too short to allow for estimation of a separate death rate, therefore, it was set equal to the birth rate before promotion), the number of cells in a papilloma at detection limit, and the mutation rate from papilloma to carcinoma cells. Papilloma cells during promotion were assumed to have a cell cycle length of 20 hours, whereas without promotion cell cycle length was assumed to last seven days (Morris and Argyris, 1983).

Using the likelihood function outlined in Appendix B(1), maximum likelihood estimates (MLE) were obtained numerically. The MLEs were: 114,183 papilloma cells due to the initiating event, the ratio of death to birth rate 1.0074 during and 1.0211 after promotion, detection limit of 352 cells (showing that a papilloma at detection limit of 1 mm in diameter contains not only intermediate cells but also consists of other types of cells), and the mutation rate from papilloma to carcinoma cells of $3.15 \cdot 10^{-8}$ per cell cycle. Application of maximum likelihood methods resulted in a good fit for papilloma data (see Figure 4) but a miserable fit for the carcinoma data (Figure 5). Although model parameters are not all identifiable from tumor incidence data alone, simultaneously evaluating number and size of intermediate and time to malignant lesion leads to identifiable model parameters.

In an attempt to explain the lack of fit of the model for carcinoma data, a two-pathway model was formulated, incorporating the hypothesis that the papilloma population is heterogeneous and only a part may go on to malignancy. However, this modification could not explain the carcinoma data either. Probably more than two rate-limiting steps are necessary for

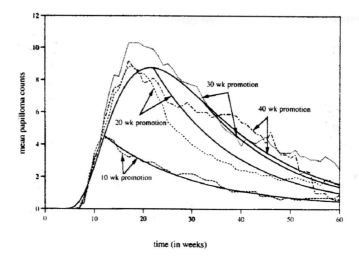

Fig. 4. Observed versus expected papilloma counts. The plots of the observed mean papilloma counts are dotted. Solid lines represent the model predictions with parameters estimated using maximum likelihood methodology.

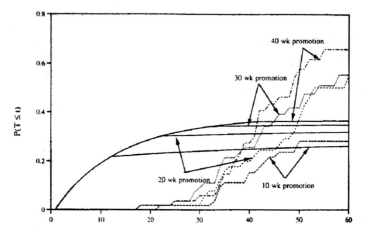

Fig. 5. Observed versus expected time to carcinoma occurrence. Solid lines represent the model predictions with parameters estimated using maximum likelihood methodology.

a carcinoma cell to be generated from a normal cell. In summary, the hypothesis that all papillomas are precursors of carcinomas and that carcinoma formation is a process involving two rate-limiting steps can be rejected.

3.2. Liver Focal Lesion Data: Testing Hypotheses about FAH Formation and Phenotype Change

Different hypotheses about the formation of preneoplastic lesions have been proposed in the scientific community. The first hypothesis assumes that single cells are transformed to the next phenotype by a mutational event. This hypothesis is depicted in the upper path of Figure 6 for the case of three different types of preneoplastic lesions. The lower path shows the alternative hypothesis, called field effect hypothesis, that all cells in a preneoplastic lesion change their phenotype more or less simultaneously rather than by mutation of single cells. The mutation hypothesis is formalized by the multistage model, whereas the field effect hypothesis is reflected in the color-shift model.

To study the mechanism of phenotype change, use was made of data from a carcinogenicity study in which rats were continuously treated with the carcinogenic compound N-nitrosomorpholine (NNM) (Weber and Bannasch, 1994) and evaluated 11, 15, 20, 27 and 37 weeks after the start of the experiment. The analysis was restricted to three types of foci representing three successive preneoplastic stages. Therefore, a multistage model with three successive types of intermediate cells was considered for the mutation hypothesis. The likelihood function for FAH data is outlined in Appendix B(2). Maximum likelihood parameter estimates for cell division rates in the multistage model were not in the range of biologically plausible values. Restricting model parameters to biologically

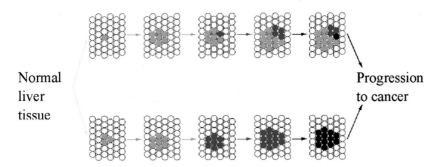

Fig. 6. Two hypotheses about the formation and phenotypic changes of liver focal lesions. The upper path shows the mutation hypothesis and the lower path shows the field effect hypothesis.

plausible values resulted in a much worsened model fit (Geisler and Kopp-Schneider, 2000).

On the other hand, a color-shift model with three different sequential phenotypes of FAH was used to formalize the field effect hypothesis. FAH of different phenotypes were allowed to grow at different rates, but growth was deterministic and rates were identical for all foci of the same phenotype (Burkholder and Kopp-Schneider, 2002). Graphical comparison of model fit showed that the color-shift model explained the liver focal lesion data slightly better than the multistage model. This is especially true for later stage FAH as shown in Figure 7.

The maximum values of the likelihood function are larger for the color-shift model than the multistage model. However, likelihood values cannot be formally compared to identify the best fitting model because the two models are not nested hierarchically. The plots suggest that for this type of preneoplastic foci the field effect hypothesis is a more likely explanation than the mutation hypothesis. This is intuitively appealing because the multistage model predicts observation of many large early stage foci and few small later stage foci while the opposite is true for the color-shift model, which predicts many small early stage foci and few large later stage foci, this being the effect observed experimentally.

3.3. Liver Focal Lesion Data: Dose-Response Analyses

As part of a long-term study of the effects of α-radioimmunotherapy in mice, an intervention study was performed to investigate the effects of α-radiation on formation and volume increase of preneoplastic liver lesions. The negative control and two dose groups treated with a single injection of α-labelled antibody ^{213}Bi-anti CD19 (^{213}Bi-CD19) were combined to obtain the dose-response relationship for FAH formation and growth. Hematoxylin and eosin (H&E) stained liver sections were evaluated for FAH six, 12 and 17 months after treatment. Five animals per group and time point were included in the study. No FAH were observed at six months, and a total of 71 FAH were observed in the entire study. Density and size distribution of focal transections was described by a multistage model, specifically a one-stage model (see Figure 1), for FAH formation and growth (see Kopp-Schneider et al., 2006). Model parameters were assumed constant for the

Stochastic Models for Preneoplastic Lesions

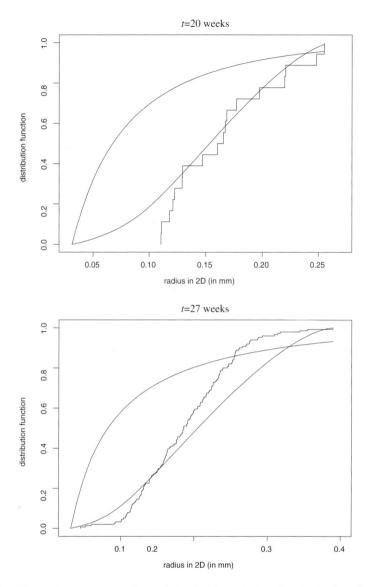

Fig. 7. Observed versus expected cumulative focal transection size distribution of type 2 foci in a hepatocarcinogenicity study with NNM 20, 27 and 37 weeks after start of treatment. Expected curves are derived from the multistage model with maximum likelihood parameter estimates restricted to biologically realistic growth rates. The dashed curves show the model fit of the multistage model whereas the long-short-dashed curves show the model fit of the color-shift model.

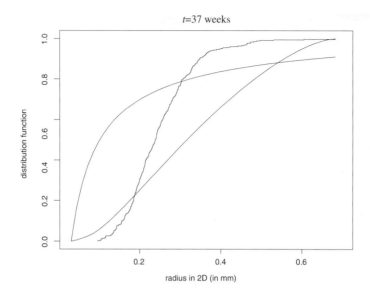

Fig. 7. (*Continued*)

duration of the study. Dose of ^{213}Bi-CD19 was incorporated into the model parameters μ (rate of transformation from normal to FAH cells) and β (growth rate of FAH cells). To assure stability in the estimation of model parameters, the death rate was kept constant proportional to the birth rate, and was assumed equal for all dose groups. As the animals were about eight weeks old when they were treated according to the study plan, the model was adapted to include a pre-treatment phase with common transformation rate and birth rate for all three groups, and the death rate was set to 0 in this phase.

Three different dose dependencies were considered for each of the two parameters μ and β. (a) The parameters were assumed to be independent of dose. (b) The parameters were different for all three dose groups. (c) The parameters for low and high dose groups had identical values distinct from those for the control group. A total of nine combinations of dose models for the two model parameters were considered. From the analysis it was apparent that although the antibody was given by single injection at the start of the experiment, the effect on FAH formation lasted for the whole duration of the experiment.

Maximum likelihood estimates (MLE) for the model parameters in all dose-response models revealed that the rates of FAH formation and the rates of FAH growth are increasing with dose, confirming what is generally presumed as the effect of α-radiation on transformation rates. The best fitting model allowed for both parameters to vary from group to group. The MLE for the rate of transformation from normal to FAH cells, μ, measured as the number of transformations per cm^3 and week, increased from 0.0 (95% CI 0.0 – 0.104) in control to 0.174 (0.075 – 0.402) in low and 0.346 (0.232 – 0.515) in high dose groups. The MLE for weekly FAH cell birth rate increased from 0.459 (0.093 – 2.259) in control to 0.600 (0.600 – 0.600) in low and 1.246 (1.071 – 1.448) in high dose groups. The MLE for the ratio of death to birth rate of FAH cells was 0.95 (0.950 – 0.951). The MLE for pre-treatment transformation rate was estimated to be 0.007 (0.003 – 0.020) (transformations per cm^3 and week), and the MLE for weekly birth rate was 0.647 (0.510 – 0.820) in all three groups for the first eight weeks. The analyses showed in addition that α-radiation has an effect on growth rates. In summary, the study suggested that treatment with ^{213}Bi-CD19 has initiating as well as promoting activity with a slightly more pronounced promoting effect. However, as shown in Figure 8, model fit is not completely satisfactory, especially for the last observation time point.

Applying a color-shift model with one color to the same data gave very similar results as to the dose-response relationship for α-radiation and its impact on formation and volume increase of FAH, although the model fit was slightly less satisfactory. This shows that the dose-response relationships for FAH formation and volume increase is distinct and different types of model can identify them, i.e., the findings are robust to misspecifications of the model.

The model-based analysis shows clear results based on just a few animals (45 animals in total) and on a small number of FAH observed in the study. This shows the power of model-based evaluation of preneoplastic lesions. A study employing 45 animals in three groups with endpoint carcinoma formation and standard statistical comparison of the proportions of carcinoma-bearing animals would at best have provided weak indications as to the carcinogenicity of the compound and would not have provided insights into the mechanism of carcinogenesis.

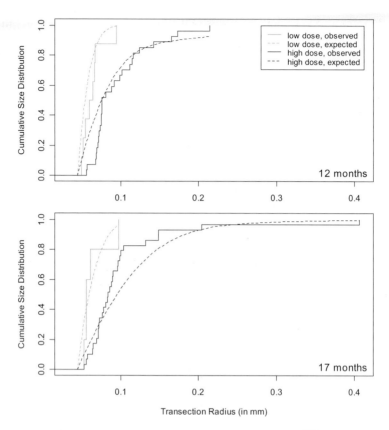

Fig. 8. Observed versus expected FAH size distribution induced by ^{213}Bi-CD19 at 12 and 17 months.

4. DISCUSSION

In this chapter we have shown the usefulness of mechanistic models for formation and growth of preneoplastic lesions in cancer research and cancer risk assessment. The models presented here can be formulated within stochastic process theory, and the distribution of the observable quantities of interest can be derived analytically or computed numerically. This allows for application of formal maximum likelihood methods to estimate parameters and identify the best fitting model. With this tool one can test which biological hypothesis formalized in a mechanistic model is more likely;

further, effects of agents on model parameters can be studied and used for quantitative risk assessment.

Whenever it is found that lesions are actual precursors of carcinomas, collection and evaluation of data on these preneoplastic lesions will lead to results obtainable in short time — weeks to a few months compared to two-year duration for carcinoma endpoint — and largely increased statistical power for quantitative risk assessment. As the appearance of preneoplastic lesions is less encumbering for experimental animals than the occurrence of carcinomas, animals can tolerate the presence of several lesions and, hence, a quantitative analysis of the number and sizes of preneoplastic lesions can be carried out. This approach provides substantially more information than the evaluation of time-to-tumor data or rates of tumor-bearing animals. In the case of liver carcinogenesis, laboratories have often confined themselves to reporting only the presence or absence of FAH in the experimental animals to avoid the time-consuming morphometric evaluation of liver sections. Kopp-Schneider (2003) showed how the ability to identify the hepatocarcinogenic potential of an agent is increased when FAH data are evaluated quantitatively, as opposed to the studies based on rates of FAH-bearing animals alone. A similar increase in power is observed for the quantitative evaluation of preneoplastic lesions in contrast to time-to-tumor data or rates of tumor-bearing animals.

Preneoplastic lesions occur much earlier than carcinomas. Therefore, studies with preneoplastic lesions as endpoints provide results earlier than carcinogenicity studies. In a large-scale interlaboratory study (Ittrich et al., 2003), it was shown that an experiment with 40 rats per gender observed over 12 weeks was enough to verify the carcinogenic potential of known hepatocarcinogens. In the case of mouse skin painting studies, Fürstenberger and Kopp-Schneider (1995) showed that treatment of animals for 20 weeks was sufficient to obtain the maximal response in terms of papilloma count (as well as carcinoma occurrence). The accumulated knowledge should be incorporated into guidelines for toxicity testing. This is especially important as large-scale toxicity tests are required by governmental authorities, e.g., the REACH program in the EU.

Quantitative risk assessment involves analysis of the dose-response relationship for an endpoint, e.g., preneoplastic lesions, induced by the agent under study. If no assumptions concerning the dose-response relationship for this endpoint are made, the only rational evaluation consists in multiple statistical tests comparing dosed groups to the background. This method was traditionally used to derive NOAELs (no observed adverse effect level) as surrogates for thresholds in risk assessment, but it strongly depends on sample size and risks to overlook the effects present at a low but still perilous level for humans (Edler *et al.*, 2002). The natural next step of refinement for processing experimental information more efficiently is the use of a statistical model to describe the dose-response relationship quantitatively. However, application of a statistical model is by its nature a fitting exercise and cannot provide insight into the mechanism of the agent under study since the model does not contain biological elements. In contrast, the use of mechanistic models for the formation and growth of preneoplastic lesions allows one to characterize the ability of the agent to alter the rate of formation and the size of preneoplastic lesions. However, any conclusion drawn from a model based analysis depends on the model adequacy.

ACKNOWLEDGMENTS

Funding for the research of Dr. Lutz Edler was provided by NATO PST.CLG 979045. Research of Dr. Iris Burkholder was supported by DFG grant KO 1886/1-1 and 1-2 and by EU-shared cost action no. FIS5-1999-00033. Research of Jutta Groos was supported by DFG grants KO 1886/1-3 and 1-5.

References

1. Armitage P and Doll R. The age distribution of cancer and a multi-stage theory of carcinogenesis. *Br. J. Cancer* 1954; **8**: 1–12.
2. Armitage P and Doll R. A two-stage theory of carcinogenesis in relation to the age distribution of human cancer. *Br. J. Cancer* 1957; **9**: 161–169.
3. Bannasch P. Pathogenesis of hepatocellular carcinoma: sequential cellular, molecular, and metabolic changes. *Prog. Liver Dis.* 1996; **14**: 161–191.
4. Bird RP, McLellan EA and Bruce WR. Aberrant crypts, putative precancerous lesions, in the study of the role of diet in the aetiology of colon cancer. *Cancer Surv.* 1989; **8**: 189–200.

5. Bird RP and Good CK (2000) The significance of aberrant crypt foci in understanding the pathogenesis of colon cancer. *Toxicol. Lett.* 2000; **112–113**: 395–402.
6. Burkholder I and Kopp-Schneider A. Incorporating phenotype-dependent growth rates into the color-shift model for preneoplastic hepatocellular lesions. *Math. Biosci.* 2002; **179**: 145–160.
7. Dewanji A, Venzon DJ and Moolgavkar SH. A stochastic model for cancer risk assessment. II. The number and size of premalignant clones. *Risk Anal.* 1989; **9**: 179–187.
8. Edler L, Poirier K, Dourson M, Kleiner J, Mileson B, Nordmann H, Renwick A, Slob W, Walton K and Würtzen G (2002) Mathematical modelling and quantitative methods. *Food Chem. Toxicol.* 2002; **40**: 283–326.
9. Enzmann H, Bomhard E, Iatropoulos M, Ahr, HJ, Schlueter G and Williams GM. Short- and intermediate-term carcinogenicity testing — a review. Part 1. The prototypes mouse skin tumour assay and rat liver focus assay. *Food Chem. Toxicol.* 1998; **36**: 979–995.
10. Geisler I and Kopp-Schneider A. A model for hepatocarcinogenesis with clonal expansion of three successive phenotypes of preneoplastic cells. *Math. Biosci.* 2000; **168**: 167–185.
11. Groos J and Kopp-Schneider A. Application of a color-shift model with heterogeneous growth to a rat hepatocarcinogenesis experiment. *Math. Biosci.* 2006; **202**: 248–268.
12. Hanin LG and Yakovlev AY, A nonidentifiability aspect of the two-stage model of carcinogenesis. *Risk Anal.* 1996; **16**: 711–715.
13. Hanin LG. Identification problem for stochastic models with application to carcinogenesis, cancer detection and radiation biology. *Disc. Dyn. Nat. Soc.* 2002; **7**: 177–189.
14. Hasegawa R and Ito N. Hepatocarcinogenesis in the rat. In: *Carcinogenesis.* Waalkes MP and Ward JM (eds.) Raven, New York, 1994, pp. 39–65.
15. Heidenreich WF. On the parameters of the clonal expansion model. *Radiat. Environ. Biophys.* 1996; **35**: 127–129.
16. Ittrich C, Deml E, Oesterle D, Küttler K, Mellert W, Enzmann H, Bannasch P, Haertel T, Mönnikes O, Schwarz M and Kopp-Schneider A (2003) Prevalidation of a rat liver foci bioassay (RLFB) based on results from 1600 rats: a study report. *Toxicol. Pathol.* 2003; **31**: 60–79.
17. Kendall DG. Birth-and-death processes, and the theory of carcinogenesis. *Biometrika* 1960; **47**: 316–330.
18. Kopp-Schneider A. Birth-death processes with piecewise constant rates. *Stat. Prob. Lett.* 1992; **13**: 121–127.

19. Kopp-Schneider A, Portier CJ and Sherman CD. The exact formula for tumor incidence in the two-stage model. *Risk Anal.* 1994; **14**, 1079–1080.
20. Kopp-Schneider A and Portier CJ. Carcinoma formation in NMRI mouse skin painting studies is a process suggesting greater than two stages. *Carcinogenesis* 1995; **16**: 53–59.
21. Kopp-Schneider A. Carcinogenesis models for risk assessment. *Stat. Methods Med. Res.* 1997; **6**: 317–340.
22. Kopp-Schneider A, Portier CJ and Bannasch P. A model for hepatocarcinogenesis treating phenotypical changes in focal hepatocellular lesions as epigenetic events. *Math. Biosci.* 1998; **148**: 181–204.
23. Kopp-Schneider A. Using a stochastic model to analyze the sequence of phenotypic changes in rat liver focal lesions. *Math. Comput. Model.* 2001; **33**: 1289–1295.
24. Kopp-Schneider A. Biostatistical analysis of focal hepatic preneoplasia. *Toxicol. Pathol.* 2003; **31**: 121–125.
25. Kopp-Schneider A, Haertel T, Burkholder I, Bannasch P, Wesch H, Groos J and Heeger S. Investigating formation and growth of α-radiation-induced foci of altered hepatocytes: a model-based approach. *Radiat. Res.* 2006; **166**: 422–430.
26. Luebeck EG and Moolgavkar SH. Stochastic analysis of intermediate lesions in carcinogenesis experiments. *Risk Anal.* 1991; **11**: 149–157.
27. Marks F and Fürstenberger G. From the normal cell to cancer. The multistep process of skin carcinogenesis. In: *Concepts and Theories in Carcinogenesis.* Maskens AP, Ebbesen P and Burny A (eds.) Exerpta Medica, Amsterdam, 1987, pp. 169–184.
28. Moolgavkar SH, Cross FT, Luebeck G and Deagle GE. A two-mutation model for radon-induced lung tumors in rats. *Radiat. Res.* 1990; **121**: 28–37.
29. Moolgavkar SH and Luebeck EG. Multistage carcinogenesis: population-based model for colon cancer. *J. Nat. Cancer Inst.* 1992; **84**: 610–618.
30. Moolgavkar SH, Luebeck EG, de Gunst M, Port RE and Schwarz M. Quantitative analysis of enzyme-altered foci in rat hepatocarcinogenesis experiments I: single agent regimen. *Carcinogenesis* 1990; **11**: 1271–1278.
31. Moolgavkar SH and Venzon DJ. Two-event model for carcinogenesis: incidence curves for childhood and adult tumors. *Math. Biosci.* 1979; **47**: 55–77.
32. Mori H, Hata K, Yamada Y, Kuno T and Hara A. Significance and role of early-lesions in experimental colorectal carcinogenesis. *Chem.-Biol. Interact.* 2005; **155**: 1–9.
33. Morris R and Argyris TS. Epidermal cell cycle and transit times during hyperplastic growth induced by abrasion or treatment with 12-*O*-tetradecanoylphorbol-13-acetate. *Cancer Res.* 1983; **43**: 4935–4942.

34. Neyman J and Scott E. Statistical aspects of the problem of carcinogenesis. In: *Fifth Berkeley Symposium on Mathematical Statistics and Probability.* University of California Press, Berkeley, CA, 1967, pp. 745–776.
35. Pitot HC. Altered hepatic foci: their role in murine hepatocarcinogenesis. *Annu. Rev. Pharmacol. Toxicol.* 1990; **30**: 465–500.
36. Portier CJ, Kohn M, Edler L, Kopp-Schneider A, Sherman CD, Maronpot RM and Lucier G. Modeling the number and size of hepatic focal lesions following exposure to 2,3,7,8-TCDD. *Toxicol. Appl. Pharmacol.* 1996; **138**, 20–30.
37. Renwick AG, Barlow SM, Hertz-Piccioto I, Boobis AR, Dybing E, Edler L, Eisenbrand G, Greig JB, Kleiner J, Lambe J, Müller DJG, Smith MR, Tritscher A, Tuijtelaars S, van den Brandt P, Walker R and Kroes R. Risk characterization of chemicals in food and diet. Final part of Food Safety in Europe (FOSIE): risk assessment of chemicals in food and diet (EC concerted action QLK1-1999-00156). *Food Chem. Toxicol.* 2003; **41**: 1211–1271.
38. Sherman CD and Portier CJ. Quantitative analysis of multiple phenotype enzyme-altered foci in rat hepatocarcinogenesis experiments: the multipath/multistage model. *Carcinogenesis* 1995; **16**: 2499–2506.
39. Solakidi S, Vorgias C and Sekeris CE. Biological carcinogenesis: theories and methods. In: *Recent Advances in Quantitative Methods in Cancer and Human Health Risk Assessment.* Edler L and Kitsos CP (eds.) Wiley, 2005.
40. Su Q, Benner A, Hofmann WJ, Otto G, Pichlmayr R and Bannasch P. Human hepatic preneoplasia: phenotypes and proliferation kinetics of foci and nodules of altered hepatocytes and their relationship to liver cell dysplasia. *Virchows Arch.* 1997; **431**: 391–406.
41. Tan WY. *Stochastic Models of Carcinogenesis.* Statistics, textbooks and monographs, Vol. 116. Marcel Dekker, New York, 1991.
42. Weber E and Bannasch P. Dose and time dependence of cellular phenotype in rat hepatic preneoplasia and neoplasia induced by continuous oral exposure to N-nitrosomorpholine. *Carcinogenesis* 1994; **15**: 1235–1242.
43. Wicksell SD. The corpuscle problem. A mathematical study of a biometrical problem. *Biometrika* 1925; **17**: 84–99.
44. Yakovlev AY and Tsodikov AD. *Stochastic Models of Tumor Latency and Their Biostatistical Applications.* World Scientific, Singapore, 1996.
45. Whittemore A and Keller JB. Quantitative theories of carcinogenesis. *SIAM Rev.* 1978; **20**: 1–30.
46. Zheng Q. On the exact hazard and survival functions of the MVK stochastic carcinogenesis model. *Risk Anal.* 1994; **14**: 1081–1084.

APPENDIX A: BASIC IDEAS OF THE COLOR-SHIFT MODEL

The color-shift-model describes the generation and the fate of focal lesions in a three-dimensional situation. Transformation of a normal cell to the first preneoplastic state is assumed to occur as a random process with an exponential waiting time. Individual normal cells are assumed to act independently. This leads to a Poisson process which generates single cells in the cube marking the center of the foci. Through clonal expansion and/or recruitment of neighboring cells, these cells form focal lesions which consist of multiple cells with the same phenotype as the original cell. We will assume that the shape of every focus is a sphere. Foci start growing from the time point of generation onward. The model includes the concept that not only single cells but whole clusters of cells are transformed to the first preneoplastic state, this is realized by assuming that the initial radius of a focus is the radius of the cluster of transformed cells. When focal lesions are generated they all have the same phenotype. The phenotype changes sequentially over time. The changes in phenotype are irreversible.

Mathematically, the color-shift model is described as follows: assume that a Poisson process generates the centers of spheres with radius r_0. The Poisson process is assumed to be homogeneous in location and to have rate $\mu(s)$ which may vary with the age (s) of the animal. Every sphere that is generated grows according to an exponential law with rate parameter B which could be a random variable. At the time point of its appearance a sphere starts in color 1 and changes its color sequentially after exponential waiting times. Hence every sphere has four attributes: (i) location X of its center, (ii) time point of appearance T, (iii) color C and (iv) radius R. It is assumed that location, X, is independent of time point of formation T, the color C and the radius R of the focus. Once a focus is generated, its color and size changes over time, i.e., $C(t)$ and $R(t)$ are both stochastic processes in time. In fact, both color and size of a focus depend upon the time elapsed since formation of the focus. It is assumed that color and radius are conditionally independent given T. However, since both depend upon time since formation, larger radii will be associated with later colors and hence a larger focus is more likely to be in a progressed state than a smaller focus. Suppose waiting time for each color change is described by the single parameter λ (per time and colony), and the number of colors for the foci is m, where m may be finite or infinite. By definition of the process, given $T = \tau$ and b as the realization of a random variable B, the radius is given by $r_0 \exp\{b(t - \tau)\}$ (with $\tau \leq t$). The joint distribution for color and size of a colony is given by:

$$f_{R(t),C(t)}(r,c) = \frac{1}{tbr(c-1)!} e^{-\frac{\lambda}{b}\ln\left(\frac{r}{r_0}\right)} \left[\frac{\lambda}{b}\ln\left(\frac{r}{r_0}\right)\right]^{c-1} \mathbf{1}_{[r_0, r_0 e^{bt}]}(r)$$

for $c < m$

and

$$f_{R(t),C(t)}(r,m)$$
$$= \frac{1}{tbr}\left\{1 - \frac{1}{(m-2)!} e^{-\frac{\lambda}{b}\ln\left(\frac{r}{r_0}\right)} \left[\frac{\lambda}{b}\ln\left(\frac{r}{r_0}\right)\right]^{m-2}\right\} \mathbf{1}_{[r_0, r_0 e^{bt}]}(r)$$

for the last color, m.

APPENDIX B: LIKELIHOOD FUNCTIONS

Appendix B(1): Likelihood Function for Skin Papilloma and Carcinoma Data

The number of papillomas, p, observed at time point t follows a Poisson distribution with rate parameter $\lambda(t)$ and the loglikelihood is given by

$$p \ln[\lambda(t)] - \ln(p!) - \lambda(t).$$

The rate parameter corresponds to the expected number of observable papilloma at time t, i.e.,

$$\lambda(t) = X(0)\mu(0)P(Y(t) > M|Y(0) = 1)$$

where $X(0)$ is the number of normal cells at the start of the experiment and $\mu(0)$ is the rate with which normal cells are transformed into papilloma cells at the start of the experiment, hence $X(0)\mu(0)$ is the expected number of papilloma cells generated by the initiating event at the start of the experiment, hence $X(0)\mu(0)$ is the expected number of papilloma cells generated by the initiating event at the start of the experiment, $Y(t)$ describes the size of a papilloma cell population subject to a linear birth-death process and M is the detection limit of a papilloma.

Skin carcinoma data are based on weekly observations. If x_i animals got a carcinoma in week i and y_i animals died in week i without a carcinoma, the loglikelihood contribution for week i is

$$x_i[F(i) - F(i-1)] - y_i[1 - F(i)]$$

where $F(t)$ corresponds to the distribution function for the first appearance of a carcinoma. Usually, time to first appearance of a carcinoma is set to time to first malignant cell. The distribution function for time to first malignant cell is either derived from a system of differential equations (see e.g., Kopp-Schneider and Portier, 1995) or from an approximate analytical solution (see e.g., Kopp-Schneider et al., 1994; Zheng, 1994).

Appendix B(2): Likelihood Function for Liver Focal Lesion Data

Every liver section obtained at time point t contributes independently to the likelihood with a factor for the observed number of focal transections in area A of the liver section and the product of size distributions for the observed radii. If n_c type c focal transections with radii $r_{i,c}$, $i = 1, \ldots, n_c$, are observed in liver section area A, the loglikelihood contribution is given by

$$\ln L = \sum_{c:\text{ foci types}} \left\{ n_c \ln(A\lambda_c(t)) - A\lambda_c(t) + \sum_{i=1}^{n_c} \ln g_{2,c}^{\varepsilon}(r_{i,c}, t) \right\}$$

where $\lambda_c(t)$ corresponds to the expected number of detectable type c focal transections per unit area in t and $g_{2,c}^{\varepsilon}$ is the conditional density function for focal transection radius given that the transection radius exceeds the detection limit of ε. For multistage models, the quantities $\lambda_c(t)$ and $g_{2,c}^{\varepsilon}$ are derived in Geisler and Kopp-Schneider (2000). For the color-shift model, the formulas are given in Kopp-Schneider et al. (1998).

Chapter 14

DRUG RESISTANCE IN CANCER MODELS

Jaroslaw Smieja

Two factors constitute major impediments to the successful chemotherapy: toxicity of chemotherapeutic agents affecting healthy cells and drug resistance. This chapter reviews how drug resistance can be incorporated into models describing dynamics of population of cancer cells. It begins with a simple example illustrating necessity of including the mechanisms of drug resistance in any dynamical model of chemotherapy. Afterwards, more complex compartmental models are introduced, including infinite-dimensional ones. The problem of optimization of chemotherapy is also stated, followed by its solution for several models.

Keywords: Drug resistance, optimization, chemotherapy.

1. INTRODUCTION

Despite a long history of mathematical modeling of cancer chemotherapy (Eisen, 1979; Martin, 1992; Swan, 1990; Swierniak *et al.*, 2003), its practical application to development of chemotherapy protocols has been arguably negligible (with minor exceptions). One of the underlying reasons for that is the emergence of resistance of cancer cells to cytotoxic agents. Recent advances in genomics and molecular biology shed light on various sources of drug resistance, opening the way for more realistic models that could ultimately facilitate development of mathematically based chemotherapy scheduling and multidrug protocols.

In the past years there has been renewed interest in these models (Fister and Panetta, 2000; Ledzewicz and Schättler, 2002) partially due to better models, but also due to a refinement of the techniques which can be used to analyze them and to estimate the necessary control parameters. Most of models described here have been introduced and analyzed by Kimmel, Swierniak and collaborators (e.g., Swierniak *et al.*, 1996, 1999, 2005), from both numerical as well as theoretical perspectives. The following sections give a review of some of these results, extend them onto a broader class of models and outline still open questions.

The most important issue that has to be dealt with before attempting any mathematical modeling concerns its goal, or more specifically, questions that the model is to answer. Only having a specific aim in mind, one can decide on necessary simplifications of the very complex models that usually arise in any biomathematical applications and methodology that can be used to solve the problems posed. The models presented in this chapter show, among others, how to include drug resistance in mathematical description, why it is crucial to do so, how mathematical analysis can help elucidate phenomena observed in experiments and, finally, how to evaluate quality of treatment protocols before they are applied clinically.

This chapter is structured as follows: First, a biological background is given, presenting current state of knowledge about biochemical processes responsible for drug resistance. Then, basic assumptions shared by all discussed cases are introduced. They are followed by a presentation of several mathematical models describing dynamics of cancer cells population under a single drug treatment, from the simplest, two-compartmental, to infinite dimensional ones. Afterwards, problems of multidrug treatment and multidrug resistance are addressed. In addition to analysis of the models, an optimization problem is stated, aiming at development of therapy protocols that destroy as many cancer cells as possible while reducing the toxic effects of chemotherapeutic agents on healthy cells. Finally, applicability of the results is discussed and open questions are presented.

2. BIOLOGICAL BACKGROUND

The ultimate task of drugs used in chemotherapy is killing cancer cells. However, drug absorption, distribution, metabolism and excretion is regulated by molecular processes taking place inside all cells. It is now widely recognized that those regulatory processes can be significantly altered by mutations that occur in cancer cells, leading to drug resistance. More specifically, three major mechanisms contributing to drug resistance have been identified (Szakacs et al., 2006; Takemura et al., 1998):

(1) Decreased uptake of water-soluble drugs which require transporters to enter cells.
(2) Various changes in cells that affect the capacity of cytotoxic drugs to kill cells, including alterations in cell cycle, increased repair of DNA damage, reduced apoptosis and altered metabolism of drugs.
(3) Increased energy-dependent efflux of hydrophobic drugs that can easily enter the cells by diffusion through the plasma membrane.

Moreover, mutation of a single gene can result in a resistance not only to a distinct cytotoxic agent, but also in cross-resistance to a whole range of drugs with different structures and cellular targets. This phenomenon is called multiple drug resistance (MDR) (Doherty and Michael, 2003; Krishna and Mayer, 2000; Liscovitch and Lavie, 2003; Ozben, 2006).

Drug resistance can be either intrinsic, exhibited by cells from the very beginning of a treatment or acquired, triggered by the chemotherapeutic agent (Biedler, 1992; Baer, 2005). From modeling perspective, acquired drug resistance determines the detailed structure of the models while intrinsic drug resistance contributes additionally to the distribution of initial conditions (number of cells in each compartment).

Regardless of the type of molecules that play a role in acquired drug resistance, it begins with a mutation of a specific gene, triggered by a drug (for an exemplary list of those genes and drugs, see e.g., Banerjee et al., 1995; Doherty and Michael, 2003; Takemura et al., 1997; Volm and Matter, 1996). This has been first considered in a point mutation model of Goldie and Coldman (1979) and then in the framework of gene amplification by Harnevo and Agur (1991, 1992, 1993). The main idea is that there exist spontaneous or induced mutations of cancer cells towards drug resistance and that the scheduling of treatment should anticipate these mutations. Depending on cell population and genes involved, this can lead either to classical irreversible mutation models (e.g., Goldie and Coldman, 1979) or models of reversible, multistage mechanisms, including a gradual increase in number of stages (Kimmel and Axelrod, 1990; Kimmel et al., 1992; Axelrod et al.,

1993). In view of recent progress in molecular biology, both approaches are justifiable.

Any mutation can alter gene expression within a cell, which results in a changed cell behavior. Gene expression is a complex process regulated at several stages in the synthesis of proteins (Lewin, 1999). Apart from the regulation of DNA transcription, the best-studied form of regulation, the expression of a gene may be controlled during RNA processing and transport, RNA translation, and the post-translational modification of proteins. The degradation of proteins and intermediate RNA products can also be regulated in the cell. The proteins performing the regulatory functions mentioned above are coded by other genes. This gives rise to genetic regulatory systems, structured by networks of regulatory interactions between DNA, RNA, proteins, and small molecules (de Jong, 2000). If, for example, a mutation disrupts such regulatory interaction at any stage, it can give rise to a single-step drug resistance, as in classical models (Coldman and Goldie, 1983; Panetta, 1998). On the other hand, if so called gene amplification process is involved additional gene copies are acquired. This leads to dramatic increase in transcription products. In that case multicompartmental models describing populations with various levels of drug resistance are more appropriate. For example, models with gene amplification predict the observed pattern of gradual loss of resistance in cancer cells placed in a non-toxic medium (Brown *et al.*, 1981; Kaufman *et al.*, 1981). The gene amplification model was extensively simulated and also resulted in recommendations for optimized therapy (Smieja and Swierniak, 2003, 2005; Smieja *et al.*, 2001). Numerous experiments proved that the process of gene amplification may be reversible (i.e., cells with increased number of gene copies tend to become extinct) whereas, in some cases, it is stable (i.e., the amplification persisted even after the selective agent causing it has been removed) (Kimmel *et al.*, 1992).

Both the multistage stepwise models of gene amplification (or, more generally, of transformations of cancer cells), and classical approaches lead to new mathematical problems. Though each of those calls for a different methodology, gene amplification models are more general, therefore more space will be devoted to them.

Regardless of the source of drug resistance, models incorporate it by dividing cell population into separate compartments, representing subpopulations of different drug sensitivity. The number of compartments and flow among them depends on the assumptions about the mechanisms causing drug resistance.

Taking into account biological background presented in this section, it is clear that any mathematical model combining cancer chemotherapy with drug resistance should be based on the following assumptions:

- The process of gaining resistance to any agent is reversible, either by nature, or with help of other drugs. Total, irreversible drug resistance is only a special case of a general model and leads to uncontrolled growth of resistant subpopulation.
- If gene amplification underlies the processes drug resistance, various levels of resistance must be taken into account (leading eventually to an infinite-dimensional model).

It should be stressed that the compartmental models shown here, though developed with cancer chemotherapy in mind, can be easily applied to describe evolution of drug resistance to antibiotics or other drugs, since in many cases the molecular mechanism underlying them are the same (Farooq and Mahajan, 2004; Tan *et al.*, 2000; Turriziani *et al.*, 2003; Wernsdorfer and Noedl, 2003).

3. PRELIMINARIES FOR MATHEMATICAL MODELS

All models presented in subsequent sections are of stochastic nature. This results from stochasticity of mutations that make cells drug resistant. Therefore, model parameters will correspond to probabilities of particular events — mutations, cell division, cell death, change of cell cycle phase, etc. All cancer cells are divided into subpopulations, grouping cells of the same type — exhibiting the same level of drug resistance or being in the same cell cycle phase (when phase-specific chemotherapy is considered). In some cases, original notation from cited papers has been changed to facilitate easier comparison of different models.

Variables N_i denote average number of cells in the ith compartment (in other words: of ith type), while variables $u(u_i)$ represent doses of drugs, or, more accurately, drugs' effect on target cells. For simplicity, no pharmacokinetics or pharmacodynamics is included in any model. These are briefly discussed in the Concluding Remarks Section at the end of this chapter.

The lifespans of all cells are independent exponentially distributed random variables with means $1/\lambda_i$ for cells of type i. It is assumed that at the end of its life the cell either undergoes a division producing two daughter cells of the same or a different type, or die (the last case will be referred to as an ineffective cell division). Parameter λ_i is assumed to be a constant, not a function of u. Though it might appear too farfetching an assumption, it is justified unless models are extended to include intracellular processes. Chemotherapeutic agents can indeed affect cell lifespans but, for models operating on average cell numbers, this influence can be incorporated in a different way.

The ultimate goal of modeling of cancer population dynamics, chemotherapy and drug resistance is to find the best possible treatment protocol. It can be formulated mathematically, as an optimization problem. Though some authors consider the (minimal) number of cancer cells at the end of therapy to be a sufficient indicator of treatment quality (e.g., Costa and Boldrini, 1997), it seems that negative cumulative effect of the administered drug(s) should be explicitly included in the performance index. Therefore, in case of a single drug the performance index to be minimised should take the following form

$$\min_u \leftarrow J = \sum_{N_i} r_i N_i(T) + r \int_0^T u(\tau) d\tau \qquad (1)$$

where T denotes time at the end of therapy, r, r_i are weighing factors. Further in the text, slight changes will be introduced to this performance index, corresponding to model modifications.

As mentioned in Section 2, acquired drug resistance implies zero initial condition in all resistant subpopulations. Intrinsic drug resistance, in turn, yields non-zero initial conditions in the resistant subpopulations. However, models presented here will be general enough to accommodate both types of drug resistance, as well as various genetical sources of resistance. The one exception would be that of gene amplification based drug resistance coupled with probabilistic distribution of initial conditions.

Regardless of the nature of drug resistance, positive feedback is a part of all models. It stems from the very nature of cancer cells, whose one of the main characteristics is a limitless replicative potential (Hanahan and Weinberg, 2000).

4. DRUG RESISTANCE AND A SINGLE CHEMOTHERAPEUTIC AGENT

4.1. A Simple, Two-Compartmental Model

The most basic model consists of only two compartments, representing sensitive and drug resistant subpopulations. It has been a subject analyzed by many researchers but arguably the most recent comprehensive studies can be found in (Ledzewicz and Schättler, 2002, 2006). Their results are briefly summarized and extended in this section.

The model based on the assumption that the mutation being the basis for drug resistance caused the cell to acquire one additional copy of a certain gene. Therefore, a sensitive mother cell produces two daughter cells, one of which remains sensitive, while the other changes into a resistant one with a probability of γ. Similarly, if a resistant cell undergoes cell division, one of the offspring remains sensitive. The other one may mutate back into a sensitive cell with a probability of d (it is assumed that no additional gene copies can be acquired).

Let us denote by N_0 and N_1 the average number of sensitive and drug resistant cells, respectively. Then, the system is described by the following set of equations

$$\begin{cases} \dot{N}_0 = (1 - 2u)\lambda_0 N_0 - \gamma(1-u)N_0 + dN_1 \\ \dot{N}_1 = \lambda_1 N_1 - dN_1 + (1-u)\gamma N_0 \end{cases} \quad (2)$$

with u satisfying

$$0 \leq u \leq u_{\max} \leq 1 \quad (3)$$

where $u = 0$ and $u = u_{\max}$ correspond to no drug being used and a full dose, respectively. If the drug being used is capable of killing all sensitive cells, then $u_{\max} = 1$.

Even such simple model can be used to show necessity of including drug resistance in the model. If it was neglected, using a constant dose treatment

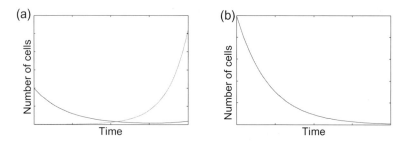

Fig. 1. Comparison of models (a) with and (b) without drug resistance. In (a) solid line represents the size of a sensitive subpopulation and dotted line the resistant one. Though α is much smaller than d in model (2), the resistant subpopulation grows uncontrollably. Both simulations started with the same initial number and lifespans of sensitive cells and identical control u.

would seemingly destroy cancer population, while in fact it would thrive (Figure 1).

Necessary conditions for the optimal control $u_{opt}(t)$ minimizing the performance index (1) can be derived using Pontryagin's Maximum Principle (Pontryagin et al., 1964). The Hamiltonian in this case takes the following form:

$$H = ru + p_0 \left[\lambda_0 N_0 (1 - 2u) - \gamma(1 - u) N_0 + dN_1 \right] \\ + p_1 \left[\lambda_1 N_1 - dN_1 + (1 - u)\gamma N_0 \right] \quad (4)$$

with adjoint variables $p(t)$ satisfying

$$\begin{cases} \dot{p}_0 = -p_0 \left[\lambda_0 (1 - 2u) - \gamma(1 - u) \right] - p_1 \gamma (1 - u) \\ \dot{p}_1 = -p_0 d - p_1 (\lambda_1 - d) \\ p_0(T) = r_0, \quad p_1(T) = r_1 \end{cases} \quad (5)$$

Since the Hamiltonian is linear with respect to u and u is bounded, we obtain

$$u_{opt}(t) = \begin{cases} 0 & \text{if } r - N_0 \left(2p_0 \lambda_0 - \gamma(p_0 - p_1) \right) > 0 \\ u_{max} & \text{if } r - N_0 \left(2p_0 \lambda_0 - \gamma(p_0 - p_1) \right) < 0 \end{cases}. \quad (6)$$

This implies that, unless singular control is also optimal, the optimal solution must be a *bang-bang* control. Then, to find optimal chemotherapy protocol, one must find only switching times between administering full dose and a rest period, in which $u = 0$. In Ledzewicz and Schättler (2006) it is shown that singular controls are not optimal if

$$\gamma N_0 < (2\lambda_0 - \gamma) N_1. \qquad (7)$$

This result can be interpreted in the following way: If the resistant subpopulation is small, then the optimal treatment should be in the form of full doses interlaced with rest periods. However, once the resistant subpopulation builds up, singular solutions might become optimal, implying administration of partial doses.

One should nevertheless remember that these solutions are optimal in the sense of a given performance index and do not necessarily provide extinction of cancer population. Indeed, in the model described by Eq. (2), even if $u \equiv 1$ on $[0, \infty)$, the total number of cancer cells can be reduced only if $\lambda_0 N_0 > \lambda_1 N_1$. This is usually true at the beginning of therapy, but due to its total resistance the subpopulation represented by N_1 eventually takes over and the cancer population will grow exponentially. Providing favorable parameter values, however, this growth might take a long time (longer than the patient's life). In another case the cancer population will actually disappear (the variables represent averages so in some cases the actual number of cells might drop to zero). These are precisely the cases when chemotherapy is successful.

Given unfavorable system asymptotic behavior, the situation in which a part of cancer population is totally drug resistant is unacceptable. Therefore it is clear that whenever possible, alternative drug, to which this subpopulation is sensitive, should be used. It leads to modeling of multidrug therapy, dealt with farther in this chapter.

The model presented in this section, although gives good qualitative insights into dynamics of cancer population behavior, confirmed by experimental research (the eventual prevalence of drug resistant subpopulation, advantages of *bang-bang* character of treatment), is based on two oversimplified assumptions, acceptable only for rough, initial analysis. First, it is assumed that the mutation responsible for drug resistance consists in acquisition of only one copy of a gene. In fact, if such gene amplification is possible, it might result in acquisition of yet additional gene copies, yielding different levels of resistance. In fact, such model is more appropriate in cases when a point mutation disable control mechanisms built into cellular signaling pathways. In the following section another model will be introduced that allows for multiple gene amplification.

4.2. Evolution of Drug Resistance Stemming from Gene Amplification

In this section certain model of cell population with evolving drug resistance caused by gene amplification or other mechanisms is presented. The model, based on results of (Axelrod *et al.*, 1994; Harnevo and Agur, 1993; Kimmel and Axelrod, 1990) is general enough to accommodate different interpretations. Among others, it is a good representation of resistance to Methotrexate (MTX) — a clinically important agent being used in the treatment of malignancies including acute lymphocytic leukemia, osteosarcoma, carcinomas of the breast, head and neck, choriocarcinoma and non-Hodgkin's lymphoma (Banerjee *et al.*, 1995). The original model and its properties were thoroughly discussed in e.g., Swierniak *et al.* (1998, 1999).

In this model gene amplification is considered the biological process leading to drug resistance. Contrary to the previous example, however, cells can gain more than one additional gene copy and the number of those determines the resistance level of a cell (for example, with increased number of DHFR or CAD gene copies). This model is a good fit for experimental data where a large number of additional gene copies has been found (e.g., Schimke *et al.*, 1984).

The compartments are numbered according to increasing numbers of gene copies and corresponding level of drug resistance. Cells of type 0, with no copies of the gene, are sensitive to the cytostatic agent. The resistant subpopulation consists of cells of types $i = 1, 2, \ldots$. Due to a mutational event a cell can acquire or lose a copy of a gene that makes it resistant to the agent. The probability of mutational event in a sensitive cell is of several orders smaller than the probability of the change in number of gene copies in a resistant cell. Since we do not limit the number of gene copies per cell, the number of different cell types is denumerably infinite. This assumption is often a source of misunderstanding. It should be stressed here that it does not postulate infinite number of gene copies — instead, it makes it possible to analyse asymptotic properties of the model and estimate e.g., stability conditions which otherwise would be always subject to an *a priori* chosen threshold in the number of gene copies (see later in this section).

Cell division and the change of the number of gene copies are stochastic processes with the following hypotheses:

(1) The lifespans of all cells are independent exponentially distributed random variables with means $1/\lambda_i$ for cells of type i.
(2) A cell of type $i \geq 1$ may mutate in a short time interval $(t, t + dt)$ into a type $i + 1$ cell with probability $b_i dt + o(dt)$ and into type $i - 1$ cell with probability $d_i dt + o(dt)$. A cell of type $i = 0$ may mutate in a short time interval $(t, t + dt)$ into a type 1 cell with probability $\alpha dt + o(dt)$, where α is several orders of magnitude smaller than any of b_i and d_i.
(3) The drug action results in fraction u_i of ineffective divisions in cells of type i (please note that there is only one drug under consideration here with different effectiveness denoted by subscript i).
(4) The process is initiated at time $t = 0$ by a finite population of cells of different types.

Under these assumptions, the model is described by the following system of ODE's

$$\begin{cases} \dot{N}_0(t) = (1 - 2u_0) \lambda_0 N_0 - \alpha N_0 + d_1 N_1 \\ \dot{N}_1(t) = (1 - 2u_1) \lambda_1 N_1 - (b_1 + d_1) N_1 + d_2 N_2 + \alpha N_0 \\ \ldots \\ \dot{N}_i(t) = (1 - 2u_i) \lambda_i N_i - (b_i + d_i) N_i + d_{i+1} N_{i+1} + b_{i-1} N_{i-1} \\ \ldots \end{cases} \quad (8)$$

In Polanski et al. (1997) and Swierniak et al. (1998, 1999), only the simplest case has been discussed in which the resistant cells are insensitive to drug's action, and there are no differences between parameters of cells of different type:

$$\begin{cases} \dot{N}_0 = (1 - 2u) \lambda N_0 - \alpha N_0 + d N_1 \\ \dot{N}_1 = \lambda N_1 - (b + d) N_1 + d N_2 + \alpha N_0 \\ \ldots \\ \dot{N}_i = \lambda N_i - (b + d) N_i + d N_{i+1} + b N_{i-1}, \quad i \geq 2 \\ \ldots \end{cases} \quad (9)$$

First, the dynamics of resistant subpopulation has been analyzed there. The resulting autonomous system, in which the sensitive cells are instantly annihilated, and there is no influx of new resistant cells led to discovery of stability conditions. It occurred that in addition to a condition that is obvious from biological point of view, i.e., $d > b$, an additional one must be satisfied, i.e., $\sqrt{d} - \sqrt{b} > \sqrt{\lambda}$. This result helped to explain the phenomena that previously escaped rigorous explanation — that even if the probability of gaining resistance is smaller than of losing it, the resistant subpopulation can grow and eventually prevail in the whole population. It must be stressed that stability conditions for a finite dimensional system would be different and depend on number of compartments. This is best illustrated in Figure 2. If one assumed that only a limited number of mutations can take place, cutting in effect model (9) into a finite-dimensional system of an arbitrarily chosen order, results of any analysis would depend on the choice of the maximum number of mutations that is allowed. Provided that an additional mutation happened (one more than the arbitrarily chosen maximum), the model would exhibit dynamics entirely different from the real system. This is illustrated in Figure 2 — where behavior of two systems, distinguished by one additional compartment only is compared. With all other parameters being the same (including value of constant control), the higher-order system is unstable, while lower-order model would suggest otherwise.

Since the autonomous system is linear, it is possible to look at it from the perspective of linear systems theory, with N_0 and N_1 representing input and output, correspondingly. The model can be decomposed into two parts, as shown in Figure 3. Such decomposition will also be used in more general models introduced in the subsequent section. The infinite dimensional part can be described then by means of a transfer function of the following form

$$K_1(s) = \frac{N_1(s)}{N_0(s)} = \frac{s - (\lambda - b - d) - \sqrt{(s - (\lambda - b - d))^2 - 4bd}}{2d}. \quad (10)$$

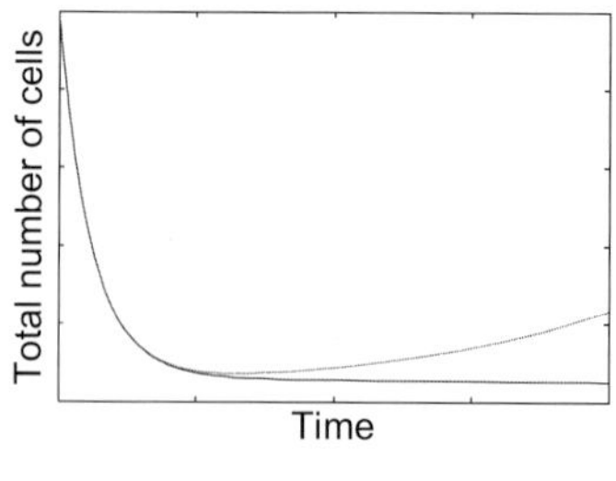

Fig. 2. Two similar finite-dimensional models of drug resistance stemming from gene amplification and described by the cut-off model (9): solid line illustrates the model with one compartment less than the one whose behavior is represented by dotted line. All parameters, except for the number of compartments are the same.

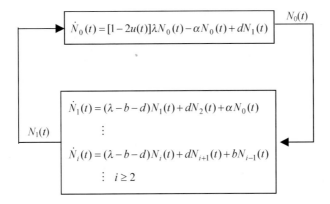

Fig. 3. Decomposition of the basic model (9).

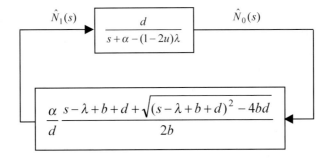

Fig. 4. Block diagram of the system (9) with constant control u.

This result, coupled with application of standard control theory methods, such as Nyquist stability criterion, helped to determine stability conditions and transform system description. Assuming constant dose of the drug, the first equation is also linear and therefore the system can be presented as in the Figure 4. It has been shown (Polanski et al., 1998) that the constant dose of the drug can lead to the extinction of drug resistant subpopulation, providing that stability conditions given above for resistant subpopulation hold and

$$u > \frac{1}{2} + \frac{\alpha}{d - b - \lambda + \sqrt{(b + d - \lambda)^2 - 4bd}}. \qquad (11)$$

This result shows if destruction of cancer cells is possible at all, taking into account the constraint (3) and gives the value of control u that should be used if this was the only goal. However, since the chemotherapy also yields negative effects, such constant control is usually unacceptable. Instead, the solution should minimize a performance index defined similarly to (1).

Due to infinite number of compartments, a special approach to optimization problem is required. First, let us modify slightly the performance index and instead of form (1) let us use the following:

$$J = \sum_{i \geq 0} N_i(T) + r \int_0^T u(\tau) d\tau. \tag{12}$$

It would be unreasonable to have a different weighing factor for each type of a cell. Therefore, all of them are to be equal, and, since in that case it is the ratio of weighing factors for summation and integral part that matters, one of them can be set to 1.

Then, one should be able to evaluate the first term in the performance index and develop an effective method to deal with optimization of a system governed by (9). It has been shown that in case of finite number of non-zero initial conditions (quite reasonable assumption taking into account model application) the system description can be transformed into one integro-differential equation (Smieja et al., 2001)

$$\dot{N}_0(t) = (1 - 2u(t))\lambda N_0(t) - \alpha N_0(t) + d\alpha \int_0^t \phi_1(t - \tau) N_0(\tau) d\tau$$

$$+ d \sum_{k \geq 1} N_k(0) \phi_k(t) \tag{13}$$

where

$$\phi_k(t) = \frac{k}{d} \left(\sqrt{\frac{d}{b}}\right)^k \frac{I_k(2\sqrt{bd}t)}{t} e^{-(b+d-\lambda)t}, \tag{14}$$

I_k – modified Bessel function of the k-th order.

Moreover, denoting $N_\Sigma(t) = \sum_{i \geq 1} N_i(t)$, the following relation holds true (Smieja et al., 2001):

$$J = N_0(T) + \sum_{k \geq 1} N_k(0) N_\Sigma^k(T) + \int_0^T \left[\alpha N_\Sigma^1(T - \tau) N_0(\tau) + r u(\tau)\right] d\tau, \tag{15}$$

where

$$N_\Sigma^k(t) = e^{\lambda t} - k e^{\lambda t} \left(\sqrt{\frac{d}{b}}\right)^k \int_0^t \frac{I_k(2\sqrt{bd}\tau)}{\tau} e^{-(b+d-\lambda)\tau} d\tau. \tag{16}$$

Such transformations, both of the system description and the performance index can be used in a more general model, introduced in the subsequent sections. Therefore, necessary conditions for optimal control will be presented there and the next several models will be introduced without detailed analysis.

The results presented above concerned a finite number of non-zero initial conditions, which means that the analysis dealt only with a subspace of all mathematically possible initial conditions. The case of infinite number of zero initial conditions, though tackled in part in e.g., Swierniak *et al.* (1998), is largely still an open problem. Though this case might initially look superficial, it has a valid biological interpretation. It allows analysis of behavior of a system, in which initial conditions are not known precisely, but given by a probability distribution. Such models would be applicable for intrinsic drug resistance or for analysis of induced drug resistance, after change of a treatment.

4.3. Partial Sensitivity of the Resistant Subpopulation

In the previous section it has been assumed for the simplified model that although multiple additional gene copies can be acquired, a single mutation leads to total resistance to a drug. Though convenient for model analysis, this is an oversimplification and it is quite straightforward to lessen this assumption. It is more reasonable to assume that only after gaining several additional gene copies the resistance is almost total (or, at least, that the higher levels of resistance are indistinguishable). Then the following set of equations is obtained (Swierniak and Smieja, 2005)

$$\begin{cases} \dot{N}_0 = 1 - 2u\lambda_0 N_0 - \alpha N_0 + d_1 N_1 \\ \dot{N}_1 = (1 - 2\mu_1 u)\lambda_1 N_1 - (b_1 + d_1)N_1 + d_2 N_2 + \alpha N_0 \\ \ldots \\ \dot{N}_{l-1} = 1 - 2\mu_{l-1} u \, \lambda_{l-1} N_{l-1} - (b_{l-1} + d_{l-1})N_{l-1} + d_l N_l + b_{l-2} N_{l-2} \\ \ldots \\ \dot{N}_i = \lambda N_i - (b+d)N_i + dN_{i+1} + bN_{i-1}, \quad i \geq l \\ \ldots \end{cases}$$

(17)

where $0 \leq \mu_i \leq 1$ μ_i are "efficiency factors," determining the effectiveness of the drug in relation to particular type of cell. Due to general assumptions about the model, presented at the beginning of this section, these factors satisfy $0 \leq \mu_i \leq \mu_{i-1} \leq 1, i = 1, 2, \ldots, l - 1$.

In addition to partial drug resistance, such models make possible taking into account other effects exerted by a chemotherapeutic agent on a cell, including, for example, modification of a cell lifespan or probability of gene amplification or deamplification.

4.4. Phase-Specific Chemotherapy

Chemotherapy introduced to the models in the preceding sections was assumed to affect all sensitive cells, regardless of a phase of a cell cycle in which those cells were. Since in general it is not true in realistic biological systems, the next logical step is to include cell cycle in the models.

The cell cycle is composed of a sequence of phases undergone by each cell from its birth to division. Starting point is a growth phase G1, after which the cell enters a phase S where DNA synthesis occurs. Then a second growth phase G2 takes place in which the cell prepares for mitosis or phase M, in which the cell divides. Each of the two daughter cells can either reenter phase G1 or for some time may simply lie dormant in a separate phase G0 until reentering G1, thus starting the entire process all over again. The simplest mathematical models which describe optimal control of cancer chemotherapy treat the entire cell cycle as one compartment (e.g., Swierniak, 1994, 1995). Actually, each drug affects cell being in particular phase and it makes sense to combine these drugs so that their cumulative effect on the cancer population would be the greatest. Very often, phase-specific chemotherapy has been considered only in the finite-dimensional case, without any regard to problems stemming from increasing drug resistance (Swan, 1990; Swierniak *et al.*, 1997). Combining infinite dimensional model of drug resistance with the phase-specific model of chemotherapy should move mathematical modeling much closer to its clinical application.

The cell cycle is modeled in the form of compartments which describe the different cell phases or combine phases of the cell cycle into clusters. The number of compartments depends on the number of processes that are taken into account. For example, killing agents (the only control considered in the previously shown models) should be applied in the G2 + M phase which makes sense from a biological standpoint for a couple of reasons. First, in mitosis M the cell becomes very thin and porous. Hence, the cell is more vulnerable to an attack while there will be a minimal effect on the normal cells. Second, chemotherapy during mitosis will prevent the creation of daughter cells. Therefore, chemotherapy usually combines a killing agent with a blocking agent which slows down the development of cells in the synthesis phase S and then releases them at the moment when another G2 + M specific anticancer drug has maximum killing potential (e.g., Brown and Thompson, 1975). This strategy may have the additional advantage of protecting the normal cells which would be less exposed to the second agent (e.g., due to less dispersion and faster transit through G2 + M). The resulting models can be either 2- or 3-compartmental, with one compartment for G2 + M phases and the other(s) for G0 + G1 combined with or separate from the S phase (Swierniak et al., 1995, Swierniak and Kimmel, 2006).

One of the major problems in chemotherapy of some leukemias is constituted by the large residuum of dormant G0 cells which are not sensitive to most cytotoxic agents. Therefore, it is reasonable to apply a drug (recruiting agent) that recruits these cells into the cycle before using a killing agent. Then, the resulting model consists of separate compartments for the G0, G1 and S + G2 + M phases and includes such a recruiting agent.

Below, an example of phase-specific chemotherapy model combined with drug resistance is presented. Since the main idea is to introduce new compartments into the existing model, the same approach can be done in any cell cycle analysis. The general framework for that will be given in the next section.

The sensitive subpopulation consists of two types of cells: type $i = 0$, being in the phase $G_1 + S$ and $i = 1$, being in the phase $G_2 + M$. The phase-specific drug (a killing agent) affects only cells of type $i = 1$. Then the following set of equations can represent the system dynamics (Swierniak and Smieja, 2005):

$$\begin{cases} \dot{N}_0 = -\lambda_0 N_0 + (2\lambda_1 - \alpha)(1 - u)\lambda_1 N_1 + dN_2 \\ \dot{N}_1 = -\lambda_1 N_1 + \lambda_0 N_0 \\ \dot{N}_2 = \lambda_2 N_2 - (b + d)N_1 + \alpha(1 - u)N_1 + bN_3 \\ \cdots \\ \dot{N}_i = \lambda N_i - (b + d)N_i + dN_{i+1} + bN_{i-1}, \quad i \geq 3 \\ \cdots \end{cases} \quad (18)$$

Since drug resistance is assumed here to be total, it does not make sense to distinguish cell cycle phases for the resistant cells.

If both killing and blocking agent actions are modeled (denoted below by u_1 and u_2, respectively), the issue of resistance is much more complex (see Section 5). However, if only the resistance to the killing agent is considered, then, denoting by $i = 0, 1, 2$ compartments of cells being in the phase G_1, S, G_2 and M, correspondingly and by $i \geq 3$ compartments of resistant cells, the following description is obtained

$$\begin{cases} \dot{N}_0 = -\lambda_0 N_0 + 2(1-\gamma)(1-u_0)\lambda_2 N_2 + dN_3 \\ \dot{N}_1 = -(1-u_1)\lambda_1 N_1 + \lambda_0 N_0 \\ \dot{N}_2 = -\lambda_2 N_2 + (1-u_1)\lambda_1 N_1 \\ \dot{N}_3 = \lambda_3 N_3 - (b+d)N_2 + 2\gamma(1-u)N_2 + bN_4 \\ \ldots \\ \dot{N}_i = \lambda N_i - (b+d)N_i + dN_{i+1} + bN_{i-1}, \quad i \geq 4 \\ \ldots \end{cases} \quad (19)$$

4.5. General Compartmental Model

All models presented in the earlier sections can be described uniformly by the following state equation (Swierniak and Smieja, 2005):

$$\dot{N} = \left(\mathbf{A} + \sum_{i=0}^{m} u_i \mathbf{B}_i \right) N, \quad (20)$$

where $N = [N_0 N_1 N_2 \ldots N_i \ldots]^T$ is an infinite dimensional state vector, and:

$$\mathbf{A} = \begin{bmatrix} \tilde{\mathbf{A}}_1 & | & \mathbf{0}_1 \\ - & - & - \\ \mathbf{0}_2 & | & \tilde{\mathbf{A}}_2 \\ & | & \end{bmatrix}, \mathbf{B} = \begin{bmatrix} \tilde{\mathbf{B}}_i & | & \mathbf{0}_1 \\ - & - & - \\ & & \mathbf{0}_3 \end{bmatrix}, \quad (21)$$

$$\tilde{\mathbf{A}}_1 = \begin{bmatrix} a_{00} & a_{01} & \cdots & a_{0,l-1} & 0 \\ a_{10} & a_{11} & \cdots & a_{1,l-1} & 0 \\ \vdots & \vdots & \cdots & \vdots & 0 \\ a_{l-1,0} & a_{l-1,1} & \cdots & a_{l-1,l-1} & a_{l-1,l} \end{bmatrix},$$

$$\tilde{\mathbf{A}}_2 = \begin{bmatrix} c_1 & a_2 & a_3 & 0 & 0 & \cdots \\ 0 & a_1 & a_2 & a_3 & 0 & 0 & \cdots \\ 0 & 0 & a_1 & a_2 & a_3 & 0 & \cdots \\ \vdots & \vdots & \ddots & \ddots & \ddots & \ddots & \ddots \end{bmatrix},$$

$$\tilde{\mathbf{B}}_i = \begin{bmatrix} b^i_{0,0} & b^i_{0,1} & \cdots & b^i_{0,l-1} \\ b^i_{1,0} & b^i_{1,1} & \cdots & b^i_{1,l-1} \\ \vdots & \vdots & \cdots & \vdots \\ b^i_{l-1,0} & b^i_{l-1,1} & \cdots & b^i_{l-1,l-1} \end{bmatrix}$$

$u(t)$ — m-dimensional control vector $u = [u_0 \ u_1 \ u_2 \ldots u_{m-1}]^T$, $\mathbf{0}_1, \mathbf{0}_2, \mathbf{0}_3$ — zero matrices of dimensions $l \times \infty$, $\infty \times l - 1$ and $\infty \times \infty$, respectively, $l > m$.

It is important to note that model parameters satisfy the following relations: $a_3 > a_1 > 0$, and $a_2 < 0$. However, full problem analysis can be done in other possible cases (e.g., when no additional conditions are to be satisfied by parameters a_1, a_3), using exactly the same line of reasoning.

The performance index to be minimized is given by

$$J = \sum_{i=0}^{l-1} N_i(T) + r_1 \sum_{i=l}^{\infty} N_i(T) + r \sum_{k=0}^{m} \int_0^T u_k(\tau) d\tau \qquad (22)$$

This general model allows even for including special cases of multidrug therapy. However, it is valid only in cases when two of the following conditions are met:

(1) each drug affects cells of different type (true in the basic model of a killing and a blocking agent treatment) and
(2) either the molecular source of resistance to each drug is the same (like in multidrug resistance (Kappelmayer et al., 2004; Nørgaard et al., 2004)) or the infinite subsystem representing gene amplification is

required for only one type of a drug (the basis for resistance to other drugs requires only a single mutation and there is only one level of resistance for each of them).

Due to these restrictions, drug resistance in other cases is discussed in Section 5 of this chapter.

As in the basic model analyzed in Section 4.2, it is convenient to present the model in the form of a block diagram shown in Figure 5 (Swierniak and Smieja, 2005). The first subsystem does not require parameters to meet any particular assumptions. The second subsystem is infinite dimensional, with tridiagonal system matrix, and does not include terms containing control variables $u_i(t)$. Since its form is exactly as in model (9), the transfer function describing relation between its output and input is given by (10), or, using the general notation that has just been introduced, by

$$K_1(s) = \frac{N_l(s)}{N_{l-1}(s)} = \frac{c_1 s - a_2 - \sqrt{(s-a_2)^2 - 4a_1 a_3}}{a_3 \quad 2a_1}. \tag{23}$$

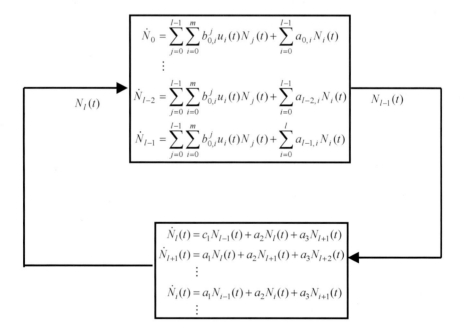

Fig. 5. Decomposition of the general model.

For constant u_is the entire system is linear. Therefore, it might be analyzed using standard linear system theory techniques, including determination of constant doses of drugs, for which the cancer population asymptotically vanishes (Smieja and Swierniak, 2003).

Taking into account that for a finite number of non-zero initial conditions

$$\sum_{i \geq l} N_i(t) = \sum_{i \geq l} N_i(0) N_\Sigma^i(t) + c_1 \int_0^t N_\Sigma^1(t-\tau) N_{l-1}(\tau) d\tau \quad (24)$$

where $N_\Sigma^i(t)$ is defined by (16) for $a_1 = b$, $a_2 = b + d - \lambda$, $a_3 = d$, $i = k = 1$ and $i = k - l + 1$ for $i \geq l, \ldots$, the performance index (22) to be minimized can be transformed into the following form

$$J = \sum_{i=0}^{l-1} N_i(T) + \int_0^T \left[r_1 c_1 N_\Sigma^l(T-\tau) N_{l-1}(\tau) + r \sum_{k=0}^m u_k(\tau) \right] d\tau \quad (25)$$

(the term $r_1 \sum_{i \geq l} N_i(0) N_\Sigma^i(T)$ has been dropped since it represents the free component and is not affected by any control u_i).

Let us further denote

$$\tilde{x} = \begin{bmatrix} N_0 \\ \vdots \\ N_{l-1} \end{bmatrix}, \quad u = \begin{bmatrix} u_0 \\ \vdots \\ u_m \end{bmatrix} \quad (26)$$

and $\mathbf{C_k} = [c_j,]$, $c_k = 1$, $c_j = 0$ for $j \neq k$, $i = 1, 2, \ldots, l-1$.

Then, the last equation in the first subsystem, influenced directly by control, as presented on Fig. 5, can be transformed into an integro-differential form:

$$\dot{N}_{l-1}(t) = \sum_{j=0}^{l-1} \sum_{i=0}^m b_{l-1,i}^j u_i(t) N_j(t) + \sum_{i=0}^{l-1} a_{l-1,i} N_i(t)$$

$$+ a_{l-1,l} \int_0^t k_1(t-\tau) N_{l-1}(\tau) d\tau + c_1 \sum_{i \geq l} N_i(0) \phi_k(t), \quad (27)$$

where $\phi_k(t)$ is defined by (14) and $k_1(t) = \phi_1(t)$ is the inverse Laplace transform of $K_1(s)$, given by (23).

Similarly, other equations can also be rewritten in the same way leading to the transformation of the model (9) into the following form:

$$\dot{\tilde{x}} = h(u, \tilde{x}) + \int_0^t \tilde{f}(\tilde{x}, t, \tau) d\tau, \quad \tilde{x}(0) = \tilde{x}_0 \qquad (28)$$

where $h(\ldots), \tilde{f}(\ldots)$-respective l-dimensional vector functions

$$h_k(u, \tilde{x}) = \sum_{j=0}^{l-1} \sum_{i=0}^{m} b_{k,i}^j u_i(t) N_j(t) + \sum_{i=0}^{l-1} a_{k,i} N_i \qquad (29)$$

$$\tilde{f}_k(\tilde{x}, t, \tau) = \begin{cases} 0 & \text{for } k < l-1 \\ a_{l-1,l} k_1(t-\tau) N_{l-1}(t) & \text{for } k = l-1 \end{cases}. \qquad (30)$$

Following assumptions made in Section 3, all control variables are bounded:

$$0 \le u_k(t) \le u_{k\,\max} \le 1. \qquad (31)$$

Due to particular form of both performance index and the equation governing the model it is possible to find the solution to the problem, applying an appropriate version of Pontryagin's maximum principle (Pontryagin *et al.*, 1962).

The necessary conditions for optimal control are as follows (Swierniak and Smieja, 2005):

$$u^{\text{opt}}(t) = \arg\min_u H(u, p, \tilde{x}), \qquad (32)$$

where $p(t)$ — adjoint vector and

$$H(u, p, \tilde{x}) = r \sum_{k=0}^{m} u_k(t) + p^T(t) h(u, \tilde{x})$$

$$+ a_{l-1,l} \int_t^T p_{l-1}(\tau) k_1(t-\tau) N_{l-1}(\tau) d\tau \qquad (33)$$

$$\dot{p}^T(t) = -\left[q^T(t) + p^T(t) h_{\tilde{x}}(u, \tilde{x})) + \int_t^T p^T(\tau) \tilde{f}_{\tilde{x}}(t-\tau) d\tau \right] \qquad (34)$$

$$q(t) = \begin{bmatrix} 0 & \ldots & 0 & r_1 c_1 N_\Sigma^l(T-t) \end{bmatrix}^T$$

$$p_i(T) = 1, \quad i = 0, 1, \ldots, l-1. \qquad (35)$$

Let us denote

$$\Phi_i(t) = \frac{\partial H(u, p, \tilde{x})}{\partial u_i}. \tag{36}$$

Taking into account constraint (31) and bilinear form of (29), it can be proved that, in order to satisfy (32), the optimal control must satisfy

$$u_i(t) = \begin{cases} 0 & \text{if } \Phi_i(t) > 0 \\ u_{i\,\max} & \text{if } \Phi_i(t) < 0 \end{cases}. \tag{37}$$

Therefore, it must be either singular or of *bang-bang* type. If singular cases can be neglected, then to find optimal number of switches and switching times, a gradient method can be developed, following the line of reasoning presented in (Smieja et al., 2001).

Bang-bang controls, which are widely used as protocols in medical treatments, are the more natural choice as candidates for optimality, and it even has been observed numerically that singular protocols actually give the worst performance (Duda, 1997; Swierniak et al., 1996). In the papers (Ledzewicz and Schättler, 2002, 2002a) with the use of high-order necessary conditions for optimality singular arcs were indeed excluded from optimality and necessary and sufficient conditions for local optimality of bang-bang controls were developed for the models of phase specific chemotherapy (without drug resitance). In Ledzewicz and Schättler (2006), it was shown that in the simplest, second order model of drug resistance, in some cases one cannot exclude singularities. The question of optimality of singular arcs in higher order systems remains an open problem.

5. MULTIDRUG THERAPY AND DRUG RESISTANCE

It is unlikely that any new therapeutic will, at least initially, be used alone in the treatment of a cancer patient. Most patients will still undergo the currently approved treatment regimen that may include multidrug chemotherapy (Kellen, 2003; Merino et al., 2004; Robert and Jarry, 2003). This is also the conclusion of analysis of even simplest models of drug resistance — if resistant population is not affected by the single chemotherapeutic agent, it might grow exponentially unless a second drug is used. Below, two examples are reviewed, both assuming two chemotherapeutical agents and drug resistance.

5.1. A Two-Compartmental Model

In Panetta (1998), the simplest possible model has been analyzed. Only two compartments have been used there, similarly as in the basic model shown here in Section 4.1. In contrast to the model given by (2), however, the mutation that renders cells resistant to the first drug is assumed to be irreversible. Therefore, second subpopulation, resistant to the first drug would increase uncontrollably unless other agent is used. Therefore it is assumed that it is sensitive to the second drug u_2. The resulting mathematical model is as follows (the original notation has been changed to reflect notation used throughout this paper):

$$\begin{cases} \dot{N}_0 = (\lambda_0 - u_1) N_0 \\ \dot{N}_1 = \gamma u_1 N_0 + (\lambda_1 - u_2) N_1 \end{cases} \tag{38}$$

where γ is a probability of a mutational event leading to drug resistance.

For this model no formal optimization has been performed. However, even here, periodic and bang-bang treatment has been carefully investigated and compared to clinical data. This model has been recently updated, showing good fit to experimental data (Panetta *et al.*, 2006). Nevertheless, the assumptions of irreversible mutation and prohibition of the case of total resistance to the second drug are restricting and to analyze more general cases other approach is required.

5.2. A Four-Compartmental Model

In Section 4.5 a general case that could also represent multidrug treatment, was discussed. However, it was assumed that each drug affects cells of a single, distinct type. Here more realistic case (at least from multidrug therapy viewpoint) with two killing agents is analyzed.

Contrary to MDR models, where once cells become resistant, they are so to various agents, here it will be assumed that it is possible to be resistant to one drug while being sensitive to another. This is the case, for example, in breast cancer treatment, where cells can be resistant to Tamoxifen and sensitive to Doxorubicin (Lilling et al., 2002).

Let the first compartment ($i = 0$) consist of cells sensitive to both drugs, the second ($i = 1$) of cells sensitive to drug u_1 but resistant to drug u_2, the third ($i = 2$) of cells sensitive to drug u_2 but resistant to drug u_1, and the fourth ($i = 3$) of cells resistant to both agents. The probability of compartment change are denoted by q_1 ($N_0 \to N_1$), $q_2 (N_0 \to N_2)$, $s_1(N_1 \to N_3)$, s_2 ($N_2 \to N_3$), r_1 ($N_2 \to N_0$ and $N_3 \to N_1$), and r_2 ($N_1 \to N_0$ and $N_3 \to N_2$). Dynamics of such system is described by the following set of equations

$$\begin{cases} \dot{N}_0 = -\lambda_0 N_0 + (1-u_1)(1-u_2)(2-q_1-q_2)\lambda_0 N_0 \\ \qquad + (1-u_1)r_2\lambda_1 N_1 + (1-u_2)r_1\lambda_2 N_2 \\ \dot{N}_1 = -\lambda_1 N_1 + (1-u_1)(2-s_1-r_2)\lambda_1 N_1 \\ \qquad + (1-u_1)(1-u_2)q_1\lambda_0 N_0 + r_1\lambda_3 N_3 \\ \dot{N}_2 = -\lambda_2 N_2 + (1-u_2)(2-s_2-r_1)\lambda_2 N_2 \\ \qquad + (1-u_1)(1-u_2)q_2\lambda_0 N_0 + r_2\lambda_3 N_3 \\ \dot{N}_3 = -\lambda_3 N_3 + (2-r_1-r_2)\lambda_3 N_3 \\ \qquad + (1-u_1)s_1\lambda_1 N_1 + (1-u_2)s_2\lambda_2 N_2 \end{cases} \quad (39)$$

It has been shown in Ledzewicz and Schättler (2006) that also in that case the optimal solution has the *bang-bang* form. Moreover, in that paper authors indicate that it is never optimal to simultaneously start administering or withdraw bot drugs. However, proving that singular solutions cannot also be optimal is much more difficult here and remains an open problem. If multiple gene amplification was behind drug resistance, with different genes for different drugs, modeling the system as in Section 4 would be impossible.

6. CONCLUDING REMARKS

In this chapter different models of evolution of drug resistance in cancer cells have been discussed. All of them are probabilistic in the sense that they assume that cell division and mutations take place at random time instants (however, deterministic modeling yields similar system description with different interpretation of parameters). It has been shown that compartmental approach is a very convenient way of dealing with such systems. First, it enables including any new findings in the biological field in models. Second, mathematical description can be identical for cases in which biological basis of drug resistance is different. And, finally, application of methodology coming from systems and control theory is straightforward in such modeling approach, allowing for effective analysis of systems dynamics and searching for optimal treatment protocols.

In all these models the attempts at finding optimal controls have been confounded by the presence of singular and periodic trajectories, and multiple solutions. Only in a few cases this has been addressed so far and remains an open problem.

All possible applications of the mathematical models of chemotherapy are contingent on our ability to estimate their parameters. Recently there has been progress in that direction, particularly concerning precise estimation of drug action in culture and estimation of cell cycle parameters of tumor cells *in vivo*. The stathmokinetic or "metaphase 14 arrest" technique consists of blocking cell division by an external agent. Flow cytometry allows precise measurements of the fractions of cells residing in different cell cycle phase. The pattern of cell accumulation in mitosis M depends on the kinetic parameters of the cell cycle and is used for estimation of these parameters. Exit dynamics from G1 and transit dynamics through S and G2 and their subcompartments can be used to characterize very precisely both unperturbed and perturbed cell cycle parameters. A true arsenal of methods have been developed to analyze the stathmokinetic data. Application of these methods allow quantification of the cell-cycle-phase action of many agents. (Swierniak and Kimmel, 2006). New techniques in molecular biology, rapidly advancing in recent years, made it possible to discern causes of drug resistance on a molecular level. Progress in analysis of tumors, such as microarray and PCR techniques to study gene expression or immunostaining of target enzymes, offer increasing promise for individualization of patient selection. Increased experience with biochemical modulators, including biologic response modifiers, has opened the possibility for selective attack on specific mechanisms of drug resistance. Sophisticated pharmacokinetic modeling and pharmacogenetic testing of metabolic phenotypes can now be done to achieve optimal dosing with less risk of toxicity. Considerations of ultimate genetic mechanisms of antimetabolite effects, especially by programmed cell death, and relationships to mechanisms of cell cycle regulation offer exciting rationales for future drug development (Spears, 1995).

What is equally important, models presented here are general enough to be applied in most cases of drug resistance, taking into account that drug actions are not specific to tumor type (Zhang *et al.*, 2006). Furthermore, they can be also applied in analysis of drug resistance in treatment of other diseases such as AIDS, malaria, etc. (Uhlemann and Krishna, 2005). However, it does not imply that those models can represent all possible cases — it is the methodology of creating compartmental models that is universal. Indeed, even in a particular type of malignancy, there can arise subtypes distinguished by different parameters of a dynamical model or even model structure (see e.g., Kager *et al.*, 2005). Therefore choice of a particular model should be preceded by a careful analysis of the biochemical sources of drug resistance (if known). A key future challenge involves determining the relative quantitative contributions of each of these mechanisms to overall resistance.

Mathematical modeling of multidrug models that take into account drug resistance is still waiting for more refined models. In particular, models that analyze systems in which additional drugs are aimed at reducing drug resistance are of the utmost importance. Such agents already do exist (Ho and Piquette-Miller, 2006; Perez-Tomas, 2006), but mathematical description of their action is another open problem.

Neither pharmacokinetics nor pharmacodynamics have been taken into account in the presented models. One of the easiest way to do it is to include them in a standard way, e.g., as the first order kinetics. However, if the goal of modeling is to analyze spatial effects in a solid tumor or include intracellular processes in investigation, then other approaches are required here, that are beyond the scope of this chapter.

ACKNOWLEDGMENTS

The idea of the paper was conceived when visiting the Mathematical Biosciences Institute, Ohio State University, Columbus, Ohio, in Fall 2003.

I would like to thank the Director of MBI, Professor Avner Friedman, for invitation and research atmosphere. This work was partially funded by KBN Grant No. 3 T11A 029 28 in 2006. I acknowledge financial support from the European Commission 6th Framework Program MRNT-CT-2004-503661.

References

1. Axelrod DE, Baggerly KA and Kimmel M. Gene amplification by unequal chromatid exchange: probabilistic modeling and analysis of drug resistance data. *J. Theor. Biol.* 1993; **168**: 151–159.
2. Baer MM. Clinical significance of multidrug resistance in AML: current insights. *Clin. Adv. Hematol. Oncol.* 2005; **3**(12): 910–912.
3. Banerjee D, Ercikan-Abali E, Waltham M, Schnieders B, Hochhauser D, Li WW, Fan J, Gorlick R, Goker E and Bertino JR. Molecular mechanisms of resistance to antifolates, a review. *Acta. Biochim. Pol.* 1995; **42**(4): 457–464.
4. Biedler JL. Genetic aspects of multidrug resistance. *Cancer* 1992; **70**(6 Suppl): 1799–1809.
5. Brown PC, Beverly SM and Schimke RT. Relationship of amplified dihydrofolate reductase genes to double minute chromosomes in unstably resistant mouse fibroblasts cell lines. *Mol. Cell. Biol.* 1981; **1**: 1077–1083.
6. Coldman AJ and Goldie JH. A model for the resistance of tumor cells to cancer chemotherapeutic agents. *Math. Biosci.* 1983; **65**: 291.
7. Costa MI and Boldrini JL. Conflicting objectives in chemotherapy with drug resistance. *Bull. Math. Biol.* 1997; **59**(4): 707–724.
8. Doherty MM and Michael M. Tumoral drug metabolism: perspectives and therapeutic implications. *Curr. Drug. Met.* 2003; **4**: 131–149.
9. Duda Z. Numerical solutions to bilinear models arising in cancer chemotherapy. *Nonlinear World* 1997; **4**: 53–72.
10. Eisen M. *Mathematical Models in Cell Biology and Cancer Chemotherapy*, Lecture Notes in Biomathematics, Vol. 30. Springer Verlag, 1979.
11. Farooq U and Mahajan RC. Drug resistance in malaria. *J. Vector Borne Dis.* 2004; **41**(3–4): 45–53.
12. Fister KR and Panetta JC. Optimal control applied to cell-cycle-specific cancer chemotherapy. *SIAM J. Appl. Math.* 2000; **60**: 1059–1072.
13. Goldie JH and Coldman AJ. A mathematical model for relating the drug sensitivity of tumors to their spontaneous mutation rate. *Cancer Treat. Rep.* 1979; **63**: 1727–1733.
14. Hanahan D and Weinberg RA. The hallmarks of cancer. *Cell* 2000; **100**: 57–70.

15. Harnevo LE and Agur Z. The dynamics of gene amplification described as a multitype compartmental model and as a branching process. *Math. Biosci.* 1991; **103**: 115–138.
16. Harnevo LE and Agur Z. Drug resistance as a dynamic process in a model for multistep gene amplification under various levels of selection stringency. *Cancer Chemother. Pharmacol.* 1992; **30**: 469–476.
17. Harnevo LE and Agur Z. Use of mathematical models for understanding the dynamics of gene amplification. *Mutat. Res.* 1993; **292**: 17–24.
18. Ho EA and Piquette-Miller M. Regulation of multidrug resistance by pro-inflammatory cytokines. *Curr. Cancer Drug Targets* 2006; **6**(4): 295–311.
19. Jong de H. Modeling and simulation of genetic regulatory systems: a literature review. *J. Comput. Biol.* 2002; **9**(1): 67–103.
20. Kager L, Cheok M, Yang W, Zaza G, Cheng Q, Panetta JC, Pui CH, Downing JR, Relling MV and Evans WE. Folate pathway gene expression differs in subtypes of acute lymphoblastic leukemia and influences methotrexate pharmacodynamics. *J. Clin. Invest.* 2005; **115**(1): 110–117.
21. Kappelmayer J, Simon A, Kiss F and Hevessy Z. Progress in defining multidrug resistance in leukemia. *Expert Rev. Mol. Diagn.* 2004; **4**(2): 209–217.
22. Kaufman RJ, Brown PC and Schimke RT. Loss and stabilization of amplified dihydrofolate reductase genes in mouse sarcoma S-180 cell lines. *Mol. Cell. Biol.* 1981; **1**: 1084–1093.
23. Kellen JA. The reversal of multidrug resistance: an update. *J. Exp. Ther. Oncol.* 2003; **3**: 5–13.
24. Kimmel M and Axelrod DE. Mathematical models of gene amplification with applications to cellular drug resistance and tumorigenicity. *Genetics* 1990; **125**: 633–644.
25. Kimmel M, Axelrod DE and Wahl GM. A branching process model of gene amplification following chromosome breakage. *Mutat. Res.* 1992; **276**: 225–240.
26. Kimmel M, Swierniak A and Polanski A. Infinite-dimensional model of evolution of drug resistance of cancer cells. *J. Math. Syst. Estimation Control* 1998; **8**: 1–16.
27. Krishna R and Mayer LD. Multidrug resistance (MDR) in cancer. Mechanisms, reversal using modulators of MDR and the role of MDR in influencing the pharmacokinetics of anticancer drugs. *Eur. J. Pharm. Sci.* 2000; **11**: 265–283.
28. Liscovitch M and Lavie Y. Cancer multidrug resistance: a review of recent drug discovery research. *Idrugs* 2002; **5**: 349–355.

29. Ledzewicz U and Schättler H. Optimal bang-bang controls for a 2-compartment model of cancer chemotherapy. *J. Opt. Theory Appl.* 2002; **114**: 609–637.
30. Ledzewicz U and Schättler H. Analysis of a cell-cycle specific model for cancer chemotherapy. *J. Biol. Syst.* 2002a; **10**: 183–206.
31. Ledzewicz U and Schättler H. Drug resistance in cancer chemotherapy as an optimal control problem. *Disc. Cont. Dyn. Syst. Ser. B* 2006; **6**(1): 129–150.
32. Lewin, B. *Genes VII.* Oxford University Press, Oxford, 1999.
33. Lilling G, Nordenberg J, Rotter V, Goldfinger N, Peller S and Sidi Y. Altered subcellular localization of p53 in estrogen-dependent and estrogen-independent breast cancer cells. *Cancer Invest.* 2002; **20**(4): 509–517.
34. Martin RB. Optimal control drug scheduling of cancer chemotherapy. *Automatica* 1992; **28**: 1113–1123.
35. Merino V, Jimenez-Torres NV and Merino-Sanjuan M. Relevance of multidrug resistance proteins on the clinical efficacy of cancer therapy. *Curr. Drug Deliv.* 2004; **1**(3): 203–212.
36. Nørgaard JM, Olesen LH and Hokland P. Changing picture of cellular drug resistance in human leukemia. *Crit. Rev. Oncol./Hematol.* 2004; **50**: 39–49.
37. Ozben T. Mechanisms and strategies to overcome multiple drug resistance in cancer. *FEBS Lett.* 2006; **580**(12): 2903–2909.
38. Panetta JC. A mathematical model of drug resistance: heterogeneous tumors. *Math. Biosci.* 1998; **147**: 41–61.
39. Panetta JC, Evans WE and Cheok MH. Mechanistic mathematical modelling of mercaptopurine effects on cell cycle of human acute lymphoblastic leukaemia cells. *Br. J. Cancer* 2006; **94**(1): 93–100.
40. Perez-Tomas R. Multidrug resistance: retrospect and prospects in anti-cancer drug treatment. *Curr. Med. Chem.* 2006; **13**(16): 1859–1876.
41. Polanski A, Kimmel M and Swierniak A. Qualitative analysis of the infinite-dimensional model of evolution of drug resistance. In: *Advances in Mathematical Population Dynamics — Molecules, Cells and Man.* World Scientific, 1997, pp. 595–612.
42. Pontryagin LS, Boltyanskii VG, Gamkrelidze RV and Mishchenko EF. *The Mathematical Theory of Optimal Processes.* MacMillan, New York, 1964.
43. Robert J and Jarry C. Multidrug resistance reversal agents. *J. Med. Chem.* 2003; **46**(23): 4805–4817.
44. Schimke RT. Gene amplification in cultured cells. *J. Biol. Chem.* 1988; **263**: 5989–5992.

45. Smieja J, Swierniak A and Duda Z. Gradient method for finding optimal scheduling in infinite dimensional models of chemotherapy. *J. Theor. Med.* 2001; **3**: 25–36.
46. Smieja J and Swierniak A. Different models of chemotherapy taking inti account drug resistance stemming from gene amplification. *Int. J. Appl. Math. Comp. Sci.* 2003; **13**: 297–306.
47. Spears CP. Clinical resistance to antimetabolites. *Hematol. Oncol. Clin. North. Am.* 1995; **9**(2): 397–413.
48. Swan GW. Role of optimal control in cancer chemotherapy. *Math. Biosci.* 1990; **101**: 237–284.
49. Swierniak A, Polanski A and Kimmel M. Optimal control problems arising in cellcycle-specific cancer chemotherapy. *Cell Prolif.* 1996; **29**: 117–139.
50. Swierniak A, Kimmel M and Polanski A. Infinite dimensional model of evolution of drug resistance of cancer cells. *J. Math. Syst. Estimation Control* 1998; **8**(1): 1–17.
51. Swierniak A, Polanski A, Kimmel M, Bobrowski A and Smieja J. Qualitative analysis of controlled drug resistance model — inverse Laplace and semigroup approach. *Control Cybern.* 1999; **28**: 61–74.
52. Swierniak A. Some control problems for simplest differential models of proliferation cycle. *Appl. Math. Comp. Sci.* 1994; **4**: 223–232.
53. Swierniak A. Cell cycle as an object of control. *J. Biol. Syst.* 1995; **3**: 41–54.
54. Swierniak A and Kimmel M. Control theory approach to cancer chemotherapy: benefiting from phase dependence and overcoming drug resistance. In: *Tutorials in Mathematical Biosciences III*. Friedman A (ed.) Lecture Notes in Mathematics (1872). Springer-Verlag, Berlin Heidelberg, 2006, pp. 185–221.
55. Swierniak A and Smieja J. Analysis and optimization of drug resistant an phase-specific cancer chemotherapy models. *Math. Biosci. Eng.* 2005; **2**(3): 657–670.
56. Swierniak A, Polanski A, Smieja J and Kimmel M. Modelling growth of drug resistant cancer populations as system with positive feedback. *Math. Comp. Model.* 2003; **37**: 1245–1252.
57. Szakacs G, Paterson JK, Ludwig JA, Booth-Genthe C and Gottesman MM. Targeting multidrug resistance in cancer. *Nat. Rev. Drug Discov.* 2006; **5**(3): 219–234.
58. Takemura Y, Kobayashi H and Miyachi H. Cellular and molecular mechanisms of resistance to antifolate drugs: new analogues and approaches to overcome the resistance. *Int. J. Hematol.* 1997; **66**(4): 459–477.
59. Tan B, Piwnica-Worms D and Ratner L. Multidrug resistance transporters and modulation. *Curr. Opin. Oncol.* 2000; **12**(5): 450–458.

60. Turriziani O, Scagnolari C, Bellomi F, Solimeo I, Focher F and Antonelli G. Cellular issues relating to the resistance of HIV to antiretroviral agents. *Scand. J. Infect. Dis. Suppl.* 2003; **35**(Suppl. 106): 45–48.
61. Uhlemann AC and Krishna S. Antimalarial multi-drug resistance in Asia: mechanisms and assessment. *Curr. Top. Microbiol. Immunol.* 2005; **295**: 39–53.
62. Volm M and Mattern J. Resistance mechanisms and their regulation in lung cancer. *Crit. Rev. Oncog.* 1996; **7**(3–4): 227–244.
63. Wernsdorfer WH and Noedl H. Molecular markers for drug resistance in malaria: use in treatment, diagnosis and epidemiology. *Curr. Opin. Infect. Dis.* 2003; **16**(6): 553–558.
64. Wheldon TE. *Mathematical Models in Cancer Chemotherapy*, Bristol Medical Science Series. Hilger, 1988.
65. Zhang W, Shannon WD, Duncan J, Scheffer GL, Scheper RJ and McLeod HL. Expression of drug pathway proteins is independent of tumour type. *J. Pathol.* 2006; **209**: 213–219.

Chapter 15

BLADDER CANCER SCREENING BY MAGNETIC RESONANCE IMAGING

Lihong Li, Zigang Wang and Zhengrong Liang

Bladder cancer is the fifth leading cause of cancer-related deaths in the United States. A common test for the cancer is urine dipsticks or standard urinalysis. It is safe, but has a very poor specificity, as low as 70%. The finding is usually at a late stage and not able to provide accurate location and staging of the lesion. As the main method of investigating bladder abnormalities, fiberoptic cystoscopy is more accurate. But it is invasive, time-consuming, expensive, uncomfortable, and has risk of urinary track infection. Recently, virtual cystoscopy (VC) has been developed as an alternative means for bladder cancer detection and evaluation. In this chapter, we present a novel mixture-based computer-aided detection (CAD) system for VC using multi-spectral (T_1- and T_2-weighted) magnetic resonance (MR) images, where a better tissue contrast can be achieved as compared to computer tomography (CT) images. As multi-spectral images are spatially registered over three-dimensional space, information extracted from multi-spectral MR images is obviously more valuable than that extracted from each (CT or MR) image individually. In addition, the urine has significantly different T_1 and T_2 relaxations compared to the bladder wall, the MR-based VC procedure can be completely non-invasive. Because bladder tumors tend to develop gradually and migrate slowly from the mucosa into the bladder wall/muscle, our focus is on the mucosa layer by mixture image segmentation. In order to obtain both geometry and texture information, we scan the bladder at two states of nearly empty and nearly full. Our CAD system utilizes fully the multi-scan and dual-state MR images for tumor detection and evaluation. Experimental results show its feasibility towards screening of bladder tumors and follow-up of recurrences.

Keywords: Bladder cancer, virtual cystoscopy, magnetic resonance.

1. INTRODUCTION

Bladder cancer is the fifth leading cause of cancer-related deaths in the United States, primarily in older men, with a 3 : 1 ratio of men to women. In the last decade, the occurrence of bladder malignances has increased by 36% (Lamm et al., 1996). Over 56,000 cases of developed bladder carcinoma and more than 12,000 deaths were reported in 2002 (Jamal et al., 2002). The lifetime probability of developing the cancer is over 3% and the probability of dying from the cancer is approximately or slightly less than 1% (Cohen et al., 1992; Steiner et al., 1992). Early asymptomatic bladder cancer may be associated with occult bleeding (microscopic hematuria) or the presence of dysplastic cells in the urine.

A common test for bladder cancer is urine dipsticks or standard urinalysis measuring the peroxidase activity of hemoglobin, which is safe and inexpensive and can be performed at home. Its sensitivity can be achieved at 90% whereas the specificity is only approximately at 70% (Shaw et al., 1985). Those findings are usually at the very late stage and unable to provide accurate location and staging information of the tumor.

As most tumors appear as small growths arising from the inner bladder-wall surface in forms of polypoid, sessile, or abnormal plaques, currently available fiberoptic cystoscopy (OC) is the most accurate diagnostic procedure for detecting and evaluating the abnormalities. This method is invasive by inserting an endoscope through the urethra into the bladder. It is also uncomfortable, expensive, lacks an objective scale and has limited field-of-view (FOV). The method also has a risk of 5% to 10% rate of urinary track infection. Patients are usually reluctant for such examination. Therefore, a minimal or non-invasive, safe, and low-cost method to evaluate the bladder would be preferred by most patients.

Recently, virtual cystoscopy (VC) techniques have been investigated as an alternative means for studying bladder abnormalities. Based on computer tomography (CT) technologies, several spiral CT scanning protocols have tested on patients and clinical VC feasibility studies have been reported (Fenlon et al., 1997; Hussain et al., 1997; Merkle et al., 1998; Vining et al., 1996). Because the earliest stages of bladder lesion development are inside the mucosa with gradual extension into bladder muscle, a desirable "visual gradient" between bladder lumen and wall is required for differentiating the associated structures (Fielding *et al.*, 2002; Prout *et al.*, 1984). However, CT images cannot provide image contrast difference between the bladder wall and the lumen filled with urine. An invasive procedure is needed to obtain the contrast through bladder insufflation with room air or CO_2 via a Foley catheter. It is still less likely to obtain image contrast inside the wall for differentiation of normal and pathologically altered tissues. Therefore, CT-based VC can only detect the lesions with significantly-developed geometry information.

By magnetic resonance (MR) imaging technologies, VC has shown its potential in evaluation of bladder lesions (Chen *et al.*, 2000; Liang *et al.*, 1999; Li *et al.*, 2003a). MR imaging has the potential to differentiate pathologically altered tissues from the normal ones inside the bladder wall. It also provides multi-spectral (T_1- and T_2-weighted) information, as compared to the CT image. Furthermore, the intrinsic T_1 and T_2 contrast of the urine against the bladder wall eliminates completely the invasive air insufflation procedure. MR-based VC shall outperform the CT-based VC in early screening, evaluation of staging and follow up recurrence without the invasive contrasting procedure and the extra radiation to the patients.

2. METHODS

Earliest stages of bladder lesion development are in the mucosa with gradual extension into bladder muscle. Morphological difference and texture variation could appear on the bladder wall when the bladder lumen is at two different states, full of urine or near empty. The dual states provide both geometric and texture information. To suppress motion artifacts, multi-scan of transverse and coronal multi-spectral MR images were acquired at empty and full states, respectively. Our computer aided diagnosis and detection (CAD) system focuses on the mucosa layer of the bladder wall for identifying bladder lesions.

2.1. MR Image Protocols

Multi-spectral MR images were acquired by a Phillips 1.5 T Edge whole-body scanner with the body coil as the transceiver. A spoiled-GRASS sequence was employed to acquire T_1-weighted transverse and coronal images with parameter of 256 × 256 matrix size, 38 cm FOV, 1.5 mm slice thickness, 3 ms T_E, 9 ms T_R, 30° flip angle, and one-scan average. Correspondingly, an axial FSE (fast spin-echo) sequence was used to collect T_2-weighted transverse and coronal images with the same acquisition location and parameters, except for 96 ms T_E, 12167 ms T_R, 90° flip angle, and two-scan average.

The advantages of MR images are (1) excellent contrast by T_1 and T_2 relaxations between urine and bladder wall, resulting in a non-invasive approach, (2) good contrast for tissues within the bladder wall, resulting in a possible detection of invasive tumors, and (3) radiation free, resulting in a possible screening modality. In our approach, after the patient voided, patient was asked to drink a cup of water before MR image scan. Images of the nearly empty state of the bladder were acquired first by an acquisition protocol of 15 minutes (including scout and location image acquisition). Waiting another 15 minutes, when the patient feels bladder full, full state MR images were acquired in 15 minutes. An entire procedure takes less than an hour, an average time course of routine MR image study.

In Figure 1, significant image contrasts were observed between urine lumen and bladder wall as well as between bladder wall and other surrounding tissues in the T_1-weighted images (Figure 1(a)). It is noted that the bladder wall becomes thicker in the near empty states (Figure 1(b)). On the other hand, T_2-weighted images provide reciprocal information with a brighten urine lumen which can be distinguished more easily from other tissues (Figures 1(c) and 1(d)).

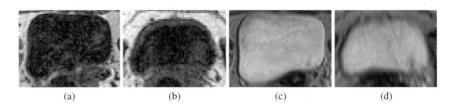

(a)　　　　　(b)　　　　　(c)　　　　　(d)

Fig. 1. (a) One slice of a T_1-weighted image at the near full state. (b) One slice of a T_1-weighted image at the near empty state, where the bladder wall is thicker than that at the near full state. (c) One slice of a T_2-weighted image at the near full state, where the urine lumen could be easily distinguished. (d) One slice of a T_2-weighted image at the near empty state.

Motion artifacts may be observed along the direction of spine in T_1-weighted transverse images as the image acquisition time was about two to five minutes. To overcome this artifact, additional coronal images were acquired correspondingly. Our multi-scan MR scheme shall provide more information for reducing motion artifact as well as the false positive (FP) in lesion detection.

2.2. Image Segmentation

We first applied Fourier-domain based interpolation to construct isotropic voxels in the three-dimensional (3D) domain (Li *et al.*, 2003). Following that, a partial volume (PV) image segmentation algorithm was developed to achieve tissue mixture segmentation, and then to extract the bladder wall from the mixture segmentation. The following is a brief description of the image segmentation algorithm.

Let $Y = \{y_{il}\}_{l=1}^{L}$ be the intensity vector of L-channel multi-spectral MR images at location i over the 3D image array of I voxels. In our study, we acquired T_1 and T_2-weighted images, i.e., $L = 2$. Assume that the images consist of K classes (or tissue types) and each class k is characterized by a Gaussian parameter vector $\theta_k(\mu_k, \nu_k)$, i.e., the mean and variance. Let M be a set of vectors $M = \{m_1, \ldots, m_2, \ldots, m_I\}$, where the mixture m_{ik} reflects the fraction of tissue type k inside voxel i. We utilized the expectation-maximization (EM) algorithm to achieve robust model parameter estimation (Li *et al.*, 2003b; Liang *et al.*, 2003). By the EM strategy, we have,

$$\mu_{kl}^{(n+1)} = \frac{\sum_i x_{ikl}^{(n)}}{\sum_i m_{ik}} \tag{1}$$

$$(\sigma_{kl}^2)^{(n+1)} = \frac{1}{I} \sum_i \frac{(x_{ikl}^2)^{(n)} - 2m_{ik}\mu_{kl}^{(n+1)}x_{ikl}^{(n)} + (m_{ik}\mu_{kl}^{(n+1)})^2}{m_{ik}} \tag{2}$$

where σ_{kl}^2 is the variance of tissue type k in image l and x_{ikl} is the contribution of tissue type k to the observation y_{il} and

$$x_{ikl}^{(n)} = m_{ik}\mu_{kl}^{(n)} + \frac{m_{ik}\sigma_{kl}^{(n)}}{\sum_{j=1}^{k} m_{ij}\sigma_{jl}^{(n)}} \cdot \left(y_{il} - \sum_{j=1}^{K} m_{ij}\mu_{jl}^{(n)} \right) \tag{3}$$

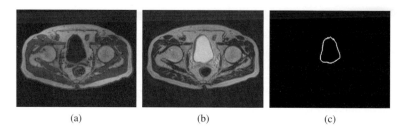

Fig. 2. (a)–(b): A slice of T_1- and T_2-weighted images in the transverse scan. (c) The extracted mucosa layer indicating the space between the bladder wall and the lumen.

$$(x_{ikl}^2)^{(n)} = (x_{ikl}^{(n)})^2 + (m_{ik}\sigma_{kl}^2)^{(n)} \cdot \frac{\sum_{j \neq k}^{K} (m_{ij}\sigma_{jl}^2)^{(n)}}{\sum_{j=1}^{K} (m_{ij}\sigma_{jl}^2)^{(n)}}. \qquad (4)$$

In order to perform the PV segmentation for $\{m_{ik}\}$, we employed the maximum *a posterior* (MAP) framework, in which a Markov random field (MRF) model-based prior was applied to integrate neighborhood information (Held *et al.*, 1997; Leahy *et al.*, 1991; Liang *et al.*, 1992, 1994). The developed framework iteratively estimates the model parameters through the EM algorithm and segments the voxels by MAP in an interleaved manner. Each voxel is then labeled as a mixture voxel (mixel) with different tissue percentages inside.

Based on the PV image segmentation, a mixture (mucosa) layer indicating the space between the bladder wall and lumen was extracted (Figure 2). The extracted mixture layer provides us both geometric and texture information, which can be further used for lesion detection. By choosing a seed point from the lumen, the bladder lumen could be extracted by means of region growing algorithm.

2.3. Interactive Visualization System

A real-time interactive visualization system was developed for facilitating physicians to diagnose 3D MR images. The system is composed of a display of the original dataset and segmentation results, fully interactive control on bladder skeleton for viewing the dual states of bladder at any angle and direction, and cut view of the bladder for inner surface inspection. At each state, we reconstructed both the inner and outer bladder wall from the mixture-based segmentation (Wang *et al.*, 2002). Registration is needed for simultaneously displaying the dual-state scans of the bladder. In this study, we calculated the central mass of the bladder and chose the central mass as a reference for the purpose of flexible registration (Chen et al., 2000). When the physician rotates the bladder in one window, the system will calculate the relative parameter, and rotate the bladder at the other three windows to display the same section of the bladder.

2.4. Detection of Bladder Lesions

From the mixture segmentation, we extracted the bladder mucosa layer, which provides us the geometric features of tissues around bladder wall. Our approach minimized the PV effects on the tissue boundaries and provided tissue growth tendency in addition to the anatomical structure of the tissues. In order to detect bladder lesions, we developed a CAD system to utilize the mixels for analyzing the features and tendency of mucosa layer for abnormities (Wang et al., 2003). A 3D geometrical feature extraction algorithm was applied to extract the principal curvature, shape index, and curvedness of each mixel. The extracted geometry information is utilized to distinguish bladder lesions from the normal bladder wall.

For describing the local shape of the surface across through bladder 3D volume, principal direction and curvature are two important geometrical attributes of the 3D surface (Dorai et al., 1997). For a given point p, we employed two quantitative measurements: the shape index SI and the curvedness R. They are defined as:

$$SI(p) = \frac{1}{2} - \arctan \frac{k_1(p) + k_2(p)}{k_1(p) - k_2(p)} \qquad (5)$$

$$R(p) = \sqrt{\frac{k_1(p)^2 + k_2(p)^2}{2}} \qquad (6)$$

where $k_1(p)$ and $k_2(p)$ are the principal curvatures, and $k_1(p) \geq k_2(p)$. SI represents what kind of type the surface is, while the R represents what "curve" the surface is.

Eight representative shapes and their corresponding shape index values are provided in Figure 3. Normal bladder wall usually has regular shapes that are described as "elliptic curvature." That is, the shape description of the inner bladder wall should be "spherical cup" or "trough." This shape

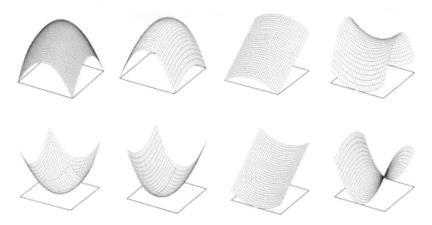

Fig. 3. Eight representative shapes on the shape index scale. Top (from left to right): spherical cap (1.0), dome (0.875), ridge (0.75), and saddle ridge (0.625). Bottom (from left to right): spherical cup (0.0), trough (0.125), rut (0.25), and saddle rut (0.375).

description can be applied to distinguish the bladder lesion from the normal bladder wall. Since this shape information only reflects a relative relation between this mixel and its neighborhood in a local region, we call it as "local" geometrical shape information. Besides the local shape information, we also integrated the corresponding "global" shape information into our CAD scheme to reduce false positive (FP) detection rate. The "global" shape index and curvedness are derived from a convolution operation on the "local" shape index and curvedness (Wang *et al.*, 2004). We evaluated the corresponding "global" shape index and found its variance remains smooth for most of the irregular bladder wall section. Therefore, all mixels whose "global" shape description is different from that of the normal bladder wall were extracted first. Based on the connectivity in the 3D space, all selected mixels were clustered into several groups, which are called as initial bladder lesion candidates.

The initial bladder lesion candidates can be further divided into three groups: real bladder lesions, noise candidates, and mimic candidates. Noise candidates are usually generated due to MR imaging scan, patient movement, or image segmentation error, which induce several little protuberance regions on the bladder wall. According to our observation, they are usually very small with a tiny spherical top section. On the other hand, the mimic candidates are generated due to normal tissue shape variations on the bladder wall. Both noise and mimic candidates are called FP candidates. In order to eliminate FP candidates, we utilized the following filtering steps in our CAD method.

Step 1. If the total voxel number of the candidate is small, this candidate will be classified as a FP candidate. Similarly, if the size of the continuous spherical top in either local or global geometrical measures is small, this candidate will be classified as a FP candidate.

Step 2. If the position of the initial lesion candidate whose "local" and "global" general shapes does not lie in the spherical shape domain, this candidate is a FP candidate.

3. RESULTS

Two patients and four healthy male volunteers were recruited in this study. Figure 4 shows the outer view of a transverse and coronal scans of bladder at

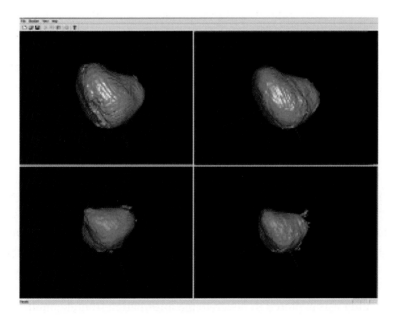

Fig. 4. The outer view of a transverse (left) and coronal (right) scans of a bladder at the two states of full (top) and near empty (bottom).

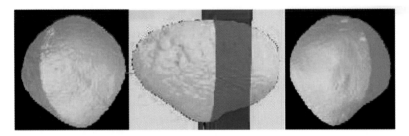

Fig. 5. In the middle is the 3D volume of a reconstructed bladder image. A plane cuts in the center vertically. On the left and right are the left and right half of the volume image, respectively. The inner and outer surfaces are clearly seen.

two states of full and near empty. Figure 5 shows a cut view of the bladder. The presented multi-scan virtual cystoscopy system provides physicians for texture analysis and visualization of bladder structure. Physicians can easily locate those locations where abnormal (morphological and pathological) tissues were observed from inner or outer view of bladder wall. Lesion detection by our CAD scheme were also performed for both healthy volunteers and patients. In the patient study as shown in Figure 6, one lesion with a size of 25–30 mm was detected at both states. A small one with 8–10 mm size was detected in the full state while missed in the near empty state. When the presented CAD scheme further examined the texture information at the corresponding location in the near empty state, this small one was detected as a candidate. The CAD results were further examined and verified by physicians using our interactive visualization system.

4. DISCUSSION AND CONCLUSIONS

The proposed MR image-based virtual cystoscopy is a non-invasive, safe, and patient-comfortable procedure. We have developed a MR image acquisition protocol to acquire multi-scans on two states of the bladder for mitigating motion artifacts and obtaining both geometric and texture information. We further developed a PV segmentation scheme to extract the bladder wall mucosa layer from the multi-spectral MR images. A flexible registration was explored to integrate the information from the multi-scans of the dual states. A display system was developed to visualize the integrated information. A CAD scheme, which combines both "local" and "global"

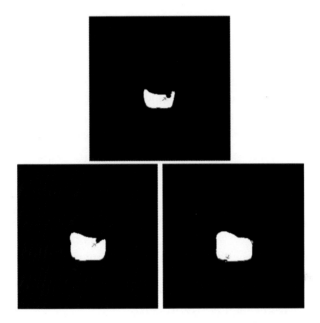

Fig. 6. One lesion with size of 25–30 mm was detected at both the near empty state (top) and full state (bottom left). A small one with size of 8–10 mm was detected in the full state (bottom right) and missed in the near empty state when only geometrical information is used. By the use of the texture information, the small tumor was detected in both states.

shape index and curvedness, was also developed to identify candidates quickly and then present them to the physician for final assessment. From the presented pilot patient study, we conclude that information extracted from multi-spectral MR images contains more valuable information than that from each image individually. The dual states of the bladder provide dynamic (both geometric and texture) information. The multi-scan (transverse and coronal) images mitigate the breathing motion artifacts. The developed CAD system with flexible registration and inner and outer wall visualization shows the feasibility towards non-invasive mass screening of bladder tumor and evaluation of following recurrence.

ACKNOWLEDGMENTS

This work was supported in part by the National Institutes of Health under Grant No. CA120917 and CA082402 and the PSC-CUNY award program under Grants No. 67677-00-36 and 68562-00-37. The authors would appreciate Drs. Wei Huang, PhD and Howard Adler, MD for assisting the patient and volunteer datasets acquisition at the Stony Brook University Hospital. The appreciation goes further to Drs. Hong Meng, MD and Donald Harrington, MD for their clinical support. The comments from Drs. Xiang Li, PhD, Dongqing Chen, PhD, and Bin Li, PhD on implementation of the MRF model in the MAP-EM algorithm shall be acknowledged.

References

1. Chen D, Li B, Huang W and Liang Z. A multi-scan MRI-based virtual cystoscopy. *SPIE Med. Imaging* 2000; **3978**: 146–152.
2. Cohen SM and Johansson SL. Epidemiology and etiology of bladder cancer. *Urol. Clin. North Am.* 1992; **19**: 421–428.
3. Dorai C and Jain AK. COSMOS — A representation scheme for 3D free-form objects. *IEEE Trans. Pattern Anal. Mach. Intell.* 1997; **45**: 2132–2138.
4. Fenlon HM, Bell TV, Ahari HK and Hussain S. Virtual cystoscopy: early clinical experience. *Radiology* 1997; **205**(1): 272–275.
5. Fielding JR, Hoyte L, Okon SA, Schreyer A, Lee J, Zou KH, Warfield S, Richie JP, Loughlin KR, O'Leary MP, Doyle CJ and Kikinis R. Tumor detection by virtual cystoscopy with color mapping of bladder wall thickness. *J. Urol.* 2002; **167**: 559–562.
6. Held K, Kops ER, Krause BJ, Wells WM, Kikinis R and Muller-Gartner H. Markov random field segmentation of brain MR images. *IEEE Trans. Med. Imaging* 1997; **16**: 876–886.
7. Hussain S, Loeffler JA, Babayan RK and Fenlon HM. Thin-section helical computer tomography of the bladder: initial clinical experience with virtual reality imaging. *Urology* 1997; **50**(5): 685–689.
8. Jemal A, Thomas A, Murray T and Thun M. Cancer statistics. *CA Cancer J. Clin.* 2002; **52**: 23–47.
9. Lamm DL and Torti FM. Bladder cancer. *CA Cancer J. Clin.* 1996; **46**: 93–112.
10. Leahy R, Hebert T and Lee R. Applications of Markov random fields in medical imaging. *Inf. Process. Med. Imaging* 1991; 1–14.
11. Li L, Li X, Lu H, Huang W, Christodoulou C, Tudorica A, Krupp LB and Liang Z. MRI volumetric analysis of multiple sclerosis: methodology and validation. *IEEE Trans. Nucl. Sci.* 2003; **50**: 1686–1692.

12. Li L, Wang Z, Harrington D, Huang W and Liang Z. A mixture-based computed aided detection system for virtual cystoscopy. In: *Proceedings of the International Society of Magnetic Resonance in Medicine*, Vol. 3, 2003a, p. 1466.
13. Li X, Eremina D, Li L and Liang Z. Partial volume segmentation of medical images. In: *IEEE Medical Imaging Conference Record*, 2003b.
14. Liang Z, Jaszczak RJ and Coleman RE. Parameter estimation of finite mixtures using the EM algorithm and information criteria with application to medical image processing. *IEEE Trans. Nucl. Sci.* 1992; **39**: 1126–1133.
15. Liang Z, MacFall JR and Harrington DP. Parameter estimation and tissue segmentation from multispectral MR images. *IEEE Trans. Med. Imaging* 1994; **13**: 441–449.
16. Liang Z, Chen D, Button T, Li H and Huang W. Feasibility studies on extracting bladder wall from MR images for virtual cystoscopy. In: *Proceedings of the International Society of Magnetic Resonance in Medicine*, Vol. 3, 1999, p. 2204.
17. Liang Z, Li X, Eremina D and Li L. An EM framework for segmentation of tissue mixtures from medical images. In: *International Conference of IEEE Engineering in Medicine and Biology*, Cancun, Mexico, 2003, pp. 682–685.
18. Merkle M, Wunderlich A, Aschoff AJ, Rilinger N, Gorich J, Bachor R, Gottfried HW, Sokiranski R, Fleiter TR and Brambs HJ. Virtual cystoscopy based on helical CT scan datasets: perspectives and limitations. *Br. J. Radiol.* 1998; **71**: 262–267.
19. Prout G. Bladder cancinoma. In: *Surgical Oncology*. McGraw Inc., New York 1984, pp. 679–697.
20. Shaw ST, Poon SY and Wong ET. Routine urinalysis: is the dipstick enough? *J. Am. Med. Assoc.* 1985; **253**: 1596–1600.
21. Steiner GD, Trump DL and Cummings KB. Metastatic bladder cancer: natural history, clinical course, and consideration for treatment. *Urol. Clin. North Am.* 1992; **19**: 735–746.
22. Vining DJ, Zagoria RJ, Liu K and Stelts D. CT cystoscopy: an innovation in bladder imaging. *AJR* 1996; **166**: 409–410.
23. Wang Z and Liang Z. Feature based rendering for 2D/3D partial volume segmentation datasets. *SPIE Med. Imaging* 2002; **4681**: 681–687.
24. Wang Z, Li L, Li X, Liang Z and Harrington DP. Skeleton based 3D computer aided detection of colonic polyps. *SPIE Med. Imaging* 2003; **5032**: 843–853.
25. Wang Z, Li L, Anderson JC, Harrington DP and Liang Z. Colonic polyp characterization and detection based on both morphological and texture features. In: *Computer Assisted Radiology and Surgery*, 2004, pp. 1004–1009.

Chapter 16

MATHEMATICAL FRAMEWORK AND WAVELETS APPLICATIONS IN PROTEOMICS FOR CANCER STUDY

Don Hong and Yu Shyr

Cancer is a proteomic disease. Though MALDI-TOF mass spectrometry allows direct measurement of the protein signature of tissue, blood, or their biological samples, and holds tremendous potential for disease diagnosis and treatment, key challenges remain in the processing of proteomic data. In this chapter, we will introduce a wavelet based mathematical framework and computational tools for proteomic data processing, feature selection, and statistical analysis in cancer study.

Keywords: Proteomics, wavelets, mass spectrometry peak detection, peak alignment, biomarker discovery, feature selection.

1. INTRODUCTION

Proteomics, the analysis of genomic complements of proteins, has attracted more and more attention to cancer researchers due to the fact that cancer is a proteomic disease and protein arrays are a breakthrough because they allow many different proteins to be tracked simultaneously. High throughput *mass spectrometry* (MS) has been motivated greatly from recent developments in both chemistry and biology. Its technology has been extended to proteomics as a tool in rapid protein identification (Chaurand *et al.*, 1999; Loo *et al.*, 1999). Comparable to the exciting development of nuclear magnetic resonance methods during the past three decades, mass spectrometry entered a phase of rapid growth in the mid-eighties beginning with the introduction of soft ionization methods, such as electrospray ionization (ESI) and matrix assisted laser desorption/ionization (MALDI). These new techniques have allowed the use of mass spectrometry in applications involving large molecules such as in biochemical, pharmaceutical, and medical research. Mass spectrometric methodology and examples of applications in biotechnology and cancer study can be found in Siuzdak (2003) and Roboz (2002). Some recent progress on automated peak identification for time-of-flight (TOF) mass spectra can be found in Hong and Shyr (2007).

Mass spectrometers are ion optical devices that produce a beam of gas-phase ions from samples. They sort the resulting mixture of ions according to their mass-to-charge (m/z) ratios or a derived property, and provide analog or digital output signals (peaks) from which the mass-to-charge ratio and intensity (abundance) of each detected ionic species may be determined. Masses are not measured directly. Mass spectrometers are m/z analyzers. The mass-to-charge ratio of an ion is obtained by dividing the mass of the ion (m), by the number of charges (z) that were acquired during the process of ionization. The mass of a particle is the sum of the atomic masses (in Dalton) of all the atoms of the elements of which it is composed.

The *mass spectrum* of a compound provides, in a graphical or tabular form, the intensities of all or a selected number of the acquired m/z values from the ionic species formed. Mass spectral peaks are observed in analog form (each peak with a height and a width) or digital form (each peak a simple line). The heart of any mass spectrometer is the mass selective analyzer. The concept of the linear time-of-flight analyzer was described by Stephens in 1946. The development of MALDI-TOF in 1988 (Karas and Hillenkamp) has paved the way for new applications, not only for biomolecules but also for synthetic polymers and polymer/biomolecule conjugates. Accordingly, the major areas of applications of mass spectrometry have been qualitative analysis and quantification.

Cancers secrete large and small molecules of numerous known and countless unknown structures. Enzymes that allow cancers to invade and metastasize, and surface molecules and compounds of unknown function often serve as critical parameters of cancer behavior. Discovery of trace compounds that could indicate the presence of early cancer is still theoretically possible, and still hoped for. Identification of such compounds in extremely small quantities in biologic fluids containing hundreds of other compounds is a classic undertaking for mass spectrometry. Coordinated immunologic assay, isotopic, spectroscopic, nuclear-magnetic resonance, and mass spectrometric analysis of such putative markers could advance the diagnostic acumen so that we might recognize pre-cancerous states, or cancers so early in their course that a cure could be readily achieved (see Henschke *et al.*, 1999; Srinivas *et al.*, 2001; and Yanagisawa *et al.*, 2003 for examples).

Mass spectrometers attempt to answer the basic questions of WHAT and HOW MUCH is present by determining ionic masses and intensities. MALDI-TOF MS is emerging as a leading technology in the proteomics revolution. Indeed, the year 2002 Nobel prizes in chemistry recognized MALDI's ability to analyze intact biological macromolecules. Though MALDI-TOF MS allows direct measurement of the protein "signature" of tissue, blood, or other biological samples, and holds tremendous potential for disease diagnosis and treatment, key challenges still remain in the processing of MALDI MS data.

The use of high-throughput mass spectrometry produces data sets comprised of spectra whose graphs are of the type shown in Figure 1. On the horizontal axis are mass/charge (m/z) values and on the vertical axis an intensity measurement that indicates a relative abundance of the particle. The analysis of such data involves inferring the existence of a peptide of a particular mass from the existence of a spike in the spectrum. The

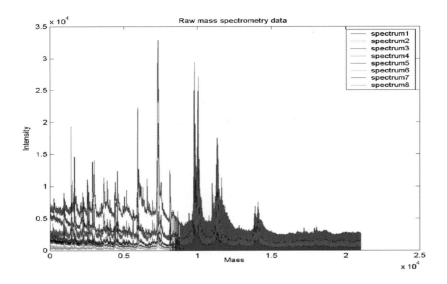

Fig. 1. Graphs of MALDI-TOF mass spectra.

data is in very high dimensional setting and there are uncertainties in peak position and as well the intensity. To identify biomarkers from these spectra, many data-analytic questions arise: What feature indicates the existence of a peptide? How does one even define a feature and subsequently extract it from a set of spectra with significant between-sample variability in intensity, background noise, and in the m/z-value at which a feature is recorded? In addition, data registration is confounded by the variability in the location and the shape/size of features when compared across samples.

To date, MALDI-TOF or SELDI (surface-enhanced laser desorption ionization), a variant of MALDI technology has been applied to search cancer biomarkers, with some success. There is also preliminary evidence that we may be able to discover patterns that can reliably distinguish cancer patients from healthy individuals (Soltys *et al.*, 2004; Waldsworth *et al.*, 2004; Yanagisawa *et al.*, 2003; Zhang *et al.*, 2004 for examples). These findings should be greeted with cautious optimism. When it has been possible to identify the protein peaks, they have often turned out to be well-known acute-phase proteins. Some authors have claimed that MS is intrinsically limited in its depth of coverage, with a dynamic range that prevents it from being able to find low-abundance proteins (Diamandis, 2004). This brings us to analysis tools. There is no consensus on the best methods to analyze mass spectra from proteomic profiling experiments. Most published studies perform data preprocessing and peak detection with software from the manufacturers of MALDI/SELDI instruments. In fact, the software is extremely conservative about calling something a peak and its baseline correction algorithm introduces substantial bias into the estimates of the size of a peak. These algorithmic weaknesses can reduce the effective sensitivity of the instrument below its true capacities and can hamper its reproducibility. Many of *ad hoc* approaches have been implemented by various groups (Coombes *et al.*, 2005; Morris *et al.*, 2005; Yu *et al.*, 2006; Chen *et al.*, 2007). It is substantial to develop a comprehensive set of mathematical and computational tools for MALDI TOF MS data analysis.

Multiscale tools such as wavelets provide promising techniques for MALDI MS data analysis. The word "wavelets" means "small waves" (the sinusoids used in Fourier analysis are "big" waves), and in short, wavelet is an oscillation that decays quickly. Mathematically, *wavelets* usually are basis functions of an L^2 space that satisfy so-called multiresolution analysis requirements (Chui, 1992; Daubechies, 1992; Hong *et al.*, 2005). In recent years, wavelets have been applied to a large variety of signal processing and image compression (Mallat, 1999). Also, there is a growing interest in using wavelets in analysis of biomedical signals and functional genomics data (see Aldrobi and Unser, 1996; de Trad *et al.*, 2002; Hirakawa *et al.*, 1999; Lio, 2003 for examples). Wavelet theory is developed now into a methodology used in many disciplines: engineering, mathematics, physics, signal processing and image compression, numerical analysis, and statistics. Wavelets are providing a rich source of useful tools for applications in time-scale types of problems. Wavelet based methods have found applications in statistics in areas such as regression, density and function estimation, modeling and forecasting in time series analysis, and spatial analysis (Donoho and Johnston, 1995, 1998; Silverman, 1999). In particular, Donoho and Johnstone found that wavelet threshold has desirable statistical optimality properties. Since then, wavelets have proved to be very useful in nonparametric statistics and time series analysis.

In the following discussions, we will focus primarily on: (a) establishing a general mathematical framework for modeling and representing MALDI MS data that allows the recovery of, as near as possible, the "true" signal from the machine data (innovative mathematical tools to be developed include non-uniform wavelets, biological diffusion maps and geometric harmonics, and shape-preserving splines); (b) designing algorithms and developing software for performing preprocessing operations such as peak alignment and detection, baseline correction and denoising, and a statistical analysis of MALDI MS data; and (c) developing tools for feature extraction and biomarker discovery. In particular, we will focus on a wavelet based novel multiscale scheme for identifying biological signatures of MALDI-TOF MS cancer data.

2. MATHEMATICAL REPRESENTATION AND PREPROCESSING OF MALDI MS DATA

In this section we model the MS signal as being composed of three distinct components: background function, true signal, and machine noise. This model allows us to address signal reconstruction and subsequent biological interpretation in a mathematically principle manner. We will explain how each component is created and how previous approaches heuristically address the modeling of each component separately. In the subsequent subsections, we will discuss the specific techniques we employ in order to model each component.

2.1. Mathematical Model for MALDI-TOF MS Data

The operation of a mass spectrometer can be divided into four main steps: sample introduction, ionization, mass analysis, and detection/data analysis. In the ionization stage, an ion-gas is produced from a given sample and in the mass analysis stage. The ions are then separated according to their mass-to-charge ratio (m/z) using electromagnetic fields. The first mass spectrometers were produced in the early 1900s (Thomson, 1913). Mass spectrometry has been used to investigate biological processes since the late 1930s; however, it is only recently that the advances in ionization technology permit the use of mass spectroscopy to study large molecules (up to 300,000 Da), such as proteins or peptide fragments that occur in biological samples. In particular, the MALDI spectrometer has become a central tool in modern protein research. In a MALDI-TOF spectrometer, the analyte is first embedded in a solid "matrix" that absorbs energy from a laser whose wavelength is matched to the matrix. The resulting intense heating of the matrix produces a gas-ion plume that is then accelerated through a potential difference V. An ion of mass m and charge z acquires a change in potential energy of zV which, to first order, is translated into a change of kinetic energy of $(1/2)\,mv^2$, exiting with a velocity determined by the ratio m/z. The ions then travel a length D to a detector where the density of ions is recorded as a function of the time of arrival t. The mass charge ratio m/z may then be expressed as a function of t of the form (Vestal and Juhasz, 1998):

$$m/z = A(t - t_0)^2, \qquad (1)$$

where A and t_0 are constants depending on instrumental parameters such as V and D.

The *mathematical processing of MS signals* can be roughly divided into two steps. First, in the "preprocessing" step, we attempt to recover from the time of arrival data, as accurately as possible, the "true" signal reflecting the mass/charge distribution of the ions originating from the sample. The preprocessing step includes registration, denoising, baseline correction, and deconvolution. In the preprocessing step, these operations are performed independently of any biological information one seeks to extract from the data. The second type of processing attempts to represent the data in a form that facilitates the extraction of biological information. This step involves operations such as dimension reduction, feature selection, clustering, and pattern recognition for classification.

A mass spectrometer has a finite resolution power mainly due to variations in the initial position and velocity contained in the ion-plume. A "pure" sample consisting of ions of a single mass/charge ratio $y = m/z$ results in an "impulse response" $k(x, y)$ where k depends on the distribution of initial position and velocity along with the value of machine parameters (such as V and D) that are chosen to minimize the resolution $(\sigma(y)/y)$ for y in some interval of interest where $\sigma(y)$ denotes some measure of the spread of $k(x, y)$, for example, standard deviation or half-width at half-maximum. Typically, k is assumed to be of the form: $k(x, y) = \exp\{-(x - y)^2/\sigma(x)\sigma(y))$, and that $\sigma(y)$ is slowly varying over intervals. As shown in (Vestal and Juhasz, 1998), k can be explicitly calculated from the mass analyzer geometry and operating voltages and from the distributions of initial ion position and velocity. Assuming a parametric form for the initial distributions, one can find parametric representations for k.

To first order (ignoring interactions between ions), a mass spectrometer is a linear device and so the output $f(x)$ in the absence of machine noise from a sample with mass/charge distribution μ is of the form:

$$f(x) = \int k(x, y) d\mu(y).$$

Specifically we propose to consider symmetric kernels k of the form:

$$k(x, y) = \frac{\gamma((x - y)^2}{(\sigma(x)\sigma(y))},$$

where: $R_+ \to R_+$ is a decreasing function with rapid decay. For fixed y, it is usually the case that $\sigma(x)$ is approximately constant for $|x - y| = O(\sigma(y))$ and so $k(x, y)$ is essentially a small symmetric bump. If the sample distribution is of the form $\mu_s = \sum_i \alpha_i \delta_{xi}$, where δ_a denotes a unit mass at a, then the observed signal is a sum of bumps of the form:

$$S(x) = \sum_i \alpha_i k(x, x_i).$$

However, a real world signal differs from this idealized scenario in several ways. First, the ion-plume contains a distribution μ_m of ionized matrix molecules having a high spectral content in the low mass region, that is, the matrix produces ions of a wide variety of masses in this range. Secondly, because of collisions or because of molecular fragmentation that occurs during the time of flight, ions are spread non-locally across the mass scale. We model this non-local scattering with a second kernel, κ, with slow decay. This suggests the following "incoherent" contribution $I(x)$ to the observed signal:

$$I(x) = \int k(x, y) d\mu_m(y) + \int \kappa(x, y) d\mu(y),$$

where $\mu = \mu_s + \mu_m$. Finally, a real MS signal contains a high frequency machine electronic machine noise $\varepsilon(x)$, and so we model an observed signal by a sum of the form $f(x) = I(x) + S(x) + \varepsilon(x)$. Generally, we shall also consider a nonlinear component in I.

Based on the above mathematical model for MALDI MS data, multi-scale deconvolution approach can be used in the statistical estimation for the MALDI MS data analysis. Mass spectrometry of proteins promises to be a very valuable tool in diagnostic applications. There are several challenges to the use of such proteomics data in classification and clustering of samples from diseased and normal patients. In particular, the number of measurements taken per sample is very large and even if considered into peak areas or peak heights, the number of potential predictors greatly exceeds the number of samples. Therefore, there has been considerable effort involved in the preprocessing of the data (see Baggerly et al., 2004; Coombes et al., 2005; Gentzel et al., 2003; Morris et al., 2005; Chen et al., 2007; Yu et al., 2006 for examples).

2.1.1. *Baseline correction and normalization*

Incoherent contribution $I(x)$ is usually approximated by a so-called *baseline* or background function. An observed MALDI MS signal $f(t)$ is often modeled as the superposition of three components:

$$f(x) = B(x) + S(x) + \varepsilon(x),$$

where, $B(x)$ is a slowly varying "baseline" that approximates the incoherent component $I(x)$, $S(x)$ is the "true" signal to be extracted, and $\varepsilon(x)$ represents a high frequency machine noise. The underlying assumption in these techniques is that B, S and ε are varying at different scales. Recently, in Coombes *et al.* (2005), Chen (2004), Chen *et al.* (2007), and Hong *et al.* (2007), wavelet-based methods was proposed in the preprocessing of SELDI and MALDI spectra data, respectively. They use baseline correction and wavelet denoising to approximate $S(x)$ and they show better peak detection than previous methods.

From experimental observations, one often includes the constraint that the baseline is non-negative and decreasing. For this purpose we are interested in the use of shape preserving splines with non-uniform knots Chen *et al.* (2007). Furthermore, we propose to use models for $I(x)$ as above to estimate its properties in order to construct better baseline approximations. In particular, we expect that $I(x)$ can be represented as $I(x) = B(x) + W(x)$, where $B(x)$ is now the projection onto a coarse space in a multiresolution and W is a component that can be represented in a wavelet basis with small coefficients. Because the bump width $\sigma(x)$ is not constant, the natural representations of $f(x)$ and $S(x)$ are in non-shift invariant spaces. We believe that using wavelets for baseline estimation and for reconstructing the true signal $S(x)$ can improve the accuracy of the appearance of the spectrum and the quality of a result from subtracting one spectrum from another.

Some simple techniques for constructing a baseline include fitting local minima with a polynomial or spline function (Chen *et al.*, 2007), or using a median filter of appropriate window size (Coombes, 2005). See Figure 2 for a comparison of a MALDI mass spectrum on raw data with and without baseline correction (Chen *et al.*, 2007).

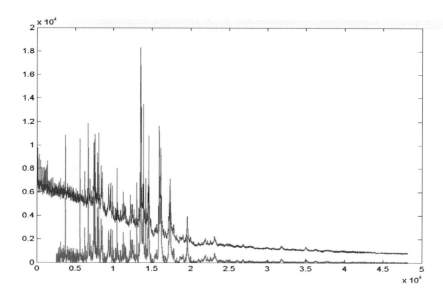

Fig. 2. MALDI MS signals with and without baseline correction.

To compare spectra in the same scale, the normalization step is inevitable. Since the spectrum after baseline correction is closer to the true distribution of the signal, we can normalize every element in the spectrum vector. In Chen *et al.* (2007), we apply an l_2 averaging formula for the normalization, which is in the energy metric.

2.1.2. *Spectra registration and peak alignment*

Spectrum data registration means that aligning the time of flight data t with m/z as accurately as possible across samples so that different samples may be compared. In the simple case of (1), this means determining the constants A and t_0 for each given sample (typically done by locating known reference mass peaks in the sample). Techniques for the registration of 2D and 3D data have been well-studied and remain an active area of research in imaging science. Many of these techniques use a multiresolution approach by first registering the signals at a "coarse" resolution and then iteratively registering the signals at successively finer resolutions. This results in algorithms both computationally faster and also more "robust" (Unser *et al.*, 1995).

The goal of the MALDI-TOF MS data preprocessing is to identify the locations and the intensities of peaks. The spectra, after all previous preprocessing steps, can be put in a matrix of column spectrum vectors of intensities.

Usually, the local maxima in each column are the peaks of each spectrum. The local maximum selection, without denoising the raw data, will generate more than one peak on a true peak interval. To filter out smaller peaks, an *ad hoc* method based on the ratio of signal and noise (S/N) is proposed in (Coombes *et al.*, 2005). In the next section, we will discuss in details on denoising for mass spectrometry data.

Now, let us discuss the cross sample alignment of MS data. For data samples from patients, the data first has to be preprocessed with the proper background subtracted, normalized, and the different fractions combined to obtain one integrated spectrum for each patient. The integrated spectrum is then binned or aligned so that the data for all patients in the sample is formatted in a matrix with one index representing the patients and the other index the peaks (discrete m/z's corresponding to the mean of the m/z of each bin).

In real application, one peak will be identified within a certain separation range (*SR*). An experimental formula for *SR* is given by $SR = 2 + (X_i/1000)$ in Daltons, where X_i is the m/z location. However, in the peak matrix, the positions of peaks of each column around the same m/z value maybe different from each other slightly (apart from two to three rows in the matrix). Therefore, we need to bin these peaks in order to correspond to the same m/z value. This is also called cross samples peak alignment. A so-called average spectrum is determined for the binning purpose in Morris *et al.* (2005). An efficient and effective binning method, called PSB, is developed in Hong *et al.* (2007) by projecting spectra to a function of number of peaks. See Figure 3 for PSB results. A so-called central spectrum idea by using local clustering techniques is introduced for binning in Chen (2004). Binning approach reduces the dimension. In Purohit *et al.* (2003), combining the binning procedure, a square root transform on the data is applied to help stabilize the variance and in turn, made a significant improvement in clustering results. A curve alignment method developed in Bar-Joseph *et al.* (2003), which combines spline interpolation with clustering can be employed as an idea for peak selection and binning in the preprocessing

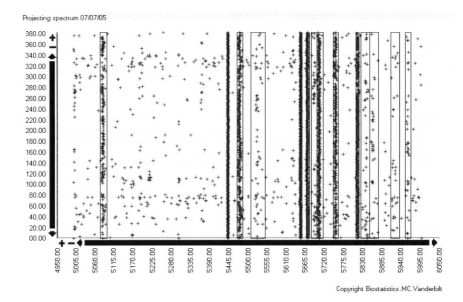

Fig. 3. A partial mass peak distribution with binned by PSB.

step. It will be interesting to compare the outcomes between the binning approach and the curve alignment method, as well as among many choices of multiscale operations.

In summary, we modeled the MALDI MS signal as a superposition of an incoherent signal created by high mass spectral content of the matrix together with non-local scattering, the true observed signal, and an electronics related noise component. We also discussed how the true observed signal is in effect a superposition of distributions of the mass values in the ion plume. In the next section, we discuss in detail the specific mathematical techniques we plan to apply to model each of these components, especially for estimation and removing the noise using wavelets.

3. MULTISCALE TOOLS

We would like to apply multiscale tools, such as wavelets in the study of MALDI MS data. In this section we first briefly introduce wavelets and then discuss how these techniques can be applied to separate the different components of an observed MALDI MS signal because of the different time-scale characteristics of the components.

3.1. Wavelets and WaveSpec Software

Wavelets are a relatively recent development in applied mathematics. Vidakovic (1999) mentioned that the first definition of wavelets could be attributed to Morlet *et al.* (1982) (Grossmann and Morlet, 1984). Now the term wavelet is usually associated with a function $\psi \in L_2(R)$ such that its translations and dilations

$$\psi_{j,k}(x) = 2^{-j/2}\psi(t/2^j - k),$$

for integers j and k constitute an orthonormal basis of $L_2(R)$. The wavelet transform is a tool that cuts up data or functions into different frequency components, and then studies each component with a resolution matched to its scale. The wavelet transform on a finite sequence of data points provides a linear mapping to the wavelet coefficients: $w_n = Wf_n$, where the matrix $W = W_{n \times n}$ is orthogonal and w_n and f_n are n-dimensional vectors. The wavelet approximation to a signal function f is built up over multiple scales and many localized positions. For the given family of scale functions and corresponding wavelet functions:

$$\phi_{J,k} = 2^{-J/2}\phi(t/2^J - k), \quad \psi_{j,k} = 2^{-j/2}\psi(t/2^j - k), \quad j = 1, 2, \ldots, J.$$

The coefficients are given by the projections:

$$s_{J,k} = \int f(t)\psi_{J,k}(t)dt, \quad d_{j,k} = \int f(t)\psi_{j,k}(t)dt$$

so that

$$f(t) = \sum_k s_{J,k}\psi_{J,k}(t) + \sum_k \sum_{j=1}^{J} d_{j,k}\psi_{j,k}(t).$$

The large J refers to the relatively small number of coefficients for the low frequency, smooth variation of f, the small j refers to the high frequency detail coefficients.

When the sample size n, the number of observations, is divisible by 2, say $n = 2^J$, then the number of coefficients, n can be grouped as $n/2$

coefficients $d_{1,k}$ at the finest level, $n/4$ coefficients $d_{2,k}$ at the next finest level, ..., $n/2^J$ coefficients $d_{J,k}$ and $n/2^J$ coefficients $s_{J,k}$ at the coarsest level. Some wavelet applications in cancer data analysis were reviewed recently in Hong and Shyr (2006).

Multiscale analysis tools such as wavelets are providing a rich source of useful tools for applications in time-scale types of problems (Sentelle *et al.*, 2002). The Fourier transform extracts details from the signal frequency, but all information about the location of a particular frequency within the signal is lost. Though window Fourier transform (WFT) can help to determine time location for nonstantionary signals, the lack of adaptivity of WFT may lead to a local under- or over-fitting. In contrast to WFTs, wavelets select widths of time slices according the local frequency in the signal. This adaptivity property of wavelets certainly can help to us to determine the location of peak difference(s) of MALDI-TOS MS protein expressions between cancerous and normal tissues in term of molecular weights.

Wavelets, as building blocks of models, are well localized in both time and scale (frequency). Signals with rapid local changes (signals with discontinuities, cusps, sharp spikes, etc.) can be precisely represented with just a few wavelet coefficients.

Wavelets can be useful in detecting patterns in DNA sequences as well. In Lio and Vannucci (2000), it was shown that wavelet variance decomposition of bacterial genome sequences can reveal the location of pathogenicity islands. The findings show that wavelet smoothing and scalogram are powerful tools to detect differences within and between genomes and to separate small (gene level) and large (putative pathogenicity islands) genomic regions that have different composition characteristics. An optimization procedure improving upon traditional Fourier analysis performance in distinguishing coding from noncoding regions in DNA sequences was introduced in Anastassiou (2000). The approach can be taken one step further by applying wavelet transforms. To find the similarities between two or more protein sequences is of great importance for protein sequence analysis. In de Trad *et al.* (2002), a comparison method based on wavelet decomposition of protein sequences and a cross-correlation study was devised that is capable of analyzing a protein sequence "hierarchically," i.e., it can examine a protein sequence at different spatial resolutions. A sequence-scale similarity vector is generated for the comparison of two sequences feasible at different spatial resolutions (scales). The cosine Fourier series and discrete wavelet transforms are applied in Morozov (2000) for describing replacement rate variation in genes and proteins, in which the profile of relative replacement rates along the length of a given sequence is defined as a function of the site number. The new models are applicable to testing biological hypotheses such as the statistical identity of rate variation profiles among homologous protein families.

Despite advances in instrument resolution and sensitivity of MS technology, the effective resolution is limited by the distribution of naturally isotopes of common elements. This isotopic envelope of molecular weights complicates analysis of spectra when two or more species differ by only a few Daltons. The species exhibit overlapping spectral signatures, and form what is here termed a "peak cluster." The resolution of such clusters would be an important advance in biomedical research in general, and cancer research in particular.

As discussed above, an observed MALDI signal consists of a true signal $S(x)$, an incoherent signal $I(x)$ and machine noise $\varepsilon(x)$. To extract the true signal we need to remove the noise and the incoherent signal from the observed data. In the wavelet representation, the noise ε is concentrated in the fine scale wavelet coefficients and the incoherent signal can be approximated by the projection onto the coarse space spanned by the functions $\phi_{J,k}$. A variety of threshold strategies can be used to remove the machine noise from the data. A baseline can be designed using a coarse approximation and a component with small coefficients in a wavelet space.

The discrete mass spectrum data provide information about the cancer tissue and normal tissue at particular molecular weights. The wavelet approximation to a signal function f is built up over multiple scales and many localized positions. A discrete wavelet transform (DWT) decomposes a signal into several vectors of wavelet coefficients. Different coefficient vectors contain information about the signal function at different scales. Coefficients at coarse scale capture gross and global features of the signal while coefficients at fine scale contain detailed information. Applying wavelet transform to MALDI-TOF MS data, the protein expression difference can be measured at different resolution scales based on a molecular weight-scale analysis. It may reveal more information than other conventional methods.

Following Donoho and Johnstone (1994, 1995), we can apply a variety of threshold techniques for MALDI MS data processing. The idea behind threshold is the removal of small (wavelet) coefficients, considered to be noise. This leaves large coefficients in the multiscale decomposition object that can then be used to estimate the signal after reconstruction. There are many ways to threshold. The universal threshold is computed as

$$\lambda = s\sqrt{(2\log M)},$$

where M is the number of data points (wavelet coefficients) and s is an estimate of the variation of the coefficients on the standard deviation scale. Probability threshold is selecting the pth quantile of the coefficients based on a given probability value p. Soft threshold is to modify the coefficients by the formula:

$$d_{jk}^{new} = sgn(d_{jk})(|d_{jk}| - \lambda)_+$$

for the thresholding scale λ. If the noise process is stationary, one effect of correlated noise is to yield an array of wavelet coefficients with variances that depend on the level j of the transform. This leads to level-dependent threshold, using for each coefficient a threshold that is proportional to its standard deviation (Johnstone and Silverman, 1997). The level-dependent threshold method applied in wavelet regression gives optimally adaptive behavior. Block threshold is to threshold the wavelet coefficients in groups (blocks) rather than individually to increasing estimation accuracy by utilizing information about neighboring coefficients. Since the high frequency components decrease as the mass weight increases, we used block threshold strategy for MALDI-TOF MS data denoising (Chen et al., 2007).

An important development in the statistical context has been the routine use of the non-decimated wavelet transform (NDWT), also called the stationary or translation-invariant wavelet transform, see (Lang et al., 1996; Nason et al., 1995; Walden and Cristan, 1998) for example. Conceptually, the NDWT is obtained by modifying the Mallat DWT algorithm: at each stage, no decimation takes place but instead the filters are padded out with alternate zeros to double their length. The effect is to yield an over determined transform with n coefficients at each of $\log_2 n$ levels. The transform contains the standard DWT for every possible choice of time origin. Johnstone and Silverman (1997) investigated the use of the NDWT in conjunction with the marginal maximum likelihood approach. The NDWT has been used for SELDI MS data analysis (Coombes, 2005). In general, in WT, hard thresholds have a better l_2 performance while soft thresholds generate better smoothness results. However, with stationary discrete wavelet transform (SDWT), since the coefficients are undecimated, hard thresholds will have both good l_2 performance and smoothness (Coombes, 2005; Lang et al., 1996).

When we do wavelet denoising, we are faced with many parameters to choose, such as the type of a mother wavelet, the decomposition level and the values of thresholds. Based on the knowledge of the wavelet analysis to the data set, we try to use the objective criteria to determine the threshold values. Basically, the choice of mother wavelet seems not matter much, while the value of thresholds does (Coombes et al., 2005). Then, setting the values of thresholds becomes a crucial topic. According to the analysis above, we would like to set the threshold values based on the data sets' properties.

It has been observed that the high frequency components of spectrum data reduce as the mass weight increases because the values of median absolute deviation (MAD) change a lot throughout different m/z segments. MAD/0.67 is a robust estimate of the non-normal variability. This phenomenon might be caused by that the machine has relative low resolution for ions of small m/z values at low m/z interval. Therefore, we should set different thresholds at different mass segments by the changing trend of the coefficients at each level (Lavielle, 1999). In this way, the denoised signal can reduce the variance in the beginning part and retain the useful information in the posterior part.

In cancer research projects carried at Vanderbilt Ingram Cancer Center (VICC), we observed that most coefficients at levels from 1 to 4 are dumped. The reason is that they are of high frequency and low energy (the proportion of the total energy of the signal is only 10^{-12}). We also need to be very cautious when manipulating the low frequency components. We believe choosing threshold values based on the exploratory data analysis will achieve better wavelet denoising performance. This denoising method performed well in the study (Chen et al., 2007).

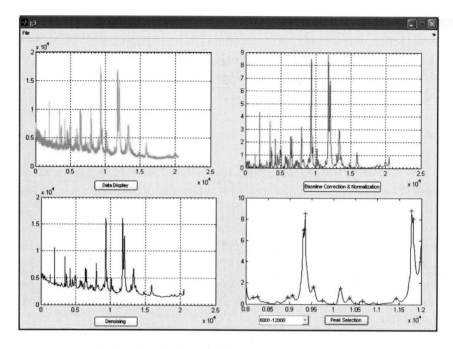

Fig. 4. Graphical user interfaces of WaveSpec software.

A software package called WaveSpec implementing the mathematical framework has been developed at Biostatistical Shared Resource of VICC. A MatLab based version of the software has been used to serve cancer research groups in VICC for MALDI-TOF MS data processing. Figure 4 shows a graphical user interface (GUI) of the software.

3.2. Diffusion Maps

Very recently, diffusion maps and geometric harmonics were introduced to understand the geometric structures of the data sets (Coifman *et al.*, 2005a, 2005b). In continuous Euclidean setting, tools from harmonic analysis, such as Fourier transforms and wavelet decompositions have proven to be highly successful in image compression, signal processing, denoising, and density estimation. In statistical data analysis, it is essential to organize graphs and data sets geometrically. Geometric diffusion is a tool for structure definition of data by extending multiscale harmonic analysis to discrete graphs and subsets of R^n.

A diffusion kernel on the data set is explicitly constructed and a diffusion map is defined by employing the spectral properties, spectrum and eigenfunctions. Also, a multiscale extension scheme is defined for decomposing empirical functions into frequency bands and showing the links between the intrinsic and extrinsic geometries of the set.

Coifman *et al.* (2005) introduced a family of diffusion maps that allow the exploration of the geometry, the statistics, and functions of the data. Diffusion maps provide a nature low-dimensional embedding of high-dimensional data that is suited for subsequent tasks such as visualization, clustering, and regression. It will be interesting to follow diffusion map's idea by emphasizing on the biological meaning and chemical structure of the mass spectrometry data set. The kernel in the model discussed above is symmetric and non-negative and we can define an associated diffusion map for such a kernel. The formalization permits the proper identification and estimation of a wavelet spectrum. Once the characteristic frequency for a particular biological function has been determined, it is possible to identify the individual mass spectrum's "hot spots" using wavelet transform that contribute mostly to the characteristic frequency and also to the protein's biological function (de Trad *et al.*, 2002). A suitable defined biological diffusion map will give potential improvements in the early detection and diagnosis of various types of cancer. It would be great to obtain a biological diffusion map for decomposing MALDI MS data into frequency bands and showing the links between the intrinsic and extrinsic biology of the data.

4. CLUSTERING AND CANCER DATA CLASSIFICATIONS

Mass spectra are intrinsically functional observations, and are well-suited to wavelet methods. We would like to apply multiscale techniques to further study preprocessed mass spectra for feature extraction. The significant difference in the findings would help the identification of protein markers.

Biomarkers are measurable molecular phenotypic parameters that characterize an organisms state of health or disease, or a response to a particular therapeutic intervention. Biomarkers are sought as instruments to help in disease risk assessment, early disease detection, and as surrogate endpoints in clinical trials (or in some cases as surrogates for environmental and other exogenous factors such as diet). Establishing/validating biomarkers include the following steps: (a) identify candidates, (b) conducting clinical assays to diagnose known disease, (c) detection of preclinical disease (pseudo-prospectively) and establishment of screen-positive rule, (d) prospective screening, establish extend and characteristics of identified disease as well as false referral rates, and (e) quantification of overall impact on disease. The biomarker selection problems maps nicely to the problem of feature selection for classification in statistics and machine learning. Feature selection is the problem of selecting a subset of variables of minimal size that can predict, classify, or diagnose a target variable of interest as well as, or better than, the full set of available predictors. In the case of MALDI-TOF MS signals, biomarkers could be individual masses, individual mass distributions or could be expressed in terms of wavelet coefficients or principal components. Selecting a minimal set of predictors with maximum accuracy is important for treating the curse of dimensionality, for reducing the cost of observing the required variables for prediction, and for gaining insight into the domain.

There are several families of methods biomarker selection on the reconstructed MALDI-TOF MS signal. *Principal component analysis* (PCA) involves a mathematical procedure that transforms a number of (possibly) correlated variables into a (smaller) number of uncorrelated variables called principal components. The first principal component accounts for as much of the variability in the data as possible, and each succeeding component accounts for as much of the remaining variability as possible. A widely used technique for the representation of sensor data is based on diagonalizing the correlation tensor of the data-set, keeping a small number of coherent structures (eigenvectors) based on principal components analysis (PCA). This approach tends to be global in character. It is possible to combine multiscale analysis and PCA to obtain proper accounting of global contributions to signal energy without loss of information on key local features. We can exploit such a combined wavelet-PCA technique in MALDI data processing.

Recently, we express MALDI MS data, after using WaveSpec preprocessing, in terms of a convex combination of dominant biological components based on principal component data for an initial investigation of

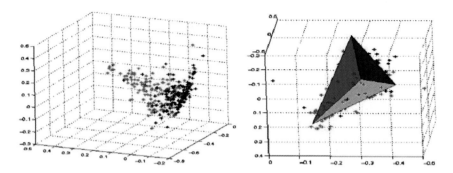

Fig. 5. MALDI-TOF MS lung cancer data plot based on selected three principal components (*left*). Simplex representation for MALDI-TOF MS lung cancer data (*right*).

feature extraction on MALDI-TOF MS data in lung cancer study (Hardin and Hong, 2006). Figure 5 shows that a convex combination (simplex) expression using super positions provides a promising tool for cancer feature extraction. The choice of analyzing wavelet basis leads to highlighting of certain features through the strengthening of a small set of coefficients, leaving the remainder at low amplitudes. It will be interesting in investigating adaptive wavelet-PCA approaches to pattern extraction from MALDI MS data, signal inversion and feature enhancement in the presence of noise.

A reliable and precise classification of tumor is essential for success in diagnosis and treatment of cancer. DNA microarrays have been used to characterize the molecular variations among tumors by monitoring gene expression profiles on a genomic scale. MALDI-TOF MS can profile proteins up to 50 kDa in size in tissue. Contributions of mass spectrometry to this infant field are largely untapped. This technology can not only directly assess peptides and proteins in sections of tumor tissue, but also can be used for high resolution image of individual biomolecules present in tissue sections. The protein profiles obtained can contain thousands of data points, necessitating sophisticated data analysis algorithms.

Clustering assigns samples to classes on the basis of their distance from objects known to be in the classes. The distance or similarity will have a large effect on the performance of the classification procedure. In orthogonal wavelet transform, the L_2 distance for the signals is equivalent to the l_2 distance for the vectors of wavelet coefficients. Therefore, we can perform clustering of MALDI MS data using wavelet coefficients with

Euclidean distance. If we view the wavelet coefficients in different scales as a microarray data set, then methods in microarray data analysis may be used for MALDI MS data analysis via wavelet coefficients as well. Shyr and Kim developed weighted flexible compound covariate methods (WFCCM) for classifying microarray data (Shyr and Kim, 2003). We can apply such methods to MALDI MS data for classification as well.

Lung cancer (Hoffman, 2002) is usually not detected early, and thus may be diagnosed at an advanced stage, where intervention or therapy is less effective. Although the incidence rate of lung cancer is lower than for breast and prostate cancer, the mortality rate of lung cancer is the highest for all cancers in both men and women. Lung cancer kills more Americans each year than the next four leading cancer killers, cancers of the colon, breast, prostate and pancreas, combined.

Precisely classifying tumors is of critical importance to cancer diagnosis and treatment. Recently, there is increasing interest in changing the basis of tumor classification from morphologic to molecular. Mass spectrometry of proteins promises to be a very valuable tool in diagnostic applications. There are several challenges to the use of such proteomics data in classification and clustering of samples from diseased and normal patients. In the following, we mention a clustering method applied to the preprocessed MALDI-TOF MS data from lung cancer patients, which was collected at VICC, using WaveSpec software (Chen, 2004).

Clustering analysis, as a multivariable statistics technique, is widely used in many different fields of study, such as engineering, genetics, medicine, psychology, and marketing. Generally, after clustering, we get the result that the profiles of objects in the same cluster are very similar and the profiles of objects in different clusters are relatively quite different.

In an example of 50 patients which will be discussed in detail later, there are many tissue mass spectra from healthy people and several groups of sick patients who have different types of cancer. We can see that even we do not know the distribution in advance, by clustering we can divide the spectra into several groups that are almost the same as the real distribution.

Generally, we build the model in the following: the initial object can be modeled as a $p \times n$ matrix for n vectors of length p. According to the characteristics of the vectors, we can cluster the matrix into several groups in the form of several submatrices: $p \times n_1, p \times n_2, \ldots$, for $\sum n_i = n$.

Hierarchical clustering and k-means clustering are two main clustering methods. Hierarchical clustering method shows us a grouping structure of the data, in the form of a cluster tree. The tree is not a single set of clusters, but rather a multi-level hierarchy, where clusters are more similar at the lower level, which allows you to decide what level or scale of clustering is most appropriate for your data.

For a $p \times n$ matrix corresponding to n vectors (objects), we use a metric (distance) to group them according to their relationship (similarity). For two vectors x and y, both having length p, some common distances are Euclidean distance: $d(x, y) = [\sum (x_i - y_i)^2]^{1/2}$, Manhattan distance: $d(x, y) = \sum |x_i - y_i|$, and correlation distance: $d(x, y) = 1 - \rho(x, y)$, where $\rho(x, y)$ is the correlation coefficient of x and y. Different distances may lead to the different cluster trees.

For n vectors, we will have $n(n - 1)/2$ pair distances. Then we need to link these newly formed clusters to other objects to create bigger clusters until all the objects in the original matrix are linked together in a hierarchical tree. There are several ways to create the cluster hierarchy tree such as shortest/longest distance, average distance and centroid distance. Matlab software has a function to display the hierarchical tree. To determine where to divide the hierarchical tree into clusters, we need to choose proper cutoff points so that we can cut the trees into several groups.

Comparing with the tree structure of hierarchical clustering, the k-means clustering method has set up the number of groups before clustering. Then all objects are grouped into k clusters, objects within each cluster are as close to each other as possible, and as far from objects in other clusters as possible. There are several member objects and a centroid, or a center in one cluster. The center for each cluster is a vector, which has the minimum sum of distances from all objects. K-means clustering uses an iterative algorithm to move objects between clusters until the sum of distances cannot be decreased any further.

Now, we have divided the $p \times n$ matrix into k clusters, but not all the elements in one cluster are different from the ones in other clusters. Therefore, we should figure out which elements are distinct in one cluster, in other words, the characteristic elements. For any two clusters, we can do pair t-test to the objects and find the rows of small p-values; or we can find the weighted average distance at a certain row:

$$w = d_B/(k_1 d_{w1} + k_2 d_{w2} + \varepsilon),$$

where d_B is the distance between cluster centers, d_{wi} is the average (Euclidean) distance among all sample pairs in one cluster, and $k_i = n_i/(n_1 + n_2)$ (Goldstein et al., 2002). Basically, the objects in the same cluster are close to each other but the distances between centers of different clusters are large. At last, if the distance of two objects is great or the p-value is small enough, then we can say that these elements are distinct. A center spectrum defined for binning scheme in (Chen, 2004) can be used for clustering as well.

As an example, we consider the MALDI TOF MS data set of 50 patients collected at VICC. The tissue sample consists of normal samples and cancer samples of Adeno, squamous, large, and other cancers. We apply the clustering analysis to the 1628×50 matrix, and then compared the results of clustering with the real data distribution.

Figures 6 and 7 show the results using the hierarchical clustering with the Euclidean distance and correlation distance, respectively.

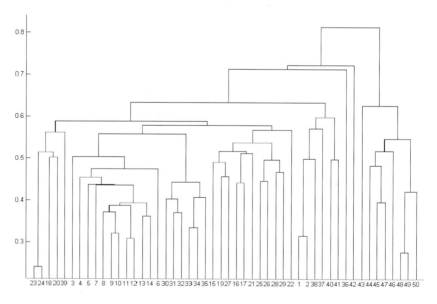

Fig. 6. Hierarchical trees by Euclidean distance.

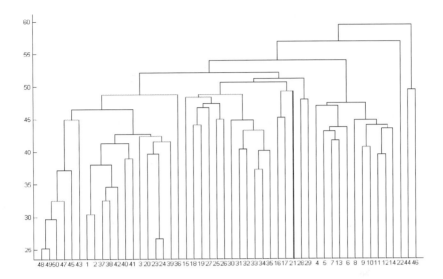

Fig. 7. Hierarchical trees by correlation distance.

Table 1. Lung cancer patients data distribution.

Labels	Cancer types
1–14	Adeno cancer
15–29	Squamous cancer
30–34	Large cancer
35–39	Meta-cancer
39–42	Other cancers
43–50	Normal

From the hierarchical trees, we can see that the clustering results match almost perfectly the real distribution data. In addition, the correlation method seems having a better performance (see Figures 6, 7, and Table 1 for comparison).

Also, when we exam the distinct elements of the normal and cancer cluster, we find that the rows: 38, 350, 356 are most significantly different and the p-values of them, which was provided by MatLab clustering program, are 0. Moreover, rows 38, 350, 356, 953, 986, and 991 are relatively distinct. That means these proteins are very likely different disease-related.

ACKNOWLEDGMENTS

The authors are grateful to Dean Billheimer, Shuo Chen, Doug Hardin, Huiming Li, Ming Li, and Jonathan B.G. Xu for their valuable discussions. This work was partially supported by Lung Cancer SPORE (Special Program of Research Excellence) (P50 CA90949), Breast Cancer SPORE (1P50 CA98131-01), GI (5P50 CA95103-02), and Cancer Center Support Grant (CCSG) (P30 CA68485) for Shyr and by National Science Foundation (IGMS 0552377), National Security Agency (#H98230-05-1-0304), and Research Enhancement Program Award from Middle Tennessee State University for Hong.

References

1. Aldroubi A and Unser M. *Wavelets in Medicine and Biology*. CRC Press, Boca Raton, FL, 1996.
2. Anastassiou D. Frequency-domain analysis of biomolecular sequences. *Bioinformatics* 2000; **16**: 1073–1081.
3. Baggerly KA, Morris JS and Coombes KR. Reproducibility of SELDI mass spectrometry patterns in serum: comparing proteomic data sets from different experiments. *Bioinformatics* 2004; **20**: 777–785.
4. Bar-Joseph Z, Gerber GK, Gifford DK, Jaakkola TS and Simon I. Continuous representations of time-series gene expression data. *J. Comput. Biol.* 2003; **10**: 341–356.
5. Chaurand P, Stoeckli M and Caprioli RM. Direct profiling of proteins in biological tissue sections by MALDI mass spectrometry. *Anal. Chem.* 1999; **71**: 5263–5270.
6. Chen S. *MALDI-TOF MS Data Processing Using Splines, Wavelets and Clustering Techniques*. Master's Thesis in Mathematical Sciences, East Tennessee State University, Johnson City, Tennessee, 2004.
7. Chen S, Hong D and Shyr Y. Wavelet-based procedures for proteomic mass spectrometry data processing. *Comput. Stat. Data Anal.* 2007; **52**: 211–220.
8. Chen HJ, Tracy ER, Cooke WE, Semmes OJ, Sasinowski M and Manos DM. Automated peak identification in a TOF-MS spectrum. In: *Quantitative Medical Data Analysis Using Mathematical Tools and Statistical Techniques*. Hong D and Shyr Y (eds.) World Scientific Publication, LLC, New Jersey, 2007, pp. 113–131.
9. Chui CK. *An Introduction to Wavelets*. Academic Press, New York, NY, 1992.

10. Coifman RR, Lafon SS, Lee AB, Maggioni M, Nadler B, Warner F and Zucker SW. Geometric diffusion as a tool for harmonic analysis and structure definition of data: diffusion maps. *Proc. Natl. Acad. Sci. USA* 2005a; **102**: 7426–7431.
11. Coifman RR, Lafon SS, Lee AB, Maggioni M, Nadler B, Warner F and Zucker SW. Geometric diffusion as a tool for harmonic analysis and structure definition of data: multiscale methods. *Proc. Natl. Acad. Sci. USA* 2005b; **102**: 7432–7437.
12. Coombes KR, Tsavachidis Morris JS, Baggerly KA, Hung MC and Kuerer HM. Improved peak detection and quantification of mass spectrometry data acquired from surface-enhanced laser desorption and ionization by denoising spectra with the undecimated discrete wavelet transform. *Proteomics* 2005; **5**: 4107–4117.
13. Daubechies I. *Ten Lectures on Wavelets*. Society for Industrial and Applied Mathematics, Philadelphia, Pennsylvania, 1992.
14. de Trad CH, Fang Q and Cosic I. Protein sequence comparison based on the wavelet transform approach. *Protein Eng.* 2002; **15**: 193–203.
15. Diamandis EP. Mass spectrometry as a diagnostic and a cancer biomarker discovery tool: opportunities and potential limitations. *Mol. Cell Proteomics* 2004; **3**: 367–378.
16. Donoho DL and Johnstone IM. Adapting to unknown smoothness via wavelet shrinkage. *J. Am. Stat. Assoc.* 1995; **90**: 1200–1224.
17. Donoho DL and Johnstone IM. Minimax estimation via wavelet shrinkage. *Ann. Stat.* 1998; **26**: 879–921.
18. Gentzel M, Kocher T, Ponnusamy S and Wilm M. Preprocessing of tandem mass spectrometric data to support automatic protein identification. *Proteomics* 2003; **8**: 1597–610.
19. Grossmann A and Morlet J. Decomposition of Hardy functions into square integrable wavelets of constant shape. *SIAM J. Math. Anal.* 1984; **15**: 723–736.
20. Hardin D and Hong D. Convex combination expression using super positions for cancer data feature extraction. *Manuscript under preparation*, 2006.
21. Henschke CI, McCauley DI, *et al.* Early lung cancer action project: overall design and findings from baseline screening. *Lancet* 1999; **354**: 99–105.
22. Hirakawa H, Muta S and Kuhara S. The hydrophobic cores of proteins predicted by wavelet analysis. *Bioinformatics* 1999; **15**: 141–148.
23. Hoffman PC, Mauer AM and Vokes EE. Lung cancer. *Lancet* 2000; **355**: 479–485.

24. Hong D, Li HM, Li M and Shyr Y. Wavelets and projecting spectrum binning for proteomic data processing. In: *Quantitative Medical Data Analysis Using Mathematical Tools and Statistical Techniques*. Hong D and Shy Y (eds.) World Scientific Publication, LLC, New Jersey, 2007, pp. 159–178.
25. Hong D and Shyr Y. Wavelet applications in cancer study. *J. Concrete Appl. Math.* 2006; **4**: 505–521.
26. Hong D and Shyr Y. *Quantitative Medical Data Analysis Using Mathematical Tools and Statistical Techniques*. World Scientific Publication, LLC, New Jersey, 2007.
27. Hong D, Wang JZ and Gardner R. *Real Analysis with an Introduction to Wavelets*. Academic Press, New York, 2005.
28. Johnstone IM and Silverman BW. Wavelet threshold estimators for data with correlated noise. *J. R. Stat. Soc. B* 1997; **59**: 319–351.
29. Karas K and Hillenkamp F. Laser desorption ionization of proteins with molecular masses exceeding 10,000 daltons. *Anal. Chem.* 1988; **60**: 2299–2301.
30. Lang M, Guo H, Odegard JE, Burrus CS and Wells RO Jr. Noise reduction using an undecimated discrete wavelet transform. *Signal Process. Lett. IEEE* 1996; **3**: 10–12.
31. Lio P. Wavelets in bioinformatics and computational biology: state of art and perspectives. *Bioinformatics* 2003; **19**: 2–9.
32. Lio P and Vannucci M. Finding pathogenicity islands and gene transfer events in genome data. *Bioinformatics* 2000; **16**: 932–940.
33. Loo JA, Dejohn DE, Du P, Stevenson TI and Ogorzalek-Loo RR. Application of mass spectrometry for target identification and characterization. *Med. Res. Rev.* 1999; **19**(4): 307–319.
34. Mallat S. *A Wavelet Tour of Signal Processing*. Academic Press, New York, 1999.
35. Morlet J, Arens G, Fourgeau E and Giard D. Wave propagation and sampling theory. *Geophysics* 1982; **47**: 203–236.
36. Morozov P, Sitnikova T, Churchill G, Ayala FJ and Rzhetsky A. A new method for characterizing replacement rate variation in molecular sequences: application of the Fourier and wavelet models to Drosophila and mammalian proteins. *Genetics* 2000; **154**: 381–395.
37. Morris JS, Coombes KR, Koomen JM, Baggerly KA and Kobayashi R. Feature extraction and quantification for mass spectrometry in biomedical applications using the mean spectrum. *Bioinformatics* 2005; **21**: 1764–1775.
38. Nason GP and Silverman BW. The stationary wavelet transform and some statistical applications. In: *Wavelets and Statistics*, Lecture Notes in Statistics,

No. 103. Antoniadis A and Oppenheim G. (eds.) Springer-Verlag, New York, 1995, pp. 281–300.
39. Purohit PV and Rocke DM. Discriminant models for high-throughput proteomics mass spectrometer data. *Proteomics* 2002; **3**: 1699–1703.
40. Roboz J. *Mass Spectrometry in Cancer Research*. CRC Press LLC, Boca Raton, Florida, 2002.
41. Sentelle S, Sentelle C and Sunton MA. Multiresolution-based segmentation of calcifications for the early detection of breast cancer. *Real-Time Imaging* 2002; **8**: 237–252.
42. Siuzdak G. *The Expanding Role of Mass Spectrometry in Biotechnology*. MCC Press, San Diego, CA, USA, 2003.
43. Silverman BW. Wavelets in statistics: beyond the standard assumptions. *Philos. Trans. R. Soc. Lond. A* 1999; **357**: 2459–2473.
44. Soltys SG, Le QT, Shi G, Tibshirani R, *et al*. The use of plasma SELDI TOF mass spectrometry proteomic patterns for detection of head and neck squamous cell cancers. *Clin. Cancer Res.* 2004; **10**: 4806–4812.
45. Srinivas PR, Srivastava S, Hanash S and Wright GL Jr. Proteomics in early detection of cancer. *Clin. Chem.* 2001; **47**: 1901–1911.
46. Thomson JJ. *Rays of Positive Electricity and Their Application to Chemical Analysis*. Longmans, Green and Co., London, 1913.
47. Vestal M and Juhasz P. Resolution and mass accuracy in matrix-assisted laser desorption ionization-time of flight. *J. Am. Soc. Mass Spectrom.* 1998; **9**: 892–911.
48. Walden AT and Cristan AC. Matching pursuit by undecimated discrete wavelet transform for non-stationary time series of arbitrary length. *Stat. Comput.* 1998; **8**: 205–219.
49. Waldsworth JT, Somers KD, Cazares LH, Malik G, *et al*. Serum protein profiles to identify head and neck cancer. *Clin. Cancer Res.* 2004; **10**: 1625–1632.
50. Yanagisawa K, Shyr Y, Xu BJ, Massion PP, Larsen PH, White BC, Roberts JR, Edgerton M, Gonzalez A, Nadaf S, Moore JH, Caprioli RM and Carbone DP. Proteomic patterns of tumour subsets in non-small-cell lung cancer. *Lancet* 2003; **362**: 433–439.
51. Yu WC, Wu BL, Lin N, Stone K, Williams K and Zhao HY. Detecting and aligning peaks in mass spectrometry data with applications to MALDI. *Comput. Biol. Chem.* 2006; **30**: 27–38.
52. Zhang Z, Bast RC, Yu Y, Li J, *et al*. Three biomarkers identified from serum proteomic analysis for the detection of early stage ovarian cancer. *Cancer Res.* 2004; **64**: 5882–5890.

Chapter 17

ADVANCED STATISTICAL METHODS FOR THE DESIGN AND ANALYSIS OF TUMOR XENOGRAFT EXPERIMENTS

Ming Tan and Hong-Bin Fang

Xenograft model is a common *in vivo* model in cancer research, where human cancer (e.g., sliced tumor tissue blocks, or tumor cells) are grafted and grown in severe combined immunodeficient (scid) nude mice. In cancer drug development, demonstrated anti-tumor activity in this model is an important step to bring a promising experimental treatment to human. These experiments provide important data on the mechanism of action of the drug and for the design of future clinical trials. For therapy with single agent, the experimental design and sample size formulae are quite well established. However, cancer therapy typically involves combination of multiple agents. Such studies should be optimally designed, so that with moderate sample size, the joint action of two drugs can be estimated and the best combinations identified. A typical outcome variable in these experiments is tumor volume measured over a period of time. The resulting data have several unique features. Since a mouse may die during the experiment or may be sacrificed when its tumor volume quadruples, then incomplete repeated measurements arise. The incompleteness or missingness is also caused by drastic tumor shrinkage (<0.01 cm^3) or random truncation. In addition, if no treatment were given to the tumor-bearing mice, the tumors would keep growing until the mice die or are sacrificed. This intrinsic growth of tumor in the absence of treatment constrains the parameters in the regression and causes further difficulties in statistical analysis. This chapter reviews the current methods of experimental design and data analysis for xenograft experiments. We describe the optimal experimental design for combination studies in xenograft models and likelihood-based methods for estimating the dose-response relationship while accounting for the special features of data such as informative censoring and model parameter constraints.

Keywords: Dose-effect, uniform, fixed ratio, optimal designs, interaction, synergism, ECM algorithm, longitudinal data, missingness, MLE, truncation, tumor xenograft model.

1. INTRODUCTION

Xenograft model is a common *in vivo* model in cancer research, where human cancer (e.g., sliced tumor tissue blocks, or tumor cells) are grafted and grown in severe combined immunodeficient (scid) nude mice. This model can be used for different purposes including elucidating the mechanism and biology of tumor and more commonly is used as a model for testing the efficacy of an experimental therapy. Since most human tumors escape clinical detection for the greater part of their growth process, it is necessary to study human tumors in an experimental situation as is in xenograft models. Because anti-tumor activity in the xenograft can be correlated well with patient response (Fiebig, 1988), demonstrated activity in this model is an important step to bring a promising compound to human. Information from xenograft models provide critical efficacy information for the design of early stage clinical trials. For example, Houghton *et al.* (1998) have successfully utilized the information to facilitate the design of clinical trial and has predicted clinical response well.

In a typical xenograft experiment, treatment is initiated when the diameter of tumor reaches certain level (e.g., 0.5 cm). Several treatment regimens are administered and the outcome variables such as tumor volumes are measured (using, e.g., the Maxcal digital caliper) at the start of the treatment and regularly in a given period of follow-up time. Measurements are transmitted directly to a computer. The renewed interest stems from humane and cost considerations as well as scientific considerations for analyzing the data efficiently and make inference properly. A general statistical guideline for the design and analysis of experiments involving animals is proposed by Festing and Altman (2002) and Rygaard and Spang-Thomsen (1997).

Although the experimental subjects are typically scid mice that are breed virtually genetically identical, their responses even to the same dose vary considerably. Such variation needs to be controlled in the experimental design and accounted for in the analysis to allow proper inference on the efficacy of the experimental therapy. For therapy with single agent, the experimental design and sample size formulae are quite well established. However, cancer therapy typically involves combination of multiple agents. It is important to examine the interaction of the two agents in xenograft model. The study of the joint action of two drugs has a long history in pharmacology and biostatistics (see reviews by Berenbaum, 1989; Greco *et al.*, 1995). A determining factor for a statistical approach to the design and analysis of combination studies is the variation in dose-effects. This variability necessitates a statistical approach so that the variation is controlled in the experimental design and accounted for in the analysis in evaluating the efficacy of the combination. The design for combination studies should be optimally designed, so that with moderate sample size, the joint action of two drugs can be estimated and the best combinations identified. An overview of the advanced methods of experimental designs will be given in Section 2.

For statistical analysis, a typical outcome variable in these experiments is tumor volume measured over a period of time, while mice are treated with an anti-cancer agent following certain schedules. However, the statistical analysis of such longitudinal data presents several challenges for a number of reasons. First, sample sizes in xenograft experiments are usually small because of cost, graft failures due to body rejection or some xeno-antigens. Therefore, small sample inference procedures are needed. Second, in xenograft experiments, the tumor growth depends on initial volumes. If no treatment were given, tumors in mice would keep growing until the tumor-bearing mice die or are sacrificed. Thus, estimating antitumor activity should adjust for the intrinsic tumor growth in the absence of treatment, which thus constrains the regression coefficients. Finally, missing data is hard to avoid in these experiments because a mouse may die of toxicity or may be sacrificed when its tumor volume reaches a set criterion (e.g., quadruples) or the tumor volume becomes unmeasurable (e.g., when it is less than 0.01 cm^3).

The rest of this chapter is organized as follows. We first describe the experimental design and sample size for combination studies of two drugs including the fixed-ratio design, Abdelbasit-Plackett optimal design and the uniform design in Section 2. We then summarize and formulate the statistical models for the analysis of xenograft experiments in Section 3. This includes likelihood estimates for dose-effects using the ECM algorithm.

Section 4 provides some methods for the comparison of two treatments while accounting for special features of xenograft data such as non-informative censoring and parameter constraints. We conclude with a discussion in Section 5.

2. DESIGN OF EXPERIMENTS FOR COMBINATION STUDIES

The joint action of two drugs is usually divided into three types: independent joint action, simple similar (additive) joint action and synergistic (or antagonistic) action. The independent joint action of two drugs can be addressed sometimes by pharmacology and biology. For example, one antihypertensive drug that lowers blood pressure by blocking angiotensin II receptors and another that exerts its effect through diuresis would have independent joint action when combined. In most situations, the joint action of two anti-cancer drugs is not independent. Study of their joint action has been based on dose-response models.

Let the dose-response curves of drugs A and B be

$$y = f_A(x_A) \quad \text{and} \quad y = f_B(x_B), \tag{1}$$

respectively, where y is the study endpoint such as tumor volume, the percent of apoptosis, etc., and x_A and x_B are the dosages of A and B, respectively. Since for most drugs higher dose results in higher toxicity, the functions f_A and f_B in (1) will be assumed monotonic. Let D_A and D_B be the doses of A and B individually that are isoeffective, then,

$$f_A(D_A) = f_B(D_B). \tag{2}$$

The **potency** ρ of B relative to A is defined by

$$\rho = D_A/D_B = f_A^{-1} f_B(D_B)/D_B, \tag{3}$$

and the dose ρD_B of A and the dose D_B of B are equivalent. The potency is a measurement of relative dose effects of the two drugs and an important parameter whose role in combination studies has only been fully recognized in Abdelbasit and Plackett (1982) and Tan et al. (2003).

To describe the joint action of two drugs A and B at a specific dose level, the additivity of Loewe (1955) is based on single drug dose-effect and is defined by the following isobole equation

$$\frac{d_A}{D_A} + \frac{d_B}{D_B} = m, \qquad (4)$$

where d_A and d_B are doses of drugs A and B in the combination required to produce an effect of given effect (e.g., the dose resulting in 50% inhibition, or p percent in general, or 50% mice died) and D_A and D_B are the doses required for each drug to individually produce the given effect. When $m = 1$, the agents at the combination (d_A, d_B) are additive (zero-interaction). For $m < 1$, the agents at the combination (d_A, d_B) are more effective than expected from their single drug dose-response curves (synergy). Conversely, the agents in combination are less effective than expected (antagonism) if $m > 1$. Graphing this for different single drug doses of D_A and D_B then creates the two-dimensional (2D) isobologram.

Another approach is the median-effect derived from enzyme kinetic arguments of Chou and Talalay (1984). This have been used often *in vitro* studies. Their method has been generalized by Greco *et al.* (1990) who derived a combination index (CI) for the mutually exclusive case of Chou and Talalay for two-drug combinations with individual components concentration-effect curves as follows.

$$\text{CI} = 1 - \frac{\alpha d_A d_B}{IC_{50,A} IC_{50,B} \left(\frac{E}{E_{\text{con}} - E}\right)^{1/2m_A + 1/2m_B}}, \qquad (5)$$

where E and E_{con} are the measured and the control effect, respectively. IC_{50} is the concentration of drug resulting in 50% inhibition, of drug A or drug B, and m_A and m_B are the slope parameters. The magnitude of α indicates the intensity of the interaction. When α is positive, Loewe synergism is indicated. When α is negative, Loewe antagonism is indicated, and when $\alpha = 0$, Loewe additivity is indicated.

The experimental design for combination studies has attracted attention of several statisticians. Under the simple similar joint action assumption, Finney (1971) proposed that the regression lines for the mixtures should be equally spaced between those for the drugs. Because this design will result in a large variance, several methods of experimental design have been developed based on different model assumptions, such as the fixed-ratio design, ray design, and uniform measures design.

2.1. Fixed-Ratio Design and Ray Design

A simple experimental design method is the fixed-ratio or ray design which was used in Tallarida (1992). In the fixed-ratio design of two drugs, the individual compounds are administered in amounts that keep the proportions of two drugs constant. The total amount in the mixture is the sum, denoted by Z

$$Z = rD_A + (1-r)D_B, \qquad (6)$$

where D_A and D_B are estimates of the doses of the individual compounds of the mixture required to obtain a certain effect level (for example, IC_{50}). r is called the mixture factor and takes values from 0 to 1.

The fixed-ratio design is desirable for several reasons. A manufactured combination product would certainly contain a constant proportion of the ingredients. Also, in experimental work, this design simplifies the analysis of the data. Finally, it has been found that synergism, when it occurs, is a function of the proportions in the combination, i.e., one proportion may be markedly synergistic while another is simply additive (Tallarida, 2000). However, at the discovery stage, we may not know which ratio is the best.

To investigate the joint action of two drugs at different ratios, the ray design has been proposed (Gennings *et al.*, 2004; Straetemans *et al.*, 2005). In ray design, the values of r are scattered uniformly in $(0, 1)$. For a given ratio r, choose M different mixtures with m replicates. Then, with a specific r, there are $M \times m$ experiments which will be used to construct dose-response curves.

2.2. Abdelbasit-Plackett Optimal Experimental Design

Assume that the single drug dose-response curves of drugs A and B are

$$y = \alpha_A + \beta \log x_A \quad \text{and} \quad y = \alpha_B + \beta \log_B, \qquad (7)$$

respectively. Then, for the simple similar action of A and B, the regression line for the mixture is

$$y = \beta\{(\log z - \mu_A) + \log[(1-\rho)\pi_A + \rho]\}, \tag{8}$$

where $\mu_A = -\alpha_A/\beta$, ρ is the potency of B relative to A, z is the total dose and π_A is the proportion of the dose of A in the total dose.

Based on the individual dose-response models (7), Abdelbasit and Plackett (1982) proposed an optimal design for the combination study of two drugs with two specific models, the first model is (Plackett and Hewlett, 1952)

$$y = \beta\{(\log z - \mu_A) + \eta^{-1}\log(\pi_A^\eta + \rho^\eta\pi_B^\eta)\}, \tag{9}$$

and the second is (Finney, 1971)

$$y = \beta\{(\log z - \mu_A) + \log(\pi_A + \rho\pi_B + \kappa(\rho\pi_A\pi_B)^{1/2})\}, \tag{10}$$

where η and κ are the parameters.

In models (9) and (10), the parameters μ_A, ρ and β are known. When $\eta = 1$ and $\kappa = 0$, the models (9) and (10) reduce to (8). For given values of the total dose z_i, Abdelbasit and Plackett (1982) proved that the asymptotic variance of η when $\eta = 1$ in (9) and the asymptotic variance of κ when $\kappa = 0$ in (10) are minimized if the proportion π_A of A in the total dose is fixed. Then, they proposed an optimal design for the combination experiment such that the total dose z is uniformly scattered in the experimental domain and the value of π_A is given by $\rho/(1+\rho)$ using Hewlett's criterion.

2.3. Uniform Experimental Design

Unfortunately, properties of the dose-response curve for a single drug are not necessarily correlated with the dose-response surface for combinations of drugs, we often do not know how the two drugs interact before the experiment nor the mechanism of the joint action. Tan et al. (2003) proposed a uniform design for an unspecific model which takes advantage of the dose-response relation of each single drug. Assume that the single drug dose-response curves are given by (7). As a generalization of models (9) and (10), the joint action of two drugs is assumed to have the non-parametric model,

$$y = \beta\{(\log z - \mu_A) + \log[(1-\rho)\pi_A + \rho]\} + f(z, \pi_A) + \varepsilon, \qquad (11)$$

where ε is the error term assumed to be normally distributed with mean 0 and variance σ^2. The function f of z and π_A is unspecified and satisfies the following orthogonal condition

$$\int_S g(z, \pi_A) f(z, \pi_A) dz d\pi_A = \mathbf{0}, \qquad (12)$$

where $\mathbf{g}(z, \pi_A) = (\log z - \mu_A, \log[(1-\rho)\pi_A + \rho])^T$ and $S = (z_L, z_H) \times (0, 1)$ is the domain of $(z, \pi_A)^T$. This orthogonality condition ensures f is independent of the first term of model (11).

The test statistic is derived by maximizing the minimum power of the statistical test for lack of fit of the model (11). Consider m mixtures of A and B are $d_1, \ldots, d_m$, and n_i experiments at the dose-level $d_i = (z_i, \pi_{ai})^T \in S$ with corresponding responses y_{ij}, $j = 1, \ldots, n_i$, $i = 1, \ldots, m$. Denote $n = n_1 + \cdots + n_m$. Let $\mathbf{y}$ be the $n \times 1$ vector with elements y_{ij} ordered lexicographically and $\mathbf{1}_k$ be the $k \times 1$ vector of one. Let Z be the $m \times 2$ matrix with ith row $(\log z_i - \mu_a, \log[(1-\rho)\pi_{ai} + \rho])$. Denote $V = UZ(Z^T U^T UZ)^{-1} Z^T U^T$, $J = U(U^T U)^{-1} U^T$ and $U = \text{diag}(\mathbf{1}_{n_1}, \ldots, \mathbf{1}_{n_m})$. The test statistic for lack of fit is the ratio of the difference between a multiple of the estimated variance (σ^2) under (11) and that under the null (additive) model to the estimated variance under the additive model. Therefore, if drugs A and B have the additive action, this test statistic can be written as

$$F = \frac{\mathbf{y}^T (J - V) \mathbf{y}/(m-2)}{\mathbf{y}^T (I - J) \mathbf{y}/(n-m)}, \qquad (13)$$

which has a central F-distribution with degrees of freedom $m-2$ and $n-m$. Otherwise, the statistic has a non-central F-distribution with the non-central parameter δ. When the m combinations of A and B are uniformly scattered in the experimental dosage region S, the non-central parameter δ is maximized,

$$\delta = \frac{n}{\sigma^2} \int_S f^2(z, \pi_A) dz d\pi_A. \qquad (14)$$

Thus, the uniform design measure maximizes the minimum power of the F-test for the additive action of drugs A and B.

To obtain the uniform experimental points, Tan et al. (2003) proposed the U-type design ($\mathbf{U}_{m,k}$). Let $\mathbf{U}_{m,2} = (u_{ij})$ be an $m \times 2$ matrix, where each column is a permutation of $\{1, 2, \ldots, m\}$. Its induced matrix, $\mathbf{V}_{m,2} = (v_{ij})$, is defined by $v_{ij} = (u_{ij} - 0.5)/m$ for $i = 1, \ldots, m; j = 1, 2$. The matrix $V_{m,2}$ can be considered as m points on C^2. Then the uniform design (UD) is to choose the m points so that the discrepancy of $V_{m,2}$ is the smallest among all possible $V_{m,2}$ and the corresponding $\mathbf{U}_{m,2}$ is called a U-type design matrix. Based on the uniform scattered points on C^2, it is easy to obtain the UD on the domain $S = (z_L, z_H) \times (0, 1)$ and the mixtures of two drugs for experiment with a linear transformation. Let $(v_{i1}, v_{i2})^T, i = 1, \ldots, m$, be m UD points on C^2. Then the m UD points on S are given by $(z_i, \pi_{Ai})^T$, where $z_i = v_{i1}(z_H - z_L) + z_L$ and $\pi_{Ai} = v_{i2}$. Hence, the ith combination of the two drugs A and B is given as follows.

$$\text{dose of } A: \pi_{Ai} z_i = v_{i2} v_{i1} z_H + v_{i2}(1 - v_{i1}) z_L,$$
$$\text{dose of } B: (1 - \pi_{Ai}) z_i = (1 - v_{i2})[v_{i1} z_H + (1 - v_{i2}) z_L],$$

which gives all the m combinations with $i = 1, \ldots, m$.

3. STATISTICAL ANALYSIS FOR TUMOR GROWTH

To analyze tumor growth in xenograft experiments, there are two common statistical approaches, linear and non-linear models. The linear model approach is more directly concerned with drawing inferences from observed growth data. In this approach one approximates the growth curve with a polynomial in time and accommodates treatment effects by including appropriate predictors in the design matrix. With models it is common to allow the variance matrix of the observations to be of arbitrary form (Gart et al., 1986; Heitjan et al., 1993). The non-linear model approach is based on mathematical modeling, such as the Gompertzian growth curve (see Leith et al., 1987; Rygaard and Spang-Thomsen, 1997), or non-parametric modeling to fit real data (Liang, 2005).

Linear models have certain advantages. One can construct valid inferences for treatment effect parameters at different time points. However, if tumor growth is actually Gompertzian, linear models rely on polynomial approximations of the mean growth curve. The approximations behave wildly outside the time period for which they are fitted, making them unsuitable for prediction. The main problem with non-linear models is that they are non-linear and therefore are more difficult to handle computationally and inferentially than linear models with limited data. Moreover, with non-linear models there is no obvious, biological basis to generalize the basic growth curves to describe different therapeutic effects of treatments.

Heitjan (1991) proposed a class of different therapeutic non-linear models, the generalized Norton-Simon models. These models accommodate time-varying treatment effects through an effective dose function and assume that unperturbed growth follows the generalized logistic form. However, those parametric models do not always fit well across the entire curves for certain tumors, especially for different tumor growth curves even in the same xenograft experiments (see Tan et al., 2005a). In addition, all of these methods ignored the special features of data arisen from xenograft experiments such as missingness and informative censoring. Tan et al. (2005b) proposed a class of semi-parametric regression models with non-decreasing intercepts for unperturbed tumor growth. In this section, we synthesize recent advances in statistical inference accounting for these challenges in xenograft experiments.

3.1. Statistical Models

Consider in general the antitumor activity of h agents. Suppose that the $m+1$ prespecified follow-up times are $t_0 < t_1 < \cdots < t_m$ for n subjects (mice). Let y_{i0} and $\mathbf{y}_i = (y_{i1}, \ldots, y_{im})^T$ be the initial tumor volume (in log scale) and an m-dimensional vector of tumor volumes (in log scale) from the ith mouse at times $t_1, \ldots, t_m$, respectively. Let $x_{ij}^{(u)}$ be the cumulative dose of the uth agent administered to the ith mouse before time t_j for $u = 1, \ldots, h$, and the interaction terms for $u = h+1, \ldots, s$, where $j = 1, \ldots, m$ and $i = 1, \ldots, n$. Denote $\mathbf{x}_i^{(u)} = (x_{i1}^{(u)}, \ldots, x_{im}^{(u)})^T$, the antitumor activity of h agents in the ith mouse is modeled by,

$$\mathbf{y}_i = \boldsymbol{\alpha} + \mathbf{X}_i \boldsymbol{\beta} + \boldsymbol{\varepsilon}_i, \quad \text{for } i = 1, \ldots, n \tag{15}$$

where $\boldsymbol{\beta} = (\beta_0, \beta_1, \ldots, \beta_s)^T$ are unknown parameter vectors, $\mathbf{X}_i = (y_{i0}\mathbf{1}_m, \mathbf{x}_i^{(1)}, \ldots, \mathbf{x}_i^{(s)})$ is an $m \times (s+1)$ the known covariate matrix, $\mathbf{1}_m$ is the m-dimensional vector with component 1. In the xenograft experiments, tumors born by the immunosuppressed mice in the control (untreated) group generally keep growing over the follow-up period because of proliferation of cancer cells. The components of the intercept $\boldsymbol{\alpha} = (\alpha_1, \ldots, \alpha_m)^T$ should show an intrinsic growth of untreated tumor. Ignoring this restriction may result in misleading inference, usually an underestimate of the treatment effect. Therefore, a reasonable assumption is that the components of $\boldsymbol{\alpha}$ are in an increasing order over the follow-up weeks, namely, satisfying the constraint,

$$\alpha_1 \leq \alpha_2 \leq \cdots \leq \alpha_m. \tag{16}$$

In the regression model (15), the first column of $\mathbf{X}_i$ is the initial tumor volume y_{i0}, treated as a covariate. The corresponding regression coefficient β_0 reflects the effect of the initial tumor volumes and is a nuisance parameter. The other columns of $\mathbf{X}_i$ consist of the cumulative doses of agents received by the ith mouse at each observation time and the corresponding interactions among the h agents. Then, the covariate matrices $\mathbf{X}_i$, $i = 1, \ldots, n$, are different for each mouse although each mouse within a treatment group has the same schedule of administration.

The error term $\boldsymbol{\varepsilon}_i$ in (15) is assumed to have the m-dimensional normal distribution with mean $\mathbf{0}$ and covariance matrix $\boldsymbol{\Sigma}_m = \sigma^2 \mathbf{R}_m$. To describe the dependence structure of $(y_{i1}, \ldots, y_{im})^T$, the covariance matrix $\boldsymbol{\Sigma}_m = \sigma^2 \mathbf{R}_m$ is generally specified in three ways. The first is the unrestricted form, where the fully unrestricted correlation matrix $\mathbf{R}_m$ has $m(m-1)/2$ unknown correlation coefficients to be estimated. Due to the limited sample size in the xenograft experiment, the problem soon becomes unidentifiable unless the follow-up times m is also small. The second is the equicorrelated (or intraclass correlation) form. In this case, the correlations are all equal and described by a single parameter ρ, that is, $\mathbf{R}_m = (1-\rho)\mathbf{I}_m + \rho\mathbf{1}_m\mathbf{1}_m^T$, where $\mathbf{I}_m$ denotes the $m \times m$ identity matrix and $\rho \in (-\frac{1}{m-1}, 1)$ denotes the unknown correlation coefficient. In the xenograft model, the equicorrelated assumption between observations at different time points ignores the length between those time points and may not hold. Therefore, a more reasonable assumption is the Toeplitz correlation structure, where a higher correlation is assumed at two points closer together, namely the correlation depends on a single parameter $\rho \in (-1, 1)$ and $\mathbf{R}_m = (r_{jl})$, and

$$r_{jl} = \mathrm{Corr}(y_{ij}, y_{il}) = \rho^{|j-l|}, \quad j, l = 1, \ldots, m. \tag{17}$$

3.2. Parameter Estimation via the ECM Algorithm

To calculate the MLEs of the parameters in model (15), denote the unknown parameters by $\theta = (\boldsymbol{\beta}, \sigma^2, \rho, \boldsymbol{\alpha})$. The complete-data $\mathbf{y}_i = (y_{i1}, \ldots, y_{im})^T$ for the ith mouse consists of three parts: the observed part $\mathbf{y}_{i,\mathrm{obs}}$ with length p_i, the informative missing part $\mathbf{y}_{i,\mathrm{inf}}$ with length q_i (i.e., each component in $\mathbf{y}_{i,\mathrm{inf}}$ is less than a known constant a), and the missing at random part $\mathbf{y}_{i,\mathrm{mis}}$ with length r_i, where $p_i + q_i + r_i = m$. Denote the observed part by $\mathbf{Y}_{\mathrm{obs}} = \{\mathbf{y}_{i,\mathrm{obs}} : i = 1, \ldots, n\}$, the informative missing part by $\mathbf{Y}_{\mathrm{inf}} = \{\mathbf{y}_{i,\mathrm{inf}} : i = 1, \ldots, n\}$, and the missing at random part by $\mathbf{Y}_{\mathrm{mis}} = \{\mathbf{y}_{i,\mathrm{mis}} : i = 1, \ldots, n\}$. We further denote the observed *data* by $\mathbf{Y}_{\mathrm{obs}}^* = \{\mathbf{Y}_{\mathrm{obs}}, \boldsymbol{\Delta}\}$, where $\boldsymbol{\Delta} = \{\boldsymbol{\delta}_i : i = 1, \ldots, n\}$, and $\boldsymbol{\delta}_i$ is a vector of indicator whose components equal 1 if the corresponding component of $\mathbf{y}_i$ is less than a, otherwise 0. Define

$$\mathcal{S}_{\boldsymbol{\alpha}} = \{(\alpha_1, \ldots, \alpha_m)^T : -\infty < \alpha_1 \leq \cdots \leq \alpha_m < +\infty\} \tag{18}$$

then, the restricted MLEs of θ subject to $\boldsymbol{\alpha} \in \mathcal{S}_{\boldsymbol{\alpha}}$ based on the complete data are determined by the following equations:

$$\begin{cases} \boldsymbol{\beta} = \left(\sum_{i=1}^n \mathbf{X}_i^T \mathbf{R}_m^{-1} \mathbf{X}_i\right)^{-1} \sum_{i=1}^n \mathbf{X}_i^T \mathbf{R}_m^{-1}(\mathbf{y}_i - \boldsymbol{\alpha}) \\ \sigma^2 = \dfrac{1}{mn} \mathrm{tr}(\mathbf{R}_m^{-1} \mathbf{B}) \\ \rho = \dfrac{\frac{1}{2}\mathrm{tr}([\mathbf{H}_m + \mathbf{H}_m^T]\mathbf{B})}{n\sigma^2 + \mathrm{tr}(\mathbf{K}_m \mathbf{B})} \end{cases} \tag{19}$$

and

$$\boldsymbol{\alpha} = \arg\min_{\boldsymbol{\alpha} \in \mathcal{S}_{\boldsymbol{\alpha}}} \mathrm{tr}(\mathbf{R}_m^{-1}\mathbf{B}), \tag{20}$$

where $\mathbf{B} = \sum_{i=1}^n (\mathbf{y}_i - \boldsymbol{\alpha} - \mathbf{X}_i\boldsymbol{\beta})(\mathbf{y}_i - \boldsymbol{\alpha} - \mathbf{X}_i\boldsymbol{\beta})^T$, $\mathbf{K}_m = \mathrm{diag}(0, \mathbf{1}_{m-2}^T, 0) = \mathrm{diag}(0, 1, \ldots, 1, 0)$ and $\mathbf{H}_m = \begin{pmatrix} 0 & 0 \\ \mathbf{I}_{m-1} & 0 \end{pmatrix}$.

The parameters θ can be estimated via (19) and (20) using the ECM (Meng and Rubin, 1993) algorithm, in which the conditional expected values are calculated by the method proposed in Tan et al. (2002) since some observations are informative censored. The standard errors of estimated parameters are equal to the square root of the diagonal elements of the inverse of the estimated information matrix (Louis, 1982). However, the restricted parameters α cause more difficulties in the computation. To obtain the solution of α in (20), a novel EM algorithm via data augmentation is given as follows.

Denote $\mathbf{V} = (v_{\ell\ell'})$ is the $m \times m$ lower-triangle matrix with $v_{\ell\ell'} = 1$ for $\ell \geq \ell'$ and 0 otherwise. With the a transformation $\alpha = \mathbf{V}\gamma$, Eq. (20) is equivalent to

$$\gamma_1 = (v_1 - \Omega_{12}\gamma_{-1})/\omega_{11}, \tag{21}$$

and

$$\gamma_{-1} = \arg\min_{\gamma_{-1} \geq 0} \left\{ \gamma_{-1}^T \Omega_{22} \gamma_{-1} - 2\gamma_{-1}^T (\mathbf{v}_{-1} - \gamma_1 \Omega_{21}) \right\}, \tag{22}$$

where $\gamma = (\gamma_1, \ldots, \gamma_m)^T$, $\gamma_{-1} = (\gamma_2, \ldots, \gamma_m)^T$, and

$$\Omega = \begin{pmatrix} \omega_{11} & \Omega_{12} \\ \Omega_{21} & \Omega_{22} \end{pmatrix} \equiv \mathbf{V}^T \mathbf{R}_m^{-1} \mathbf{V},$$

$$\mathbf{v} = \begin{pmatrix} v_1 \\ \mathbf{v}_{-1} \end{pmatrix} \equiv \mathbf{V}^T \mathbf{R}_m^{-1} \sum_{i=1}^n (\mathbf{y}_i - \mathbf{X}_i \boldsymbol{\beta}).$$

From the Cholesky decomposition of Ω_{22} such that $\Omega_{22} = \mathbf{A}^T \mathbf{A}$ such that $\mathbf{A} = (a_{ik})$ is the upper triangular matrix with positive diagonal elements. Let $\boldsymbol{\mu} = (\mu_2, \ldots, \mu_m)^T = \mathbf{A}(\mathbf{v}_{-1} - \alpha_1 \Omega_{21})$, we have

$$\gamma_{-1} = \arg\min_{\alpha_{-1} \geq 0} \|\boldsymbol{\mu} - \mathbf{A}\boldsymbol{\alpha}_{-1}\|^2.$$

Given the current estimate $\gamma_{-1}^{(t)} = (\gamma_2^{(t)}, \ldots, \gamma_m^{(t)})^T$, the E-step calculates

$$S_k^{(t)} = \sum_{i=2}^m a_{ik} \left[a_{ik}\gamma_k^{(t)} + \frac{\mu_i - \sum_{\ell=2}^m a_{i\ell}\gamma_\ell^{(t)}}{m-1} \right] / \sum_{i=2}^m a_{ik}^2,$$

for $k = 2, \ldots, m$, and the M-step updates

$$\gamma_k^{(t+1)} = \max\{0, S_k^{(t)}\}, \quad k = 2, \ldots, m.$$

4. COMPARISON OF TREATMENT EFFECTS

To compare two treatments in xenograft experiments, a common method is to test the treatment difference at a fixed points using a t-test or a Mann-Whitney test, an ANOVA F-test or a Kruskal-Wallis test, indicating all the times at which differences were significant (e.g., Kasprzyk et al., 1992). For example, a test may be performed at the final measurement time or the final time when a substantial fraction of the animals were alive (Sakaguchi et al., 1992). The method to analyze animal survival times such as time to tumor volume quadruple in addition to tumor volumes was also used in literature (Houghton et al., 2000). Based on a comparison of those methods, Heitjan et al. (1993) indicated that the methods commonly used are deficient in that they have either low power or misleading type I error rates and proposed a multivariate method to improving the efficiency of testing. However, this multivariate test method does not account for missing data and large sample sizes are necessary.

A major goal in the xenograft model is to assess the effectiveness of treatment regimens, e.g., the mean tumor sizes at different time points as opposed to a growth curve analysis where the nonlinearity of tumor responses over the follow-up period and the limited amount of data in xenograft models preclude a growth curve characterization. When the goal of the study is to compare two treatment groups, a test can be derived as a special case in the random-effects model. However, a simpler approach is to use a modified t-test and the Bayesian hypothesis testing.

Consider two treatment groups with m prespecified follow-up times $t_1 < t_2 < \cdots < t_m$. Let $\mathbf{Y}_i^{(k)} = (Y_{i1}^{(k)}, \ldots, Y_{im}^{(k)})^T$ be an $m \times 1$ vector of outcomes which are the tumor volumes (in log-scale) from the ith subject in the kth group, $i = 1, 2, \ldots, n_k$, $k = 1, 2$, and $n = n_1 + n_2$. Assume that $\mathbf{Y}_i^{(k)}$ has a multivariate normal distribution with mean vector and covariance matrix of the toeplitz form

$$E(\mathbf{Y}_i^{(k)}) = \boldsymbol{\mu}^{(k)} \hat{=} (\mu_1^{(k)}, \ldots, \mu_m^{(k)})^T,$$

$$\mathrm{Cov}(\mathbf{Y}_i^{(k)}) = \boldsymbol{\Sigma}_m = \sigma^2 \mathbf{R}_m,$$

(23)

respectively, $k = 1, 2$, where $\mathbf{R}_m$ is defined in (17).

Our goal is to test hypotheses

$$H_0: \mathbf{d}^T \boldsymbol{\mu}^{(1)} = \mathbf{d}^T \boldsymbol{\mu}^{(2)} \quad \text{versus} \quad H_1: \mathbf{d}^T \boldsymbol{\mu}^{(1)} < \mathbf{d}^T \boldsymbol{\mu}^{(2)},$$

where $\mathbf{d} = (d_1, \ldots, d_m)^T$ is a known contrast vector which is chosen based on the scientific goal of the study. For example, if we want to find out if a different dosing schedule can decrease the total tumor volumes (or area under the tumor growth curve) further than does another schedule in the xenograft models, the contrast vector $\mathbf{d}$ would be the m-dimensional vector with unit component. If our goal is to compare tumor response after the first course of treatment (consisting of multiple doses of a drug), the components of the contrast vector $\mathbf{d}$ should be weighted appropriately. To test the hypotheses H_0, Tan et al. (2002) proposed two approaches, one is a heuristic t-test and a Bayesian hypothesis test.

4.1. Quasi *t*-Test Based on the EM Algorithm

If we had the complete observations of $\mathbf{Y}_i^{(k)}$, where $\mathbf{Y}_i^{(k)} \sim N_m(\boldsymbol{\mu}^{(k)}, \sigma^2 \mathbf{R}_m)$, $i = 1, \ldots, n_k$, $k = 1, 2$, we could simply use the t-statistic to test H_0,

$$t = \frac{\mathbf{d}^T \bar{\mathbf{Y}}^{(1)} - \mathbf{d}^T \bar{\mathbf{Y}}^{(2)}}{\sqrt{\mathbf{d}^T (S^{(1)} + S^{(2)}) \mathbf{d}}} \sqrt{\frac{n_1 n_2 (n_1 + n_2 - 2)}{n_1 + n_2}}, \tag{24}$$

where $\bar{\mathbf{Y}}^{(k)} = \frac{1}{n_k} \sum_{i=1}^{n_k} \mathbf{Y}_i^{(k)}$ and

$$S^{(k)} = \sum_{i=1}^{n_k} \mathbf{Y}_i^{(k)} \mathbf{Y}_i^{(k)T} - n_k \bar{\mathbf{Y}}^{(k)} \bar{\mathbf{Y}}^{(k)T},$$

for $k = 1, 2$. If H_0 holds, then the t-statistic in (24) has t-distribution with degree of freedom $n_1 + n_2 - 2$.

When there are missing data, the ECM algorithm can be applied to obtain maximum likelihood estimates of the parameters of interest $\boldsymbol{\phi} = (\boldsymbol{\mu}^{(1)}, \boldsymbol{\mu}^{(2)}, \sigma^2, \rho)^T$ (cf. Section 3.2 above). Denote the MLE of $\boldsymbol{\phi}$ by $\hat{\boldsymbol{\phi}}$. Then, the t-statistic in (24) is calculated with $\mathbf{Y}_i^{(k)}$ and $\mathbf{Y}_i^{(k)} \mathbf{Y}_i^{(k)T}$ being replaced by the corresponding conditional expectations

$$E(\mathbf{Y}_i^{(k)} | \mathbf{Y}_{\text{obs}}^*, \hat{\boldsymbol{\phi}}) \quad \text{and} \quad E(\mathbf{Y}_i^{(k)} \mathbf{Y}_i^{(k)T} | \mathbf{Y}_{\text{obs}}^*, \hat{\boldsymbol{\phi}}), \tag{25}$$

respectively, where $\mathbf{Y}_{\text{obs}}^*$ is defined as in Section 3.2.

4.2. Bayesian Test

The Bayesian hypothesis testing requires calculating the observed posterior probability of the one-sided alternative hypothesis $H_1 : \mathbf{d}^T \boldsymbol{\mu}^{(1)} < \mathbf{d}^T \boldsymbol{\mu}^{(2)}$, namely,

$$\Pr\{H_1|\mathbf{Y}^*_{\text{obs}}\} = \Pr\{\mathbf{d}^T(\boldsymbol{\mu}^{(1)} - \boldsymbol{\mu}^{(2)}) < 0|\mathbf{Y}^*_{\text{obs}}\}$$

$$\doteq \frac{1}{L}\sum_{\ell=1}^{L} I(\mathbf{d}^T(\boldsymbol{\mu}^{(1,\ell)} - \boldsymbol{\mu}^{(2,\ell)}) < 0), \qquad (26)$$

where $\{(\boldsymbol{\mu}^{(1,\ell)}, \boldsymbol{\mu}^{(2,\ell)}) : \ell = 1, \ldots, L\}$ is an iid sample from the observed posterior density $f(\boldsymbol{\mu}^{(1)}, \boldsymbol{\mu}^{(2)}|\mathbf{Y}^*_{\text{obs}})$ and $I(\cdot)$ denotes the indicator function. If the posterior probability of H_1 is greater than or equal to certain level (e.g., 95%), we reject the null hypothesis. We chose posterior probability (over the Bayes factor) for its ease of interpretation to biologists. For incomplete data, the samples from the observed posterior density $f(\boldsymbol{\mu}^{(1)}, \boldsymbol{\mu}^{(2)}|\mathbf{Y}^*_{\text{obs}})$ can be obtained using the IBF sampler (Tan et al., 2002).

With an analysis of a real study on xenograft models for two new anti-cancer agents temozolomide and irinotecan, Tan et al. (2002) shows that these two test approaches are in concordance. The main advantage of the presented methods is that they are valid for small samples, a typical case in animal studies for cancer drug development.

5. SUMMARY AND DISCUSSION

In this chapter, we reviewed recent developments of statistical methods for xenograft experiments, including the experimental design for combination studies and statistical models accounting for the special features of data from tumor xenografts.

Several experimental designs for detecting synergy in combination of two drugs are described. Abdelbasit-Plackett optimal design is available only for some specific models. The fixed-ratio design is based on the assumption that the synergism is a function of the proportions in the combination. It is shown that applications at individual combinations using the fixed-ratio design results in too many false synergistic combinations and a modified large sample procedure was proposed (Dawson et al., 2000). The ray design has primarily been applied to *in vitro* studies and needs a large sample size but may have a lower power to detect synergy in combination of two drugs. While these methods provide an overall test for synergism and may detect a consistent small but clinical significant interaction, it may miss an apparent interaction at a particular combination. The uniform design proposed by Tan et al. (2003) is suitable for a general model for joint action of two drugs where no specific parametric forms of the synergistic/antagonistic effect are assumed. The optimality is derived from the properties of uniform measures and by minimizing the variability in modeling the dose-effect while allocating the combinations reasonably to extract maximum information on the joint action. Clearly, the uniform design is an improvement over the design of Abdelbasit and Plackett (1982), the fixed-ratio and the ray designs. Thus, the uniform design avoids the arbitrariness in fixing one dimension. The model allows the exploration of the response surface with moderate sample sizes. The method is also flexible in that if we can only afford to do very limited *in vivo* combination experiments, we can narrow the dose range of interest and thus reduce the number of uniform points (doses).

We described a class of multivariate models to characterize the dose-response relationship in xenograft experiments with incomplete (missing at random and informative censoring) longitudinal data and the monotonicity of model parameters. The maximum likelihood method with ECM algorithm is proposed for model estimation. Although we have focused on the Toeplitz correlation structure, the simpler compound symmetry structure may be justified for within-cluster correlation in some cases. On the other hand, when the sample sizes are moderate, more complicated covariance structures may be incorporated in models (15) and (16). For small sample sizes, the Bayesian approach is an appealing alternative (see Fang et al., 2004; Tan et al., 2005).

Although we have focused on the type of xenograft experiments in our developmental therapeutics programs, we believe that the model formulation and estimation methods can be adopted to other animal experiments in research. Because the mechanism of the compound in development is better understood and some drugs are designed based on molecular targets resulted from the vast progresses in molecular and cellular biology and genetics in the last decade, coupled with the need for protecting human research subjects, tumor xenograft models play an important role in the translational research of bringing laboratory advances to clinic. The statistical models and methods proposed here serve as a basis for further development of methods to fully utilize the costly data and provide relevant information for the design of subsequent clinical trials.

ACKNOWLEDGMENTS

This research was supported in part by US National Institutes of Health grant CA106767.

References

1. Abdelbasit KM and Plackett RL. Experimental design for joint action. *Biometrics* 1982; **38**: 171–179.
2. Berenbaum MC. What is synergy? *Pharmacol. Rev.* 1989; **41**: 93–141.
3. Chou TC and Talalay P. Quantitative analysis of dose-effect relationships: the combined effects of multiple drugs or enzyme inhibitors. *Adv. Enzyme Regul.* 1984; **22**: 27–55.
4. Dawson KS, Carter WH Jr and Gennings C. A statistical test for detecting and characterizing/departures from additivity in drug/chemical combinations. *J. Agric. Biol. Environ. Stat.* 2000; **5**: 342–359.
5. Fang HB, Tian GL and Tan M. Hierarchical models for tumor xenograft experiments in drug development. *J. Biopharm. Stat.* 2004; **14**(4): 1–15.
6. Festing MFW and Altman DG. Guidelines for the design and statistical analysis of experiments using laboratory animals. *Inst. Lab. Anim. Res. J.* 2002; **43**(4): 244–258.
7. Fiebig HH, *et al.* Comparison of tumor response in nude mice and in the patients. In: *Human Tumour Xenografts in Anticancer Drug Development*. Veronesi U (ed.) Springer-Verlag, Berlin, 1988, pp. 25–30.
8. Finney DJ. *Probit Analysis*, 3rd ed. Cambridge University Press, London, 1971.
9. Gart JJ, Krewski D, Lee PN, Tarone RE and Wahrendorf J. *Statistical Methods in Cancer Research. Volume III: The Design and Analysis of Long-Term Animal Experiments*. International Agency for Research on Cancer, Lyon, France, 1986.

10. Gennings C, Carter WH Jr, Carney EW, Charles GD, Gollapudi BB and Carchman RA. A novel flexible approach for evaluating fixed ratio mixtures of full and partial agonists. *Toxicol. Sci.* 2004; **80**: 134–150.
11. Greco WR, Bravo G and Parsons JC. The search for synergy: a critical review from a response surface perspective. *Pharmacol. Rev.* 1995; **47**: 331–385.
12. Greco WR, Park HS and Rustum YM. An application of a new approach for the quantitation of drug synergism to the combination of cis-diamminedichloroplatinum and 1-β-D-arabinofur-anosylcytosine. *Cancer Res.* 1990; **50**: 5318–5327.
13. Heitjan DF. Generalized Norton-Simon models of tumour growth. *Stat. Med.* 1991; **10**: 1075–1088.
14. Heitjan DF, Manni A and Santen R. Statistical analysis of *in vivo* tumor growth experiments. *Cancer Res.* 1993 **53**: 6042–6050.
15. Houghton PJ, Stewart CF, Cheshire PJ, Richmond LB, Kirstein MN, Poquette CA, Tan M, Friedman HS and Brent TP. Antitumor activity of temozolomide combined with irinotecan is partly independent of O^6-methylguanine-DNA methyltransferase and mismatch repair phenotypes in xenograft models. *Clin. Cancer Res.* 2000; **6**: 4110–4118.
16. Houghton PJ, Stewart CF, Thompson J, Santana VM, Furman WL and Friedman HS. Extending principles learned in model systems to clinical trials design. *Oncology* 1998; **12**(8 Suppl. 6): 84–93.
17. Kasprzyk PG, Song SU, Di Fiore PP and King CR. Therapy of an animal model of human gastric cancer using a combination of anti-*erb*B-2 monoclonal antibodies. *Cancer Res.* 1992; **52**: 2771–2776.
18. Leith JT, Michelson S, Faulkner LE and Bliven SF. Growth properties of artificial heterogeneous human colon tumors. *Cancer Res.* 1987; **47**: 1045–1051.
19. Liang H. Modeling antitumor activity in xenograft tumor treatment. *Biom. J.* 2005; **47**: 358–368.
20. Loewe S. Isobols of dose-effect relations in the combination of pentylenetetrazole and phenobarbital. *J. Pharmacol. Exp. Ther.* 1955; **114**: 185–191.
21. Louis TA. Finding observed information using the EM algorithm. *J. R. Stat. Soc. B* 1982; **44**: 98–130.
22. Meng XL and Rubin DB. Maximum likelihood estimation via the ECM algorithm: a general framework. *Biometrika* 1993; **80**: 267–278.
23. Plackett RL and Hewlett PS. Quantal response to mixtures of poisons. *J. R. Stat. Soc. B* 1952; **14**: 141–163.
24. Rygaard K and Spang-Thomsen M. Quantitation and Gompertzian analysis of tumor growth. *Breast Cancer Res. Treat.* 1997; **46**: 303–312.

25. Sakaguchi Y, Maehara Y, Baba H, Kusumoto T, Sugimachi K and Newman R. Flavone acetic acid increases the antitumor effect of hyperthermia in mice. *Cancer Res.* 1992; **52**: 3306–3309.
26. Straetemans R, O'Brien T, Wouters L, van Dun J, Janicot M, Bijnens L, Burzykowski T and Aerts M. Design and analysis of drug combination experiments. *Biom. J.* 2005; **47**: 299–308.
27. Tallarida RJ. Statistical analysis of drug combinations for synergism. *Pain* 1992; **49**: 93–97.
28. Tallarida RJ. *Drug Synergism and Dose-Effect Data Analysis.* Chapman and Hall/CRC, New York, 2000.
29. Tan M, Fang HB, Tian GL and Houghton PJ. Small sample inference for incomplete longitudinal data with truncation and censoring in tumor xenograft models. *Biometrics* 2002; **58**: 612–620.
30. Tan M, Fang HB, Tian GL and Houghton PJ. Experimental design and sample size determination for testing synergy in drug combination studies based on uniform measures. *Stat. Med.* 2003; **22**: 2091–2100.
31. Tan M, Fang HB and Tian GL. Statistical analysis for tumor xenograft experiments in drug development. In: *Contemporary Multivariate Analysis and Experimental Design.* Fan J and Li G (eds.) World Scientific Publishing Co., Singapore, 2005a, pp. 351–368.
32. Tan M, Fang HB, Tian GL and Houghton PJ. Repeated-measures models with constrained parameters for incomplete data in tumor xenograft experiments. *Stat. Med.* 2005b; **24**: 109–119.

Chapter 18

ANALYSIS OF OCCULT TUMOR STUDIES

Shesh N. Rai

The primary motivation for this work is drawn from problems arising in the analysis of incomplete data. Although data can be incomplete in many ways, we are most interested in those situations where an intermediate event, which is of prime importance, cannot be observed. Such types of data are also referred as interval censored data. For example, in the analysis of data from occult tumor trials, the time of tumor onset is not known. Due to incompleteness of the data analyses become very complex and many assumptions are often required to develop a basis for inference concerning the tumor incidence. We briefly discuss the assumptions made that lead to many different analyses of occult tumor trial data in past two decades. From the prospect of reducing the sample size in occult tumor trials, some models have been proposed. These semi-parametric models assume relationships in tumor-bearing and tumor-free animals and have impact on tumor onset rates and potency measures of carcinogenic substances that we have explored further.

Keywords: Interval censored data, carcinogenicity experiments, occult tumor trials, tumorigenicity experiments, EM algorithm.

1. INTRODUCTION

Tumorigenicity (also known as carcinogenicity) experiments are commonly used to screen chemicals, drugs, and food additives for carcinogenic effects. Experiments of this type have three different, though related, purposes; these are:

(a) to estimate the effect of tumor presence on the death rate,
(b) to estimate the rate of tumor development, which is assumed to be irreversible, and

(c) to estimate carcinogenic potency, i.e., the magnitude of the dose effect of the substance of interest.

In addition, with respect to (a) one is interested in estimating how tumor presence alters the rate of death in tumor-bearing subjects; in other words, how lethal is the tumor? Generally, the lethality of a tumor can be classified as incidental, lethal or intermediate. An incidental tumor does not alter the death rate in tumor-bearing subjects, i.e, the death rates in tumor-bearing and tumor-free animals are the same. On the other hand, if an experimental animal dies almost immediately after tumor onset, the corresponding tumor is known to be one of the lethal type. Tumors which are neither lethal nor incidental are known as tumors of intermediate lethality.

Another interesting aspect of this type of experiment is the problem of estimating the rate at which tumor develops in a specific environment on specific sets of subjects. However, the most important focus of this type of experiment is to compare a potentially carcinogenic agent to its absence in relation to the rate of tumor onset.

We now turn our attention to those trials which involve occult tumors, i.e., the presence of a tumor is determined only at the time of postmortem; that is why these are known as occult tumor trials or studies. A typical experiment involves about 600 experimental animals of both sexes in each of two strains randomized to a control group and one or more exposure groups. In most of these experiments, the animals, which are usually mice or rats, are maintained in a controlled environment and dosed with the potential carcinogen according to the experimental protocol. During the experiment, animals may be selected for interim sacrifice according to the protocol in order to determine their tumor status; that is why these studies are also called survival/sacrifice experiments. At the conclusion of the study, all surviving animals are killed for humane reasons and to discover their tumor status as well; this is known as the terminal sacrifice. In experiments which involve smaller sample sizes, often only a terminal sacrifice is performed.

Occult tumor studies represent an important source of information concerning the possible carcinogenic effect of potentially hazardous substances such as chemicals, drugs and food additives. Commonly referred names for such studies are tumorigenicity experiments, carcinogenicity experiments, occult tumor trials or studies and animal survival/sacrifice experiments.

Hence onward, we will refer to occult tumor studies. Such studies furnish data that typically include the administered dose of the suspected carcinogen, the age of the animal at death and indicators of the presence or absence, at death, of various tumors. These data are frequently both grouped and incomplete, and are invariably difficult to analyze (Sun, 2006).

In the remaining sections of this article we introduce the statistical framework within which occult tumor studies data are analyzed by reviewing the key developments which have appeared in the scientific literature.

The non-parametric analysis of occult tumor studies depends on the availability of information collected from numerous interim sacrifices. However, many occult tumor studies involve only a terminal kill at the conclusion of the study. We address the problem of providing a satisfactory basis for non-parametric estimation of the tumor incidence function in these situations. Rai *et al.* (2000) have proposed two semi-parametric models which require virtually no interim sacrifice information and yet yield unique parameter estimates. The key aspect in each of these models is the simplicity of the postulated relationship between the death rates for tumor-free and tumor-bearing animals. We briefly outline the methods for estimating the parameters in various versions of these models. Two data sets from two experiments are analyzed to demonstrate various aspects of their use.

In Section 2, we summarize recent results along with literature review. We define preliminary notation in Section 3 for parametric and non-parametric estimation. Interval estimation is given in Section 4. Testing tumor lethality and carcinogenic effect is addressed in Section 5. The methods demonstrated on two data sets are of focus in the penultimate section. Final section contains the discussion.

2. A REVIEW OF THE LITERATURE

The study of the time to onset and rate of progression of an occult tumor induced by a potential carcinogen is an area of research which has attracted a great deal of attention. Hoel and Walburg (1972) were the first investigators to consider the analysis of data from occult tumor studies, and made a useful distinction between rapidly lethal tumors and incidental tumors.

In the former, the time to death following tumor onset is short, and therefore time to death is a good proxy for the time of tumor onset. Accordingly, an analysis based on time to death with tumor is indicated. In the incidental tumor case, the tumor has no effect on the death rate and the proportion of deaths with tumor provides an estimate of the tumor prevalence at that time. Most tumors, however, are neither lethal nor incidental; consequently, neither of these methods is appropriate. Under the assumption that the cause of death can be identified, i.e., whether death is principally due to tumor or due to competing risks, Kodell and Nelson (1980) consider a semi-Markov model with Weibull transition intensity functions. They estimate the parameters by maximizing the likelihood function for the model, and illustrate their results by analyzing an occult tumor study involving the toxic substance benzidine dihydrochloride. Kalbfleisch, Krewski and Van Ryzin (1983) provide a thorough review of the field, and describe the construction of the full likelihood function in detail. Other related articles include Kodell, Shaw and Johnson (1982), Dinse and Lagakos (1982), Turnbull and Mitchell (1984) and Portier (1986); in these papers, the authors are interested in estimating the tumor onset distribution based on a multiple decrement analysis.

McKnight and Crowley (1984) and Dewanji and Kalbfleisch (1986) both provide an extensive survey of non-parametric methods of estimation in occult tumor studies. McKnight and Crowley argue that the tumor incidence rate should be the principal quantity of interest in occult tumor studies, and propose a non-parametric estimator of this quantity. Dewanji and Kalbfleisch derive a non-parametric estimate of this rate using the EM algorithm. In both papers, information from numerous interim sacrifices is essential in estimating this key rate. Some other results which also require numerous interim sacrifices may be found in the papers by Williams and Portier (1992, 1993).

From the perspective of an experimentalist, interim sacrifices represent an undesirable aspect of the non-parametric approach, because they frequently inflate the size and cost of a proposed trial. One approach which reduces the necessity for interim sacrifices involves the use of parametric models for the death rates experienced by tumor-free and tumor-bearing animals. Although many different parametric models might be worth considering, two particular forms arise quite naturally in this occult tumor context. In the first, which corresponds to a cause-separable hazards model, the rates of death for tumor-bearing and tumor-free animals differ by a constant. We shall refer to this approach as the additive model. Dinse (1991) adopts this solution to the problem of analyzing data from occult tumor studies and claims that when the time to tumor onset is considered to be a continuous random variable and the time to death is modeled as a discrete process, only a terminal sacrifice of the surviving experimental units is necessary to ensure unique estimates for the parameters in the model. However, in Dinse's approach a crucial simplification of the complete data likelihood function which yields a closed-form solution at the maximization step of the EM algorithm is based on the assumption that the cause of death can be identified. Although the context of observation is a fairly common assumption on which the analysis of occult tumor studies is based, the cause of death is not always known, or it may be uncertain (see Finkelstein and Ryan (1987), Lagakos and Ryan (1985) or Kodell, Shaw and Johnson (1982)). For example, in an empirical investigation of the ED_{01} data, Lagakos and Ryan (1985) found the cause of death information to be inadequate for several tumor types occurring in that experiment. Consequently, it seems unwise to assume that reliable information of this type is likely to be available.

In the second parametric form, which is known as the multiplicative model, the rate of death for tumor-bearing animals is a constant multiple of the corresponding hazard for tumor-free animals. Lindsey and Ryan (1993) investigate this approach, assuming that the key random variables are continuous, and show that when the hazard rate for death without tumor is piecewise constant, unique parameter estimates are obtained when the only sacrifice in the experiment is a terminal one.

Ryan and Orav (1988) propose the use of covariate information, tumor grade, to estimate the tumor prevalence function. However, in all the papers mentioned above, markers such as tumor growth and tumor size have not been incorporated into the analysis of the data.

Non-parametric statistical analyses of occult tumor studies require the results of many interim sacrifices (see McKnight and Crowley, 1984; Dewanji and Kalbfleisch, 1986). From the perspective of an experimentalist, interim sacrifices represent an undesirable aspect of the non-parametric approach, because they frequently inflate the size and cost of a proposed trial. One approach which reduces the necessity for interim sacrifices involves the use of parametric models for the death rates experienced by tumor-free and tumor-bearing animals. Although many different semi-parametric models might be possible, two particular forms are considered quite naturally in this occult tumor context. Testing the effect of carcinogen was also straightforward using the likelihood ratio test in these two settings; however, alternative test procedure is recently developed by Kim, Ahn and Moon (2007).

In the first parametric form, which is referred as the constant risk ratio (CRR) model (Dinse, 1991, 1993; Lindsey and Ryan, 1993) and multiplicative failure rate (MFR) model (Rai and Matthews, 1993), are based on the Cox proportional hazards model (Cox, 1972), the rate of death for tumor-bearing animals is a constant multiple of the corresponding hazard for tumor-free animals, a direct use of the Cox proportional model for the continuous and a discrete scale. In the second, which corresponds to a cause-separable hazards model, the rates of death for tumor-bearing and tumor-free animals differ by a constant, referred as constant risk difference (CRD) model (Dinse, 1991, 1993) and additive failure rate (AFR) model by Rai and Matthews (1993); when the death process is a discrete random variable, there is not any preferred version of the AFR model proposed in the occult tumor trials.

Dinse (1991) considers that the death rates for tumor-free and tumor-bearing animals differ by a constant in case of the death and tumor onset processes are continuous and derives a relationship in these two types of death rates when the deaths are recorded on a discrete scale. This derived relationship between the death rates in tumor-free and tumor bearing animals is based on some assumptions which may not hold in the occult tumor trials. Rai and Matthews (1993) propose a very simple form for the AFR model which is based on a cause-separable hazards model.

In analyzing data from occult tumor studies, another question to consider concerns the nature of the time scales on which deaths and tumor onset occur. Various authors differ in their resolution of this issue. For example, Dewanji and Kalbfleisch (1986) and Rai and Matthews (1997) assume that the tumor onset and death occur on a discrete scale, whereas Dewanji, Krewski and Goddard (1993) and Lindsey and Ryan (1993) adopt a continuous scale for both types of events. Rai, Matthews and Krewski (2000), consider a mixed scale models in which the time to tumor onset is assumed to be a continuous random variable and the time to death is modelled as a discrete process. Under the semi-parametric assumptions proposed in Rai and Matthews (1997), the mixed scale model does not require an interim sacrifice for the estimation.

3. PRELIMINARY CONSIDERATIONS

Consider an occult tumor trial in which the presence of tumor is not clinically observable, i.e., tumor status can be determined only at necropsy. At the time of observation, an experimental unit can occupy any of the three states illustrated in Figure 1. Let the stochastic process $\{X(t)\}$ identify the state occupied by an animal at time t.

For simplicity, we suppose that n animals in state 1 at time $t = 0$ are randomly selected as experimental units and are observed for the duration of the trial. Let the random variable T denote the time of death and U the time of tumor onset. We also assume that the development of a tumor is an irreversible event, and therefore transitions from state 2 to state 1 do not occur, as illustrated in Figure 1. From time to time, experimental units are sacrificed to determine their status. These animals are chosen for

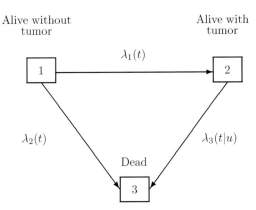

Fig. 1. An illness-death model involving three states. State 1 corresponds to animals which are alive without tumor. Tumor-bearing animals which are alive are in state 2, which is frequently unobservable in occult tumor studies. State 3 is an absorbing state and corresponds to death.

sacrifice independent of their health status, etc., to ensure that sacrifices can be regarded as independent of the times of the events of interest.

The intensities shown in Figure 1 are defined as the limits

$$\lambda_1(t) = \lim_{\Delta t \to 0} Pr\{X(t + \Delta t) = 2 | X(t) = 1\}/\Delta t, \quad (1)$$

$$\lambda_2(t) = \lim_{\Delta t \to 0} Pr\{T \in [t, t + \Delta t) | X(t) = 1\}/\Delta t, \quad (2)$$

and

$$\lambda_3(t|u) = \lim_{\Delta t \to 0} Pr\{T \in [t, t + \Delta t) | X(t) = 2, U = u\}/\Delta t, \quad (3)$$

for $u \leq t$; otherwise $\lambda_3(t|u) = 0$.

We now define various quantities of interest, as functions of the three hazard rates $\lambda_1(t)$, $\lambda_2(t)$ and $\lambda_3(t|u)$. The marginal distribution of T, the time to failure, can be determined from the survivor function

$$S(t) = E_1\left[\exp\left\{-\int_u^t \lambda_3(v|u)dv\right\}\right]$$

where the expectation E_1 is taken with respect to the distribution of U, the onset time. The pseudo-survival functions corresponding to the intensities $\lambda_1(.)$, $\lambda_2(.)$ and $\lambda_3(t|u)$ are

$$Q_i(t) = \exp\left\{-\int_0^t \lambda_i(v)dv\right\}$$

for $i = 1, 2$ and

$$Q_3(t|u) = \exp\left\{-\int_u^t \lambda_3(v|u)dv\right\}, \quad (4)$$

whereas

$$Q(t) = \exp\left\{-\int_0^t (\lambda_1(v) + \lambda_2(v))dv\right\} = Q_1(t)Q_2(t)$$

denotes the probability that the time to the first event — tumor onset or death without tumor — exceeds t. Note that

$$S(t) = Pr\{X(t) = 1\} + Pr\{X(t) = 2\}$$
$$= Q(t) + \int_0^t \lambda_1(u) Q(u) Q_3(t|u) du.$$

Following the work of McKnight and Crowley (1984), we can define an average hazard associated with the transition from state 2 to state 3 as

$$\lambda^{D|T}(t) = \lim_{\Delta t \to 0} Pr\{T \in [t, t + \Delta t) | X(t) = 2\}/\Delta t$$
$$= \lim_{\Delta t \to 0} \frac{Pr\{T \in [t, t + \Delta t), X(t) = 2\}/\Delta t}{Pr\{X(t) = 2\}}$$
$$= \frac{\int_0^t \lambda_1(u) Q(u) \lambda_3(t|u) Q_3(t|u) du}{\int_0^t \lambda_1(u) Q(u) Q_3(t|u) du}.$$

Similarly, the tumor prevalence function, which is the proportion of live animals with tumor in the population, is defined to be

$$\pi(t) = Pr\{X(t) = 2 | T \geq t\}$$
$$= \frac{Pr\{X(t) = 2\}}{Pr\{X(t) = 1\} + Pr\{X(t) = 2\}}$$
$$= \frac{\int_0^t \lambda_1(u) Q(u) Q_3(t|u) du}{S(t)}.$$

Finally, we define the lethality function, $l(t)$, to be either the relative death rate, $\lambda^{D|T}(t)/\lambda_2(t)$, or the difference of death rates, $\lambda^{D|T}(t) - \lambda_2(t)$, for different forms of $\lambda_3(t|u)$. With respect to the analysis of data from an occult tumor study, there are two special cases, based on lethality assumptions, which are easily handled; these correspond to tumors which are rapidly lethal and, at the other extreme, tumors which are incidental. In the case of tumors which are rapidly lethal, $\lambda_3(t|u)$ is very large, and death occurs immediately after tumor onset. In this situation, the time of death can be regarded as a surrogate for the time of tumor development. On the other hand, if the tumor is incidental, tumor development has no effect on the death rate and $\lambda_3(t|u) = \lambda_2(t)$. In this case, the analysis of data from the trial can be based on the fact that the proportion of deaths at time t which involve tumor-bearing animals provides an estimate of the tumor prevalence in the population at that time.

3.1. Constructing the Likelihood Function

In this section, we outline a general framework for constructing the likelihood function associated with any of the models to which we will have cause to refer in subsequent chapters. Let θ represent the full parametric vector associated with a particular model. Thus, θ includes parameters which specify the general form of the transition intensities, c.f. Figure 1. In general, the data arising from an occult tumor study trial will consist of the time and type of failure, i.e., natural death or sacrifice, and an indication of tumor presence or absence at autopsy. Let t_i be the realization of the random variable T for the ith experimental animal, $i = 1, 2, \ldots, n$. If C_i represents the contribution to the likelihood due to the ith animal, then the likelihood function for θ is $L(\theta) = \prod_{i=1}^{n} C_i$. Table 1 identifies the various types of observations which occur and the corresponding contribution to the likelihood.

The terms $A(t)$ and $B(t)$ which appear in Table 1 represent the integrals

$$A(t) = \int_0^t \lambda_1(u) Q(u) \lambda_3(t|u) Q_3(t|u) du$$

and

$$B(t) = \int_0^t \lambda_1(u) Q(u) Q_3(t|u) du,$$

Table 1. Likelihood contributions for data from an occult tumor trial.

Observation type	Outcome	Likelihood contribution
Death without tumor	$T = t, X(t^-) = 1$	$\lambda_2(t) Q(t)$
Sacrifice without tumor	$T > t, X(t) = 1$	$Q(t)$
Death with tumor	$T = t, X(t^-) = 2$	$A(t)$
Sacrifice with tumor	$T > t, X(t) = 2$	$B(t)$

respectively. These same terms can also be expressed in other familiar forms. Let the random variable $W = T - U$ denote the time between tumor onset and subsequent failure. For $t \geq u$, i.e., $w \geq 0$, we can write

$$Pr\{W \in [w, w+dw)|U = u\} = \lambda_3(w+u|u)\overline{G}(w|u)dw$$

where

$$\overline{G}(w|u) = Pr\{W \geq w|U = u\} = Q_3(u+w|u)$$

is the probability of survival to time t given tumor onset at u. Then,

$$A(t) = \int_0^t \lambda_3(w+u|u) f(u)\overline{G}(w|u)du$$

and

$$B(t) = \int_0^t f(u)\overline{G}(w|u)du,$$

where $f(.)$ is the (sub)density function of U, i.e., $f(u) = \lambda_1(u)Q(u)$.

Particular versions of $A(t)$ and $B(t)$ will be derived explicitly for different forms of $\lambda_3(t|u)$ in subsequent chapters. When the time scale on which one or both events corresponding to tumor onset or death are recorded is discrete, the expressions for the intensities specified in (1) – (3) reduce to conditional probabilities; appropriate versions of these intensities will be introduced as required. For parametric estimation, when a simpler form for these intensities (such as constant hazard rates, Weibull hazard rates, or combination, etc.) are assumed the likelihood construction is straightforward, however, estimation is somewhat cumbersome and we do not consider further discussion here.

3.2. Non-Parametric Settings

In an occult tumor trial, tumor status can be determined only at necropsy. At the time of observation, an experimental unit can occupy any of the three states illustrated in Figure 1 (cf. Kalbfleisch et al., 1983).

Let $X(t)$ indicate the absence ($X(t) = 0$) or presence ($X(t) = 1$) of a tumor at time t. For simplicity, we suppose that n tumor-free animals are randomly selected as experimental units at time $t = 0$ and are observed for the duration of the trial. Let the random variable T denote the time of death and U the time of tumor onset. We assume that the development of a tumor is an irreversible event, so that transitions from state 2 to state 1 do not occur, as illustrated in Figure 1. From time to time, animals may be sacrificed to determine their tumor status. Since animals are chosen for sacrifice are selected at random, the sacrifices can be regarded as independent of the times of the events of interest.

The observed data for each animal consist of the time of death or sacrifice and an indicator of tumor presence or absence. Suppose there are M distinct death times, denoted by $t_1 < \cdots < t_M$, and let $I_j = (t_{j-1}, t_j]$, $j = 1, 2, \ldots, M$, where, for completeness, we define $t_0 = 0$ and t_M is the time of terminal sacrifice. Without loss of generality, set $t_j = j$ for $j = 0, 1, \ldots, M$. The argument of Kaplan and Meier (1958) can be used to show that, without the imposition of distributional restrictions, the likelihood is maximized when the death rates place mass only at the observed times of death. Following Dinse (1991), we can treat death as a discrete process and T as a discrete random variable, and we identify the range of T as $\{1, 2, \ldots, M\}$. In addition, we suppose that the random variable U, which denotes the time of tumor onset, is continuous. Under this assumption, we will show that the terms in the complete data likelihood which involve the tumor incidence rate, $\lambda_1(t)$, and the rate of death without tumor, $\lambda_2(t)$, are separable.

For animal i, there is an associated time of sacrifice, Y_i, which is chosen in advance of the experiment with $Pr(Y_i = j) = q_j$ ($j = 1, 2, \ldots, M$) with $\sum q_j = 1$. This random sacrifice model is analogous to a random censorship model. The event $Y_i = j$ corresponds to a planned sacrifice of animal i at time j^+, so that sacrifices at j are presumed to follow other events that may occur at j. At time M the experiment is terminated and all surviving animals are sacrificed. For this reason, such experiments are referred to as survival/sacrifice experiments.

Let u and t be realizations of U and T respectively. The transition intensities specified in Figure 1 are defined as

$$\lambda_1^*(u) = Pr\{X(u) = 1 | X(u^-) = 0, T \geq u\},$$
$$\lambda_2(t) = Pr\{T = t | X(t) = 0, T \geq t\} \quad \text{and}$$
$$\lambda_3(t|u) = Pr\{T = t | X(t) = 1, U = u, T \geq t \geq u\}$$

for $t = 1, 2, \ldots, M$ and $u \geq 0$. Furthermore, we define the tumor incidence rate

$$\lambda_1(j) = Pr\{U \in I_j | T > j-1, X(j-1) = 0\}$$
$$= 1 - \exp\left\{-\int_{j-1}^{j} \lambda_1^*(u) du\right\}$$
$$\approx \int_{j-1}^{j} \lambda_1^*(u) du,$$

for interval $j = 1, \ldots, M$. By assuming that U is a continuous random variable, we exclude the possibility that tumor onset and death with tumor occur simultaneously.

Although different semi-parametric models for the intensity $\lambda_3(t|u)$ can be considered, we concentrate on the two simple forms

$$\frac{\lambda_3(t|u)}{1 - \lambda_3(t|u)} = \frac{\lambda_2(t)}{1 - \lambda_2(t)} e^{\gamma}, \qquad (5)$$

and

$$\lambda_3(t|u) = \lambda_2(t) + \gamma, \qquad (6)$$

proposed by Rai (1993) and Rai and Matthews (1997) for $t = 1, 2, \ldots, M$. Note that, the former model, which we refer to as the multiplicative failure (MFR) model corresponds to the discrete version of the proportional hazards model (Cox, 1972), while Dinse's (1991) the constant risk ratio (CRR) model is closer to the continuous proportional hazards model. The latter formulation, which we call the additive failure rate (AFR) model, represents a cause-separable hazards model for $\lambda_3(t|u)$. For notational convenience, we refer to both models by $\lambda_3(t)$. Note that these models assume that the death rate does not depend on the tumor onset time. In the class of separable models for such type of experiments, our AFR model may be more comparable to the Cox proportional hazards model on a discrete scale (MFR).

One of the major differences between the CRD and AFR models can be understood as follows. Hypothetically assume that in tumor-bearing animals there are two types (competing causes) of deaths: deaths due to tumor and deaths due to causes other than tumor. The death rate due to causes other than tumor is the death rate in tumor-free animals. Both the CRD and AFR models consider that the death rates in tumor-bearing and tumor-free animals differ by a constant. However, in the CRD model the time of death is measured on a continuous scale, while in the AFR model the time of deaths are recorded on a discrete scale (for a detailed comparison of the AFR and CRD models see Rai, 1997).

Since both the tumor incidence rate and the death rate without tumor are completely unspecified and events may occur at any time $t = 1, \ldots, M$, we must estimate the $2M$ values of $\lambda_1(t)$ and $\lambda_2(t)$. If $\lambda_3(t)$ was also unspecified, corresponding to a fully non-parametric model of the experiment, we would have to estimate $3M$ parameters; however, the parametric forms for $\lambda_3(t)$ prescribed in Eqs. (1) and (2) reduce that total to $2M + 1$.

We conclude this section by defining some additional notation which will be used in later sections of the chapter. Let

$$Q(t) = Pr\{T > t, X(t) = 0\}$$
$$= Pr\{\min(U, T) > t\}$$
$$= \prod_{j=1}^{t}\{1 - \lambda_1(j)\} \prod_{j=1}^{t}\{1 - \lambda_2(j)\}$$

and

$$Q_3(t|u) = Pr\{T > t, X(t) = 1 | U \in I_u\}$$
$$= \prod_{j \geq u}^{t}\{1 - \lambda_3(j)\},$$

for $u \leq t = 1, 2, \ldots, M$. The function $Q(t)$ represents the probability that an experimental animal is alive and tumor-free at time t, while $Q_3(t|u)$ corresponds to the conditional probability that an animal which developed a tumor in I_u is still alive at time t.

3.3. Fitting the Semi-Parametric Model

In what follows, we describe an algorithm for maximizing the likelihood of the data, based on the semi-parametric model outlined in Section 2. This method of estimation is an algorithm of the EM type which was first described by Dempster, Laird, and Rubin (1977). As with all such algorithms, maximum likelihood estimation of the parameters in the model using the observed data (the incomplete data problem) is accomplished by maximizing the conditional expectation, given the data, of the likelihood function generated by the corresponding complete data formulation.

A Complete Data Formulation

In a complete data formulation of the problem of estimating the parameters of the semi-parametric model, we assume that the time of tumor onset is known to belong to one of the intervals I_j, and that sacrifice is simply a right-censoring of the multistate process. In that case, the data from a sample of n experimental animals can be summarized as the observed values of the counting processes

$$N_1(j) = \#\{i | U_i \in I_j, T_i \geq j, Y_i \geq j\},$$
$$N_2(t) = \#\{i | X_i(t) = 0, T_i = t, Y_i \geq t\},$$
$$N_3(t) = \#\{i | X_i(t) = 1, T_i = t, Y_i \geq t\},$$
$$Y_0(t) = \#\{i | X_i(t) = 0, Y_i = t, T_i > t\},$$
$$Y_1(t) = \#\{i | X_i(t) = 1, Y_i = t, T_i > t\}$$

$(j, t = 1, \ldots, M)$. Here, the subscripts on the random variables U, $X(.)$, T and Y identify the values of these variables associated with the ith experimental animal. The quantity $N_1(j)$ represents the number of animals with tumor onset time in I_j, whereas $N_2(t)$ identifies the number of tumor-free animals which die at time t. Likewise, $N_3(t)$ indicates the number of tumor-bearing animals which die at time t. The random variables $Y_1(t)$ and $Y_0(t)$ summarize the number of animals with and without tumor, respectively, which are sacrificed at t; thus $n = \sum_{j=1}^{M} \{N_1(j) + N_2(j) + Y_0(j)\}$.

In addition, let

$$R_1(t) = \#\{i | U_i > t - 1, T_i \geq t, Y_i \geq t\}$$
$$= \sum_{j \geq t}^{M} \{N_1(j) + N_2(j) + Y_0(j)\}, \qquad (7)$$
$$R_2(t) = \#\{i | U_i \geq t, T_i \geq t, Y_i \geq t\}$$
$$= R_1(t) - N_1(t), \qquad (8)$$

and

$$R_3(t) = \#\{i|U_i < t, T_i \geq t, Y_i \geq t\}$$
$$= \sum_{j \leq t}\{N_1(j) - N_3(j-1) - Y_1(j-1)\} \quad (9)$$

($t = 1, \ldots, M$). The variable $R_1(t)$ represents the number of animals at risk of developing a tumor in I_t; likewise, $R_2(t)$ and $R_3(t)$ specify the numbers of tumor-free and tumor-bearing animals, respectively, which are at risk of death at t.

The complete data may be divided into two groups, depending on the type of information available. The first group corresponds to animals in state 1; these either remain in state 1 or move into one of the other two states, i.e., state 2 or state 3. Thus, at any time t, i.e., at the end of I_t, $N_1(t)$ animals out of a total $R_1(t)$ move to state 2. At t the remaining animals, i.e., $R_2(t) = R_1(t) - N_1(t)$, are at risk of moving from state 1 to state 3. Of these, $N_2(t)$ die and $R_2(t) - N_2(t)$ remain in state 1. For such animals, the contribution to the likelihood function at time t is

$$L_1(t) \propto \lambda_1(t)^{N_1(t)}\{1 - \lambda_1(t)\}^{R_1(t)-N_1(t)} \times \lambda_2(t)^{N_2(t)}\{1 - \lambda_2(t)\}^{R_2(t)-N_2(t)}.$$

The second group corresponds to animals in state 2; these either remain in state 2 or move into state 3. If there are $R_3(t)$ animals at risk in state 2 at time t, $N_3(t)$ move into state 3 and the remaining $R_3(t) - N_3(t)$ stay in state 2. Thus, animals in state 2 at time t contribute

$$L_2(t) \propto \lambda_3(t)^{N_3(t)}\{1 - \lambda_3(t)\}^{R_3(t)-N_3(t)} \quad (10)$$

to the likelihood function.

Combining these two groups, we obtain the likelihood function for the semi-parametric model based on the complete data, viz.,

$$L \propto \prod_{t=1}^{M} L_1(t)L_2(t). \quad (11)$$

An explicit version of this likelihood function is obtained by replacing $\lambda_3(t)$ with various model-specific forms.

The Incomplete Data Problem

Since tumor information can be obtained only at autopsy, the complete data are not available; instead, the observations consist of $N_2(t)$, $Y_0(t)$, $N_3(t)$, and $Y_1(t)$. Let $\theta^T = (\lambda_1(j), \lambda_2(j), \gamma; j = 1, \ldots, M)$ be the vector of parameters in the semi-parametric model. A modified version of the EM algorithm (Rai and Matthews, 1993) provides a simple method for estimating θ in two steps: E (expectation) and M1 (one-step maximization). Starting with an initial estimate of θ, say $\theta^{(0)}$, these steps are applied in a strictly alternating sequence until the parameter estimates converge and the log likelihood function of the observed data is maximized.

Suppose that $\theta^{(i-1)}$ represents the value of θ which was obtained at iteration $i-1$ of the algorithm. At the next E-step, we have to evaluate the conditional expectation

$$N_1^{(i)}(j) = E\{N_1(j) | N_3(t), Y_1(t), t = j, j+1, \ldots, M, \theta = \theta^{(i-1)}\}$$

for $j = 1, \ldots, M$. To compute this expectation, the conditional probability

$$P^{(i-1)}(j|t) = Pr\{U \in I_j | T = t, X(t) = 1\}$$
$$= \frac{\lambda_1(j) Q(j-1) Q_3(t-1|j)}{\sum_{l=1}^{t} \lambda_1(l) Q(l-1) Q_3(t-1|l)}$$
$$= Pr\{U \in I_j | Y = t, X(t) = 1\} \quad (12)$$

for $j \leq t$, is required. Note that $Q_3(t-1|t) = 1$, i.e., the probability of surviving at least up to time $t-1$ given that the animal develops a tumor in the interval I_t is one. The dependence of the right-hand side of expression (8) on $(i-1)$ has been suppressed for notational convenience. Since we have previously assumed that $\lambda_3(t|u)$, the rate of death in tumor-bearing animals, does not depend on U, the time of tumor onset, the common factor of $\lambda_3(t)$ or $\{1 - \lambda_3(t)\}$ in both the numerator and denominator of the above expression cancels. It follows that, at the conclusion of iteration $i-1$, the conditional probability that tumor onset occurred in interval I_j, given that the animal died at time t, is equal to the conditional probability of the same event, given that the animal was sacrificed at time t. Consequently,

$$N_1^{(i)}(j) = \sum_{t \geq j} \{N_3(t) + Y_1(t)\} P^{(i-1)}(j|t).$$

At the ith iteration, the expected values of the risk sets $R_1(.)$, $R_2(.)$ and $R_3(.)$ are evaluated by substituting the quantities $N_1^{(i)}(.)$, $N_2(.)$, $N_3(.)$, $Y_0(.)$ and $Y_1(.)$ in Eqs. (3), (4) and (5), respectively. These expectations are required in order to proceed with the succeeding maximization step.

At the next step of the algorithm we maximize the complete data likelihood specified in Eq. (7) with respect to the parameter vector θ to obtain the updated value $\theta^{(i)}$. It can easily be seen that the maximum likelihood estimates for the parameters $\lambda_1(j)$ are

$$\hat{\lambda}_1^{(i)}(j) = \begin{cases} N_1^{(i-1)}(j)/R_1^{(i-1)}(j), & \text{if } R_1^{(i-1)}(j) \neq 0 \\ 0 & \text{otherwise} \end{cases}$$

($j = 1, \ldots, M$). Unfortunately, there are no closed-form expressions for the updated parameter estimates $\hat{\lambda}_2^{(i)}(.)$ and $\hat{\gamma}^{(i)}$; consequently, a numerical method of maximizing L with respect to $\lambda_2(.)$ and γ at iteration i is required. The one-step maximization described by Rai and Matthews (1993) provides an effective method of reducing the computational burden at this stage in the derivation of the maximum likelihood estimates.

Parameter Identifiability

Before deriving the estimated variance of $\hat{\theta}$, it is appropriate to summarize our experience concerning the problem of identifiability. As noted previously, the MFR and AFR models discussed in this chapter each involve $2M + 1$ unknown parameters. In order to uniquely estimate θ, we therefore require at least $2M + 1$ independent observations. According to the design of the experiment, only two types of event can occur at any time: death with tumor or death without tumor. The data on death information form a multinomial table with $2M$ cell frequencies which specifies the number of deaths with and without tumor at M time points. Given that there is only one terminal sacrifice at M, the data on sacrifice information form a binomial table with two cell frequencies which specifies the number of sacrificed animals with and without tumor. Thus, we will have $2M + 1$ independent information in the data to use in estimating $2M + 1$ parameters in either semi-parametric model. The results of numerous simulations are encouraging; however, a theoretical basis for believing that unique parameter estimates can always be obtained in a single dose group has not yet been established. One of the examples discussed here is specifically chosen because the study design incorporated only a terminal sacrifice.

4. INTERVAL ESTIMATION

Once the maximum likelihood estimate of θ has been obtained, it is natural to consider point estimates, and subsequently interval estimates, of functions of θ. For example, the cumulative tumor incidence rate at time t,

$$\Lambda_1(t) = \sum_{j=1}^{t} \lambda_1(j),$$

the subdistribution function,

$$F_1(t) = \sum_{j=1}^{t} \lambda_1(j) Q(j-1),$$

or the prevalence of the disease among surviving animals at time t,

$$\pi(t) = \frac{\sum_{j=1}^{t} \lambda_1(j) Q(j-1) Q_3(t|j)}{Q(t) + \sum_{j=1}^{t} \lambda_1(j) Q(j-1) Q_3(t|j)},$$

for $t = 1, \ldots, M$ may be of interest.

Several different approaches might be suggested which would enable us to avoid deriving the observed information directly. For example, we could use one of the resampling methods described by Efron and Tibshirani (1986) to construct a confidence interval for the parameter of interest.

A second approach is based on the technique suggested by Turnbull and Mitchell (1984) and others. This method involves specifying a fixed value c for $g(\theta) = c$, the parameter of interest. The EM1 algorithm is then used to obtain the constrained maximum of the likelihood function when $g(\theta) = c$, based on the incomplete data. Inverting χ_1^2, which is the asymptotic sampling distribution of twice the difference of the constrained and the unconstrained maximized values of the log likelihood function, generates a confidence interval for $g(\theta)$.

Alternatively, the observed information can be computed indirectly, using the approach outlined by Louis (1982). In the remainder of this section, we describe this method in detail.

For ease of exposition, we require some additional notation. Thus, let x represent the complete data and y the corresponding incomplete data. Then $I_y(\hat{\theta})$ denotes the observed information based on y, $I_x(\hat{\theta})$ the observed information based on the complete data x, and

$$I_{x|y}(\hat{\theta}) = E\{S_x(\theta)S_x^T(\theta)|y\}|_{\theta=\hat{\theta}},$$

where $S_x(\theta)$ is the score vector from the complete data formulation of the problem and T stands for matrix transpose. Then, as Louis (1982) demonstrates,

$$I_y(\hat{\theta}) = I_x(\hat{\theta}) - I_{x|y}(\hat{\theta}).$$

To compute $I_{x|y}(\hat{\theta})$ we use the approach described in Dewanji and Kalbfleisch (1986). Let $S_{j,t}^B$ and $S_{j,t}^C$ be the score vectors from the complete data formulation corresponding to the events $(U \in I_j, T = t, Y \geq t)$ and $(U \in I_j, Y = t, T > t)$, respectively, for $j, t = 1, \ldots, M$. Then

$$S_{j,t}^B = \frac{\partial}{\partial \theta} \log\{\lambda_1(j)Q(j-1)\lambda_3(t)Q_3(t-1|j)\}$$

and

$$S_{j,t}^C = \frac{\partial}{\partial \theta} \log\{\lambda_1(j)Q(j-1)Q_3(t|j)\};$$

expressions for the components of $S_{j,t}^B$ and $S_{j,t}^C$ are given in the end of this section.

In addition, let

$$\bar{S}_{j,t}^B = \frac{1}{N_3(t)} \sum_{j=1}^{t} N_3(t|j) S_{j,t}^B$$

and

$$\bar{S}_{j,t}^C = \frac{1}{Y_1(t)} \sum_{j=1}^{t} Y_1(t|j) S_{j,t}^C,$$

where

$$N_3(t|j) = N_3(t) P(j|t)$$

and

$$Y_1(t|j) = Y_1(t) P(j|t)$$

are the imputed frequencies of death and sacrifice at time t which are generated at the concluding cycle of the EM1 algorithm for animals with tumor onset in interval I_j. The conditional probability $P(j|t)$ corresponds to the limiting value of the expression specified in Eq. (8).

Following Dewanji and Kalbfleisch (1986), we define $I_{\hat{x}}(\hat{\theta})$ to be the observed information based on the complete data likelihood and the complete data. The latter, which we denote by $\hat{x}$, are generated at the final cycle of the EM1 algorithm. Similarly, let the quantities $\hat{N}_3(t)$, $\hat{Y}_1(t)$, $\hat{S}_{j,t}^B$, $\hat{S}_{j,t}^C$, $\bar{\hat{S}}_{j,t}^B$ and $\bar{\hat{S}}_{j,t}^C$ denote the values of $N_3(t)$, $Y_1(t)$, $S_{j,t}^B$, $S_{j,t}^C$, $\bar{S}_{j,t}^B$ and $\bar{S}_{j,t}^C$, respectively, evaluated at the maximum likelihood estimate $\hat{\theta}$ and $\hat{x}$. Then the observed information based on the incomplete data is equal to

$$I_y(\hat{\theta}) = I_{\hat{x}}(\hat{\theta}) - \sum_{t=1}^{M}\sum_{j=1}^{t}\{\hat{N}_3(t|j)(\hat{S}_{j,t}^B - \bar{\hat{S}}_{j,t}^B)(\hat{S}_{j,t}^B - \bar{\hat{S}}_{j,t}^B)^T$$
$$- \hat{Y}_1(t|j)(\hat{S}_{j,t}^C - \bar{\hat{S}}_{j,t}^C)(\hat{S}_{j,t}^C - \bar{\hat{S}}_{j,t}^C)^T\}.$$

Variance estimates for particular quantities of interest can now be obtained by the delta method (see Rao, 1973, p. 388). The estimated variance of $g(\hat{\theta})$ is

$$(\partial g/\partial \theta) I_y^{-1}(\theta)(\partial g/\partial \theta)^T|_{\theta=\hat{\theta}},$$

provided g is a continuous and differentiable function of θ.

Components of $S_{j,t}^B$ and $S_{j,t}^C$

The components of the score vectors $S_{j,t}^B$ and $S_{j,t}^C$ which correspond to the events

$$(U \in I_j, T = t, Y \geq t) \quad \text{and} \quad (U \in I_j, Y = t, T > t) \quad \text{are}$$
$$S_{j,t}^B = (\xi_{11}, \ldots, \xi_{1M}, \xi_{21}, \ldots, \xi_{2M}, \xi_3)^T$$

where

$$\xi_{1k} = \begin{cases} -1/\{1 - \lambda_1(k)\} & k < j \\ 1/\lambda_1(k) & k = j \\ 0 & k > j, \end{cases}$$

$$\xi_{2k} = \begin{cases} -1/\{1 - \lambda_2(k)\} & k < j \\ \partial/\partial\lambda_2(k)[\log\{1 - \lambda_3(k)\}] & j \leq k < t \\ \partial/\partial\lambda_2(k)[\log \lambda_3(k)] & k = t \\ 0 & k > t, \end{cases}$$

$$\xi_3 = \partial/\partial\gamma \ [\log \lambda_3(k) + \sum_{k=j}^{t-1} \log\{1 - \lambda_3(k)\}],$$

and

$$S_{j,t}^C = (\zeta_{11}, \ldots, \zeta_{1M}, \zeta_{21}, \ldots, \zeta_{2M}, \zeta_3)^T$$

where

$$\zeta_{1k} = \begin{cases} -1/\{1 - \lambda_1(k)\} & k < j \\ 1/\lambda_1(k) & k = j \\ 0 & k > j, \end{cases}$$

$$\zeta_{2k} = \begin{cases} -1/\{1 - \lambda_2(k)\} & k < j \\ \partial/\partial\lambda_2(k)[\log\{1 - \lambda_3(k)\}] & j \leq k \leq t \\ 0 & k > t, \end{cases}$$

and

$$\zeta_3 = \partial/\partial\gamma \left[\sum_{k=j}^{t} \log\{1 - \lambda_3(k)\}\right].$$

5. TESTING TUMOR LETHALITY AND CARCINOGENIC EFFECT

Two particular issues arise in occult tumor studies which naturally prompt consideration in the framework of hypothesis testing. These are tumor lethality — the effect of tumor presence on the mortality rate — and the possible carcinogenic effect of a substance, which is the primary reason for the occult tumor study. This effect is evaluated by studying the association between the rate of tumor incidence and different dose levels of the possible carcinogen. In what follows, we consider each of these issues for the various models used in this chapter.

Testing Tumor Lethality

In Section 2, we described two semi-parametric models (c.f. Eqs. (1) and (2)) in which the change in the death rates in tumor-bearing and tumor-free animals is indexed by a single parameter γ, the lethality parameter. It would be desirable to be able to test for the possible departure of the MFR and AFR models from the assumed parametric form, which is independent of time. Thus, we suppose that

$$\gamma = \gamma_0 + \gamma_1 t$$

in both the MFR and AFR models. A method for testing $\gamma_1 = 0$ with arbitrary γ_0 and $\gamma = c$, where c is a known scalar quantity, is outlined below. The latter case is of particular interest since if we let $\gamma = c$ in each of the MFR and AFR models, then incidental tumors correspond to $c = 0$ and lethal tumors to $c^{-1} = 0$. This notion of incidental and lethal tumors was first introduced by Hoel and Walburg (1972).

Following the arguments of Dewanji and Kalbfleisch (1986) and Louis (1982), the score statistic corresponding to $\gamma = C$, based on the observed data, is

$$\hat{U}_\gamma = \sum_{t=1}^{M} \hat{w}_t \{N_3(t) - \hat{\lambda}_3(t)\hat{R}_3(t)\},$$

where $\hat{w}_t$ is a weighting factor determined by the choice of model. For the case represented by $C = \gamma_0 + \gamma_1 t$, where $\gamma_1 = 0$ and γ_0 is unknown, $\hat{w}_t = t$

(MFR) or $t[\hat{\lambda}_3(t)\{1 - \hat{\lambda}_3(t)\}]^{-1}$ (AFR). Similarly, if $C = c$, where c is a known scalar quantity, the weighting factors are 1 and $[\hat{\lambda}_3(t)\{1 - \hat{\lambda}_3(t)\}]^{-1}$ for the MFR and the AFR models, respectively. The quantities $\hat{w}_t$, $\hat{\lambda}_3(t)$ and $\hat{R}_3(t)$ specified in the above expressions are computed using the maximum likelihood estimates and the expected value of $R_3(t)$ obtained on the final cycle of the EM1 algorithm, under the assumption that $\gamma = C$.

An estimated variance for this score statistic can be based on the observed information, which may be evaluated using the method described in Section 3.

Testing for a Carcinogenic Effect

In most occult tumor studies, several different levels or dosages of a potential carcinogen are assayed. The modelling framework which we have described above is sufficiently flexible that dose effects can be incorporated into one or more components of both the MFR and AFR models. At the most general level, we might suppose that the different levels of the carcinogen affect the rates of tumor onset and of death differently, i.e., the parameters $\lambda_1(.)$, $\lambda_2(.)$ and γ would differ according to the dose of the carcinogen which had been administered. For convenience, we will refer to this rather general model of an occult tumor study as Model I.

In many experiments, the possible toxicity of the carcinogenic substance, i.e., the effect of dose on the death rates $\lambda_2(.)$ and $\lambda_3(.)$, may be of little interest. We will refer to a model in which these death rates do not depend on dose as Model II. On the other hand, since the tumor incidence rate $\lambda_1(.)$ is of primary interest, it is natural to try and model the effect of dose on $\lambda_1(.)$ using only a few parameters. Such a framework, which we will refer to as Model III, might be

$$\frac{\lambda_1(t, z)}{1 - \lambda_1(t, z)} = \frac{\lambda_1(t, 0)}{1 - \lambda_1(t, 0)} \exp(z^T \beta) = p_0(t) \exp(z^T \beta)$$

for $t = 1, 2, \ldots, M$. In this formulation, the covariate z represents the dose group, suitably coded, and the regression coefficient β summarizes the dose effect of the carcinogenic substance on $\lambda_1(t; z)$, the rate of tumor onset in I_t for animals in dose group z.

In the most parsimonious semi-parametric framework, none of the parameters differ according to dose and the observed data from all dose groups are pooled to estimate $\lambda_1(t)$, $\lambda_2(t)$ and γ. In Section 6, we will refer to this simple description of an occult tumor study as Model IV.

Under Model III, the complete data consist of $N_i(t; z)$ and $R_i(t; z)$, $i = 1, 2, 3$ as well as $Y_0(t; z)$ and $Y_1(t; z)$, the corresponding information regarding sacrificed animals. For simplicity, we consider the case involving only two dose groups corresponding to the values $z = 0$ and $z = 1$. Generalizations to the situation involving more than two groups are straightforward. The likelihood function for $\theta^T = (\lambda_1(t), \lambda_2(t), \gamma, \beta; t = 1, \ldots, M)$ is equal to

$$L \propto \prod_{i=1}^{t} L_1(t) L_2(t)$$

where $L_2(t)$ corresponds to the expression specified in Eq. (6), but $L_1(t)$ is given by

$$\{p_0(t)\}^{N_1(t)} \exp\{\beta N_1(t; 1)\} \{1 + e^{\beta} p_0(t)\}^{-R_1(t;1)} \{1 + p_0(t)\}^{-R_1(t;0)}$$

where $N_1(t) = N_1(t; 0) + N_1(t; 1)$ denotes the total number of animals with tumor onset time in I_t. The score statistic for testing the hypothesis $\beta = 0$, based on the likelihood function L, reduces to the log-rank statistic

$$U_\beta = \sum_{t=1}^{M} \left\{ N_1(t; 1) - \frac{N_1(t) R_1(t; 1)}{R_1(t)} \right\},$$

where $R_1(t) = R_1(t; 0) + R_1(t; 1)$ denotes the total number of animals at risk of developing a tumor in I_t. Following the argument of Dewanji and Kalbfleisch (1986), the score statistic for testing the hypothesis $\beta = 0$, based on the incomplete data, is

$$\hat{U}_\beta = \sum_{t=1}^{M} \left\{ \hat{N}_1(t; 1) - \frac{\hat{N}_1(t) \hat{R}_1(t; 1)}{\hat{R}_1(t)} \right\}.$$

The quantities $\hat{N}_1(t; 1)$, $\hat{N}_1(t)$, $\hat{R}_1(t; 1)$ and $\hat{R}_1(t)$ specified in the above expression are computed using the expected values of $N_1(t; 0)$, $N_1(t; 1)$, $R_1(t; 0)$ and $R_1(t; 1)$ which were obtained on the final cycle of the EM1 algorithm, under the assumption that $\beta = 0$, i.e., the tumor incidence rates in both dose groups are the same. An estimated variance for this score statistic can be based on the observed information, which may be evaluated using the method described in Section 4. If $\hat{V}_\beta$ denotes the estimated variance of $\hat{U}_\beta$, we can compare the observed value of $\hat{U}_\beta^2 \hat{V}_\beta^{-1}$ to the χ_1^2 sampling distribution in order to test the hypothesis of interest.

6. TWO EXAMPLES

Many occult tumor studies incorporate only a single sacrifice which occurs at the end of the study, i.e., the terminal kill. The models which we have described in this chapter are naturally suited to the analysis of experimental designs of this type. In this section we analyze the data from two different occult tumor studies in order to illustrate the use of the proposed methods in this commonly-occurring situation.

NTP Study of Benzyl Acetate and Liver Tumors

In this first example we use both the MFR and AFR models to analyze data on liver tumors in male mice from the study of benzyl acetate carried out by the National Toxicology Program. The route of the exposure was gavage. Fifty mice received corn oil alone (the vehicle control group), while 50 mice in each of two additional groups were exposed to 500 (low-dose) and 1000 (high-dose) milligrams of benzyl acetate (in corn oil) per kilogram of body weight, respectively. The data, which are summarized in the first three columns of Table 2, are taken from Dinse (1991); for the purposes of

Table 2. Observed and expected frequencies of the number of deaths occurring in the control ($i = 0$) and high dose ($i = 1$) groups of the National Toxicology Program study of benzyl acetate.

Age in Weeks	Observed Frequencies		Expected frequencies			
			MFR model		AFR model	
	a_0/b_0^*	a_1/b_1	a_0/b_0	a_1/b_1	a_0/b_0	a_1/b_1
0–33	0/4	0/2	0.001/4.001	0.001/2.001	0.001/4.001	0.001/2.001
33–66	0/2	1/1	0.001/2.001	1.001/1.002	0.001/2.001	0.994/0.995
66–84	1/3	1/1	1.000/3.000	1.001/1.002	1.000/3.000	1.007/1.008
84–94	1/2	3/5	1.000/2.000	3.001/5.001	1.000/2.000	2.999/4.999
94–104	1/1	2/2	1.000/1.001	2.001/2.001	1.000/1.001	2.001/2.002
104+	7/38	16/39	7.000/38.00	15.999/39.00	7.000/38.00	16.00/39.00

*a_i denotes the number of deaths with tumor out of b_i total deaths, in the ith dose group.

this example, we have chosen to analyze only the information derived from the control and high dose groups. The five time intervals corresponding to Weeks 0–33, 33–66, 66–84, 84–94, and 94–104+ of the study are based on the available information concerning the times of death; the terminal sacrifice was performed at 104 weeks. These intervals are chosen based on the death times and are different from Rai, Matthews and Krewski (2000). For further details concerning this data set, see Dinse (1991) and National Toxicology Program (1986).

Due to the form of both the MFR and AFR models, difficulties in estimating the parameters $\lambda_2(.)$ and γ arise whenever the observed number of deaths with tumor is zero and the corresponding number of deaths without tumor is positive, or *vice versa*. As Table 2 indicates, these complications occur several times in both groups of mice (Weeks 0–33, 33–66, 66–84 and 94–104), and lead to maximum likelihood estimates of $\lambda_2(.)$ and γ which are not permissible. To resolve this problem, we replaced an observed frequency of zero by δ, say, where δ is small in comparison to the total number of deaths observed in any particular interval. To investigate the sensitivity of the parameter estimates to different choices of δ, we used the values 0.0001, 0.001, 0.01, and 0.1 to replace observed zeros in the control group data and maximized the likelihood of the adjusted frequencies. The estimates of the cumulative tumor incidence rate which we obtained in each case, and which are of primary importance in occult tumor studies, hardly differed. However, since boundary problems during maximization of the likelihood still occurred when δ was less than 0.0001, we replaced all the observed counts of zero in Table 2 by $\delta = 0.001$ before deriving the results reported below.

To evaluate the goodness-of-fit of both semi-parametric models which were used to analyze the data, we calculated model-specific expected numbers of deaths with and without tumor for both groups of mice. These values, which were obtained from the E-step of the final cycle of the EM1 algorithm, are presented for comparison with the corresponding observed frequencies in Table 2. The expected frequencies for both groups of mice and for both the MFR and AFR models are in good agreement with their observed counterparts. The Kaplan-Meier estimate of the survival function for each group of mice and the semi-parametric estimates obtained using either model are almost identical (the figures are not produced here as these convey the same message using the expected frequency given in Table 2). Thus, both models appear to fit the observed data very well. However, the

Table 3. Estimated parameters for the NTP liver tumor data using the basic versions of the mixed scale models.

Age in weeks	Control dose				High dose			
	MFR model		AFR model		MFR model		AFR model	
	$\Lambda_1(.)$	$\sum \lambda_2(.)$	$\Lambda_1(.)$	$\sum \lambda_2(.)$	$\Lambda_1(.)$	$\sum \lambda_2(.)$	$\Lambda_1(.)$	$\sum \lambda_2(.)$
0–33	0.000	0.080	0.000	0.080	0.000	0.040	0.000	0.040
33–66	0.000	0.124	0.000	0.124	0.104	0.040	0.186	0.040
66–84	0.023	0.170	0.128	0.176	0.104	0.040	0.215	0.040
84–94	0.047	0.195	0.179	0.205	0.104	0.087	0.443	0.108
94–104	0.252	0.195	0.254	0.205	0.543	0.087	0.600	0.108
γ	8.396		0.125		9.137		0.111	

estimates of $\Lambda_1(.)$, the cumulative tumor incidence rates, which are displayed in Table 3 for both dose groups do not exhibit the same degree of agreement. In the case of the MFR model, a major increase in $\hat{\Lambda}_1(.)$ occurs in the last observation interval (94–104 weeks), whereas in the AFR model the estimated values of $\Lambda_1(.)$ are distributed more broadly throughout the duration of the experiment, and the largest increase in $\hat{\Lambda}_1(.)$ occurs prior to the final observation interval. The fact that the estimate of $\Lambda_1(.)$ in the high-dose group does not change under the MFR model from week 33 to week 94 is rather surprising since tumors were found in 23 animals out of 32 subsequent to week 33.

Table 4 summarizes the values of the maximized log likelihoods and the numbers of parameters estimated for each semi-parametric formulation (MFR or AFR) and choice of model (I–IV). Despite the basic structural difference between the MFR and AFR models, corresponding values in Table 4 for these two models are almost identical. This result might have been anticipated, in view of the excellent agreement between the observed and expected frequencies displayed in Table 2; however, the degree of concordance exhibited in Table 4 is still surprising. Since Models I–IV are hierarchical, various hypotheses concerning the possible effect of the high dose of benzyl acetate can be tested using the likelihood ratio statistic. Due to the similarity of the maximized log-likelihoods, the conclusions will be the same for both the MFR and AFR models. For example, a comparison of Model I and Model II indicates that the data do not contradict the simpler

Table 4. Values of the maximized log likelihood and the corresponding number of parameters estimated, by semi-parametric formulation (MFR and AFR) and various choices of the mixed scale models, for the NTP liver tumor data.

Model	Number of parameters	Maximized log likelihood	
		MFR	AFR
I	22	−138.960	−138.960
II	16	−142.361	−142.340
III	12	−142.946	−142.843
IV	11	−146.009	−146.009

model in which $\lambda_2(.)$ and γ are the same in both dose groups; the observed value of the likelihood ratio statistic used to test this hypothesis that the effect of benzyl acetate is not toxic is 6.8, and corresponds to a significance level of 0.66. The additional simplification represented by Model III, in which the effect of the 1000 mg dose of benzyl acetate on tumor incidence is modelled by the same proportionality factor, e^β, throughout the two-year duration of the study is also not contradicted by the data. However, based on a comparison of the maximized log-likelihoods for Models III and IV, the data provide strong evidence against the hypothesis that the high-dose of benzyl acetate is not carcinogenic ($p = 0.013$). Although Dinse (1991) considers all three dose groups in an analysis of the NTP benzyl acetate study, it is satisfying to note that the final model which he reports includes a dose effect for the tumor incidence function, but not for the death rates with or without tumor.

The same data set is also analyzed by Rai, Matthews and Krewski (2000) but with different intervals. Although the conclusion for toxicity effect and carcinogenicity effects remains the same, the estimates of the onset rates and death rates are drastically different. This suggests further investigation is needed to estimate the tumor onset rate and the tumor prevalence rate.

Low-level Ionizing Radiation and the Occurrence of Glomerulosclerosis

Data for this first example is given Table 2, which is extracted from Table 2 in Dewanji and Kalbfleisch (1986). The data represent a summary, in intervals of 100 days, of the information gleaned from deaths and sacrifices concerning the presence or absence of the disease glomerulosclerosis. Further information concerning this comparative assay regarding the occurrence of glomerulosclerosis following exposure to ionizing radiation may be found in the report of Berlin, Brodsky and Clifford (1979).

The results of fitting various versions of the basic MFR and AFR models to these data are summarized in Figures 2 – 4 and Tables 5 – 8. Consider first Figure 2, which displays the Kaplan-Meier estimate of the survival function for both the control and irradiated groups of mice. The corresponding estimates of the survival function which were derived from the MFR (M) and AFR (A) models are virtually identical to the Kaplan-Meier estimates, both for a constant value of γ, i.e., $\gamma_1 = 0$, and for a more general time-dependent version, i.e., $\gamma = \gamma_0 + \gamma_1 t$ (denoted by GM and GA in the case of the MFR and AFR models, respectively). On this basis, it appears that both types of semi-parametric models appear to fit the observed data very well. On the other hand, a comparison of the Kaplan-Meier estimates and those derived from both the MFR and AFR models involving proportional dose effects (MP and AP, respectively) suggests that the MP and AP models uniformly underestimate the probability of survival in the control group and overestimate the same probability for the irradiated group of mice. Similar remarks apply to the estimates of the probability of survival based on MFR and AFR models incorporating death rates which are constant for both dose groups (MC and AC, respectively).

Estimates of the cumulative tumor incidence rate in each of the dose groups for various MFR and AFR models are presented in Table 6 and displayed, graphically, in Figure 2. As a basis for comparison, the corresponding non-parametric estimates obtained by Dewanji and Kalbfleisch (1986), and denoted by the symbol DK, are also included in Table 6 and Figure 2. While agreement between the estimates based on various MFR and AFR models and the non-parametric estimates is generally quite good in the control group, the estimated cumulative rate of tumor onset in the irradiated group of mice is higher in all versions of both types of models, relative to the non-parametric estimates, in the latter half of the observation period.

Estimates of the tumor prevalence function, $\pi(t)$, based on various semi-parametric models are summarized in Table 7 and plotted in Figure 4. The

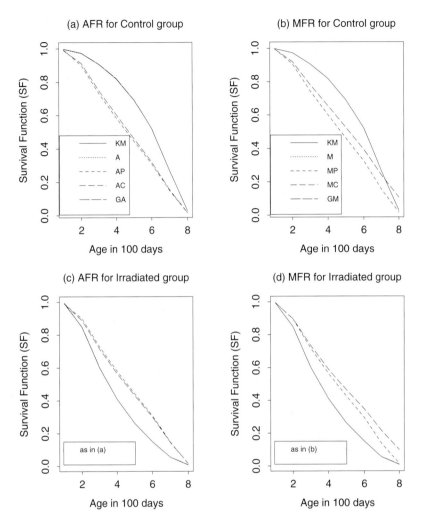

Fig. 2. Estimated survival functions for the glomerulosclerosis data based on various mixed scale models.

corresponding values derived by Dewanji and Kalbfleisch (1986) are once again included for comparison purposes. Relative to the non-parametric estimates of $\pi(t)$, both the MFR and AFR models appear to estimate a higher prevalence rate in both groups of mice.

The general impression conveyed by Figures 2 to 4 is that more parsimonious models involving a proportional dose effect or death rates which

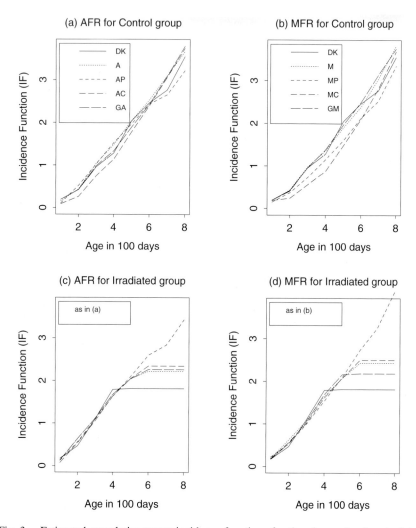

Fig. 3. Estimated cumulative tumor incidence functions for the glomerulosclerosis data based on various mixed scale models.

are identical for both the irradiated and control groups of mice are less satisfactory in terms of fitting the observed data. This impression can be confirmed on the basis of likelihood ratio tests derived from the maximized log likelihoods obtained when either semi-parametric model (MFR or AFR) is fitted.

Analysis of Occult Tumor Studies

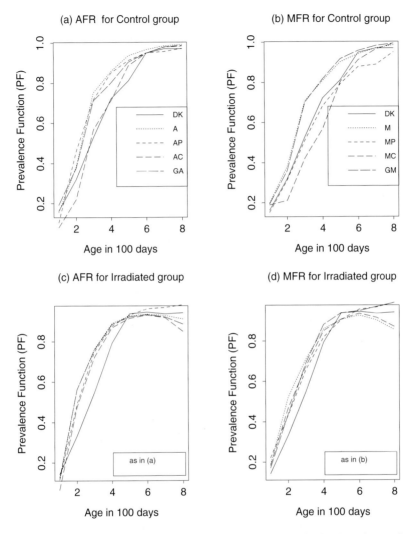

Fig. 4. Estimated prevalence functions for the glomerulosclerosis data based on various mixed scale models.

Table 8(a) summarizes the values of the maximized log likelihoods and the numbers of parameters estimated for each semi-parametric model (MFR and AFR) that was fitted to the data. Half of the models involve the time-dependent version $\gamma = \gamma_0 + \gamma_1 t$; in the remaining models, we only fitted the basic semi-parametric formulation represented by $\gamma_1 = 0$. The values in

Table 5. Age at death and the occurrence of glomerulosclerosis due to low-level ionizing radiation.

Age in days	Control group		Irradiated group	
	Deaths	Sacrifice	Deaths	Sacrifice
0–100	1/2*	14/72†	4/8	11/67
101–200	16/25	24/65	144/205	25/68
201–300	55/63	44/63	257/322	50/66
301–400	71/76	31/40	204/228	41/43
401–500	98/104	37/39	150/162	38/39
501–600	127/131	32/33	99/104	30/31
601–700	177/179	32/33	65/71	26/27
701–	138/139	16/16	9/11	2/2

*Number of deaths with tumor/total number of deaths.
†Number of sacrifices with tumor/total number of sacrifices.

Table 6. Estimated cumulative tumor incidence rates for the glomerulosclerosis data based on various mixed scale models.

Age in days	Control group					Irradiated group				
	DK*	M	A	GM	GA	DK	M	A	GM	GA
0–100	0.195	0.198	0.160	0.195	0.193	0.166	0.190	0.121	0.173	0.130
101–200	0.420	0.432	0.433	0.410	0.427	0.469	0.623	0.637	0.538	0.642
201–300	0.961	0.980	1.040	0.962	0.984	1.104	1.077	1.119	1.052	1.117
301–400	1.266	1.360	1.481	1.380	1.333	1.794	1.649	1.676	1.692	1.670
401–500	2.034	1.894	2.044	1.958	1.894	1.794	2.074	2.047	2.163	2.044
501–600	2.442	2.385	2.529	2.497	2.429	1.794	2.431	2.220	2.183	2.262
601–700	2.750	3.035	3.094	3.130	3.086	1.794	2.431	2.220	2.183	2.262
701–	3.537	3.813	3.672	3.747	3.789	1.794	2.431	2.220	2.183	2.262

*DK = Dewanji and Kalbfleischapproach, M = MFR model, A = AFR model, GM = generalized MFR model, GA = generalized AFR model.

the table provide a basis for examining the adequacy of the simpler model for γ with respect to the glomerulosclerosis data. In the case of the MFR model, it appears that the hypothesis $\gamma_1 = 0$ is not contradicted by the data for the control group of mice ($p = 0.230$); however, the same test yields a significance level of 0.001 when applied to the results of fitting the MFR model to the irradiated mice. Somewhat curiously, the reverse of the above

Table 7. Estimated prevalence functions for the glomerulosclerosis data based on various mixed scale models.

Age in days	Control group					Irradiated group				
	DK*	M	A	GM	GA	DK	M	A	GM	GA
0–100	0.194	0.198	0.157	0.194	0.193	0.164	0.194	0.116	0.171	0.127
101–200	0.369	0.387	0.381	0.360	0.375	0.368	0.553	0.561	0.438	0.562
201–300	0.698	0.715	0.752	0.701	0.713	0.758	0.728	0.763	0.692	0.759
301–400	0.775	0.814	0.858	0.817	0.802	0.939	0.868	0.889	0.880	0.885
401–500	0.949	0.904	0.937	0.917	0.905	0.945	0.910	0.926	0.938	0.922
501–600	0.970	0.940	0.966	0.958	0.949	0.937	0.928	0.935	0.949	0.933
601–700	0.970	0.965	0.985	0.983	0.978	0.941	0.895	0.928	0.969	0.921
701–	1.000	0.963	0.992	0.994	0.984	1.000	0.775	0.911	0.988	0.886

*DK = Dewanji and Kalbfleisch approach, M = MFR model, A = AFR model, GM = generalized MFR model, GA = generalized AFR model.

Table 8(a). Values of the maximized log likelihood and the corresponding number of parameters estimated when the glomerulosclerosis data were fitted with the mixed scale MFR and AFR models involving $\gamma = \gamma_0 + \gamma_1 t$.

Dose group	$\gamma_1 = 0$			$\gamma_1 \neq 0$		
	Number of parameters	Maximized log likelihood		Number of parameters	Maximized log likelihood	
		MFR	AFR		MFR	AFR
Control	17	−1778.50	−1782.06	18	−1777.78	−1778.41
Irradiated	17	−2886.39	−2886.79	18	−2880.87	−2886.72
Pool dose	17	−4929.36	−4934.24	18	−4921.23	−4932.47

remarks summarize the conclusions based on the likelihood ratio tests when the AFR model is fitted to the same data set.

The final table (Table 8(b)) in this section indicates the aggregate values of the maximized log likelihoods for the combined control and irradiated groups of mice, as well as the number of parameters estimated, for each version of the mixed scale models which was fitted to the glomerulosclerosis data. Likelihood ratio tests, based on differences in these maximized log likelihoods, can be used to investigate the plausibility of different methods

Table 8(b). Values of the maximized log likelihood and the corresponding number of parameters estimated, by semi-parametric formulation (MFR and AFR) and various choices of the mixed scale models, for the glomerulosclerosis data.

Model	Number of parameters	Maximized log likelihood	
		MFR model	AFR model
I	34	−4664.89	−4668.85
II	25	−4897.33	−4914.35
III	18	−4920.99	−4933.53
IV	17	−4929.36	−4934.24

of modelling the effect of dose on the tumor incidence rate and the death rates. The principal conclusion to which these tests point is the fact that simplified models involving a proportional dose effect or death rates which are the same for the both groups of mice in the experiment do not appear to fit the data particularly well.

The same data set is also analyzed by Rai and Matthews (1997) but with different model setting. In the previous analysis we considered the tumor onset process and death process both to be recorded on the discrete scale. When comparing two results based on two settings, the conclusions remain the same in terms of estimates of survival, cumulative incidence and prevalence functions, and toxicity and carcinogenicity effects. Unlike the first example, this data set has many interim sacrifices. Thus, irrespective of which Markov model for relating death rate in tumor-bearing animals and tumor-free animals are selected or likelihood constructions are based on a discrete or mixed scale model, when there are a lot of interim sacrifice information, inference is very robust.

7. DISCUSSION

The analysis of data from an occult tumor study is a complex statistical problem, particularly because the event of interest, tumor onset, generally is not observed. Apart from a very few experiments such as the ED_{01} study (Cairns, 1980) which involve multiple interim sacrifices, realistic approaches to the estimation of tumor incidence rates and carcinogenic potency will necessarily involve modeling assumptions. The various models which we have described in this chapter, and which were used to analyze two examples example in Section 6, were based on a few key assumptions concerning the transition intensities $\lambda_1(.)$, $\lambda_2(.)$ and $\lambda_3(.)$. In particular, we assumed that the time to tumor onset was a continuous random variable, whereas the time scale on which deaths are recorded is discrete. This combination results in a model involving both discrete and continuous time scales, i.e., a mixed scale model, and confers both advantages and disadvantages from a modeling perspective. An obvious alternative to such a mixed scale model would be one in which both time scales are discrete;we shall refer to such a combination as a discrete scale model. The methods described by Dewanji and Kalbfleisch (1986) and Rai and Matthews (1997) are examples of discrete scale models.

There is considerable merit in deciding, *a priori*, to adopt a discrete scale for the endpoint death since a certain arbitrariness in the grouping of observed deaths may be avoided (Lindsey and Ryan, 1993). In addition, as we remarked in Section 2, provided we wish to treat death, with or without tumor, non-parametrically, the use of a discrete time scale is a natural consequence of the argument of Kaplan and Meier (1958). In addition, choosing to specify the time of death as a discrete random variable reduces the problem of estimation to one which is multinomial in character. This reduction has important consequences for ease of estimation of both the model parameters and their asymptotic variances.

A further benefit which arises through selection of the mixed scale model concerns the form of the likelihood function. More specifically, the complete data formulation which results from a mixed scale model can be factored into two components;one of these involves parameters related to tumor onset, while the remaining factor concerns the death rates with and without tumor. Such a separation is not achieved in the corresponding complete data likelihood generated by the discrete scale model.

It is worth noting that the choice of a mixed scale model induces a natural preference or ordering of the two events, tumor onset and death, which is not present in corresponding discrete scale models. In effect, because tumor onset occurs on a continuous scale whereas the time to death is a discrete random variable, tumor-free animals are at risk of developing tumor at any time, whereas the same animals are only at risk of death immediately prior to the next point on the discrete scale of the random variable T after remaining tumor-free for the intervening interval. This natural ordering has implications for the resulting estimates of the tumor incidence rate, and the death rates with and without tumor, which we believe are deserving of further study. For example, in the mixed scale model it is possible for animals to die with tumor at the first value in the range of T, i.e., the time of the first observed death or sacrifice, whereas the same event is impossible in a discrete scale model. Consequently, at least one interim sacrifice is required in order to obtain unique parameter estimates in discrete scale models corresponding to the mixed scale models which we have considered, whereas no interim sacrifice is required in the case of the mixed scale models. Thus, the mixed scale models are recommended for the analysis of such data sets.

An important assumption on which the method of analysis argued in this chapter is based is the Markov condition, viz. $\lambda_3(t|u) = \lambda_3(t)$. Although this assumption has no biological motivation, there is anecdotal evidence (see Dewanji and Kalbfleisch (1986)) to suggest that its impact on the estimation of the tumor incidence rate seems to be minimal. The assumption underlying the semi-parametric models which we have described concerns the relationship between $\lambda_2(t)$, the death rate without tumor, and $\lambda_3(t)$, the corresponding rate of death for tumor-bearing animals. By adopting a relatively simple parametric relationship between these two rates, *viz.*, the MFR model specified in Eq. (1) or the AFR model summarized in Eq. (2), we were able to achieve a substantial reduction in the number of parameters required to model the observed data. The MFR model is based on the proportional hazards model for a discrete time scale which was first introduced by Cox (1972). However, there is no similar basis for the AFR model. Since MFR model is the natural extension, and results are very similar, this is recommendation for analyzing such data sets.

The idea of using information from a marker process (size of tumor or grade of tumor) in the analysis of occult tumor studies is not explored much (Rai and Matthews, 1995). Semi-parametric models proposed here can be extended to strengthen inference with this additional information. For example, let $Z(t, u)$ be a stochastic covariate. Given that the time of tumor onset is u, the transition rate from state 2 to state 3 specifies the relationship between the marker process $\{Z(s)|0 \leq s \leq t\}$ and the hazard of failure at time t, and can be

$$\lambda_3(t|u, z(s) : 0 \leq s \leq t) = \lim_{\Delta t \to 0} Pr\{T \in [t, t + \Delta t)|X(t) = 2,$$
$$U = u, Z(s) = z(s) : 0 \leq s \leq t\}/\Delta t.$$

Suppose that $Z(t) = k$, and let the death rate at time t in tumor-bearing animals depend only on k, the size of the tumor at time t, and not on the sample path of the marker prior to t. Note that in the present context $k = 0$ for all $u < t$. In the sequel we will denote $\lambda_3(t|u, z(s) : 0 \leq s \leq t, z(t) = k)$ by $\lambda_3(t|u, k)$. One of the different forms for this intensity function which naturally suggest themselves is

$$\lambda_3(t|u, k) = \lambda_2(t) \exp(\gamma k).$$

The form for $\lambda_3(t|u, k)$ specified is the familiar proportional hazards model, which specifies that the failure rate for an animal which has a tumor of size k is a constant multiple, $\exp(\gamma k)$, of the baseline death rate at all times. Another alternative for this intensity function can be

$$\lambda_3(t|u, k) = \lambda_2(t) + \gamma k.$$

The above corresponds to a cause-separable hazards model for $\lambda_3(t|u, k)$, where $\lambda_2(t)$ is the hazard associated with competing causes of death and γk is the extra risk of failure at time t which is due to a tumor of size k. However, since the exact marker process will never be observed, it may be difficult, if not impossible, to assess the adequacy of any particular model involving a marker process such as tumor growth.

The two examples described in Section 6 appear to indicate that these more parsimonious, semi-parametric models may be as capable of adequately representing the information gathered in an occult tumor study as non-parametric models which necessarily involve many more parameters. The same non-parametric models also require revised experimental designs in which substantial amounts of interim sacrifice information is collected in order to estimate the tumor incidence rate and indications of carcinogenic potency.

Establishing the model adequacy is a challenging problem in this type of incomplete data. An *ad-hoc* assessment of the overall goodness-of-fit can be obtained by comparing the survival function from either semi-parametric model with that based on the Kaplan-Meier estimate. However, additional research is required to rigorously establish methods of differentiating between the semi-parametric models with and withour marker process information that we have described with respect to any particular set of data arising from an occult tumor study.

References

1. Berlin B, Brodsky J and Clifford P. Testing disease dependence in survival experiments with serial sacrifice. *J. Am. Stat. Assoc.* 1979; **74**: 5–14.
2. Cairns T. The ED_{01} study: introduction, objectives and experimental design. *J. Environ. Pathol. Toxicol.* 1980; **3**: 1–7.
3. Cox DR. Regression models and life-tables (with discussion). *J. R. Stat. Soc. B* 1972; **34**: 187–220.
4. Dempster AP, Laird NM and Rubin DB. Maximum likelihood estimation from incomplete data via the EM algorithm. *J. R. Stat. Soc. B* 1977; **39**: 1–22.
5. Dewanji A, Krewski D and Goddard, MJ. A Weibull model for the estimation of tumorigenic potency. *Biometrics* 1993; **49**: 367–377.
6. Dewanji A and Kalbfleisch JD. Non-parametric methods for survival/sacrifice experiments. *Biometrics* 1986; **42**: 325–341.
7. Dinse GE. Constant risk differences in the analysis of animal tumorigenicity data. *Biometrics* 1991; **47**: 685–700.
8. Dinse GE. Simple parametric analysis of animal tumorigenicity data. *J. Am. Stat. Assoc.* 1988; **83**: 638–649.
9. Dinse GE and Lagakos SW. Non-parametric estimation of lifetime and disease onset distributions from incomplete observations. *Biometrics* 1982; **38**: 921–932.
10. Efron B and Tibshirani R. Bootstrap methods for standard errors, confidence intervals and other measure of statistical accuracy. *Stat. Sci.* 1986; **1**: 54–77.
11. Finkelstein DM and Ryan LM. Estimating carcinogenicpotency from a rodent tumorigenicity experiment. *Appl. Stat.* 1987; **36**: 121–133.
12. Hoel DG and Walburg HE. Statistical analysis of survival experiments. *J. Nat. Cancer Inst.* 1972; **49**: 361–372.
13. Kalbfleisch JD, Krewski DR and Van Ryzin J. Dose-response models for time-to-response toxic data. *Can. J. Stat.* 1983; **11**: 25–49.
14. Kaplan EL and Meier P. Non-parametric estimation from incomplete observation. *J. Am. Stat. Assoc.* 1958; **53**: 457–481.

15. Kim W, Ahn H and Moon H. A dose-response test via closed-form solutions for constrained MLEs in survival/sacrifice experiments. *Stat. Med.* 2007; **26**: 694–708.
16. Kodell RL and Nelson CJ. An illness-death model for study of the carcinogenic process using survival/sacrifice data. *Biometrics* 1980; **36**: 267–277.
17. Kodell RL, Shaw GW and Johnson AM. Non-parametric joint estimates for disease resistance and survival function in survival/sacrifice experiments. *Biometrics* 1982; **38**: 43–58.
18. Lagakos SW and Ryan LM. Statistical analysis of disease onset and lifetime data from tumorigenicity experiments. *Environ. Health Perspect.* 1985; **63**: 211–216.
19. Lindsey J and Ryan LM. A multiplicative 3-state model for rodent tumorigenicity experiments. *Appl. Stat.* 1993; **42**: 283–300.
20. Louis TA. Finding the observed information matrix using the EM Algorithm. *J. R. Stat. Soc. B* 1982; **44**: 226–233.
21. McKnight B and Crowley J. Tests for the differences in tumor incidence based on animal carcinogenesis experiments. *J. Am. Stat. Assoc.* 1984; **79**: 639–648.
22. National Toxicology Program. *Report on the NTP ad hoc Panel on Chemical Carcinogenesis Testing and Evaluation*. National Institute of Environmental Health Sciences, Research Triangle Park, NC, 1986.
23. Portier JC. Estimating the tumor onset distribution in animal carcinogenicity-experiments. *Biometrika* 1986; **73**: 371–378.
24. Rai SN. On semi-parametric models in occult tumour experiments. *Biom. J.* 1997; **39**: 909–918.
25. Rai SN and Matthews DE. Improving the E-M algorithm. *Biometrics* 1993; **49**: 587–591.
26. Rai SN and Matthews DE. Discrete scale models for survival/sacrifice experiments. *Appl. Stat.* 1997; **46**: 93–103.
27. Rai SN, Matthews DE and Krewski DR. Mixed-scale scale models for survival/sacrifice experiments. *Can. J. Stat.* 2000; **28**: 65–80.
28. Rao CR. *Linear Statistical Inference and Its Applications*. Wiley, New York, 1973.
29. Ryan LM and Orav EJ. On the use of covariates for rodent bioassay and screening experiments. *Biometrika* 1988; **75**: 631–637.
30. Sun J. *The Statistical Analysis of Interval-Censored Failure Time Data*. Springer, New York, 2006.

31. Turnbull BW and Mitchell TJ. Non-parametric estimation of the distribution of time to onset and time to death for specific disease in survival/sacrifice experiments. *Biometrics* 1984; **40**: 41–50.
32. Williams PL and Portier CJ. Analytic expressions for maximum likelihood estimators in a non-parametric model of tumor incidence and death. *Comm. Stat.* 1986; **21**: 711–732.
33. Williams PL and Portier CJ. Explicit solutions for constrained maximum likelihood estimators in survival/sacrifice experiments. *Biometrika* 1993; **79**: 711–729.

Index

Absorbed dose, 291
Age-specific incidence, 29–30, 93–94
AIC (Akaike information criterion), 78, 365, 370
Anaerobic respiration, 225–226
Ankylosing spondylitis, 126–128
APC-(beta)catenin-Tcf-myc pathway, 61, 63–64, 66, 347, 377
Apoptosis, 27, 47–48, 55, 95–97, 99, 106, 328, 335–336
ARF-MDM2-p53 pathway, 51
Armitage-Doll model, 31, 39–41, 113–119, 121, 124, 133–134
Arsenic, 115

Bang-bang control, 432–433, 447, 448, 449
Base excision repair (BER), 303–307
Baseline hazard, 30–31, 35
Bayesian procedure; see Generalized Bayesian procedure
Beta-distributed growth rate, 406
BIC (Bayesian information criterion), 78, 365, 370
Binomial distribution, 70, 383
Bioelectrical conductivity, 225–226
Biological endpoint, 324, 328, 336, 338
Biomarkers, 489
Birth and death process, 25, 28, 402–403, 405
Bootstrap, 19
British physician data, 76, 94
Bystander effect, 137–138

Canadian National Dose Registry cohort data, 94
Cancer natural history, 149–151, 153–156, 160–161, 165, 167
Cancer risk assessment, 375, 382, 384–385
Carcinogenesis, 45–46, 55, 225–226
Cell cycle and carcinogenesis model, 55–56, 95–102, 104–108
Cellular kinetic parameter, 28, 34, 38
Checkpoint delay, 95–96, 98, 106
Chemotherapy, 174–175, 200–202, 205–206, 212–215, 218–220, 222–223
Chernoff inequality, 12
Chinese tin miners cohort data, 93–94
Chromosome aberration, 125, 310–314
Chromosomal abnormalities, 225–226
Chromosome copy number aberration (CNA), 14
Chromosomal instability, 136–138
Cigarette smoking, 115–116, 121, 126
CIS (CIN, Chromosomal instability), 51, 58, 62–63, 346, 350, 378–379
Clastogenic factor, 326
Cluster analysis, 492
Colon cancer; see Human colon cancer
Color-shift model, 401, 405–407, 411–413, 415
Comparative genomic hybridization (CGH), 14
Compartment, 26–27
Computer aided detection, 457
Computer tomography, 457
Conditioned medium, 333
Confidence interval, 388, 539

563

Cox regression model, 78, 83
Cox proportional hazards model, 526, 533
CPS-I, CPS-II data, American Cancer Society cohorts, 94–95
Cumulative hazard, 30

Data augmentation, 345, 358, 363
Differentiation, 27, 95–99, 105–106
Direct effect, 297–298
DNA damage, 58, 95–97, 107, 328, 333, 335, 297–303
DNA damage, complex, 95–97, 105–107
DNA damage, endogenous, 95–97, 107
DNA damage, simple, 95–97, 105–107
DNA damage pathway, 58, 95–97, 105–107
DNA repair, 47, 58, 95–97, 105–107
DNA repair, complex damage (slow), 95–97, 105–107
DNA repair, simple damage (fast), 95–97, 105–107
Dose-effect, 505, 517–520
Dose response curve (Dose response relationship), 388, 397–398, 407, 412, 415, 417–418
Double strand break (DSB), 297–298, 301–302

ECM algorithm, 503, 512, 515, 517, 519
ED_{01} 525, 556
EM (Expectation-maximization) algorithm, 345–346, 384, 461, 521, 524, 525
EM1 algorithm, 537, 539, 541, 544, 545, 547
Environmental agent, 375–376, 394
Environmental carcinogen, 225–226
Epigenetic, 48–49, 55, 57, 59, 334
Epigenetic carcinogenesis, 225
Extended multi-event model of carcinogenesis, 45, 64, 370, 377
Extracellular signaling, 330
Evolution process; *see* Micro-evolution process

FAP (Familial adenomatous polyposis) pathway, 52, 55, 65–66, 345–347, 349, 359, 361–362, 365, 368, 370
Fiberoptic cystoscopy, 458
Field effect hypothesis, 407, 411–412
Fixed ratio design, 503, 506, 516–517, 519
FLOH (Familial LOH) pathway, 65–66, 346–347, 349, 351, 356, 361–362, 365, 368
Foci of altered hepatocyte, 399, 407
Frailty, 30–32, 35, 38

GA algorithm (Genetic algorithm), 345–346, 364, 375, 386, 388, 394
Gamma distribution, 30–31
Gap junction, 324, 330–333, 225–226, 251–254
Gene amplification, 427–430, 433–434, 440, 443, 449
Generalized Bayesian procedure, 45–46, 75, 345, 361, 363
Generalized MVK model, 123–128
Genetic change, 48–49, 53
Genomic instability, 112–113, 129–138, 325, 328–329, 330
Gibbs sampling methods; *see* Multi-level Gibbs sampling procedures
G1 cell cycle state (early, late), 95–98, 100, 106, 108
G1/S checkpoint, 96, 98, 102, 106
G2 cell cycle state (early, late), 95–97, 101, 104, 106
G2/M checkpoint, 96–98, 106
Glomerulosclerosis, 550–554
Gompertzian growth, 186, 188, 195–196, 210, 212, 214, 218

Haploinsufficiency, 62, 348, 350–351, 378
Hazard function, 25–26, 29, 31–32, 34–35, 39–41
Hazard rate, 26
HNPCC (Hereditary non-polyposis colon cancer) pathway, 55, 66, 345, 347, 350–351, 356, 359–360, 362, 365, 368–370
Homeostatic regulation, 97–98, 102, 106

Index 565

Homogeneous model, 164, 167
Homologous recombination (HR), 308–309
Hypermethylation, 59, 225–226, 260–261, 264
Hypomethylation, 59, 225–226, 260–261, 264
Human bladder cancer, 375, 389–390, 394
Human colon cancer, 92, 114–115, 121–122, 126–127, 129–137, 345–347
H2AX, 327–328

Immortalization, 49, 60
Incidence function, 523, 549, 552
Indirect effect, 297–298
Incomplete gamma function, 31
Initial value problem, 33–35, 37–40
Initiator, 376, 392
Initiated cell, 92–94
Instantaneous seeding model, 164–165, 167
Intercellular communication, 225–226
International Radiation Study of Cervical Cancer (IRSCC), 126–128
Intracellular signaling, 325
Ionizing radiation (See Radiation, ionizing)

Japanese atomic bomb survivors, 109–112, 116–119, 121, 124–125, 127–128, 133, 137

Kaplan-Meier estimate, 547, 550, 559

Leukemia, 110, 112, 116, 118–119, 124–128, 134, 173–174, 177, 185–186, 189, 191, 193, 195–197, 203, 210–211, 213–223
Likelihood, 154, 160, 539, 541, 544–545, 547–549, 552, 555–557
Linear energy transfer (LET), 292–293, 295
Lineal energy, 294
Linear-No threshold model, 324
Linear-quadratic (LQ) model, 311, 319–321

LOH (Loss of Heterozygosity), 51–53, 59, 61, 63, 65, 82, 345–347, 349–352, 356, 359, 361–362, 365, 368–370, 378
Longitudinal data, 503, 517, 520
Lung cancer, 46, 65, 76, 92–95, 121–123, 125, 127

Magnetic resonance, 457
MALDI, 471
MAPK pathway, 51, 54
Markov random field, 462
Mass spectrometry, 471
Mass spectrometer, 472
Mass spectrum, 472
Mathematical processing of MS signals, 476
Maximum a posteriori, 462
Maximum principle, 432, 446
M cell cycle phase, 95–97, 101, 104–105
M phase arrest, 96, 105–106
Methylation, 57–60
Micro-evolution process, 46
Microdosimetry, 294–296
Micronucleus, 328
Micronutrient, 225–226, 238
Mid-S checkpoint, 96, 104, 106
Missingness, 501, 510
Mixture models, 45, 64–65
Maximum likelihood estimate (MLE), 512, 538–539, 541, 544, 547
MMR (Mis-match repair genes), 53–55, 58–59, 62–64, 346–347, 349–351, 360, 379
Microsatellite instability; *see* MSI
Model identifiability, 149, 165, 167
Mouse intestinal crypts model, 95–96, 104–108
MSI (Microsatellite instability) pathway, 58–59, 61–63, 65–66, 136–137, 345–346, 349–351, 356, 359–360, 362, 368–370, 379
Multidrug protocols (treatment, therapy), 425–426, 433, 443, 447–448
Multidrug resistance (MDR), 426–427, 443, 448

Multi-level Gibbs sampling, 45, 75, 346, 363, 375–376, 384, 394
Multinomial distribution, 67–69, 358, 381
Multiple myeloma, 173, 201–203, 212–213, 215–217, 222
Multiple pathways (Multiple pathway model), 59, 61, 65, 128–133, 345–346
Multiscale tools, 474
Multistage model of carcinogenesis (also Multistage cancer model), 26, 28, 31–32, 48–50, 64, 92–95, 345, 352, 375, 379–380, 397, 400–402, 405–407, 411–413
Multistage model of cell cycle and carcinogenesis, 95–102, 104–108
Mutation hypothesis, 407, 411–412
MVK two-stage model of carcinogenesis (Two-mutation (MVK) model), 67, 92–95, 119–122
Myc, 52, 54–55

NAD, 225, 242–251
Necrosis, 27
Non-homologous end joining (NHEJ), 334, 308–309
Nyquist stability criterion, 437

Observation model in carcinogenesis, 45, 72, 375, 385
Occult (undetectable) metastasis, 149, 150, 151, 154, 161, 521–561
Oncogene, 53, 257–258
Oncogenetic tree, 3, 22
Oncogenetic tree-error model, 5
Oncogenetic tree-reconstruction, 5–6, 9, 11–13, 15–17, 18–19
Oncogenetic tree-skewness, 6–7, 11
Oncogenetic tree-timed, 4, 7, 11
Oncogenetic tree-untimed, 4, 6–7, 11
Optimal control, 200–203, 210, 212, 218–219, 222–223
Optimal design, 503, 507, 516

Parameter estimation, 152, 154, 161
Partial differential equation (PDE), 32–33, 36–37, 39
Partial volume, 461

Phase-specific chemotherapy, 429, 440
PI3k-Akt pathway, 51, 54
P53-Bax pathway, 63, 350–351, 379
P53; see Tumor suppressor gene
Point mutation, 307–308
Poisson distribution, 67, 69, 160, 381, 383
Poisson process, 27, 36, 41, 155–156, 401–402, 404–405, 422
Posterior distribution of parameters in models of carcinogenesis, 74, 362
Power law of cancer incidence, 32
Prevalence function, 525, 529, 550, 553, 555–556
Predictive inference, 45–46, 375, 383, 386–387
Preneoplastic lesion, 397–398, 400–401, 406–407, 411, 415–418
Primary tumor, 149–158, 161–164, 167
Principal component analysis (PCA), 490
Prior distributions of parameters in models of carcinogenesis, 74, 361–362
Probability distributions of cancer tumors in carcinogenesis, 69–70, 355, 381
Probability distribution of state variables in carcinogenesis, 69–72, 381, 384
Probability distribution of cancer incidence data in carcinogenesis, 356–357
Probability generating function (p.g.f.), 32–34, 36–37, 39
Promoter, 376, 392
Promotion time, 93–94, 149–151, 156, 158–159, 161–162, 164
Proportional hazard model, 30; see also Cox proportional hazard model
Proteomics, 471
Proto-oncogene, 257

Radiation, ionizing, 94, 96–102, 107, 549–550, 554
Radiation, protracted exposure, inverse dose-rate effect, 93–94
Radon daughters, 93–94, 121–123, 125–126
Radon daughters, interaction with tobacco and arsenic, 93–94

Ras, 50–55, 57, 348, 378
Reactive oxygen species, 326, 329–330, 332
Reporter cell, 334
Recombinational repair, 334
Repair (*See* DNA Repair)
Risk assessment, 397–399, 401, 407, 416–418; *see also* Cancer risk assessment

S cell cycle state (early, late), 95–97, 100, 102, 104, 106–107
Secondary metastasis, 157
SEER (SEER database), 26, 92, 345–346, 356, 360, 364
Sequential nature of carcinogenesis, 50–51
Shape index, 463
Single strand break (SSB), 297–298, 301–302
Singular control, 432–433, 447, 449
Sister-chromatid exchange, 328, 330
Skin papilloma, 398, 401, 404
Smoking, 46, 76, 93–95
Specific energy, 294–295
Spectrum data registration, 480
State space model in carcinogenesis, 45, 104–107, 345, 361, 375, 383–384, 394
State space model for extended multi-event model, 70–71
State space model for extended multiple pathways model, 361
State space model for cancer risk assessment, 384
State space model for cell cycle and carcinogenesis, 95–98, 104–107
Stem cell, 45–49, 92, 98, 104
Statistical model of cancer incidence data, 45–46, 93–95, 356, 375
Stochastic difference equation for state variables, 375, 380
Stochastic differential equation for state variables, 45–46, 67–68, 107–108
Stochastic differentiation, 45–48, 64, 346
Stochastic equation for state variables in carcinogenesis, 107–108, 353

Stochastic four-stage model of carcinogenesis, 92, 245
Stochastic model of carcinogenesis, 45–46, 64, 66–67, 92, 107–108, 345–346, 361, 375, 377
Stochastic model of cell cycle and carcinogenesis, 95–98, 107–108
Stochastic proliferation, 45–46, 48, 64, 346
Stochastic system model, 45, 71, 365, 375, 384
Stochastic three-stage model of carcinogenesis, 390
Survival function, 25, 29, 32, 34, 38–40, 528, 547, 550–551, 559
Synergism, 505–506, 516–517, 519–520

Telomere, 60–61
TGF-β signaling pathway, 52, 61, 63, 350–351, 377–379
Tumor latency, 155, 161
Truncation, 501, 520
Tumor suppressor genes, 53
  APC, 49–55, 59, 61–64, 66, 347–349, 351, 360, 377–379
  P53, 48, 50–52, 54, 56–57, 61–63, 329–330, 348, 350–351, 378–379
  Rb, 54–55, 58, 61
  Smad (Smad2/Smad3, Smad2/Smad4), 50, 52, 61–62, 347–348, 351, 377–379
Tumor xenograft model, 503, 505, 507, 509, 511, 513, 515, 517, 519

Uniform design, 503, 506–507, 509, 517, 520
Uranium miners, 121–123, 125–126

Viral infection, 225, 254
Virtual cystoscopy, 457
Vitamin, 225, 238

Wavelets, 474, 483
Weighted bootstrap method, 75
Wicksell transformation, 404
Wnt signalling pathway, 59, 63, 350–351, 379